Student Solutions Manual

to accompany

Elementary Linear Algebra

Third Edition

Student Solutions Manual

to accompany

Elementary Linear Algebra

Third Edition

Stephen Andrilli
La Salle University

and

David Hecker
Saint Joseph's University

ELSEVIER
ACADEMIC PRESS

Amsterdam • Boston • Heidelberg • London
New York • Oxford • Paris • San Diego
San Francisco • Singapore • Sydney • Tokyo

Acquisitions Editor: Barbara A. Holland
Associate Editor: Tom Singer
Project Manager: Troy Lilly
Marketing Manager: Linda Beattie
Marketing Manager: Clare Fleming
Cover Design: Eric DeCicco
Printer: Quebecor World Book Services, Martinsburg

Elsevier Academic Press
200 Wheeler Road, Burlington, MA 01803, USA
525 B Street, Suite 1900, San Diego, California 92101-4495, USA
84 Theobald's Road, London WC1X 8RR, UK

This book is printed on acid-free paper. ∞

Library of Congress Cataloging-in-Publication Data
Application submitted

British Library Cataloguing in Publication Data
A catalogue record for this book is available from the British Library

ISBN: 0-12-058622-3

For all information on all Academic Press publications
visit our website at www.academicpressbooks.com

Printed in the United States of America

03 04 05 06 07 08 9 8 7 6 5 4 3 2 1

Transferred to Digital Printing 2007

Dedication

To all the students who have used the various editions of our book over the years

Table of Contents

Preface

This *Student Solutions Manual* is designed to accompany the third edition of *Elementary Linear Algebra*, by Andrilli and Hecker. It contains detailed solutions for all of the exercises in the textbook marked with a star (★) or a triangle (▶). In the triangle exercises, the student is typically asked to prove a theorem that appears in the textbook.

The solutions presented are generally self-contained, except that a comment may appear at the beginning of an exercise that applies to the solutions for each part of that exercise. For example, the solution to Exercise 1 in Section 1.1 begins with a description of the general strategy used to solve the problems in each of parts (a) and (c). Then, the solution is given to each part separately, following the previously stated strategy. Therefore, you should always check for a comment at the heading of a problem before jumping to the part in which you are interested.

We hope you find these solutions helpful. You can find other useful information at our web site:

http://www.sju.edu/~dhecker/linalg.html

Stephen Andrilli
David Hecker
August 2003

Solutions to Selected Exercises

Chapter 1

Section 1.1

(1) In each part, to find the vector, subtract the corresponding coordinates of the initial vector from the terminal vector.

(a) In this case, $[5, -1] - [-4, 3] = [9, -4]$. And so, the desired vector is $[9, -4]$. The distance between the two points $= \|[9, -4]\| = \sqrt{9^2 + (-4)^2} = \sqrt{97}$.

(c) In this case, $[0, -3, 2, -1, -1] - [1, -2, 0, 2, 3] = [-1, -1, 2, -3, -4]$. And so, the desired vector is $[-1, -1, 2, -3, -4]$. The distance between the two points $= \|[-1, -1, 2, -3, -4]\| =$ $\sqrt{(-1)^2 + (-1)^2 + 2^2 + (-3)^2 + (-4)^2} = \sqrt{31}$

(2) In each part, the terminal point is found by adding the coordinates of the given vector to the corresponding coordinates of the initial point $(1, 1, 1)$.

(a) $[1, 1, 1] + [2, 3, 1] = [3, 4, 2]$ (see Figure 1), so the terminal point $= (3, 4, 2)$.

(c) $[1, 1, 1] + [0, -3, -1] = [1, -2, 0]$ (see Figure 2), so the terminal point $= (1, -2, 0)$.

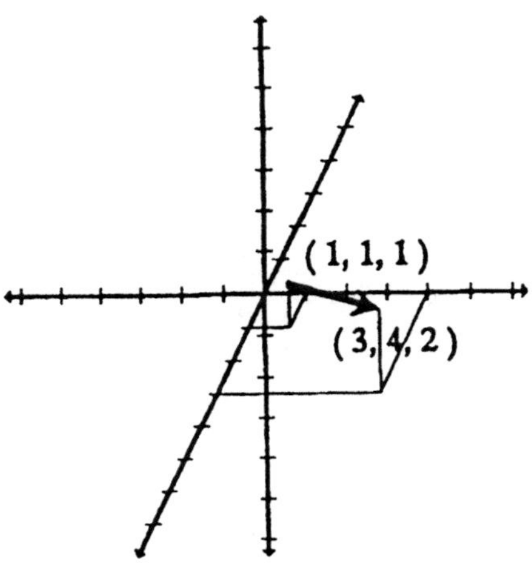

Figure 1

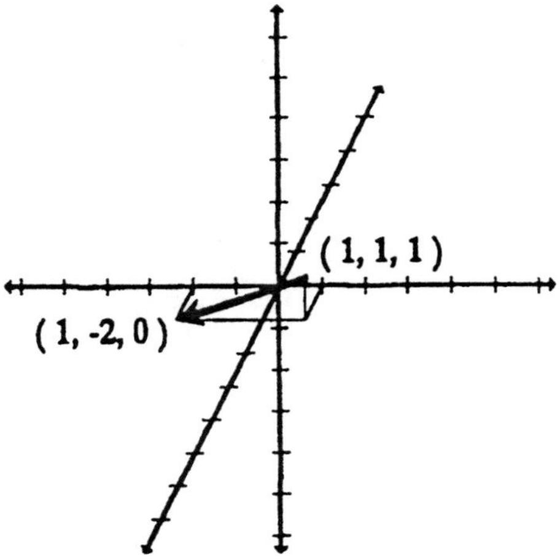

Figure 2

(3) In each part, the initial point is found by subtracting the coordinates of the given vector from the corresponding coordinates of the given terminal point.

(a) $[6, -9] - [-1, 4] = [7, -13]$, so the initial point is $(7, -13)$.

(c) $[2, -1, -1, 5, 4] - [3, -4, 0, 1, -2] = [-1, 3, -1, 4, 6]$, so the initial point $= (-1, 3, -1, 4, 6)$.

(4) (a) First, we find the vector $\mathbf{v}$ having the given initial and terminal point by subtracting: $[10, -10, 11] - [-4, 7, 2] = [14, -17, 9] = \mathbf{v}$. Next, the desired point is found by adding $\frac{2}{3}\mathbf{v}$ (which

is $\frac{2}{3}$ the length of $\mathbf{v}$) to the vector for the initial point $(-4,7,2)$:
$[-4,7,2]+\frac{2}{3}\mathbf{v}=[-4,7,2]+\frac{2}{3}[14,-17,9]=[-4,7,2]+\left[\frac{28}{3},-\frac{34}{3},6\right]=\left[\frac{16}{3},-\frac{13}{3},8\right]$,
so the desired point is $\left(\frac{16}{3},-\frac{13}{3},8\right)$.

(5) Let $\mathbf{v}$ represent the given vector. Then the desired unit vector $\mathbf{u}$ equals $\frac{1}{\|\mathbf{v}\|}\mathbf{v}$.

(a) $\mathbf{u}=\frac{1}{\|\mathbf{v}\|}\mathbf{v}=\frac{1}{\|[3,-5,6]\|}[3,-5,6]=\frac{1}{\sqrt{3^2+(-5)^2+6^2}}[3,-5,6]=\left[\frac{3}{\sqrt{70}},-\frac{5}{\sqrt{70}},\frac{6}{\sqrt{70}}\right]$;
$\mathbf{u}$ is shorter than $\mathbf{v}$ since $\|\mathbf{u}\|=1\leq\sqrt{70}=\|\mathbf{v}\|$.

(c) $\mathbf{u}=\frac{1}{\|\mathbf{v}\|}\mathbf{v}=\frac{1}{\|[0.6,-0.8]\|}[0.6,-0.8]=\frac{1}{\sqrt{(0.6)^2+(-0.8)^2}}[0.6,-0.8]=[0.6,-0.8]$. Neither vector is longer because $\mathbf{u}=\mathbf{v}$.

(6) Two nonzero vectors are parallel if and only if one is a scalar multiple of the other.

(a) $[12,-16]$ and $[9,-12]$ are parallel because $\frac{3}{4}[12,-16]=[9,-12]$.

(c) $[-2,3,1]$ and $[6,-4,-3]$ are not parallel. To show why not, suppose they are. Then there would be a $c\in\mathbb{R}$ such that $c[-2,3,1]=[6,-4,-3]$. Comparing first coordinates shows that $-2c=6$, or $c=-3$. However comparing second coordinates shows that $3c=-4$, or $c=-\frac{4}{3}$ instead. But c can not have both values.

(7) (a) $3[-2,4,5]=[3(-2),3(4),3(5)]=[-6,12,15]$

(c) $[-2,4,5]+[-1,0,3]=[(-2+(-1)),(4+0),(5+3)]=[-3,4,8]$

(e) $4[-1,0,3]-5[-2,4,5]=[4(-1),4(0),4(3)]-[5(-2),5(4),5(5)]=[-4,0,12]-[-10,20,25]=[(-4-(-10)),(0-20),(12-25)]=[6,-20,-13]$

(8) (a) $\mathbf{x}+\mathbf{y}=[-1,5]+[2,-4]=[(-1+2),(5-4)]=[1,1]$, $\mathbf{x}-\mathbf{y}=[-1,5]-[2,-4]=[(-1-2),(5-(-4))]=[-3,9]$, $\mathbf{y}-\mathbf{x}=[2,-4]-[-1,5]=[(2-(-1)),((-4)-5)]=[3,-9]$ (see Figure 3, next page)

(c) $\mathbf{x}+\mathbf{y}=[2,5,-3]+[-1,3,-2]=[(2+(-1)),(5+3),((-3)+(-2))]=[1,8,-5]$,
$\mathbf{x}-\mathbf{y}=[2,5,-3]-[-1,3,-2]=[(2-(-1)),(5-3),((-3)-(-2))]=[3,2,-1]$,
$\mathbf{y}-\mathbf{x}=[-1,3,-2]-[2,5,-3]=[((-1)-2),(3-5),((-2)-(-3))]=[-3,-2,1]$ (see Figure 4, next page)

(10) In each part, consider the center of the clock to be the origin.

(a) At 12 PM, the tip of the minute hand is at $(0,10)$. At $12:15$ PM, the tip of the minute hand is at $(10,0)$. To find the displacement vector, we subtract the vector for the initial point from the vector for the terminal point, yielding $[10,0]-[0,10]=[10,-10]$.

(b) At 12 PM, the tip of the minute hand is at $(0,10)$. At $12:40$ PM, the minute hand makes a 210° angle with the positive x-axis. So, as shown in Figure 1.10 in the textbook, the minute hand makes the vector $\mathbf{v}=[\|\mathbf{v}\|\cos\theta,\|\mathbf{v}\|\sin\theta]=[10\cos(210^\circ),10\sin(210^\circ)]=\left[10\left(-\frac{\sqrt{3}}{2}\right),10\left(-\frac{1}{2}\right)\right]=[-5\sqrt{3},-5]$. To find the displacement vector, we subtract the vector for the initial point from the vector for the terminal point, yielding $[-5\sqrt{3},-5]-[0,10]=[-5\sqrt{3},-15]$.

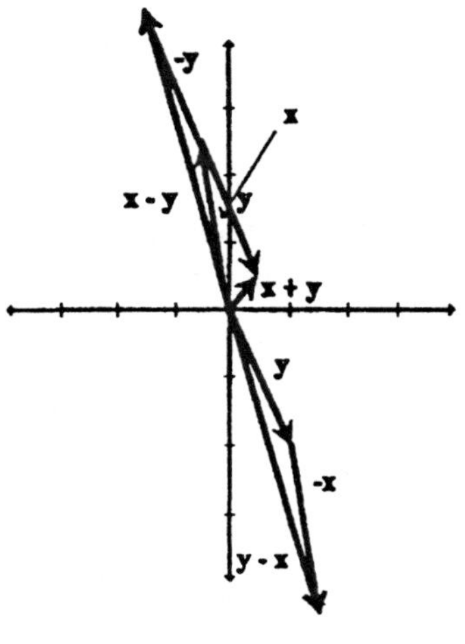

Scale: Each tickmark =2 units

Figure 3

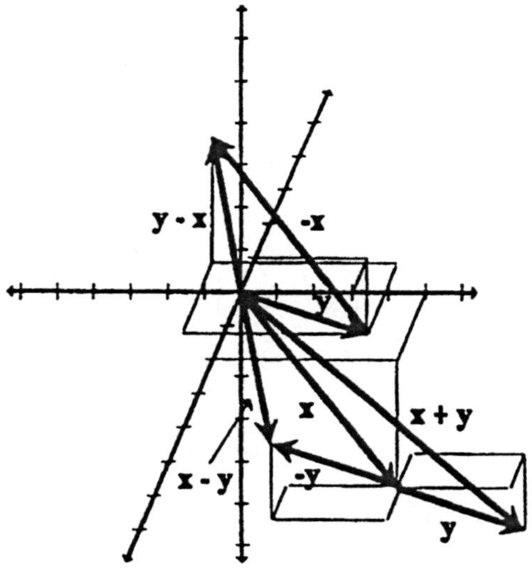

Figure 4

(13) As shown in Figure 1.10 in the textbook, for any vector $\mathbf{v} \in \mathbb{R}^2$, $\mathbf{v} = [\|\mathbf{v}\|\cos\theta, \|\mathbf{v}\|\sin\theta]$, where θ is the angle $\mathbf{v}$ makes with the positive x-axis. Now "southwest" corresponds to 225°, "east" corresponds to 0°, and northwest corresponds to 135°. Hence, if $\mathbf{v}_1$, $\mathbf{v}_2$, and $\mathbf{v}_3$ are the vectors corresponding to the three given movements, then
$\mathbf{v}_1 = [1\cos(225°), 1\sin(225°)] = \left[-\frac{\sqrt{2}}{2}, -\frac{\sqrt{2}}{2}\right]$, $\mathbf{v}_2 = [0.5\cos(0°), 0.5\sin(0°)] = [0.5, 0]$, and
$\mathbf{v}_3 = [0.2\cos(135°), 0.2\sin(135°)] = \left[0.2\left(-\frac{\sqrt{2}}{2}\right), 0.2\frac{\sqrt{2}}{2}\right]$. Hence, the result of all three putts is
$\mathbf{v}_1+\mathbf{v}_2+\mathbf{v}_3 = \left[-\frac{\sqrt{2}}{2}, -\frac{\sqrt{2}}{2}\right]+[0.5,0]+\left[0.2\left(-\frac{\sqrt{2}}{2}\right), 0.2\frac{\sqrt{2}}{2}\right] = [0.5-0.6\sqrt{2}, -0.4\sqrt{2}] \approx [-0.3485, -0.5657]$.

(15) As shown in Figure 1.10 in the textbook, for any vector $\mathbf{v} \in \mathbb{R}^2$, $\mathbf{v} = [\|\mathbf{v}\|\cos\theta, \|\mathbf{v}\|\sin\theta]$, where θ is the angle $\mathbf{v}$ makes with the positive x-axis. Now "northwestward" corresponds to 135°, and "southward" corresponds to 270°. Hence, if $\mathbf{v}_1$ corresponds to the rower, and $\mathbf{v}_2$ corresponds to the current, then $\mathbf{v}_1 = [4\cos(135°), 4\sin(135°)] = \left[4\left(-\frac{\sqrt{2}}{2}\right), 4\frac{\sqrt{2}}{2}\right] = [-2\sqrt{2}, 2\sqrt{2}]$ and $\mathbf{v}_2 = [3\cos(270°), 3\sin(270°)] = [0, -3]$. Therefore, the rower's net velocity $= \mathbf{v}_1+\mathbf{v}_2 = [-2\sqrt{2}, 2\sqrt{2}]+[0,-3] = [-2\sqrt{2}, -3+2\sqrt{2}]$. The resultant speed $= \|[-2\sqrt{2}, -3+2\sqrt{2}]\| = \sqrt{(-2\sqrt{2})^2+(-3+2\sqrt{2})^2} = \sqrt{25-12\sqrt{2}} \approx 2.83$ km/hr.

(17) If $\mathbf{v}_1$ corresponds to the rower, and $\mathbf{v}_2$ corresponds to the current and $\mathbf{v}_3$ corresponds to the resultant velocity, then $\mathbf{v}_3 = \mathbf{v}_1 + \mathbf{v}_2$, or $\mathbf{v}_1 = \mathbf{v}_3 - \mathbf{v}_2$. Now, as shown in Figure 1.10 in the textbook, for any vector $\mathbf{v} \in \mathbb{R}^2$, $\mathbf{v} = [\|\mathbf{v}\|\cos\theta, \|\mathbf{v}\|\sin\theta]$, where θ is the angle $\mathbf{v}$ makes with the positive x-axis. Also, "westward" corresponds to 180°, and "northeastward" corresponds to 45°. Thus,
$\mathbf{v}_2 = [2\cos(45°), 2\sin(45°)] = \left[2\frac{\sqrt{2}}{2}, 2\frac{\sqrt{2}}{2}\right] = [\sqrt{2}, \sqrt{2}]$ and $\mathbf{v}_3 = [8\cos(180°), 8\sin(180°)] = [-8, 0]$.
Hence, $\mathbf{v}_1 = \mathbf{v}_3 - \mathbf{v}_2 = [-8, 0] - [\sqrt{2}, \sqrt{2}] = [-8-\sqrt{2}, -\sqrt{2}]$.

(18) Let $\mathbf{f}_R$ and $\mathbf{a}_R$ represent the resultant force and the resultant acceleration, respectively. Then $\mathbf{f}_R = \mathbf{f}_1 + \mathbf{f}_2 + \mathbf{f}_3$. Now, $\mathbf{f}_1 = 4\frac{[3,-12,4]}{\|[3,-12,4]\|} = \frac{4}{\sqrt{3^2+(-12)^2+4^2}}[3,-12,4] = \frac{4}{13}[3,-12,4] = \left[\frac{12}{13}, -\frac{48}{13}, \frac{16}{13}\right]$,

$\mathbf{f}_2 = 2\frac{[0,-4,-3]}{\|[0,-4,-3]\|} = \frac{2}{\sqrt{0^2+(-4)^2+(-3)^2}}[0,-4,-3] = \frac{2}{5}[0,-4,-3] = \left[0,-\frac{8}{5},-\frac{6}{5}\right]$, and

$\mathbf{f}_3 = 6\frac{[0,0,1]}{\|[0,0,1]\|} = \frac{6}{1}[0,0,1] = [0,0,6]$. Therefore, $\mathbf{f}_R = \mathbf{f}_1+\mathbf{f}_2+\mathbf{f}_3 = \left[\frac{12}{13},-\frac{48}{13},\frac{16}{13}\right]+\left[0,-\frac{8}{5},-\frac{6}{5}\right]+[0,0,6] = \left[\frac{12}{13},-\frac{344}{65},\frac{392}{65}\right]$. Finally, $\mathbf{f}_R = m\mathbf{a}_R$. Since $m = 20$, acceleration $= \mathbf{a}_R = \frac{1}{m}\mathbf{f}_R = \frac{1}{20}\left[\frac{12}{13},-\frac{344}{65},\frac{392}{65}\right] \approx [0.0462,-0.2646,0.3015]$.

(20) As shown in Figure 1.10 in the textbook, for any vector $\mathbf{v} \in \mathbb{R}^2$, $\mathbf{v} = [\|\mathbf{v}\|\cos\theta, \|\mathbf{v}\|\sin\theta]$, where θ is the angle $\mathbf{v}$ makes with the positive x-axis. Thus, if we let $r = \|\mathbf{a}\|$ and $s = \|\mathbf{b}\|$, then $\mathbf{a} = [r\cos(135°), r\sin(135°)] = \left[-r\frac{\sqrt{2}}{2}, r\frac{\sqrt{2}}{2}\right]$ and $\mathbf{b} = [s\cos(60°), s\sin(60°)] = \left[s\frac{1}{2}, s\frac{\sqrt{3}}{2}\right]$. Now, since the weight is not moving, $\mathbf{a}+\mathbf{b}+\mathbf{g} = \mathbf{0}$. That is, $\left[-r\frac{\sqrt{2}}{2}, r\frac{\sqrt{2}}{2}\right] + \left[s\frac{1}{2}, s\frac{\sqrt{3}}{2}\right] + [0,-mg] = [0,0]$. Using the first coordinates gives us $-r\frac{\sqrt{2}}{2} + s\frac{1}{2} = 0$. Hence, $s = \sqrt{2}r$. Using the second coordinates gives us $r\frac{\sqrt{2}}{2} + s\frac{\sqrt{3}}{2} - mg = 0$. Substituting in $s = \sqrt{2}r$ yields $r\frac{\sqrt{2}}{2} + \sqrt{2}r\frac{\sqrt{3}}{2} - mg = 0$. Solving for r produces $r = \frac{2mg}{\sqrt{2}(1+\sqrt{3})}$, and $s = \sqrt{2}r = \frac{2mg}{1+\sqrt{3}}$. Plugging these values in to the formulas for $\mathbf{a}$ and $\mathbf{b}$ gives $\mathbf{a} = \left[\frac{-mg}{1+\sqrt{3}}, \frac{mg}{1+\sqrt{3}}\right]$ and $\mathbf{b} = \left[\frac{mg}{1+\sqrt{3}}, \frac{mg\sqrt{3}}{1+\sqrt{3}}\right]$.

(22) Let $\mathbf{x} = [x_1, x_2, \ldots, x_n]$. Then $\|c\mathbf{x}\| = \sqrt{(cx_1)^2+\cdots+(cx_n)^2} = \sqrt{c^2(x_1^2+\cdots+x_n^2)}$ $= |c|\sqrt{x_1^2+\cdots+x_n^2} = |c|\,\|\mathbf{x}\|$.

(23) In each part, let $\mathbf{x} = [x_1, x_2, \ldots, x_n]$, $\mathbf{y} = [y_1, y_2, \ldots, y_n]$, and $c \in \mathbb{R}$.

(b) $\mathbf{x}+(-\mathbf{x}) = [x_1,\ldots,x_n] + [-x_1,\ldots,-x_n] = [(x_1+(-x_1)),\ldots,(x_n+(-x_n))] = [0,\ldots,0] = \mathbf{0}$. Also, $(-\mathbf{x})+\mathbf{x} = \mathbf{x}+(-\mathbf{x})$ (by part (1) of Theorem 1.3) $= \mathbf{0}$, by the above.

(c) $c(\mathbf{x}+\mathbf{y}) = c([(x_1+y_1),\ldots,(x_n+y_n)]) = [c(x_1+y_1),\ldots,c(x_n+y_n)] = [(cx_1+cy_1),\ldots,(cx_n+cy_n)] = [cx_1,\ldots,cx_n] + [cy_1,\ldots,cy_n] = c\mathbf{x}+c\mathbf{y}$

(24) If $c = 0$, we are done. Otherwise, $c\mathbf{x} = \mathbf{0} \Rightarrow (\frac{1}{c})(c\mathbf{x}) = \frac{1}{c}(\mathbf{0}) \Rightarrow (\frac{1}{c}\cdot c)\mathbf{x} = \mathbf{0}$ (by part (7) of Theorem 1.3) $\Rightarrow \mathbf{x} = \mathbf{0}$. Thus either $c = 0$ or $\mathbf{x} = \mathbf{0}$.

(26) (a) False. The length of $[a_1, a_2, a_3]$ is $\sqrt{a_1^2+a_2^2+a_3^2}$. The formula given in the problem is missing the square root. So, for example, the length of $[0,3,4]$ is actually 5, *not* $0^2+3^2+4^2 = 25$.

(b) True. This is easily proved using parts (1) and (2) of Theorem 1.3.

(c) True. $[2,0,-3] = 2[1,0,0] + (-3)[0,0,1]$.

(d) False. Two nonzero vectors are parallel if and only if one is a scalar multiple of the other. To show why $[3,-5,2]$ and $[6,-10,5]$ are not parallel, suppose they are. Then there would be a $c \in \mathbb{R}$ such that $c[3,-5,2] = [6,-10,5]$. Comparing first coordinates shows that $3c = 6$, or $c = 2$. But this value of c must work in all coordinates. However, it does not work in the third coordinate, since $(2)(2) \neq 5$.

(e) True. Multiply both sides of $d\mathbf{x} = \mathbf{0}$ by $\frac{1}{d}$.

(f) False. Parallel vectors can be in opposite directions. For example, $[1,0]$ and $[-2,0]$ are parallel, but are not in the same direction.

(g) False. The properties of vectors in Theorem 1.3 are independent of the location of the initial points of the vectors.

Section 1.2

(1) In each part, we use the formula $\cos\theta = \frac{\mathbf{x}\cdot\mathbf{y}}{\|\mathbf{x}\|\|\mathbf{y}\|}$.

(a) $\cos\theta = \frac{\mathbf{x}\cdot\mathbf{y}}{\|\mathbf{x}\|\|\mathbf{y}\|} = \frac{[-4,3]\cdot[6,-1]}{\|[-4,3]\|\|[6,-1]\|} = \frac{(-4)(6)+(3)(-1)}{\sqrt{(-4)^2+3^2}\sqrt{6^2+(-1)^2}} = \frac{-27}{5\sqrt{37}}$. And so, using a calculator yields $\theta = \arccos(-\frac{27}{5\sqrt{37}})$, or about 152.6°, or 2.66 radians.

(c) $\cos\theta = \frac{\mathbf{x}\cdot\mathbf{y}}{\|\mathbf{x}\|\|\mathbf{y}\|} = \frac{[7,-4,2]\cdot[-6,-10,1]}{\|[7,-4,2]\|\|[-6,-10,1]\|} = \frac{(7)(-6)+(-4)(-10)+(2)(1)}{\sqrt{7^2+(-4)^2+2^2}\sqrt{(-6)^2+(-10)^2+1^2}} = \frac{0}{\sqrt{69}\sqrt{137}} = 0$. And so, $\theta = \arccos(0)$, which is 90°, or $\frac{\pi}{2}$ radians.

(4) When a force $\mathbf{f}$ moves an object by a displacement $\mathbf{d}$, the work performed is $\mathbf{f}\cdot\mathbf{d}$.

(b) First, we need to compute $\mathbf{f}$. A unit vector in the direction of $\mathbf{f}$ is $\mathbf{u} = \frac{1}{\|[-2,4,5]\|}[-2,4,5] = \frac{1}{\sqrt{(-2)^2+4^2+5^2}}[-2,4,5] = \frac{1}{3\sqrt{5}}[-2,4,5]$. Thus, since $\|\mathbf{f}\| = 26$, $\mathbf{f} = 26\mathbf{u} = \frac{26}{3\sqrt{5}}[-2,4,5] = \left[\frac{-52\sqrt{5}}{15}, \frac{104\sqrt{5}}{15}, \frac{26\sqrt{5}}{3}\right]$, where we have rationalized the denominators.
Next, we need to compute $\mathbf{d}$. A unit vector in the direction of $\mathbf{d}$ is $\mathbf{v} = \frac{1}{\|[-1,2,2]\|}[-1,2,2] = \frac{1}{\sqrt{(-1)^2+2^2+2^2}}[-1,2,2] = [-\frac{1}{3}, \frac{2}{3}, \frac{2}{3}]$. Thus, since $\|\mathbf{d}\| = 10$, $\mathbf{d} = 10\mathbf{v} = [-\frac{10}{3}, \frac{20}{3}, \frac{20}{3}]$.
Hence, the work performed is
$\mathbf{f}\cdot\mathbf{d} = \left[\frac{-52\sqrt{5}}{15}, \frac{104\sqrt{5}}{15}, \frac{26\sqrt{5}}{3}\right]\cdot[-\frac{10}{3}, \frac{20}{3}, \frac{20}{3}] = \left(\frac{-52\sqrt{5}}{15}\right)(-\frac{10}{3}) + \left(\frac{104\sqrt{5}}{15}\right)(\frac{20}{3}) + \left(\frac{26\sqrt{5}}{3}\right)(\frac{20}{3})$
$= \frac{1}{9}\left(104\sqrt{5} + 416\sqrt{5} + 520\sqrt{5}\right) = \frac{1040\sqrt{5}}{9}$, or 258.4 joules.

(6) In all parts, let $\mathbf{x} = [x_1, x_2, \ldots, x_n]$, $\mathbf{y} = [y_1, y_2, \ldots, y_n]$, and $\mathbf{z} = [z_1, z_2, \ldots, z_n]$.
Part (1): $\mathbf{x}\cdot\mathbf{y} = [x_1, x_2, \ldots, x_n]\cdot[y_1, y_2, \ldots, y_n] = x_1y_1 + \cdots + x_ny_n = y_1x_1 + \cdots + y_nx_n = [y_1, y_2, \ldots, y_n]\cdot[x_1, x_2, \ldots, x_n] = \mathbf{y}\cdot\mathbf{x}$
Part (2): $\mathbf{x}\cdot\mathbf{x} = [x_1, x_2, \ldots, x_n]\cdot[x_1, x_2, \ldots, x_n] = x_1x_1 + \cdots + x_nx_n = x_1^2 + \cdots + x_n^2$. Now $x_1^2 + \cdots + x_n^2$ is a sum of squares, each of which must be nonnegative. Hence, the sum is also nonnegative, and so its square root is defined. Thus, $0 \le \mathbf{x}\cdot\mathbf{x} = x_1^2 + \cdots + x_n^2 = \left(\sqrt{x_1^2 + \cdots + x_n^2}\right)^2 = \|\mathbf{x}\|^2$.
Part (3): Suppose $\mathbf{x}\cdot\mathbf{x} = 0$. From part (2), $0 = x_1^2 + \cdots + x_n^2 \ge x_i^2$, for each i, since the sum of the remaining squares (without x_i^2) is nonnegative. Hence, $0 \ge x_i^2$ for each i. But $x_i^2 \ge 0$, because it is a square. Hence each $x_i = 0$. Therefore, $\mathbf{x} = \mathbf{0}$.
Next, suppose $\mathbf{x} = \mathbf{0}$. Then $\mathbf{x}\cdot\mathbf{x} = [0, \ldots, 0]\cdot[0, \ldots, 0] = (0)(0) + \cdots + (0)(0) = 0$.
Part (4): $c(\mathbf{x}\cdot\mathbf{y}) = c([x_1, x_2, \ldots, x_n]\cdot[y_1, y_2, \ldots, y_n]) = c(x_1y_1 + \cdots + x_ny_n) = cx_1y_1 + \cdots + cx_ny_n = [cx_1, cx_2, \ldots, cx_n]\cdot[y_1, y_2, \ldots, y_n] = (c\mathbf{x})\cdot\mathbf{y}$.
Next, $c(\mathbf{x}\cdot\mathbf{y}) = c(\mathbf{y}\cdot\mathbf{x})$ (by part (1)) $= (c\mathbf{y})\cdot\mathbf{x}$ (by the above) $= \mathbf{x}\cdot(c\mathbf{y})$, by part (1).
Part (6): $(\mathbf{x}+\mathbf{y})\cdot\mathbf{z} = ([x_1, x_2, \ldots, x_n] + [y_1, y_2, \ldots, y_n])\cdot[z_1, z_2, \ldots, z_n]$
$= [x_1 + y_1, x_2 + y_2, \ldots, x_n + y_n]\cdot[z_1, z_2, \ldots, z_n] = (x_1 + y_1)z_1 + (x_2 + y_2)z_2 + \cdots + (x_n + y_n)z_n$
$= (x_1z_1 + x_2z_2 + \cdots + x_nz_n) + (y_1z_1 + y_2z_2 + \cdots + y_nz_n)$.
Also, $(\mathbf{x}\cdot\mathbf{z}) + (\mathbf{y}\cdot\mathbf{z}) = ([x_1, x_2, \ldots, x_n]\cdot[z_1, z_2, \ldots, z_n]) + ([y_1, y_2, \ldots, y_n]\cdot[z_1, z_2, \ldots, z_n])$
$= (x_1z_1 + x_2z_2 + \cdots + x_nz_n) + (y_1z_1 + y_2z_2 + \cdots + y_nz_n)$. Hence, $(\mathbf{x}+\mathbf{y})\cdot\mathbf{z} = (\mathbf{x}\cdot\mathbf{z}) + (\mathbf{y}\cdot\mathbf{z})$.

(7) No. Consider $\mathbf{x} = [1,0]$, $\mathbf{y} = [0,1]$, and $\mathbf{z} = [1,1]$. Then $\mathbf{x}\cdot\mathbf{z} = [1,0]\cdot[1,1] = (1)(1) + (0)(1) = 1$ and $\mathbf{y}\cdot\mathbf{z} = [0,1]\cdot[1,1] = (0)(1) + (1)(1) = 1$, so $\mathbf{x}\cdot\mathbf{z} = \mathbf{y}\cdot\mathbf{z}$. But $\mathbf{x} \ne \mathbf{y}$.

(13) θ_1 is the angle between $\mathbf{x}$ and $[1,0,0]$. So $\cos\theta_1 = \frac{[a,b,c]\cdot[1,0,0]}{\|[a,b,c]\|\|[1,0,0]\|} = \frac{a(1)+b(0)+c(0)}{\sqrt{a^2+b^2+c^2}\sqrt{1^2+0^2+0^2}} = \frac{a}{\sqrt{a^2+b^2+c^2}}$. Similarly, $\cos\theta_2 = \frac{[a,b,c]\cdot[0,1,0]}{\|[a,b,c]\|\|[0,1,0]\|} = \frac{a(0)+b(1)+c(0)}{\sqrt{a^2+b^2+c^2}\sqrt{0^2+1^2+0^2}} = \frac{b}{\sqrt{a^2+b^2+c^2}}$, and $\cos\theta_3 = \frac{[a,b,c]\cdot[0,0,1]}{\|[a,b,c]\|\|[0,0,1]\|} = \frac{a(0)+b(0)+c(1)}{\sqrt{a^2+b^2+c^2}\sqrt{0^2+0^2+1^2}} = \frac{c}{\sqrt{a^2+b^2+c^2}}$.

(14) Position the cube so that one of its corners is at the origin and the three edges from that corner are along the positive x-, y, and z-axes. Then the corner diagonally opposite from the origin is at the point (s,s,s). The vector representing this diagonal is $\mathbf{d} = [s,s,s]$.

(a) The length of the diagonal $= \|\mathbf{d}\| = \sqrt{s^2+s^2+s^2} = \sqrt{3s^2} = \sqrt{3}s$.

(b) The angle $\mathbf{d}$ makes with a side of the cube is equal to the angle between $\mathbf{d}$ and $[s,0,0]$, which is $\arccos\left(\frac{\mathbf{d}\cdot[s,0,0]}{\|\mathbf{d}\|\|[s,0,0]\|}\right) = \arccos\left(\frac{s(s)+s(0)+s(0)}{(\sqrt{3}s)(\sqrt{s^2+0^2+0^2})}\right) = \arccos\left(\frac{s^2}{\sqrt{3}s^2}\right) = \arccos(\frac{\sqrt{3}}{3}) \approx 54.7°$, or 0.955 radians.

(15) (a) $\mathbf{proj_a b} = \left(\frac{\mathbf{a}\cdot\mathbf{b}}{\|\mathbf{a}\|^2}\right)\mathbf{a} = \left(\frac{[2,1,5]\cdot[1,4,-3]}{\|[2,1,5]\|^2}\right)[2,1,5] = \left(\frac{(2)(1)+(1)(4)+(5)(-3)}{2^2+1^2+5^2}\right)[2,1,5] = \frac{-9}{30}[2,1,5]$ $= [-\frac{3}{5},-\frac{3}{10},-\frac{3}{2}]$. $\mathbf{b}-\mathbf{proj_a b} = [1,4,-3]-[-\frac{3}{5},-\frac{3}{10},-\frac{3}{2}] = [\frac{8}{5},\frac{43}{10},-\frac{3}{2}]$. $\mathbf{proj_a b}$ is orthogonal to $(\mathbf{b}-\mathbf{proj_a b})$ because $\mathbf{proj_a b}\cdot(\mathbf{b}-\mathbf{proj_a b}) = [-\frac{3}{5},-\frac{3}{10},-\frac{3}{2}]\cdot[\frac{8}{5},\frac{43}{10},-\frac{3}{2}] =$ $(-\frac{3}{5})(\frac{8}{5})+(-\frac{3}{10})(\frac{43}{10})+(-\frac{3}{2})(-\frac{3}{2}) = 0$.

(c) $\mathbf{proj_a b} = \left(\frac{\mathbf{a}\cdot\mathbf{b}}{\|\mathbf{a}\|^2}\right)\mathbf{a} = \left(\frac{[1,0,-1,2]\cdot[3,-1,0,-1]}{\|[1,0,-1,2]\|^2}\right)[1,0,-1,2] =$ $\left(\frac{(1)(3)+(0)(-1)+(-1)(0)+(2)(-1)}{1^2+0^2+(-1)^2+2^2}\right)[1,0,-1,2] = \frac{1}{6}[1,0,-1,2] = [\frac{1}{6},0,-\frac{1}{6},\frac{1}{3}]$.
$\mathbf{b}-\mathbf{proj_a b} = [3,-1,0,-1]-[\frac{1}{6},0,-\frac{1}{6},\frac{1}{3}] = [\frac{17}{6},-1,\frac{1}{6},-\frac{4}{3}]$. $\mathbf{proj_a b}$ is orthogonal to $(\mathbf{b}-\mathbf{proj_a b})$ because $\mathbf{proj_a b}\cdot(\mathbf{b}-\mathbf{proj_a b}) = [\frac{1}{6},0,-\frac{1}{6},\frac{1}{3}]\cdot[\frac{17}{6},-1,\frac{1}{6},-\frac{4}{3}] =$ $(\frac{1}{6})(\frac{17}{6})+(0)(-1)+(-\frac{1}{6})(\frac{1}{6})+(\frac{1}{3})(-\frac{4}{3}) = 0$.

(17) Let $\mathbf{a} = [a,b,c]$. Then $\mathbf{proj_i a} = \left(\frac{\mathbf{i}\cdot\mathbf{a}}{\|\mathbf{i}\|^2}\right)\mathbf{i} = \left(\frac{[1,0,0]\cdot[a,b,c]}{\|[1,0,0]\|^2}\right)[1,0,0] = \left(\frac{1a+0b+0c}{1^2+0^2+0^2}\right)[1,0,0] = a\mathbf{i}$. Similarly, $\mathbf{proj_j a} = \left(\frac{\mathbf{j}\cdot\mathbf{a}}{\|\mathbf{j}\|^2}\right)\mathbf{j} = \left(\frac{[0,1,0]\cdot[a,b,c]}{\|[0,1,0]\|^2}\right)[0,1,0] = \left(\frac{0a+1b+0c}{0^2+1^2+0^2}\right)[0,1,0] = b\mathbf{j}$, and $\mathbf{proj_k a} = \left(\frac{\mathbf{k}\cdot\mathbf{a}}{\|\mathbf{k}\|^2}\right)\mathbf{k} = \left(\frac{[0,0,1]\cdot[a,b,c]}{\|[0,0,1]\|^2}\right)[0,0,1] = \left(\frac{0a+0b+1c}{0^2+0^2+1^2}\right)[0,0,1] = c\mathbf{k}$.

(18) In each part, if $\mathbf{v}$ is the given vector, then by Theorem 1.10, $\mathbf{x} = \mathbf{proj_v x}+(\mathbf{x}-\mathbf{proj_v x})$, where $\mathbf{proj_v x}$ is parallel to $\mathbf{v}$ and $(\mathbf{x}-\mathbf{proj_v x})$ is orthogonal to $\mathbf{v}$.

(a) In this case, $\mathbf{x} = [-6,2,7]$ and $\mathbf{v} = [2,-3,4]$. Then $\mathbf{proj_v x} = \left(\frac{\mathbf{v}\cdot\mathbf{x}}{\|\mathbf{v}\|^2}\right)\mathbf{v} =$ $\left(\frac{[2,-3,4]\cdot[-6,2,7]}{\|[2,-3,4]\|^2}\right)[2,-3,4] = \left(\frac{(2)(-6)+(-3)(2)+(4)(7)}{2^2+(-3)^2+4^2}\right)[2,-3,4] = \left(\frac{10}{29}\right)[2,-3,4] = [\frac{20}{29},-\frac{30}{29},\frac{40}{29}]$.
And so $\mathbf{proj_v x}$ is clearly parallel to $\mathbf{v}$ since $\mathbf{proj_v x} = (\frac{10}{29})\mathbf{v}$. Also, $(\mathbf{x}-\mathbf{proj_v x}) = [-6,2,7]-[\frac{20}{29},-\frac{30}{29},\frac{40}{29}] = [-\frac{194}{29},\frac{88}{29},\frac{163}{29}]$. We can easily check that $\mathbf{proj_v x}+(\mathbf{x}-\mathbf{proj_v x}) = [\frac{20}{29},-\frac{30}{29},\frac{40}{29}]+[-\frac{194}{29},\frac{88}{29},\frac{163}{29}] = [-6,2,7] = \mathbf{x}$, and that $\mathbf{proj_v x}$ and $(\mathbf{x}-\mathbf{proj_v x})$ are orthogonal, since $\mathbf{proj_v x}\cdot(\mathbf{x}-\mathbf{proj_v x}) = [\frac{20}{29},-\frac{30}{29},\frac{40}{29}]\cdot[-\frac{194}{29},\frac{88}{29},\frac{163}{29}]$ $= (\frac{20}{29})(-\frac{194}{29})+(-\frac{30}{29})(\frac{88}{29})+(\frac{40}{29})(\frac{163}{29}) = 0$.

(c) In this case, $\mathbf{x} = [-6, 2, 7]$ and $\mathbf{v} = [3, -2, 6]$. Then $\mathbf{proj_v x} = \left(\frac{\mathbf{v}\cdot\mathbf{x}}{\|\mathbf{v}\|^2}\right)\mathbf{v} =$ $\left(\frac{[3,-2,6]\cdot[-6,2,7]}{\|[3,-2,6]\|^2}\right)[3,-2,6] = \left(\frac{(3)(-6)+(-2)(2)+(6)(7)}{3^2+(-2)^2+6^2}\right)[3,-2,6] = \left(\frac{20}{49}\right)[3,-2,6] = \left[\frac{60}{49}, -\frac{40}{49}, \frac{120}{49}\right]$. And so, $\mathbf{proj_v x}$ is clearly parallel to $\mathbf{v}$ since $\mathbf{proj_v x} = \left(\frac{20}{49}\right)\mathbf{v}$. Also, $(\mathbf{x} - \mathbf{proj_v x}) = [-6,2,7] - \left[\frac{60}{49}, -\frac{40}{49}, \frac{120}{49}\right] = \left[-\frac{354}{49}, \frac{138}{49}, \frac{223}{49}\right]$. We can easily check that $\mathbf{proj_v x} + (\mathbf{x} - \mathbf{proj_v x}) = \left[\frac{60}{49}, -\frac{40}{49}, \frac{120}{49}\right] + \left[-\frac{354}{49}, \frac{138}{49}, \frac{223}{49}\right] = [-6,2,7] = \mathbf{x}$, and that $\mathbf{proj_v x}$ and $(\mathbf{x} - \mathbf{proj_v x})$ are orthogonal, since $\mathbf{proj_v x}\cdot(\mathbf{x} - \mathbf{proj_v x}) = \left[\frac{60}{49}, -\frac{40}{49}, \frac{120}{49}\right]\cdot\left[-\frac{354}{49}, \frac{138}{49}, \frac{223}{49}\right]$ $= \left(\frac{60}{49}\right)\left(-\frac{354}{49}\right) + \left(-\frac{40}{49}\right)\left(\frac{138}{49}\right) + \left(\frac{120}{49}\right)\left(\frac{223}{49}\right) = 0$.

(23) (a) True by part (4) of Theorem 1.5.

(b) True. Theorem 1.6 (the Cauchy-Schwarz Inequality) states that $|\mathbf{x}\cdot\mathbf{y}| \le \|\mathbf{x}\|\,\|\mathbf{y}\|$. Since $\mathbf{x}\cdot\mathbf{y} \le |\mathbf{x}\cdot\mathbf{y}|$, we have $\mathbf{x}\cdot\mathbf{y} \le \|\mathbf{x}\|\,\|\mathbf{y}\|$. But then, because $\|\mathbf{x}\| > 0$, we can divide both sides by $\|\mathbf{x}\|$ to obtain the desired inequality.

(c) False. For example, if $\mathbf{x} = [1,0,0]$ and $\mathbf{y} = [0,1,0]$, then $\|\mathbf{x}-\mathbf{y}\| = \|[1,-1,0]\| =$ $\sqrt{1^2+(-1)^2+0^2} = \sqrt{2}$, and $\|\mathbf{x}\| - \|\mathbf{y}\| = \|[1,0,0]\| - \|[0,1,0]\| =$ $\sqrt{1^2+0^2+0^2} - \sqrt{0^2+1^2+0^2} = 1 - 1 = 0$. However, $\sqrt{2} \not\le 0$.

(d) False. Theorem 1.8 shows that if $\theta > \frac{\pi}{2}$, then $\mathbf{x}\cdot\mathbf{y} < 0$. (Remember, θ is defined to be $\le \pi$.)

(e) True. If $i \neq j$, then $\mathbf{e}_i\cdot\mathbf{e}_j = 0(0)+\cdots+\underbrace{1(0)}_{i\text{th terms}}+\cdots+\underbrace{0(1)}_{j\text{th terms}}+0(0)+\cdots+0(0) = 0+\cdots+0 = 0$.

(f) False. $\mathbf{proj_a b} = \left(\frac{\mathbf{a}\cdot\mathbf{b}}{\|\mathbf{a}\|^2}\right)\mathbf{a}$, and so $\mathbf{proj_a b}$ is parallel to $\mathbf{a}$, since it is a scalar multiple of $\mathbf{a}$. Thus, if $\mathbf{proj_a b} = \mathbf{b}$, $\mathbf{a}$ and $\mathbf{b}$ are parallel, not perpendicular. For a particular example, suppose $\mathbf{a} = [1,0]$ and $\mathbf{b} = [2,0]$. Then $\mathbf{proj_a b} = \left(\frac{\mathbf{a}\cdot\mathbf{b}}{\|\mathbf{a}\|^2}\right)\mathbf{a} = \left(\frac{[1,0]\cdot[2,0]}{\|[1,0]\|}\right)[1,0] = \left(\frac{1(2)+0(0)}{\sqrt{1^2+0^2}}\right)[1,0] = 2[1,0] = [2,0] =$ $\mathbf{b}$. However, $\mathbf{a} = [1,0]$ and $\mathbf{b} = [2,0]$ are parallel (since $\mathbf{b} = 2\mathbf{a}$) and *not* perpendicular (since $\mathbf{a}\cdot\mathbf{b} = [1,0]\cdot[2,0] = 1(2)+0(0) = 2 \neq 0$.

Section 1.3

(1) (b) Let $m = \max\{|c|, |d|\}$. Then $\|c\mathbf{x} \pm d\mathbf{y}\| \le m(\|\mathbf{x}\| + \|\mathbf{y}\|)$.
The proof is as follows:

$$\begin{aligned}
\|c\mathbf{x} \pm d\mathbf{y}\| &= \|c\mathbf{x} + (\pm d\mathbf{y})\| \\
&\le \|c\mathbf{x}\| + \|\pm d\mathbf{y}\| && \text{by the Triangle Inequality} \\
&= |c|\,\|\mathbf{x}\| + |\pm d|\,\|\mathbf{y}\| && \text{by Theorem 1.1} \\
&\le m\,\|\mathbf{x}\| + m\,\|\mathbf{y}\| && \text{because } |c| \le m \text{ and } |\pm d| \le m \\
&= m\,(\|\mathbf{x}\| + \|\mathbf{y}\|)
\end{aligned}$$

(2) (b) The converse is: If an integer has the form $3k+1$, then it also has the form $6j-5$, where j and k are integers. For a counterexample, consider that the number $4 = 3k+1$ with $k = 1$, but there is no integer j such that $4 = 6j-5$, since this equation implies $j = \frac{3}{2}$, which is not an integer.

(5) (a) Consider $\mathbf{x} = [1,0,0]$ and $\mathbf{y} = [1,1,0]$. Then $\mathbf{x}\cdot\mathbf{y} = [1,0,0]\cdot[1,1,0] = 1 = \|[1,0,0]\|^2 = \|\mathbf{x}\|^2$, but $\mathbf{x} \neq \mathbf{y}$.

(b) If $\mathbf{x} \neq \mathbf{y}$, then $\mathbf{x} \cdot \mathbf{y} \neq \|\mathbf{x}\|^2$.

(c) Yes. Using the counterexample from part (a), $\mathbf{x} \neq \mathbf{y}$, but $\mathbf{x} \cdot \mathbf{y} = \|\mathbf{x}\|^2$.

(8) (a) Contrapositive: If $\mathbf{x} = \mathbf{0}$, then $\mathbf{x}$ is not a unit vector.
Converse: If $\mathbf{x}$ is nonzero, then $\mathbf{x}$ is a unit vector.
Inverse: If $\mathbf{x}$ is not a unit vector, then $\mathbf{x} = \mathbf{0}$.

(c) (Let $\mathbf{x}$, $\mathbf{y}$ be nonzero vectors.)
Contrapositive: If $\mathbf{proj_y x} \neq \mathbf{0}$, then $\mathbf{proj_x y} \neq \mathbf{0}$.
Converse: If $\mathbf{proj_y x} = \mathbf{0}$, then $\mathbf{proj_x y} = \mathbf{0}$.
Inverse: If $\mathbf{proj_x y} \neq \mathbf{0}$, then $\mathbf{proj_y x} \neq \mathbf{0}$.

(10) (b) Converse: Let $\mathbf{x}$ and $\mathbf{y}$ be vectors in $\mathbb{R}^n$. If $\|\mathbf{x}+\mathbf{y}\| \geq \|\mathbf{y}\|$, then $\mathbf{x} \cdot \mathbf{y} = 0$. The original statement is true, but the converse is false in general.
Proof of the original statement:

$$\begin{aligned} \|\mathbf{x}+\mathbf{y}\|^2 &= (\mathbf{x}+\mathbf{y}) \cdot (\mathbf{x}+\mathbf{y}) && \text{by part (2) of Theorem 1.5} \\ &= \|\mathbf{x}\|^2 + 2(\mathbf{x} \cdot \mathbf{y}) + \|\mathbf{y}\|^2 && \text{using Theorem 1.5} \\ &= \|\mathbf{x}\|^2 + \|\mathbf{y}\|^2 && \text{since } \mathbf{x} \cdot \mathbf{y} = 0 \\ &\geq \|\mathbf{y}\|^2 && \text{by part (2) of Theorem 1.5} \end{aligned}$$

Taking the square root of both sides yields $\|\mathbf{x}+\mathbf{y}\| \geq \|\mathbf{y}\|$.
Counterexample to converse: let $\mathbf{x} = [1, 0]$ and $\mathbf{y} = [1, 1]$. Then $\|\mathbf{x}+\mathbf{y}\| = \|[2, 1]\| = \sqrt{5}$ and $\|\mathbf{y}\| = \|[1, 1]\| = \sqrt{2}$. Hence $\|\mathbf{x}+\mathbf{y}\| \geq \|\mathbf{y}\|$. However, $\mathbf{x} \cdot \mathbf{y} = [1, 0] \cdot [1, 1] = 1 \neq 0$.

(18) Step 1 cannot be reversed, because y could equal $-(x^2+2)$ and then the converse would not hold.
Step 2 cannot be reversed, because y^2 could equal x^4+4x^2+c with $c > 4$. For a specific counterexample to the converse, set $y = \sqrt{x^4+4x^2+5}$.
Step 3 is reversible. To prove the converse, multiply both sides of $\frac{dy}{dx} = \frac{4x^3+8x}{2y}$ by $2y$.
Step 4 cannot be reversed, because in general y does not have to equal x^2+2. In particular, y could equal x^2+5.
Step 5 can be reversed by multiplying $2x$ by $\frac{2(x^2+2)}{2(x^2+2)}$.
Step 6 cannot be reversed, since $\frac{dy}{dx}$ could equal $2x + c$ for some $c \neq 0$.

(19) (a) Negation: For every unit vector $\mathbf{x}$ in $\mathbb{R}^3$, $\mathbf{x} \cdot [1, -2, 3] \neq 0$. (Note that the existential quantifier has changed to a universal quantifier.)

(c) Negation: $\mathbf{x} = \mathbf{0}$ or $\|\mathbf{x}+\mathbf{y}\| \neq \|\mathbf{y}\|$, for all vectors $\mathbf{x}$ and $\mathbf{y}$ in $\mathbb{R}^n$. (The "and" has changed to an "or," and the existential quantifier has changed to a universal quantifier.)

(e) Negation: There is an $\mathbf{x} \in \mathbb{R}^3$ such that for every nonzero $\mathbf{y} \in \mathbb{R}^3$, $\mathbf{x} \cdot \mathbf{y} \neq 0$. (Note that the universal quantifier on $\mathbf{x}$ has changed to an existential quantifier, and the existential quantifier on $\mathbf{y}$ has changed to a universal quantifier.)

(20) (a) When negating the conclusion "$\mathbf{x} = \mathbf{0}$ or $\|\mathbf{x}-\mathbf{y}\| > \|\mathbf{y}\|$" of the original statement, we must change the "or" to "and."
Contrapositive: If $\mathbf{x} \neq \mathbf{0}$ and $\|\mathbf{x}-\mathbf{y}\| \leq \|\mathbf{y}\|$, then $\mathbf{x} \cdot \mathbf{y} \neq 0$.
Converse: If $\mathbf{x} = \mathbf{0}$ or $\|\mathbf{x}-\mathbf{y}\| > \|\mathbf{y}\|$, then $\mathbf{x} \cdot \mathbf{y} = 0$.
Inverse: If $\mathbf{x} \cdot \mathbf{y} \neq 0$, then $\mathbf{x} \neq \mathbf{0}$ and $\|\mathbf{x}-\mathbf{y}\| \leq \|\mathbf{y}\|$.

(25) (a) False. We must also verify that the reversed steps are truly reversible. That is, we must also supply valid reasons for the forward steps thus generated, as in the proof of Result 2.

(b) True. "If not B then not A" is the contrapositive of "If A then B." A statement and its contrapositive are always logically equivalent.

(c) True. We saw that "A only if B" is an equivalent form for "If A then B," whose converse is "If B then A."

(d) False. "A is a necessary condition for B" is an equivalent form for "If B then A." As such, it does not include the statement "If A then B" that is included in the "A if and only if B" statement.

(e) False. "A is a necessary condition for B" and "B is a sufficient condition for A" are both equivalent forms for "If B then A." As such, neither includes the statement "If A then B" that is included in the "A if and only if B" statement.

(f) False. The inverse of a statement is the contrapositive of the converse. Hence, the converse and the inverse are logically equivalent, and always have the same truth values.

(g) False. The problem only describes the inductive step. The base step of the proof is also required.

(h) True. This rule is given in the section for negating statements with quantifiers.

(i) False. The "and" must change to an "or." Thus, the negation of "A and B" is "not A or not B."

Section 1.4

(1) (a) $$\begin{bmatrix} -4 & 2 & 3 \\ 0 & 5 & -1 \\ 6 & 1 & -2 \end{bmatrix} + \begin{bmatrix} 6 & -1 & 0 \\ 2 & 2 & -4 \\ 3 & -1 & 1 \end{bmatrix} = \begin{bmatrix} (-4+6) & (2+(-1)) & (3+0) \\ (0+2) & (5+2) & ((-1)+(-4)) \\ (6+3) & (1+(-1)) & ((-2)+1) \end{bmatrix}$$
$$= \begin{bmatrix} 2 & 1 & 3 \\ 2 & 7 & -5 \\ 9 & 0 & -1 \end{bmatrix}$$

(c) $$4\begin{bmatrix} -4 & 2 & 3 \\ 0 & 5 & -1 \\ 6 & 1 & -2 \end{bmatrix} = \begin{bmatrix} 4(-4) & 4(2) & 4(3) \\ 4(0) & 4(5) & 4(-1) \\ 4(6) & 4(1) & 4(-2) \end{bmatrix} = \begin{bmatrix} -16 & 8 & 12 \\ 0 & 20 & -4 \\ 24 & 4 & -8 \end{bmatrix}$$

(e) Impossible. **C** is a 2×2 matrix, **F** is a 3×2 matrix, and **E** is a 3×3 matrix. Hence, 3**F** is 3×2 and $-$**E** is 3×3. Thus, **C**, 3**F**, and $-$**E** have different sizes. However, in order to add matrices, they must be the same size.

(g) $$2\begin{bmatrix} -4 & 2 & 3 \\ 0 & 5 & -1 \\ 6 & 1 & -2 \end{bmatrix} - 3\begin{bmatrix} 3 & -3 & 5 \\ 1 & 0 & -2 \\ 6 & 7 & -2 \end{bmatrix} - \begin{bmatrix} 6 & -1 & 0 \\ 2 & 2 & -4 \\ 3 & -1 & 1 \end{bmatrix} =$$
$$\begin{bmatrix} 2(-4) & 2(2) & 2(3) \\ 2(0) & 2(5) & 2(-1) \\ 2(6) & 2(1) & 2(-2) \end{bmatrix} + \begin{bmatrix} -3(3) & -3(-3) & -3(5) \\ -3(1) & -3(0) & -3(-2) \\ -3(6) & -3(7) & -3(-2) \end{bmatrix} + \begin{bmatrix} -6 & -(-1) & -0 \\ -2 & -2 & -(-4) \\ -3 & -(-1) & -1 \end{bmatrix}$$
$$= \begin{bmatrix} -8 & 4 & 6 \\ 0 & 10 & -2 \\ 12 & 2 & -4 \end{bmatrix} + \begin{bmatrix} -9 & 9 & -15 \\ -3 & 0 & 6 \\ -18 & -21 & 6 \end{bmatrix} + \begin{bmatrix} -6 & 1 & 0 \\ -2 & -2 & 4 \\ -3 & 1 & -1 \end{bmatrix}$$
$$= \begin{bmatrix} (-8)+(-9)+(-6) & 4+9+1 & 6+(-15)+0 \\ 0+(-3)+(-2) & 10+0+(-2) & (-2)+6+4 \\ 12+(-18)+(-3) & 2+(-21)+1 & (-4)+6+(-1) \end{bmatrix} = \begin{bmatrix} -23 & 14 & -9 \\ -5 & 8 & 8 \\ -9 & -18 & 1 \end{bmatrix}$$

(i) $\begin{bmatrix} -4 & 2 & 3 \\ 0 & 5 & -1 \\ 6 & 1 & -2 \end{bmatrix}^T + \begin{bmatrix} 3 & -3 & 5 \\ 1 & 0 & -2 \\ 6 & 7 & -2 \end{bmatrix}^T = \begin{bmatrix} -4 & 0 & 6 \\ 2 & 5 & 1 \\ 3 & -1 & -2 \end{bmatrix} + \begin{bmatrix} 3 & 1 & 6 \\ -3 & 0 & 7 \\ 5 & -2 & -2 \end{bmatrix}$

$= \begin{bmatrix} ((-4)+3) & (0+1) & (6+6) \\ (2+(-3)) & (5+0) & (1+7) \\ (3+5) & ((-1)+(-2)) & ((-2)+(-2)) \end{bmatrix} = \begin{bmatrix} -1 & 1 & 12 \\ -1 & 5 & 8 \\ 8 & -3 & -4 \end{bmatrix}$

(l) Impossible. Since $\mathbf{C}$ is a 2×2 matrix, both $\mathbf{C}^T$ and $2\mathbf{C}^T$ are 2×2. Also, since $\mathbf{F}$ is a 3×2 matrix, $-3\mathbf{F}$ is 3×2. Thus, $2\mathbf{C}^T$ and $-3\mathbf{F}$ have different sizes. However, in order to add matrices, they must be the same size.

(n) $(\mathbf{B} - \mathbf{A})^T = \begin{bmatrix} 10 & 2 & -3 \\ -3 & -3 & -2 \\ -3 & -3 & 3 \end{bmatrix}$ and $\mathbf{E}^T = \begin{bmatrix} 3 & 1 & 6 \\ -3 & 0 & 7 \\ 5 & -2 & -2 \end{bmatrix}$.

Thus $(\mathbf{B} - \mathbf{A})^T + \mathbf{E}^T = \begin{bmatrix} 13 & 3 & 3 \\ -6 & -3 & 5 \\ 2 & -5 & 1 \end{bmatrix}$, and so $\left((\mathbf{B} - \mathbf{A})^T + \mathbf{E}^T\right)^T = \begin{bmatrix} 13 & -6 & 2 \\ 3 & -3 & -5 \\ 3 & 5 & 1 \end{bmatrix}$.

(2) For a matrix to be square, it must have the same number of rows as columns. A matrix $\mathbf{X}$ is diagonal if and only if it is square and $x_{ij} = 0$ for $i \neq j$. A matrix $\mathbf{X}$ is upper triangular if and only if it is square and $x_{ij} = 0$ for $i > j$. A matrix $\mathbf{X}$ is lower triangular if and only if it is square and $x_{ij} = 0$ for $i < j$. A matrix $\mathbf{X}$ is symmetric if and only if it is square and $x_{ij} = x_{ji}$ for all i, j. A matrix $\mathbf{X}$ is skew-symmetric if and only if it is square and $x_{ij} = -x_{ji}$ for all i, j. Notice that this implies that if $\mathbf{X}$ is skew-symmetric, then $x_{ii} = 0$ for all i. Finally, the transpose of an $m \times n$ matrix $\mathbf{X}$ is the $n \times m$ matrix whose (i, j)th entry is x_{ji}. We now check each matrix in turn.

$\mathbf{A}$ is a 3×2 matrix, so is not square. Therefore it can not be diagonal, upper triangular, lower triangular, symmetric, or skew-symmetric. Finally, $\mathbf{A}^T = \begin{bmatrix} -1 & 0 & 6 \\ 4 & 1 & 0 \end{bmatrix}$.

$\mathbf{B}$ is a 2×2 matrix, and so is square. Because $b_{12} = b_{21} = 0$, $\mathbf{B}$ is diagonal, upper triangular, lower triangular, and symmetric. $\mathbf{B}$ is not skew-symmetric because $b_{11} \neq 0$. Finally, $\mathbf{B}^T = \mathbf{B}$.

$\mathbf{C}$ is a 2×2 matrix, and so is square. Because $c_{12} \neq 0$, $\mathbf{C}$ is neither diagonal, nor lower triangular. Because $c_{21} \neq 0$, $\mathbf{C}$ is not upper triangular. $\mathbf{C}$ is not symmetric because $c_{12} \neq c_{21}$. $\mathbf{C}$ is not skew-symmetric because $c_{11} \neq 0$. Finally, $\mathbf{C}^T = \begin{bmatrix} -1 & -1 \\ 1 & 1 \end{bmatrix}$.

$\mathbf{D}$ is a 3×1 matrix, so is not square. Therefore it can not be diagonal, upper triangular, lower triangular, symmetric, or skew-symmetric. Finally, $\mathbf{D}^T = \begin{bmatrix} -1 & 4 & 2 \end{bmatrix}$.

$\mathbf{E}$ is a 3×3 matrix, and so is square. Because $e_{13} \neq 0$, $\mathbf{E}$ is neither diagonal, nor lower triangular. Because $e_{31} \neq 0$, $\mathbf{E}$ is not upper triangular. $\mathbf{E}$ is not symmetric because $e_{31} \neq e_{13}$. $\mathbf{E}$ is not skew-symmetric because $e_{22} \neq 0$. Finally, $\mathbf{E}^T = \begin{bmatrix} 0 & 0 & -6 \\ 0 & -6 & 0 \\ 6 & 0 & 0 \end{bmatrix}$.

$\mathbf{F}$ is a 4×4 matrix, and so is square. Because $f_{14} \neq 0$, $\mathbf{F}$ is neither diagonal, nor lower triangular. Because $f_{41} \neq 0$, $\mathbf{F}$ is not upper triangular. $\mathbf{F}$ is symmetric because $f_{ij} = f_{ji}$ for every i, j. $\mathbf{F}$ is not skew-symmetric because $f_{32} \neq -f_{23}$. Finally, $\mathbf{F}^T = \mathbf{F}$.

$\mathbf{G}$ is a 3×3 matrix, and so is square. Because $g_{ij} = 0$ whenever $i \neq j$, $\mathbf{G}$ is diagonal, lower triangular, upper triangular, and symmetric. $\mathbf{G}$ is not skew-symmetric because $g_{11} \neq 0$. Finally, $\mathbf{G}^T = \mathbf{G}$.

H is a 4×4 matrix, and so is square. Because $h_{14} \neq 0$, **H** is neither diagonal, nor lower triangular. Because $h_{41} \neq 0$, **H** is not upper triangular. **H** is not symmetric because $h_{12} \neq h_{21}$. **H** is skew-symmetric because $h_{ij} = -h_{ji}$ for every i, j. Finally, $\mathbf{H}^T = -\mathbf{H}$.

J is a 4×4 matrix, and so is square. Because $j_{12} \neq 0$, **J** is neither diagonal, nor lower triangular. Because $j_{21} \neq 0$, **J** is not upper triangular. **J** is symmetric because $j_{ik} = j_{ki}$ for every i, k. **J** is not skew-symmetric because $j_{32} \neq -j_{23}$. Finally, $\mathbf{J}^T = \mathbf{J}$.

K is a 4×4 matrix, and so is square. Because $k_{14} \neq 0$, **K** is neither diagonal, nor lower triangular. Because $k_{41} \neq 0$, **K** is not upper triangular. **K** is not symmetric because $k_{12} \neq k_{21}$. **K** is not skew-symmetric because $k_{11} \neq 0$. Finally, $\mathbf{K}^T = \begin{bmatrix} 1 & -2 & -3 & -4 \\ 2 & 1 & -5 & -6 \\ 3 & 5 & 1 & -7 \\ 4 & 6 & 7 & 1 \end{bmatrix}$.

L is a 3×3 matrix, and so is square. Because $l_{13} \neq 0$, **L** is neither diagonal, nor lower triangular. Because $l_{ij} = 0$ whenever $i > j$, **L** is upper triangular. **L** is not symmetric because $l_{31} \neq l_{13}$. **L** is not skew-symmetric because $l_{11} \neq 0$ (or because $l_{21} \neq -l_{12}$). Finally, $\mathbf{L}^T = \begin{bmatrix} 1 & 0 & 0 \\ 1 & 1 & 0 \\ 1 & 1 & 1 \end{bmatrix}$.

M is a 3×3 matrix, and so is square. Because $m_{31} \neq 0$, **M** is neither diagonal, nor upper triangular. Because $m_{ij} = 0$ whenever $i < j$, **M** is lower triangular. **M** is not symmetric because $m_{31} \neq m_{13}$. **M** is not skew-symmetric because $m_{31} \neq -m_{13}$. Finally, $\mathbf{M}^T = \begin{bmatrix} 0 & 1 & 1 \\ 0 & 0 & 1 \\ 0 & 0 & 0 \end{bmatrix}$.

N is a 3×3 matrix, and so is square. Because $n_{ij} = 0$ whenever $i \neq j$, **N** is diagonal, lower triangular, upper triangular, and symmetric. **N** is not skew-symmetric because $n_{11} \neq 0$. Finally, $\mathbf{N}^T = \mathbf{N} = \mathbf{I}_3$.

P is a 2×2 matrix, and so is square. Because $p_{12} \neq 0$, **P** is neither diagonal, nor lower triangular. Because $p_{21} \neq 0$, **P** is not upper triangular. **P** is symmetric because $p_{12} = p_{21}$. **P** is not skew-symmetric because $p_{12} \neq -p_{21}$. Finally, $\mathbf{P}^T = \mathbf{P}$.

Q is a 3×3 matrix, and so is square. Because $q_{31} \neq 0$, **Q** is neither diagonal, nor upper triangular. Because $q_{ij} = 0$ whenever $i < j$, **Q** is lower triangular. **Q** is not symmetric because $q_{31} \neq q_{13}$. **Q** is not skew-symmetric because $q_{11} \neq 0$ (or because $q_{31} \neq -q_{13}$). Finally, $\mathbf{Q}^T = \begin{bmatrix} -2 & 4 & -1 \\ 0 & 0 & 2 \\ 0 & 0 & 3 \end{bmatrix}$.

R is a 3×2 matrix, so is not square. Therefore it can not be diagonal, upper triangular, lower triangular, symmetric, or skew-symmetric. Finally, $\mathbf{R}^T = \begin{bmatrix} 6 & 3 & -1 \\ 2 & -2 & 0 \end{bmatrix}$.

(3) Just after Theorem 1.13 in the textbook, the desired decomposition is described as $\mathbf{A} = \mathbf{S} + \mathbf{V}$, where $\mathbf{S} = \frac{1}{2}(\mathbf{A} + \mathbf{A}^T)$ is symmetric and $\mathbf{V} = \frac{1}{2}(\mathbf{A} - \mathbf{A}^T)$ is skew-symmetric.

(a) $\mathbf{S} = \frac{1}{2}(\mathbf{A} + \mathbf{A}^T) = \frac{1}{2}\left(\begin{bmatrix} 3 & -1 & 4 \\ 0 & 2 & 5 \\ 1 & -3 & 2 \end{bmatrix} + \begin{bmatrix} 3 & 0 & 1 \\ -1 & 2 & -3 \\ 4 & 5 & 2 \end{bmatrix}\right) =$

$$\tfrac{1}{2}\left(\begin{bmatrix} (3+3) & ((-1)+0) & (4+1) \\ (0+(-1)) & (2+2) & (5+(-3)) \\ (1+4) & ((-3)+5) & (2+2) \end{bmatrix}\right) =$$

$$\tfrac{1}{2}\begin{bmatrix} 6 & -1 & 5 \\ -1 & 4 & 2 \\ 5 & 2 & 4 \end{bmatrix} = \begin{bmatrix} \tfrac{1}{2}(6) & \tfrac{1}{2}(-1) & \tfrac{1}{2}(5) \\ \tfrac{1}{2}(-1) & \tfrac{1}{2}(4) & \tfrac{1}{2}(2) \\ \tfrac{1}{2}(5) & \tfrac{1}{2}(2) & \tfrac{1}{2}(4) \end{bmatrix} = \begin{bmatrix} 3 & -\tfrac{1}{2} & \tfrac{5}{2} \\ -\tfrac{1}{2} & 2 & 1 \\ \tfrac{5}{2} & 1 & 2 \end{bmatrix}$$, which is symmetric, since $s_{ij} = s_{ji}$ for all i, j.

$$\mathbf{V} = \tfrac{1}{2}(\mathbf{A} - \mathbf{A}^T) = \tfrac{1}{2}\left(\begin{bmatrix} 3 & -1 & 4 \\ 0 & 2 & 5 \\ 1 & -3 & 2 \end{bmatrix} - \begin{bmatrix} 3 & 0 & 1 \\ -1 & 2 & -3 \\ 4 & 5 & 2 \end{bmatrix}\right)$$

$$= \tfrac{1}{2}\left(\begin{bmatrix} (3-3) & ((-1)-0) & (4-1) \\ (0-(-1)) & (2-2) & (5-(-3)) \\ (1-4) & ((-3)-5) & (2-2) \end{bmatrix}\right) =$$

$$\tfrac{1}{2}\begin{bmatrix} 0 & -1 & 3 \\ 1 & 0 & 8 \\ -3 & -8 & 0 \end{bmatrix} = \begin{bmatrix} \tfrac{1}{2}(0) & \tfrac{1}{2}(-1) & \tfrac{1}{2}(3) \\ \tfrac{1}{2}(1) & \tfrac{1}{2}(0) & \tfrac{1}{2}(8) \\ \tfrac{1}{2}(-3) & \tfrac{1}{2}(-8) & \tfrac{1}{2}(0) \end{bmatrix} = \begin{bmatrix} 0 & -\tfrac{1}{2} & \tfrac{3}{2} \\ \tfrac{1}{2} & 0 & 4 \\ -\tfrac{3}{2} & -4 & 0 \end{bmatrix}$$, which is skew-symmetric, since $v_{ij} = -v_{ji}$ for all i, j.

Note that $\mathbf{S} + \mathbf{V} = \begin{bmatrix} 3 & -\tfrac{1}{2} & \tfrac{5}{2} \\ -\tfrac{1}{2} & 2 & 1 \\ \tfrac{5}{2} & 1 & 2 \end{bmatrix} + \begin{bmatrix} 0 & -\tfrac{1}{2} & \tfrac{3}{2} \\ \tfrac{1}{2} & 0 & 4 \\ -\tfrac{3}{2} & -4 & 0 \end{bmatrix} =$

$$\begin{bmatrix} (3+0) & (-\tfrac{1}{2}+(-\tfrac{1}{2})) & (\tfrac{5}{2}+\tfrac{3}{2}) \\ (-\tfrac{1}{2}+\tfrac{1}{2}) & (2+0) & (1+4) \\ (\tfrac{5}{2}+(-\tfrac{3}{2})) & (1+(-4)) & (2+0) \end{bmatrix} = \begin{bmatrix} 3 & -1 & 4 \\ 0 & 2 & 5 \\ 1 & -3 & 2 \end{bmatrix} = \mathbf{A}.$$

(5) (d) The matrix (call it **A**) must be a square zero matrix; that is, $\mathbf{O}_n$, for some n. First, since **A** is diagonal, it must be square (by definition of a diagonal matrix). To prove that $a_{ij} = 0$ for all i, j, we consider two cases: First, if $i \neq j$, then $a_{ij} = 0$ since **A** is diagonal. Second, if $i = j$, then $a_{ij} = a_{ii} = 0$ since **A** is skew-symmetric ($a_{ii} = -a_{ii}$ for all i, implying $a_{ii} = 0$ for all i). Hence, every entry $a_{ij} = 0$, and so **A** is a zero matrix.

(11) Part (1): Let $\mathbf{B} = \mathbf{A}^T$ and $\mathbf{C} = (\mathbf{A}^T)^T$. Then $c_{ij} = b_{ji} = a_{ij}$. Since this is true for all i and j, we have $\mathbf{C} = \mathbf{A}$; that is, $(\mathbf{A}^T)^T = \mathbf{A}$.
Part (3): Let $\mathbf{B} = c(\mathbf{A}^T)$, $\mathbf{D} = c\mathbf{A}$ and $\mathbf{F} = (c\mathbf{A})^T$. Then $f_{ij} = d_{ji} = ca_{ji} = b_{ij}$. Since this is true for all i and j, we have $\mathbf{F} = \mathbf{B}$; that is, $(c\mathbf{A})^T = c(\mathbf{A}^T)$.

(14) (a) Trace (**B**) $= b_{11} + b_{22} = 2 + (-1) = 1$, trace (**C**) $= c_{11} + c_{22} = (-1) + 1 = 0$, trace (**E**) $= e_{11} + e_{22} + e_{33} = 0 + (-6) + 0 = -6$, trace (**F**) $= f_{11} + f_{22} + f_{33} + f_{44}$ $= 1+0+0+1 = 2$, trace (**G**) $= g_{11}+g_{22}+g_{33} = 6+6+6 = 18$, trace (**H**) $= h_{11}+h_{22}+h_{33}+h_{44} =$ $0+0+0+0 = 0$, trace (**J**) $= j_{11}+j_{22}+j_{33}+j_{44} = 0+0+1+0 = 1$, trace (**K**) $= k_{11}+k_{22}+k_{33}+k_{44}$ $= 1+1+1+1 = 4$, trace (**L**) $= l_{11} + l_{22} + l_{33} = 1+1+1 = 3$, trace (**M**) $= m_{11} + m_{22} + m_{33} =$ $0+0+0 = 0$, trace (**N**) $= n_{11} + n_{22} + n_{33} = 1+1+1 = 3$, trace (**P**) $= p_{11} + p_{22} = 0+0 = 0$, trace (**Q**) $= q_{11} + q_{22} + q_{33} = (-2) + 0 + 3 = 1$.

(c) No; consider matrices $\mathbf{L}$ and $\mathbf{N}$ in Exercise 2. From part (a) of this exercise, trace($\mathbf{L}$) $= 3 =$ trace($\mathbf{N}$). However, $\mathbf{L} \neq \mathbf{N}$.

(15) (a) False. The main diagonal entries of a 5×6 matrix $\mathbf{A}$ are a_{11}, a_{22}, a_{33}, a_{44}, and a_{55}. Thus, there are 5 entries on the main diagonal. (There is no entry a_{66} because there is no 6th row.)

(b) True. Suppose $\mathbf{A}$ is lower triangular. Then $a_{ij} = 0$ whenever $i < j$. If $\mathbf{B} = \mathbf{A}^T$, then $b_{ij} = a_{ji} = 0$, if $i > j$, and so $\mathbf{B}$ is upper triangular.

(c) False. The square zero matrix $\mathbf{O}_n$ is both skew-symmetric and diagonal.

(d) True. This follows from the definition of a skew-symmetric matrix.

(e) True. $\left(c\left(\mathbf{A}^T + \mathbf{B}\right)\right)^T = \left(c\left(\mathbf{A}^T + \mathbf{B}\right)^T\right)$ (by part (3) of Theorem 1.12) $= \left(c\left(\left(\mathbf{A}^T\right)^T + \mathbf{B}^T\right)\right)$ (by part (2) of Theorem 1.12) $= c\left(\mathbf{A} + \mathbf{B}^T\right)$ (by part (1) of Theorem 1.12) $= c\mathbf{A} + c\mathbf{B}^T$ (by part (5) of Theorem 1.11) $= c\mathbf{B}^T + c\mathbf{A}$ (by part (1) of Theorem 1.11).

Section 1.5

(1) (b) $\mathbf{BA} = \begin{bmatrix} -5 & 3 & 6 \\ 3 & 8 & 0 \\ -2 & 0 & 4 \end{bmatrix} \begin{bmatrix} -2 & 3 \\ 6 & 5 \\ 1 & -4 \end{bmatrix} =$

$$\begin{bmatrix} ((-5)(-2) + (3)(6) + (6)(1)) & ((-5)(3) + (3)(5) + (6)(-4)) \\ ((3)(-2) + (8)(6) + (0)(1)) & ((3)(3) + (8)(5) + (0)(-4)) \\ ((-2)(-2) + (0)(6) + (4)(1)) & ((-2)(3) + (0)(5) + (4)(-4)) \end{bmatrix} = \begin{bmatrix} 34 & -24 \\ 42 & 49 \\ 8 & -22 \end{bmatrix}$$

(c) Impossible; (number of columns of $\mathbf{J}$) $= 1 \neq 2 =$ (number of rows of $\mathbf{M}$)

(e) $\mathbf{RJ} = \begin{bmatrix} -3 & 6 & -2 \end{bmatrix} \begin{bmatrix} 8 \\ -1 \\ 4 \end{bmatrix} = [(-3)(8) + (6)(-1) + (-2)(4)] = [-38]$

(f) $\mathbf{JR} = \begin{bmatrix} 8 \\ -1 \\ 4 \end{bmatrix} \begin{bmatrix} -3 & 6 & -2 \end{bmatrix} = \begin{bmatrix} (8)(-3) & (8)(6) & (8)(-2) \\ (-1)(-3) & (-1)(6) & (-1)(-2) \\ (4)(-3) & (4)(6) & (4)(-2) \end{bmatrix} = \begin{bmatrix} -24 & 48 & -16 \\ 3 & -6 & 2 \\ -12 & 24 & -8 \end{bmatrix}$

(g) Impossible; (number of columns of $\mathbf{R}$) $= 3 \neq 1 =$ (number of rows of $\mathbf{T}$)

(j) Impossible; (number of columns of $\mathbf{F}$) $= 2 \neq 4 =$ (number of rows of $\mathbf{F}$)

(l) $\mathbf{E}^3 = \mathbf{E}(\mathbf{E}^2)$. Let $\mathbf{W} = \mathbf{E}^2 = \begin{bmatrix} 1 & 1 & 0 & 1 \\ 1 & 0 & 1 & 0 \\ 0 & 0 & 0 & 1 \\ 1 & 0 & 1 & 0 \end{bmatrix}^2$. Then

$w_{11} = (1)(1) + (1)(1) + (0)(0) + (1)(1) = 3$
$w_{12} = (1)(1) + (1)(0) + (0)(0) + (1)(0) = 1$
$w_{13} = (1)(0) + (1)(1) + (0)(0) + (1)(1) = 2$
$w_{14} = (1)(1) + (1)(0) + (0)(1) + (1)(0) = 1$
$w_{21} = (1)(1) + (0)(1) + (1)(0) + (0)(1) = 1$
$w_{22} = (1)(1) + (0)(0) + (1)(0) + (0)(0) = 1$
$w_{23} = (1)(0) + (0)(1) + (1)(0) + (0)(1) = 0$
$w_{24} = (1)(1) + (0)(0) + (1)(1) + (0)(0) = 2$
$w_{31} = (0)(1) + (0)(1) + (0)(0) + (1)(1) = 1$
$w_{32} = (0)(1) + (0)(0) + (0)(0) + (1)(0) = 0$

$w_{33} = (0)(0) + (0)(1) + (0)(0) + (1)(1) = 1$
$w_{34} = (0)(1) + (0)(0) + (0)(1) + (1)(0) = 0$
$w_{41} = (1)(1) + (0)(1) + (1)(0) + (0)(1) = 1$
$w_{42} = (1)(1) + (0)(0) + (1)(0) + (0)(0) = 1$
$w_{43} = (1)(0) + (0)(1) + (1)(0) + (0)(1) = 0$
$w_{44} = (1)(1) + (0)(0) + (1)(1) + (0)(0) = 2$

Let $\mathbf{Y} = \mathbf{E}^3 = \mathbf{E}\mathbf{E}^2 = \mathbf{E}\mathbf{W} =$

$$\begin{bmatrix} 1 & 1 & 0 & 1 \\ 1 & 0 & 1 & 0 \\ 0 & 0 & 0 & 1 \\ 1 & 0 & 1 & 0 \end{bmatrix} \begin{bmatrix} 3 & 1 & 2 & 1 \\ 1 & 1 & 0 & 2 \\ 1 & 0 & 1 & 0 \\ 1 & 1 & 0 & 2 \end{bmatrix}. \text{ So,}$$

$y_{11} = (1)(3) + (1)(1) + (0)(1) + (1)(1) = 5$
$y_{12} = (1)(1) + (1)(1) + (0)(0) + (1)(1) = 3$
$y_{13} = (1)(2) + (1)(0) + (0)(1) + (1)(0) = 2$
$y_{14} = (1)(1) + (1)(2) + (0)(0) + (1)(2) = 5$
$y_{21} = (1)(3) + (0)(1) + (1)(1) + (0)(1) = 4$
$y_{22} = (1)(1) + (0)(1) + (1)(0) + (0)(1) = 1$
$y_{23} = (1)(2) + (0)(0) + (1)(1) + (0)(0) = 3$
$y_{24} = (1)(1) + (0)(2) + (1)(0) + (0)(2) = 1$
$y_{31} = (0)(3) + (0)(1) + (0)(1) + (1)(1) = 1$
$y_{32} = (0)(1) + (0)(1) + (0)(0) + (1)(1) = 1$
$y_{33} = (0)(2) + (0)(0) + (0)(1) + (1)(0) = 0$
$y_{34} = (0)(1) + (0)(2) + (0)(0) + (1)(2) = 2$
$y_{41} = (1)(3) + (0)(1) + (1)(1) + (0)(1) = 4$
$y_{42} = (1)(1) + (0)(1) + (1)(0) + (0)(1) = 1$
$y_{43} = (1)(2) + (0)(0) + (1)(1) + (0)(0) = 3$
$y_{44} = (1)(1) + (0)(2) + (1)(0) + (0)(2) = 1$

Hence, $\mathbf{E}^3 = \mathbf{Y} = \begin{bmatrix} 5 & 3 & 2 & 5 \\ 4 & 1 & 3 & 1 \\ 1 & 1 & 0 & 2 \\ 4 & 1 & 3 & 1 \end{bmatrix}$.

(n) $\mathbf{D}(\mathbf{FK}) = \mathbf{D}\left(\begin{bmatrix} 9 & -3 \\ 5 & -4 \\ 2 & 0 \\ 8 & -3 \end{bmatrix} \begin{bmatrix} 2 & 1 & -5 \\ 0 & 2 & 7 \end{bmatrix}\right) =$

$$\mathbf{D}\begin{bmatrix} ((9)(2) + (-3)(0)) & ((9)(1) + (-3)(2)) & ((9)(-5) + (-3)(7)) \\ ((5)(2) + (-4)(0)) & ((5)(1) + (-4)(2)) & ((5)(-5) + (-4)(7)) \\ ((2)(2) + (0)(0)) & ((2)(1) + (0)(2)) & ((2)(-5) + (0)(7)) \\ ((8)(2) + (-3)(0)) & ((8)(1) + (-3)(2)) & ((8)(-5) + (-3)(7)) \end{bmatrix}$$

$$= \begin{bmatrix} -1 & 4 & 3 & 7 \\ 2 & 1 & 7 & 5 \\ 0 & 5 & 5 & -2 \end{bmatrix} \begin{bmatrix} 18 & 3 & -66 \\ 10 & -3 & -53 \\ 4 & 2 & -10 \\ 16 & 2 & -61 \end{bmatrix} = \mathbf{Z}. \quad \text{Then,}$$

$z_{11} = (-1)(18) + (4)(10) + (3)(4) + (7)(16) = 146$
$z_{12} = (-1)(3) + (4)(-3) + (3)(2) + (7)(2) = 5$

$z_{13} = (-1)(-66) + (4)(-53) + (3)(-10) + (7)(-61) = -603$
$z_{21} = (2)(18) + (1)(10) + (7)(4) + (5)(16) = 154$
$z_{22} = (2)(3) + (1)(-3) + (7)(2) + (5)(2) = 27$
$z_{23} = (2)(-66) + (1)(-53) + (7)(-10) + (5)(-61) = -560$
$z_{31} = (0)(18) + (5)(10) + (5)(4) + (-2)(16) = 38$
$z_{32} = (0)(3) + (5)(-3) + (5)(2) + (-2)(2) = -9$
$z_{33} = (0)(-66) + (5)(-53) + (5)(-10) + (-2)(-61) = -193$

Hence, $\mathbf{D}(\mathbf{FK}) = \mathbf{Z} = \begin{bmatrix} 146 & 5 & -603 \\ 154 & 27 & -560 \\ 38 & -9 & -193 \end{bmatrix}$.

(2) (a) No. $\mathbf{LM} = \begin{bmatrix} 10 & 9 \\ 8 & 7 \end{bmatrix}\begin{bmatrix} 7 & -1 \\ 11 & 3 \end{bmatrix} = \begin{bmatrix} ((10)(7)+(9)(11)) & ((10)(-1)+(9)(3)) \\ ((8)(7)+(7)(11)) & ((8)(-1)+(7)(3)) \end{bmatrix}$

$= \begin{bmatrix} 169 & 17 \\ 133 & 13 \end{bmatrix}$, but $\mathbf{ML} = \begin{bmatrix} 7 & -1 \\ 11 & 3 \end{bmatrix}\begin{bmatrix} 10 & 9 \\ 8 & 7 \end{bmatrix}$

$= \begin{bmatrix} ((7)(10)+(-1)(8)) & ((7)(9)+(-1)(7)) \\ ((11)(10)+(3)(8)) & ((11)(9)+(3)(7)) \end{bmatrix} = \begin{bmatrix} 62 & 56 \\ 134 & 120 \end{bmatrix}$. Hence, $\mathbf{LM} \neq \mathbf{ML}$.

(c) No. Since **A** is a 3×2 matrix and **K** is a 2×3 matrix, **AK** is a 3×3 matrix, while **KA** is a 2×2 matrix. Therefore **AK** can not equal **KA**, since they are different sizes.

(d) Yes. $\mathbf{NP} = \begin{bmatrix} 0 & 0 \\ 0 & 0 \end{bmatrix}\begin{bmatrix} 3 & -1 \\ 4 & 7 \end{bmatrix} = \begin{bmatrix} ((0)(3)+(0)(4)) & ((0)(-1)+(0)(7)) \\ ((0)(3)+(0)(4)) & ((0)(-1)+(0)(7)) \end{bmatrix} = \begin{bmatrix} 0 & 0 \\ 0 & 0 \end{bmatrix}$.

Also, $\mathbf{PN} = \begin{bmatrix} 3 & -1 \\ 4 & 7 \end{bmatrix}\begin{bmatrix} 0 & 0 \\ 0 & 0 \end{bmatrix} = \begin{bmatrix} ((3)(0)+(-1)(0)) & ((3)(0)+(-1)(0)) \\ ((4)(0)+(7)(0)) & ((4)(0)+(7)(0)) \end{bmatrix} = \begin{bmatrix} 0 & 0 \\ 0 & 0 \end{bmatrix}$.

Hence, $\mathbf{NP} = \mathbf{PN}$, and so **N** and **P** commute.

(3) (a) (2nd row of **BG**) = (2nd row of **B**)**G** = $\begin{bmatrix} 3 & 8 & 0 \end{bmatrix}\begin{bmatrix} 5 & 1 & 0 \\ 0 & -2 & -1 \\ 1 & 0 & 3 \end{bmatrix}$. The entries obtained from this are:

(1st column entry) $= ((3)(5) + (8)(0) + (0)(1)) = 15$,
(2nd column entry) $= ((3)(1) + (8)(-2) + (0)(0)) = -13$,
(3rd column entry) $= ((3)(0) + (8)(-1) + (0)(3)) = -8$.

Hence, (2nd row of **BG**) $= [15, -13, -8]$.

(c) (1st column of **SE**) = **S**(1st column of **E**) $= \begin{bmatrix} 6 & -4 & 3 & 2 \end{bmatrix}\begin{bmatrix} 1 \\ 1 \\ 0 \\ 1 \end{bmatrix}$

$= [(6)(1) + (-4)(1) + (3)(0) + (2)(1)] = [4]$

(4) (a) Valid, by Theorem 1.14, part (1).

(b) Invalid. The equation claims that all pairs of matrices that can be multiplied in both orders commute. However, parts (a) and (c) of Exercise 2 illustrate two different pairs of matrices that do not commute (see above).

(c) Valid, by Theorem 1.14, part (1).

(d) Valid, by Theorem 1.14, part (2).

(e) Valid, by Theorem 1.16.

(f) Invalid. For a counterexample, consider the matrices **L** and **M** from Exercises 1 through 3. Using our computation of **ML** from Exercise 2(a), above, $\mathbf{L}(\mathbf{ML}) = \begin{bmatrix} 10 & 9 \\ 8 & 7 \end{bmatrix}\begin{bmatrix} 62 & 56 \\ 134 & 120 \end{bmatrix} =$

$$\begin{bmatrix} ((10)(62)+(9)(134)) & ((10)(56)+(9)(120)) \\ ((8)(62)+(7)(134)) & ((8)(56)+(7)(120)) \end{bmatrix} = \begin{bmatrix} 1826 & 1640 \\ 1434 & 1288 \end{bmatrix},$$

but $\mathbf{L}^2\mathbf{M} = (\mathbf{LL})\mathbf{M} = \left(\begin{bmatrix} 10 & 9 \\ 8 & 7 \end{bmatrix}\begin{bmatrix} 10 & 9 \\ 8 & 7 \end{bmatrix}\right)\mathbf{M}$

$$= \begin{bmatrix} ((10)(10)+(9)(8)) & ((10)(9)+(9)(7)) \\ ((8)(10)+(7)(8)) & ((8)(9)+(7)(7)) \end{bmatrix}\mathbf{M} = \begin{bmatrix} 172 & 153 \\ 136 & 121 \end{bmatrix}\begin{bmatrix} 7 & -1 \\ 11 & 3 \end{bmatrix}$$

$$= \begin{bmatrix} ((172)(7)+(153)(11)) & ((172)(-1)+(153)(3)) \\ ((136)(7)+(121)(11)) & ((136)(-1)+(121)(3)) \end{bmatrix} = \begin{bmatrix} 2887 & 287 \\ 2283 & 227 \end{bmatrix}.$$

(g) Valid, by Theorem 1.14, part (3).

(h) Valid, by Theorem 1.14, part (2).

(i) Invalid. For a counterexample, consider the matrices **A** and **K** from Exercises 1 through 3. Note that **A** is a 3×2 matrix and **K** is a 2×3 matrix. Therefore, **AK** is a 3×3 matrix. Hence, $(\mathbf{AK})^T$ is a 3×3 matrix. However, $\mathbf{A}^T$ is a 2×3 matrix and $\mathbf{K}^T$ is a 3×2 matrix. Thus, $\mathbf{A}^T\mathbf{K}^T$ is a 2×2 matrix, and so can not equal $(\mathbf{AK})^T$.
The equation is also false in general for square matrices, where it is not the sizes of the matrices that cause the problem. For example, let $\mathbf{A} = \begin{bmatrix} 1 & 1 \\ 0 & 0 \end{bmatrix}$, and let $\mathbf{K} = \begin{bmatrix} 1 & -1 \\ 0 & 0 \end{bmatrix}$. Then,

$$(\mathbf{AK})^T = \left(\begin{bmatrix} 1 & 1 \\ 0 & 0 \end{bmatrix}\begin{bmatrix} 1 & -1 \\ 0 & 0 \end{bmatrix}\right)^T = \begin{bmatrix} ((1)(1)+(1)(0)) & ((1)(-1)+(1)(0)) \\ ((0)(1)+(0)(0)) & ((0)(-1)+(0)(0)) \end{bmatrix}^T =$$

$$\begin{bmatrix} 1 & -1 \\ 0 & 0 \end{bmatrix}^T = \begin{bmatrix} 1 & 0 \\ -1 & 0 \end{bmatrix}, \text{ while } \mathbf{A}^T\mathbf{K}^T = \begin{bmatrix} 1 & 0 \\ 1 & 0 \end{bmatrix}\begin{bmatrix} 1 & 0 \\ -1 & 0 \end{bmatrix}$$

$$= \begin{bmatrix} ((1)(1)+(0)(-1)) & ((1)(0)+(0)(0)) \\ ((1)(1)+(0)(-1)) & ((1)(0)+(0)(0)) \end{bmatrix} = \begin{bmatrix} 1 & 0 \\ 1 & 0 \end{bmatrix}.$$

However, there are some rare instances for which the equation is true, such as when $\mathbf{A} = \mathbf{K}$, or when either **A** or **K** equals $\mathbf{I}_n$. Another example for which the equation holds is for $\mathbf{A} = \mathbf{G}$ and $\mathbf{K} = \mathbf{H}$, where **G** and **H** are the matrices used in Exercises 1, 2, and 3.

(j) Valid, by Theorem 1.14, part (3), and Theorem 1.16.

(5) To find the total in salaries paid by Outlet 1, we must compute the total amount paid to executives, the total amount paid to salespersons, and the total amount paid to others, and then add these amounts. But each of these totals is found by multiplying the number of that type of employee working at Outlet 1 by the salary for that type of employee. Hence, the total salary paid at Outlet 1 is

$$\underbrace{(3)(30000)}_{\text{Executives}} + \underbrace{(7)(22500)}_{\text{Salespersons}} + \underbrace{(8)(15000)}_{\text{Others}} = \underbrace{367500}_{\text{Salary Total}}.$$

Note that this is the $(1,1)$ entry obtained when multiplying the two given matrices. A similar analysis shows that multiplying the two given matrices gives all the desired salary and fringe benefit totals. In particular, if

$$\mathbf{W} = \begin{bmatrix} 3 & 7 & 8 \\ 2 & 4 & 5 \\ 6 & 14 & 18 \\ 3 & 6 & 9 \end{bmatrix} \begin{bmatrix} 30000 & 7500 \\ 22500 & 4500 \\ 15000 & 3000 \end{bmatrix}, \text{ then}$$

$w_{11} = (3)(30000) + (7)(22500) + (8)(15000) = 367500,$
$w_{12} = (3)(7500) + (7)(4500) + (8)(3000) = 78000,$
$w_{21} = (2)(30000) + (4)(22500) + (5)(15000) = 225000,$
$w_{22} = (2)(7500) + (4)(4500) + (5)(3000) = 48000,$
$w_{31} = (6)(30000) + (14)(22500) + (18)(15000) = 765000,$
$w_{32} = (6)(7500) + (14)(4500) + (18)(3000) = 162000,$
$w_{41} = (3)(30000) + (6)(22500) + (9)(15000) = 360000,$
$w_{42} = (3)(7500) + (6)(4500) + (9)(3000) = 76500.$

$$\text{Hence, we have: } \mathbf{W} = \begin{array}{c} \\ \text{Outlet 1} \\ \text{Outlet 2} \\ \text{Outlet 3} \\ \text{Outlet 4} \end{array} \begin{array}{c} \begin{array}{cc} \text{Salary} & \text{Fringe Benefits} \end{array} \\ \begin{bmatrix} \$367500 & \$78000 \\ \$225000 & \$48000 \\ \$765000 & \$162000 \\ \$360000 & \$76500 \end{bmatrix} \end{array}.$$

(6) To compute the tonnage of a particular chemical applied to a given field, for each type of fertilizer you must multiply the percent concentration of the chemical in the fertilizer by the number of tons of that fertilizer applied to the given field. After this is done for each fertilizer type, we add these results to get the total tonnage of the chemical that was applied. For example, to compute the number of tons of potash applied to field 2, we compute

$$\underbrace{0.05}_{\text{\% Potash in Fert. 1}} \cdot \underbrace{2}_{\text{Tons of Fert. 1}} + \underbrace{0.05}_{\text{\% Potash in Fert. 2}} \cdot \underbrace{1}_{\text{Tons of Fert. 2}} + \underbrace{0.20}_{\text{\% Potash in Fert. 3}} \cdot \underbrace{1}_{\text{Tons of Fert. 3}} = \underbrace{0.35}_{\text{Tons of Potash}}.$$

Note that this is the dot product (3rd column of $\mathbf{A}$)·(2nd column of $\mathbf{B}$). Since this is the dot product of two columns, it is not in the form of an entry of a matrix product. To turn this into a form that looks like an entry of a matrix product, we consider the transpose of $\mathbf{A}^T$, and express it as (3rd row of $\mathbf{A}^T$)·(2nd column of $\mathbf{B}$). Thus, this is the $(3,2)$ entry of $\mathbf{A}^T\mathbf{B}$. A similar analysis shows that computing all of the entries of $\mathbf{A}^T\mathbf{B}$ produces all of the information requested in the problem. Hence, to solve the problem, we compute

$$\mathbf{Y} = \mathbf{A}^T\mathbf{B} = \begin{bmatrix} 0.10 & 0.25 & 0.00 \\ 0.10 & 0.05 & 0.10 \\ 0.05 & 0.05 & 0.20 \end{bmatrix} \begin{bmatrix} 5 & 2 & 4 \\ 2 & 1 & 1 \\ 3 & 1 & 3 \end{bmatrix}.$$

Solving for each entry produces

$y_{11} = (0.10)(5) + (0.25)(2) + (0.00)(3) = 1.00,$
$y_{12} = (0.10)(2) + (0.25)(1) + (0.00)(1) = 0.45,$
$y_{13} = (0.10)(4) + (0.25)(1) + (0.00)(3) = 0.65,$
$y_{21} = (0.10)(5) + (0.05)(2) + (0.10)(3) = 0.90,$
$y_{22} = (0.10)(2) + (0.05)(1) + (0.10)(1) = 0.35,$
$y_{23} = (0.10)(4) + (0.05)(1) + (0.10)(3) = 0.75,$

$y_{31} = (0.05)(5) + (0.05)(2) + (0.20)(3) = 0.95,$
$y_{32} = (0.05)(2) + (0.05)(1) + (0.20)(1) = 0.35,$
$y_{33} = (0.05)(4) + (0.05)(1) + (0.20)(3) = 0.85.$

$$\text{Hence,}\quad \mathbf{Y} = \begin{array}{c} \\ \text{Nitrogen} \\ \text{Phosphate} \\ \text{Potash} \end{array} \begin{array}{c} \begin{array}{ccc} \text{Field 1} & \text{Field 2} & \text{Field 3} \end{array} \\ \left[\begin{array}{ccc} 1.00 & 0.45 & 0.65 \\ 0.90 & 0.35 & 0.75 \\ 0.95 & 0.35 & 0.85 \end{array}\right] \end{array} \text{(in tons).}$$

(7) (a) An example: $\mathbf{A} = \begin{bmatrix} 1 & 1 \\ 0 & -1 \end{bmatrix}$. Note that $\mathbf{A}^2 = \begin{bmatrix} 1 & 1 \\ 0 & -1 \end{bmatrix}\begin{bmatrix} 1 & 1 \\ 0 & -1 \end{bmatrix}$

$$= \begin{bmatrix} ((1)(1) + (1)(0)) & ((1)(1) + (1)(-1)) \\ ((0)(1) + (-1)(0)) & ((0)(1) + (-1)(-1)) \end{bmatrix} = \mathbf{I}_2.$$

(b) One example: Suppose $\mathbf{A} = \begin{bmatrix} 1 & 1 & 0 \\ 0 & -1 & 0 \\ 0 & 0 & 1 \end{bmatrix}$. Let $\mathbf{W} = \mathbf{A}^2 = \begin{bmatrix} 1 & 1 & 0 \\ 0 & -1 & 0 \\ 0 & 0 & 1 \end{bmatrix}\begin{bmatrix} 1 & 1 & 0 \\ 0 & -1 & 0 \\ 0 & 0 & 1 \end{bmatrix}$.

Computing each entry yields

$w_{11} = (1)(1) + (1)(0) + (0)(0) = 1,$
$w_{12} = (1)(1) + (1)(-1) + (0)(0) = 0,$
$w_{13} = (1)(0) + (1)(0) + (0)(1) = 0,$
$w_{21} = (0)(1) + (-1)(0) + (0)(0) = 0,$
$w_{22} = (0)(1) + (-1)(-1) + (0)(0) = 1,$
$w_{23} = (0)(0) + (-1)(0) + (0)(1) = 0,$
$w_{31} = (0)(1) + (0)(0) + (1)(0) = 0,$
$w_{32} = (0)(1) + (0)(-1) + (1)(0) = 0,$
$w_{33} = (0)(0) + (0)(0) + (1)(1) = 1.$

Hence, $\mathbf{A}^2 = \mathbf{W} = \mathbf{I}_3$.

(c) Consider $\mathbf{A} = \begin{bmatrix} 0 & 0 & 1 \\ 1 & 0 & 0 \\ 0 & 1 & 0 \end{bmatrix}$. Let $\mathbf{Y} = \mathbf{A}^2 = \begin{bmatrix} 0 & 0 & 1 \\ 1 & 0 & 0 \\ 0 & 1 & 0 \end{bmatrix}\begin{bmatrix} 0 & 0 & 1 \\ 1 & 0 & 0 \\ 0 & 1 & 0 \end{bmatrix}$. First, we compute each entry of $\mathbf{Y}$.

$y_{11} = (0)(0) + (0)(1) + (1)(0) = 0,$
$y_{12} = (0)(0) + (0)(0) + (1)(1) = 1,$
$y_{13} = (0)(1) + (0)(0) + (1)(0) = 0,$
$y_{21} = (1)(0) + (0)(1) + (0)(0) = 0,$
$y_{22} = (1)(0) + (0)(0) + (0)(1) = 0,$
$y_{23} = (1)(1) + (0)(0) + (0)(0) = 1,$
$y_{31} = (0)(0) + (1)(1) + (0)(0) = 1,$
$y_{32} = (0)(0) + (1)(0) + (0)(1) = 0,$
$y_{33} = (0)(1) + (1)(0) + (0)(0) = 0.$

And so $\mathbf{Y} = \mathbf{A}^2 = \begin{bmatrix} 0 & 1 & 0 \\ 0 & 0 & 1 \\ 1 & 0 & 0 \end{bmatrix}$. Hence, $\mathbf{A}^3 = \mathbf{A}^2\mathbf{A} = \mathbf{Y}\mathbf{A} = \begin{bmatrix} 0 & 1 & 0 \\ 0 & 0 & 1 \\ 1 & 0 & 0 \end{bmatrix}\begin{bmatrix} 0 & 0 & 1 \\ 1 & 0 & 0 \\ 0 & 1 & 0 \end{bmatrix}$.

Letting this product equal $\mathbf{Z}$, we compute each entry as follows:

$z_{11} = (0)(0) + (1)(1) + (0)(0) = 1,$
$z_{12} = (0)(0) + (1)(0) + (0)(1) = 0,$

$z_{13} = (0)(1) + (1)(0) + (0)(0) = 0,$
$z_{21} = (0)(0) + (0)(1) + (1)(0) = 0,$
$z_{22} = (0)(0) + (0)(0) + (1)(1) = 1,$
$z_{23} = (0)(1) + (0)(0) + (1)(0) = 0,$
$z_{31} = (1)(0) + (0)(1) + (0)(0) = 0,$
$z_{32} = (1)(0) + (0)(0) + (0)(1) = 0,$
$z_{33} = (1)(1) + (0)(0) + (0)(0) = 1.$

Therefore, $\mathbf{A}^3 = \mathbf{Z} = \mathbf{I}_3$.

(8) (a) This is the dot product of the 3rd row of **A** with the 4th column of **B**. Hence, the result is the 3rd row, 4th column entry of **AB**.

(c) This is (2nd column of **A**)·(3rd row of **B**) = (3rd row of **B**)·(2nd column of **A**) = 3rd row, 2nd column entry of **BA**.

(9) (a) ((3, 2) entry of **AB**) = (3rd row of **A**)·(2nd column of **B**) = $\sum_{k=1}^{n} a_{3k}b_{k2}$

(11) Several crucial steps in the following proofs rely on Theorem 1.5 for their validity.

Proof of Part (2): The (i,j) entry of $\mathbf{A}(\mathbf{B}+\mathbf{C})$
= (ith row of **A**)·(jth column of $(\mathbf{B}+\mathbf{C})$)
= (ith row of **A**)·(jth column of **B** + jth column of **C**)
= (ith row of **A**)·(jth column of **B**)
+ (ith row of **A**)·(jth column of **C**)
= (i,j) entry of **AB** + (i,j) entry of **AC**
= (i,j) entry of $(\mathbf{AB}+\mathbf{AC})$.

Proof of Part (3): The (i,j) entry of $(\mathbf{A}+\mathbf{B})\mathbf{C}$
= (ith row of $(\mathbf{A}+\mathbf{B})$)·(jth column of **C**)
= ((ith row of **A**) + (ith row of **B**)) · (jth column of **C**)
= (ith row of **A**)·(jth column of **C**)
+ (ith row of **B**)·(jth column of **C**)
= (i,j) entry of **AC** + (i,j) entry of **BC**
= (i,j) entry of $(\mathbf{AC}+\mathbf{BC})$.

For the first equation in part (4), the (i,j) entry of $c(\mathbf{AB})$
= c ((ith row of **A**)·(jth column of **B**))
= (c(ith row of **A**))·(jth column of **B**)
= (ith row of $c\mathbf{A}$)·(jth column of **B**)
= (i,j) entry of $(c\mathbf{A})\mathbf{B}$.

Similarly, the (i,j) entry of $c(\mathbf{AB})$
= c ((ith row of **A**)·(jth column of **B**))
= (ith row of **A**)·(c(jth column of **B**))
= (ith row of **A**)·(jth column of $c\mathbf{B}$)
= (i,j) entry of $\mathbf{A}(c\mathbf{B})$.

(16) Proof of Part (1): We use induction on the variable t. Base Step: $\mathbf{A}^{s+0} = \mathbf{A}^s = \mathbf{A}^s\mathbf{I} = \mathbf{A}^s\mathbf{A}^0$. Inductive Step: Assume $\mathbf{A}^{s+t} = \mathbf{A}^s\mathbf{A}^t$ for some $t \geq 0$. We must prove $\mathbf{A}^{s+(t+1)} = \mathbf{A}^s\mathbf{A}^{t+1}$. But $\mathbf{A}^{s+(t+1)} = \mathbf{A}^{(s+t)+1} = \mathbf{A}^{s+t}\mathbf{A} = (\mathbf{A}^s\mathbf{A}^t)\mathbf{A}$ (by the inductive hypothesis) $= \mathbf{A}^s(\mathbf{A}^t\mathbf{A}) = \mathbf{A}^s\mathbf{A}^{t+1}$.

Proof of Part (2): Again, we use induction on the variable t. Base Step: $(\mathbf{A}^s)^0 = \mathbf{I} = \mathbf{A}^0 = \mathbf{A}^{s0}$. Inductive Step: Assume $(\mathbf{A}^s)^t = \mathbf{A}^{st}$ for some integer $t \geq 0$. We must prove $(\mathbf{A}^s)^{t+1} = \mathbf{A}^{s(t+1)}$. But $(\mathbf{A}^s)^{t+1} = (\mathbf{A}^s)^t\mathbf{A}^s$ (by definition) $= \mathbf{A}^{st}\mathbf{A}^s$ (by the inductive hypothesis) $= \mathbf{A}^{st+s}$ (by part (1))

$= \mathbf{A}^{s(t+1)}$. Finally, reversing the roles of s and t in the proof above shows that $(\mathbf{A}^t)^s = \mathbf{A}^{ts}$, which equals $\mathbf{A}^{st}$.

(23) (a) Consider any matrix $\mathbf{A}$ of the form $\begin{bmatrix} 1 & 0 \\ x & 0 \end{bmatrix}$. Then $\mathbf{A}^2 = \begin{bmatrix} 1 & 0 \\ x & 0 \end{bmatrix}\begin{bmatrix} 1 & 0 \\ x & 0 \end{bmatrix} =$

$\begin{bmatrix} ((1)(1)+(0)(x)) & ((1)(0)+(0)(0)) \\ ((x)(1)+(0)(x)) & ((x)(0)+(0)(0)) \end{bmatrix} = \mathbf{A}$. So, for example, $\begin{bmatrix} 1 & 0 \\ 1 & 0 \end{bmatrix}$ is idempotent.

(24) (b) Consider $\mathbf{A} = \begin{bmatrix} 1 & 2 & -1 \\ 2 & 4 & -2 \end{bmatrix}$ and $\mathbf{B} = \begin{bmatrix} 1 & -2 \\ 0 & 1 \\ 1 & 0 \end{bmatrix}$. Then $\mathbf{AB} = \begin{bmatrix} 1 & 2 & -1 \\ 2 & 4 & -2 \end{bmatrix}\begin{bmatrix} 1 & -2 \\ 0 & 1 \\ 1 & 0 \end{bmatrix} =$

$\begin{bmatrix} ((1)(1)+(2)(0)+(-1)(1)) & ((1)(-2)+(2)(1)+(-1)(0)) \\ ((2)(1)+(4)(0)+(-2)(1)) & ((2)(-2)+(4)(1)+(-2)(0)) \end{bmatrix} = \mathbf{O}_2$.

(25) A 2×2 matrix $\mathbf{A}$ that commutes with every other 2×2 matrix must have the form $\mathbf{A} = c\mathbf{I}_2$. To prove this, note that if $\mathbf{A}$ commutes with every other 2×2 matrix, then $\mathbf{AB} = \mathbf{BA}$, where $\mathbf{B} = \begin{bmatrix} 0 & 1 \\ 0 & 0 \end{bmatrix}$. But

$\mathbf{AB} = \begin{bmatrix} a_{11} & a_{12} \\ a_{21} & a_{22} \end{bmatrix}\begin{bmatrix} 0 & 1 \\ 0 & 0 \end{bmatrix} = \begin{bmatrix} ((a_{11})(0)+(a_{12})(0)) & ((a_{11})(1)+(a_{12})(0)) \\ ((a_{21})(0)+(a_{22})(0)) & ((a_{21})(1)+(a_{22})(0)) \end{bmatrix} = \begin{bmatrix} 0 & a_{11} \\ 0 & a_{21} \end{bmatrix}$ and

$\mathbf{BA} = \begin{bmatrix} 0 & 1 \\ 0 & 0 \end{bmatrix}\begin{bmatrix} a_{11} & a_{12} \\ a_{21} & a_{22} \end{bmatrix} = \begin{bmatrix} ((0)(a_{11})+(1)(a_{21})) & ((0)(a_{12})+(1)(a_{22})) \\ ((0)(a_{11})+(0)(a_{21})) & ((0)(a_{12})+(0)(a_{22})) \end{bmatrix} = \begin{bmatrix} a_{21} & a_{22} \\ 0 & 0 \end{bmatrix}$.

Setting these equal shows that $a_{21} = 0$ and $a_{11} = a_{22}$. Let $c = a_{11} = a_{22}$. Then $\mathbf{A} = \begin{bmatrix} c & a_{12} \\ 0 & c \end{bmatrix}$.

Let $\mathbf{D} = \begin{bmatrix} 0 & 0 \\ 1 & 0 \end{bmatrix}$. Then $\mathbf{AD} = \begin{bmatrix} c & a_{12} \\ 0 & c \end{bmatrix}\begin{bmatrix} 0 & 0 \\ 1 & 0 \end{bmatrix} =$

$\begin{bmatrix} ((c)(0)+(a_{12})(1)) & ((c)(0)+(a_{12})(0)) \\ ((0)(0)+(c)(1)) & ((0)(0)+(c)(0)) \end{bmatrix} = \begin{bmatrix} a_{12} & 0 \\ c & 0 \end{bmatrix}$ and $\mathbf{DA} = \begin{bmatrix} 0 & 0 \\ 1 & 0 \end{bmatrix}\begin{bmatrix} c & a_{12} \\ 0 & c \end{bmatrix} =$

$\begin{bmatrix} ((0)(c)+(0)(0)) & ((0)(a_{12})+(0)(c)) \\ ((1)(c)+(0)(0)) & ((1)(a_{12})+(0)(c)) \end{bmatrix} = \begin{bmatrix} 0 & 0 \\ c & a_{12} \end{bmatrix}$.

Setting $\mathbf{AD} = \mathbf{DA}$ shows that $a_{12} = 0$, and so $\mathbf{A} = c\mathbf{I}_2$.
Finally, we must show that $c\mathbf{I}_2$ actually does commute with every 2×2 matrix. If $\mathbf{M}$ is a 2×2 matrix, then $(c\mathbf{I}_2)\mathbf{M} = c(\mathbf{I}_2\mathbf{M}) = c\mathbf{M}$. Similarly, $\mathbf{M}(c\mathbf{I}_2) = c(\mathbf{MI}_2) = c\mathbf{M}$. Hence $c\mathbf{I}_2$ commutes with $\mathbf{M}$.

(27) (a) True. This is a "boxed" fact given in the section.

(b) True. $\mathbf{D}(\mathbf{A}+\mathbf{B}) = \mathbf{DA}+\mathbf{DB}$ (by part (2) of Theorem 1.14) $= \mathbf{DB}+\mathbf{DA}$ (by part (1) of Theorem 1.1)

(c) True. This is part (4) of Theorem 1.14.

(d) False. Let $\mathbf{D} = \begin{bmatrix} 1 & 1 \\ 0 & -1 \end{bmatrix}$, and let $\mathbf{E} = \begin{bmatrix} 1 & -2 \\ 0 & 1 \end{bmatrix}$. Now $\mathbf{DE} = \begin{bmatrix} 1 & 1 \\ 0 & -1 \end{bmatrix}\begin{bmatrix} 1 & -2 \\ 0 & 1 \end{bmatrix} =$

$\begin{bmatrix} ((1)(1)+(1)(0)) & ((1)(-2)+(1)(1)) \\ ((0)(1)+(-1)(0)) & ((0)(-2)+(-1)(1)) \end{bmatrix} = \begin{bmatrix} 1 & -1 \\ 0 & -1 \end{bmatrix}$. Hence,

$$(\mathbf{DE})^2 = \begin{bmatrix} 1 & -1 \\ 0 & -1 \end{bmatrix} \begin{bmatrix} 1 & -1 \\ 0 & -1 \end{bmatrix} = \begin{bmatrix} ((1)(1)+(-1)(0)) & ((1)(-1)+(-1)(-1)) \\ ((0)(1)+(-1)(0)) & ((0)(-1)+(-1)(-1)) \end{bmatrix} = \mathbf{I}_2.$$

But, $\mathbf{D}^2 = \begin{bmatrix} 1 & 1 \\ 0 & -1 \end{bmatrix} \begin{bmatrix} 1 & 1 \\ 0 & -1 \end{bmatrix} = \begin{bmatrix} ((1)(1)+(1)(0)) & ((1)(1)+(1)(-1)) \\ ((0)(1)+(-1)(0)) & ((0)(1)+(-1)(-1)) \end{bmatrix} = \mathbf{I}_2$. Also,

$$\mathbf{E}^2 = \begin{bmatrix} 1 & -2 \\ 0 & 1 \end{bmatrix} \begin{bmatrix} 1 & -2 \\ 0 & 1 \end{bmatrix} = \begin{bmatrix} ((1)(1)+(-2)(0)) & ((1)(-2)+(-2)(1)) \\ ((0)(1)+(1)(0)) & ((0)(-2)+(1)(1)) \end{bmatrix} = \begin{bmatrix} 1 & -4 \\ 0 & 1 \end{bmatrix}.$$

Hence, $\mathbf{D}^2\mathbf{E}^2 = \mathbf{I}_2\mathbf{E}^2 = \mathbf{E}^2 \neq \mathbf{I}_2$. Thus, $(\mathbf{DE})^2 \neq \mathbf{D}^2\mathbf{E}^2$. The problem here is that $\mathbf{D}$ and $\mathbf{E}$ do not commute. For, *if they did*, we would have $(\mathbf{DE})^2 = (\mathbf{DE})(\mathbf{DE}) = \mathbf{D}(\mathbf{ED})\mathbf{E} = \mathbf{D}(\mathbf{DE})\mathbf{E} = (\mathbf{DD})(\mathbf{EE}) = \mathbf{D}^2\mathbf{E}^2$.

(e) False. See the answer for part (i) of Exercise 4 for a counterexample. We know from Theorem 1.16 that $(\mathbf{DE})^T = \mathbf{E}^T\mathbf{D}^T$. So if $\mathbf{D}^T$ and $\mathbf{E}^T$ do not commute, then $(\mathbf{DE})^T \neq \mathbf{D}^T\mathbf{E}^T$.

(f) False. See the answer for part (b) of Exercise 24 for a counterexample. $\mathbf{DE} = \mathbf{O}$ whenever every row of $\mathbf{D}$ is orthogonal to every column of $\mathbf{E}$, but neither matrix is forced to be zero for $\mathbf{DE} = \mathbf{O}$ to be true.

Chapter 2

Section 2.1

(1) In each part, we first set up the augmented matrix corresponding to the given system of linear equations. Then we perform the row operations designated by the Gaussian elimination method. Finally, we use the final augmented matrix obtained to give the solution set for the original system.

(a) First augmented matrix: $\left[\begin{array}{ccc|c} -5 & -2 & 2 & 14 \\ 3 & 1 & -1 & -8 \\ 2 & 2 & -1 & -3 \end{array}\right]$. The first pivot is the (1,1) entry. We make this a "1" by performing the following type (I) row operation:

Type (I) operation: $\langle 1\rangle \leftarrow -\frac{1}{5}\langle 1\rangle$: $\left[\begin{array}{ccc|c} \textcircled{1} & \frac{2}{5} & -\frac{2}{5} & -\frac{14}{5} \\ 3 & 1 & -1 & -8 \\ 2 & 2 & -1 & -3 \end{array}\right]$

We now target the (2,1) entry, and make it "0." We do this using the following type (II) row operation:

Type (II) operation: $\langle 2\rangle \leftarrow (-3)\times\langle 1\rangle + \langle 2\rangle$:

Side Calculation

$$\begin{array}{rccc|c} (-3)\times \text{(row 1)} & -3 & -\frac{6}{5} & \frac{6}{5} & \frac{42}{5} \\ \text{(row 2)} & 3 & 1 & -1 & -8 \\ \hline \text{(sum)} & 0 & -\frac{1}{5} & \frac{1}{5} & \frac{2}{5} \end{array}$$

Resultant Matrix

$$\left[\begin{array}{ccc|c} \textcircled{1} & \frac{2}{5} & -\frac{2}{5} & -\frac{14}{5} \\ 0 & -\frac{1}{5} & \frac{1}{5} & \frac{2}{5} \\ 2 & 2 & -1 & -3 \end{array}\right]$$

Next, we target the (3,1) entry.
Type (II) operation: $\langle 3\rangle \leftarrow (-2)\times\langle 1\rangle + \langle 3\rangle$:

Side Calculation

$$\begin{array}{rccc|c} (-2)\times \text{(row 1)} & -2 & -\frac{4}{5} & \frac{4}{5} & \frac{28}{5} \\ \text{(row 3)} & 2 & 2 & -1 & -3 \\ \hline \text{(sum)} & 0 & \frac{6}{5} & -\frac{1}{5} & \frac{13}{5} \end{array}$$

Resultant Matrix

$$\left[\begin{array}{ccc|c} \textcircled{1} & \frac{2}{5} & -\frac{2}{5} & -\frac{14}{5} \\ 0 & -\frac{1}{5} & \frac{1}{5} & \frac{2}{5} \\ 0 & \frac{6}{5} & -\frac{1}{5} & \frac{13}{5} \end{array}\right]$$

Now we move the pivot to the (2,2) entry. We make this entry a "1" using a type (I) row operation.

Type (I) operation: $\langle 2\rangle \leftarrow -5\,\langle 2\rangle$: $\left[\begin{array}{ccc|c} 1 & \frac{2}{5} & -\frac{2}{5} & -\frac{14}{5} \\ 0 & \textcircled{1} & -1 & -2 \\ 0 & \frac{6}{5} & -\frac{1}{5} & \frac{13}{5} \end{array}\right]$

Now we target the (3,2) entry.
Type (II) operation: $\langle 3\rangle \leftarrow (-\frac{6}{5})\times\langle 2\rangle + \langle 3\rangle$:

Side Calculation

$$\begin{array}{rccc|c} (-\frac{6}{5})\times \text{(row 2)} & 0 & -\frac{6}{5} & \frac{6}{5} & \frac{12}{5} \\ \text{(row 3)} & 0 & \frac{6}{5} & -\frac{1}{5} & \frac{13}{5} \\ \hline \text{(sum)} & 0 & 0 & 1 & 5 \end{array}$$

Resultant Matrix

$$\left[\begin{array}{ccc|c} 1 & \frac{2}{5} & -\frac{2}{5} & -\frac{14}{5} \\ 0 & \textcircled{1} & -1 & -2 \\ 0 & 0 & 1 & 5 \end{array}\right]$$

The pivot moves to the (3,3) entry. However, the (3,3) entry already equals 1. Since there are no rows below the third row, we are finished performing row operations. The final augmented matrix corresponds to the following linear system:

$$\begin{cases} x_1 + \frac{2}{5}x_2 - \frac{2}{5}x_3 = -\frac{14}{5} \\ \quad x_2 - x_3 = -2 \\ \quad x_3 = 5 \end{cases}$$

The last equation clearly gives $x_3 = 5$. Substituting this value into the second equation yields $x_2 - 5 = -2$, or $x_2 = 3$. Substituting both values into the first equation produces $x_1 + \frac{2}{5}(3) - \frac{2}{5}(5) = -\frac{14}{5}$, which leads to $x_1 = -2$. Hence, the solution set for the system is $\{(-2, 3, 5)\}$.

(c) First augmented matrix: $\left[\begin{array}{ccc|c} 3 & -2 & 4 & -54 \\ -1 & 1 & -2 & 20 \\ 5 & -4 & 8 & -83 \end{array}\right]$. The first pivot is the (1,1) entry. We make this a "1" by performing the following type (I) row operation:

Type (I) operation: $\langle 1\rangle \leftarrow \frac{1}{3}\langle 1\rangle$: $\left[\begin{array}{ccc|c} ① & -\frac{2}{3} & \frac{4}{3} & -18 \\ -1 & 1 & -2 & 20 \\ 5 & -4 & 8 & -83 \end{array}\right]$

We now target the (2,1) entry, and make it "0." We do this using the following type (II) row operation:

Type (II) operation: $\langle 2\rangle \leftarrow (1) \times \langle 1\rangle + \langle 2\rangle$:

Side Calculation

$$\begin{array}{rccc|c} (1)\times \text{ (row 1)} & 1 & -\frac{2}{3} & \frac{4}{3} & -18 \\ \text{(row 2)} & -1 & 1 & -2 & 20 \\ \hline \text{(sum)} & 0 & \frac{1}{3} & -\frac{2}{3} & 2 \end{array}$$

Resultant Matrix

$$\left[\begin{array}{ccc|c} ① & -\frac{2}{3} & \frac{4}{3} & -18 \\ 0 & \frac{1}{3} & -\frac{2}{3} & 2 \\ 5 & -4 & 8 & -83 \end{array}\right]$$

Next, we target the (3,1) entry.

Type (II) operation: $\langle 3\rangle \leftarrow (-5) \times \langle 1\rangle + \langle 3\rangle$:

Side Calculation

$$\begin{array}{rccc|c} (-5)\times \text{ (row 1)} & -5 & \frac{10}{3} & -\frac{20}{3} & 90 \\ \text{(row 3)} & 5 & -4 & 8 & -83 \\ \hline \text{(sum)} & 0 & -\frac{2}{3} & \frac{4}{3} & 7 \end{array}$$

Resultant Matrix

$$\left[\begin{array}{ccc|c} ① & -\frac{2}{3} & \frac{4}{3} & -18 \\ 0 & \frac{1}{3} & -\frac{2}{3} & 2 \\ 0 & -\frac{2}{3} & \frac{4}{3} & 7 \end{array}\right]$$

Now we move the pivot to the (2,2) entry. We make this entry a "1" using a type (I) row operation.

Type (I) operation: $\langle 2\rangle \leftarrow 3\,\langle 2\rangle$: $\left[\begin{array}{ccc|c} 1 & -\frac{2}{3} & \frac{4}{3} & -18 \\ 0 & ① & -2 & 6 \\ 0 & -\frac{2}{3} & \frac{4}{3} & 7 \end{array}\right]$

Now we target the (3,2) entry.

Type (II) operation: $\langle 3\rangle \leftarrow (\frac{2}{3}) \times \langle 2\rangle + \langle 3\rangle$:

Side Calculation

$$\begin{array}{rccc|c} (\frac{2}{3})\times \text{ (row 2)} & 0 & \frac{2}{3} & -\frac{4}{3} & 4 \\ \text{(row 3)} & 0 & -\frac{2}{3} & \frac{4}{3} & 7 \\ \hline \text{(sum)} & 0 & 0 & 0 & 11 \end{array}$$

Resultant Matrix

$$\left[\begin{array}{ccc|c} 1 & -\frac{2}{3} & \frac{4}{3} & -18 \\ 0 & ① & -2 & 6 \\ 0 & 0 & 0 & 11 \end{array}\right]$$

The pivot moves to the (3,3) entry. However, the (3,3) entry equals 0. Because there are no rows below the third row, we can not switch with a row below the third row to produce a nonzero pivot. Also, since there are no more columns before the augmentation bar, we are finished performing row operations. The final augmented matrix corresponds to the following linear system:

$$\begin{cases} x_1 - \frac{2}{3}x_2 + \frac{4}{3}x_3 = -18 \\ \qquad\; x_2 - 2x_3 = 6 \\ \qquad\qquad\; 0 = 11 \end{cases}$$

The last equation states that $0 = 11$. Since this equation can not be satisfied, the original linear system has no solutions. The solution set is empty.

(e) First augmented matrix: $\left[\begin{array}{ccccc|c} 6 & -12 & -5 & 16 & -2 & -53 \\ -3 & 6 & 3 & -9 & 1 & 29 \\ -4 & 8 & 3 & -10 & 1 & 33 \end{array}\right]$. The first pivot is the (1,1) entry. We make this a "1" by performing the following type (I) row operation:

Type (I) operation: $\langle 1\rangle \leftarrow \frac{1}{6}\langle 1\rangle$: $\left[\begin{array}{ccccc|c} \textcircled{1} & -2 & -\frac{5}{6} & \frac{8}{3} & -\frac{1}{3} & -\frac{53}{6} \\ -3 & 6 & 3 & -9 & 1 & 29 \\ -4 & 8 & 3 & -10 & 1 & 33 \end{array}\right]$

We now target the (2,1) entry, and make it "0." We do this using the following type (II) row operation:

Type (II) operation: $\langle 2\rangle \leftarrow (3) \times \langle 1\rangle + \langle 2\rangle$:

Side Calculation:

$$\begin{array}{r|rrrrr|r} (3)\times\text{(row 1)} & 3 & -6 & -\frac{5}{2} & 8 & -1 & -\frac{53}{2} \\ \text{(row 2)} & -3 & 6 & 3 & -9 & 1 & 29 \\ \hline \text{(sum)} & 0 & 0 & \frac{1}{2} & -1 & 0 & \frac{5}{2} \end{array}$$

Resultant Matrix: $\left[\begin{array}{ccccc|c} \textcircled{1} & -2 & -\frac{5}{6} & \frac{8}{3} & -\frac{1}{3} & -\frac{53}{6} \\ 0 & 0 & \frac{1}{2} & -1 & 0 & \frac{5}{2} \\ -4 & 8 & 3 & -10 & 1 & 33 \end{array}\right]$

Next, we target the (3,1) entry.

Type (II) operation: $\langle 3\rangle \leftarrow (4) \times \langle 1\rangle + \langle 3\rangle$:

Side Calculation:

$$\begin{array}{r|rrrrr|r} (4)\times\text{(row 1)} & 4 & -8 & -\frac{10}{3} & \frac{32}{3} & -\frac{4}{3} & -\frac{106}{3} \\ \text{(row 3)} & -4 & 8 & 3 & -10 & 1 & 33 \\ \hline \text{(sum)} & 0 & 0 & -\frac{1}{3} & \frac{2}{3} & -\frac{1}{3} & -\frac{7}{3} \end{array}$$

Resultant Matrix: $\left[\begin{array}{ccccc|c} \textcircled{1} & -2 & -\frac{5}{6} & \frac{8}{3} & -\frac{1}{3} & -\frac{53}{6} \\ 0 & 0 & \frac{1}{2} & -1 & 0 & \frac{5}{2} \\ 0 & 0 & -\frac{1}{3} & \frac{2}{3} & -\frac{1}{3} & -\frac{7}{3} \end{array}\right]$

We attempt to move the pivot to the (2,2) entry; however, there is a "0" there. Also, the entry below this is also 0, so we can not switch rows to make the (2,2) entry nonzero. Thus, we move the pivot horizontally to the next column – to the (2,3) entry. We make this pivot a "1" using the following type (I) row operation:

Type (I) operation: $\langle 2\rangle \leftarrow 2\,\langle 2\rangle$: $\left[\begin{array}{ccccc|c} 1 & -2 & -\frac{5}{6} & \frac{8}{3} & -\frac{1}{3} & -\frac{53}{6} \\ 0 & 0 & \textcircled{1} & -2 & 0 & 5 \\ 0 & 0 & -\frac{1}{3} & \frac{2}{3} & -\frac{1}{3} & -\frac{7}{3} \end{array}\right]$

Now we target the (3,3) entry.

Type (II) operation: $\langle 3\rangle \leftarrow (\frac{1}{3}) \times \langle 2\rangle + \langle 3\rangle$:

Side Calculation:

$$\begin{array}{r|rrrrr|r} (\frac{1}{3}) \times \text{(row 2)} & 0 & 0 & \frac{1}{3} & -\frac{2}{3} & 0 & \frac{5}{3} \\ \text{(row 3)} & 0 & 0 & -\frac{1}{3} & \frac{2}{3} & -\frac{1}{3} & -\frac{7}{3} \\ \hline \text{(sum)} & 0 & 0 & 0 & 0 & -\frac{1}{3} & -\frac{2}{3} \end{array}$$

Resultant Matrix: $\left[\begin{array}{ccccc|c} 1 & -2 & -\frac{5}{6} & \frac{8}{3} & -\frac{1}{3} & -\frac{53}{6} \\ 0 & 0 & \textcircled{1} & -2 & 0 & 5 \\ 0 & 0 & 0 & 0 & -\frac{1}{3} & -\frac{2}{3} \end{array}\right]$

We attempt to move the pivot diagonally to the (3,4) entry, however that entry is also 0. Since there are no rows below with which to swap, we must move the pivot horizontally to the (3,5) entry. We make this pivot a "1" by performing the following type (I) row operation:

Type (I) operation: $\langle 3\rangle \leftarrow -3\,\langle 3\rangle$: $\left[\begin{array}{ccccc|c} 1 & -2 & -\frac{5}{6} & \frac{8}{3} & -\frac{1}{3} & -\frac{53}{6} \\ 0 & 0 & 1 & -2 & 0 & 5 \\ 0 & 0 & 0 & 0 & \textcircled{1} & 2 \end{array}\right]$

At this point, we are finished performing row operations. The linear system corresponding to the final augmented matrix is:

$$\left\{\begin{array}{rcrcrcrcrcr} x_1 & - & 2x_2 & - & \frac{5}{6}x_3 & + & \frac{8}{3}x_4 & - & \frac{1}{3}x_5 & = & -\frac{53}{6} \\ & & & & x_3 & - & 2x_4 & & & = & 5 \\ & & & & & & & & x_5 & = & 2 \end{array}\right.$$

The last equation clearly gives us $x_5 = 2$. The variable x_4 corresponds to a nonpivot column, and hence is an independent variable. Let $x_4 = d$. Then the second equation yields $x_3 - 2d = 5$, or $x_3 = 2d + 5$. The variable x_2 also corresponds to a nonpivot column, making it an independent variable as well. We let $x_2 = b$. Finally, plugging the all these values into the first equation produces $x_1 - 2b - \frac{5}{6}(2d+5) + \frac{8}{3}d - \frac{1}{3}(2) = -\frac{53}{6}$. Solving this equation for x_1 gives $x_1 = 2b - d - 4$. Hence, the general solution set for the original system is $\{(2b-d-4, b, 2d+5, d, 2) \mid b, d \in \mathbb{R}\}$. To find three particular solutions, we plug in three different sets of values for b and d. Using $b = 0$, $d = 0$ yields the solution $(-4, 0, 5, 0, 2)$. Letting $b = 1$, $d = 0$ gives us $(-2, 1, 5, 0, 2)$. Finally, setting $b = 0$, $d = 1$ produces $(-5, 0, 7, 1, 2)$.

(g) First augmented matrix: $\left[\begin{array}{rrr|r} 4 & -2 & -7 & 5 \\ -6 & 5 & 10 & -11 \\ -2 & 3 & 4 & -3 \\ -3 & 2 & 5 & -5 \end{array}\right]$. The first pivot is the (1,1) entry. We make this a "1" by performing the following type (I) row operation:

Type (I) operation: $\langle 1\rangle \leftarrow \frac{1}{4}\,\langle 1\rangle$: $\left[\begin{array}{rrr|r} \textcircled{1} & -\frac{1}{2} & -\frac{7}{4} & \frac{5}{4} \\ -6 & 5 & 10 & -11 \\ -2 & 3 & 4 & -3 \\ -3 & 2 & 5 & -5 \end{array}\right]$

We now target the (2,1) entry, and make it "0." We do this using the following type (II) row operation:
Type (II) operation: $\langle 2\rangle \leftarrow (6) \times \langle 1\rangle + \langle 2\rangle$:

Side Calculation	Resultant Matrix
$\begin{array}{rrrr\|r} (6)\times \text{(row 1)} & 6 & -3 & -\frac{21}{2} & \frac{15}{2} \\ \text{(row 2)} & -6 & 5 & 10 & -11 \\ \hline \text{(sum)} & 0 & 2 & -\frac{1}{2} & -\frac{7}{2} \end{array}$	$\left[\begin{array}{rrr\|r} \textcircled{1} & -\frac{1}{2} & -\frac{7}{4} & \frac{5}{4} \\ 0 & 2 & -\frac{1}{2} & -\frac{7}{2} \\ -2 & 3 & 4 & -3 \\ -3 & 2 & 5 & -5 \end{array}\right]$

Next, we target the (3,1) entry.
Type (II) operation: $\langle 3\rangle \leftarrow (2) \times \langle 1\rangle + \langle 3\rangle$:

Side Calculation	Resultant Matrix
$\begin{array}{rrrr\|r} (2)\times \text{(row 1)} & 2 & -1 & -\frac{7}{2} & \frac{5}{2} \\ \text{(row 3)} & -2 & 3 & 4 & -3 \\ \hline \text{(sum)} & 0 & 2 & \frac{1}{2} & -\frac{1}{2} \end{array}$	$\left[\begin{array}{rrr\|r} \textcircled{1} & -\frac{1}{2} & -\frac{7}{4} & \frac{5}{4} \\ 0 & 2 & -\frac{1}{2} & -\frac{7}{2} \\ 0 & 2 & \frac{1}{2} & -\frac{1}{2} \\ -3 & 2 & 5 & -5 \end{array}\right]$

To finish the first column, we now target the (4,1) entry.
Type (II) operation: $\langle 4\rangle \leftarrow (3) \times \langle 1\rangle + \langle 4\rangle$:

Side Calculation	Resultant Matrix
$\begin{array}{rrrr\|r} (3)\times \text{(row 1)} & 3 & -\frac{3}{2} & -\frac{21}{4} & \frac{15}{4} \\ \text{(row 4)} & -3 & 2 & 5 & -5 \\ \hline \text{(sum)} & 0 & \frac{1}{2} & -\frac{1}{4} & -\frac{5}{4} \end{array}$	$\left[\begin{array}{rrr\|r} \textcircled{1} & -\frac{1}{2} & -\frac{7}{4} & \frac{5}{4} \\ 0 & 2 & -\frac{1}{2} & -\frac{7}{2} \\ 0 & 2 & \frac{1}{2} & -\frac{1}{2} \\ 0 & \frac{1}{2} & -\frac{1}{4} & -\frac{5}{4} \end{array}\right]$

Now we move the pivot to the (2,2) entry. We make this entry a "1" using a type (I) row operation.

Type (I) operation: $\langle 2\rangle \leftarrow \frac{1}{2}\langle 2\rangle$: $\left[\begin{array}{rrr|r} 1 & -\frac{1}{2} & -\frac{7}{4} & \frac{5}{4} \\ 0 & \textcircled{1} & -\frac{1}{4} & -\frac{7}{4} \\ 0 & 2 & \frac{1}{2} & -\frac{1}{2} \\ 0 & \frac{1}{2} & -\frac{1}{4} & -\frac{5}{4} \end{array}\right]$

Now we target the (3,2) entry.
Type (II) operation: $\langle 3\rangle \leftarrow (-2) \times \langle 2\rangle + \langle 3\rangle$:

Side Calculation	Resultant Matrix
$\begin{array}{rrrr\|r} (-2)\times \text{(row 2)} & 0 & -2 & \frac{1}{2} & \frac{7}{2} \\ \text{(row 3)} & 0 & 2 & \frac{1}{2} & -\frac{1}{2} \\ \hline \text{(sum)} & 0 & 0 & 1 & 3 \end{array}$	$\left[\begin{array}{rrr\|r} 1 & -\frac{1}{2} & -\frac{7}{4} & \frac{5}{4} \\ 0 & \textcircled{1} & -\frac{1}{4} & -\frac{7}{4} \\ 0 & 0 & 1 & 3 \\ 0 & \frac{1}{2} & -\frac{1}{4} & -\frac{5}{4} \end{array}\right]$

To finish the second column, we target the (4,2) entry.
Type (II) operation: $\langle 4\rangle \leftarrow (-\frac{1}{2}) \times \langle 2\rangle + \langle 4\rangle$:

Side Calculation	Resultant Matrix
$\begin{array}{rrrr\|r} (-\frac{1}{2})\times \text{(row 2)} & 0 & -\frac{1}{2} & \frac{1}{8} & \frac{7}{8} \\ \text{(row 4)} & 0 & \frac{1}{2} & -\frac{1}{4} & -\frac{5}{4} \\ \hline \text{(sum)} & 0 & 0 & -\frac{1}{8} & -\frac{3}{8} \end{array}$	$\left[\begin{array}{rrr\|r} 1 & -\frac{1}{2} & -\frac{7}{4} & \frac{5}{4} \\ 0 & \textcircled{1} & -\frac{1}{4} & -\frac{7}{4} \\ 0 & 0 & 1 & 3 \\ 0 & 0 & -\frac{1}{8} & -\frac{3}{8} \end{array}\right]$

The pivot now moves to the (3,3) entry. Luckily, the pivot already equals 1. Therefore, we next

target the (4,3) entry.
Type (II) operation: $\langle 4\rangle \leftarrow (\frac{1}{8}) \times \langle 3\rangle + \langle 4\rangle$:

Side Calculation

$$\begin{array}{rcccc|c} (\frac{1}{8})\times \text{ (row 3)} & 0 & 0 & \frac{1}{8} & \frac{3}{8} \\ \text{(row 4)} & 0 & 0 & -\frac{1}{8} & -\frac{3}{8} \\ \hline \text{(sum)} & 0 & 0 & 0 & 0 \end{array}$$

Resultant Matrix

$$\left[\begin{array}{ccc|c} 1 & -\frac{1}{2} & -\frac{7}{4} & \frac{5}{4} \\ 0 & 1 & -\frac{1}{4} & -\frac{7}{4} \\ 0 & 0 & \textcircled{1} & 3 \\ 0 & 0 & 0 & 0 \end{array}\right]$$

This finishes the third column. Since there are no more columns before the augmentation bar, we are finished performing row operations. The linear system corresponding to the final matrix is:

$$\begin{cases} x_1 - \frac{1}{2}x_2 - \frac{7}{4}x_3 = \frac{5}{4} \\ \qquad x_2 - \frac{1}{4}x_3 = -\frac{7}{4} \\ \qquad\qquad x_3 = 3 \\ \qquad\qquad 0 = 0 \end{cases}$$

The last equation, $0 = 0$, provides no information regarding the solution set for the system because it is true for every value of x_1, x_2, and x_3. Thus, we can ignore it. The third equation clearly gives $x_3 = 3$. Plugging this value into the second equation yields $x_2 - \frac{1}{4}(3) = -\frac{7}{4}$, or $x_2 = -1$. Substituting the values we have for x_1 and x_2 into the first equation produces $x_1 - \frac{1}{2}(-1) - \frac{7}{4}(3) = \frac{5}{4}$, or $x_1 = 6$. Hence, the original linear system has a unique solution. The full solution set is $\{(6, -1, 3)\}$.

(2) (a) The system of linear equations corresponding to the given augmented matrix is

$$\begin{cases} x_1 - 5x_2 + 2x_3 + 3x_4 - 2x_5 = -4 \\ \qquad x_2 - x_3 - 3x_4 - 7x_5 = -2 \\ \qquad\qquad x_4 + 2x_5 = 5 \\ \qquad\qquad 0 = 0 \end{cases}$$

The last equation, $0 = 0$, provides no information regarding the solution set for the system because it is true for every value of x_1, x_2, x_3, x_4, and x_5. So, we can ignore it. The column corresponding to the variable x_5 is not a pivot column, and so x_5 is an independent variable. Let $x_5 = e$. Substituting e for x_5 into the third equation yields $x_4 + 2e = 5$, or $x_4 = -2e + 5$. The column corresponding to the variable x_3 is not a pivot column, and so x_3 is another independent variable. Let $x_3 = c$. Substituting the expressions we have for x_3, x_4, and x_5 into the second equation produces $x_2 - c - 3(-2e + 5) - 7e = -2$. Solving for x_2 and simplifying gives $x_2 = c + e + 13$. Finally, we plug all these expressions into the first equation to get
$x_1 - 5(c + e + 13) + 2c + 3(-2e + 5) - 2e = -4$, which simplifies to $x_1 = 3c + 13e + 46$. Hence, the complete solution set is $\{(3c + 13e + 46, c + e + 13, c, -2e + 5, e) \mid c, e \in \mathbb{R}\}$.

(c) The system of linear equations corresponding to the given augmented matrix is

$$\begin{cases} x_1 + 4x_2 - 8x_3 - x_4 + 2x_5 - 3x_6 = -4 \\ \qquad x_2 - 7x_3 + 2x_4 - 9x_5 - x_6 = -3 \\ \qquad\qquad x_5 - 4x_6 = 2 \\ \qquad\qquad 0 = 0 \end{cases}.$$

The last equation, $0 = 0$, provides no information regarding the solution set for the system because it is true for every value of x_1, x_2, x_3, x_4, x_5 and x_6. So, we can ignore it. The column corresponding to the variable x_6 is not a pivot column, and so x_6 is an independent variable. Let $x_6 = f$. Substituting f for x_6 into the third equation yields $x_5 - 4f = 2$, or $x_5 = 4f + 2$. The columns corresponding to the variables x_3 and x_4 are not pivot columns, and so x_3 and x_4 are also independent variables. Let $x_3 = c$ and $x_4 = d$. Substituting the expressions we have for x_3, x_4, x_5 and x_6 into the second equation produces $x_2 - 7c + 2d - 9(4f + 2) - f = -3$. Solving for x_2 and simplifying gives $x_2 = 7c - 2d + 37f + 15$. Finally, we plug all these expressions into the first equation to get $x_1 + 4(7c - 2d + 37f + 15) - 8c - d + 2(4f + 2) - 3f = -4$, which simplifies to $x_1 = -20c + 9d - 153f - 68$. Hence, the complete solution set is
$\{(-20c + 9d - 153f - 68, 7c - 2d + 37f + 15, c, d, 4f + 2, f) \mid c, d, f \in \mathbb{R}\}$

(3) Let x represent the number of nickels, y represent the number of dimes, and z represent the number of quarters. The fact that the total value of all of the coins is \$16.50 gives us the following equation: $0.05x + 0.10y + 0.25z = 16.50$, or, $5x + 10y + 25z = 1650$.
Next, using the fact that there are twice as many dimes as quarters gives us $y = 2z$, or, $y - 2z = 0$. Finally, using that the total number of nickels and quarters is 20 more than the number of dimes produces $x + z = y + 20$, or $x - y + z = 20$. Putting these three equations into a single system yields

$$\begin{cases} x - y + z = 20 \\ \phantom{x - {}} y - 2z = 0 \\ 5x + 10y + 25z = 1650 \end{cases},$$

where we have arranged the equations in an order that will make the row operations easier. The augmented matrix for this system is

$$\left[\begin{array}{ccc|c} 1 & -1 & 1 & 20 \\ 0 & 1 & -2 & 0 \\ 5 & 10 & 25 & 1650 \end{array}\right].$$

The (1,1) entry is the first pivot. Since this already equals 1, we continue by targeting nonzero entries in the first column. We get a zero in the (3,1) entry by performing a type (II) operation:
Type (II) operation: $\langle 3\rangle \leftarrow (-5) \times \langle 1\rangle + \langle 3\rangle$:

Side Calculation

$(-5)\times$ (row 1)	-5	5	-5	-100
(row 3)	5	10	25	1650
(sum)	0	15	20	1550

Resultant Matrix

$$\left[\begin{array}{ccc|c} \textcircled{1} & -1 & 1 & 20 \\ 0 & 1 & -2 & 0 \\ 0 & 15 & 20 & 1550 \end{array}\right]$$

Next, the pivot moves to the (2,2) entry. Since this already equals 1, we continue with the second column by targeting the (3,2) entry.
Type (II) operation: $\langle 3\rangle \leftarrow (-15) \times \langle 2\rangle + \langle 3\rangle$:

Side Calculation

$(-15)\times$ (row 2)	0	-15	30	0
(row 3)	0	15	20	1550
(sum)	0	0	50	1550

Resultant Matrix

$$\left[\begin{array}{ccc|c} 1 & -1 & 1 & 20 \\ 0 & \textcircled{1} & -2 & 0 \\ 0 & 0 & 50 & 1550 \end{array}\right]$$

The pivot now moves to the (3,3) entry. We must perform a type (I) operation to turn the pivot into a "1."

Type (I) operation: $\langle 3\rangle \leftarrow \frac{1}{50}\langle 3\rangle$: $\left[\begin{array}{ccc|c} 1 & -1 & 1 & 20 \\ 0 & 1 & -2 & 0 \\ 0 & 0 & \textcircled{1} & 31 \end{array}\right]$

This ends the row operations. The final matrix corresponds to the linear system

$$\left\{\begin{array}{rcrcrcr} x & - & y & + & z & = & 20 \\ & & y & - & 2z & = & 0 \\ & & & & z & = & 31 \end{array}\right. .$$

The third equation in this system yields $z = 31$. Plugging this value into the second equation produces $y-2(31) = 0$, or $y = 62$. Finally, we substitute these values into the first equation to get $x-62+31 = 20$, or $x = 51$. Hence, there are 51 nickels, 62 dimes, and 31 quarters.

(4) First, plug each of the three given points into the equation $y = ax^2 + bx + c$. This yields the following linear system:

$$\left\{\begin{array}{rcrcrcrll} 18 & = & 9a & + & 3b & + & c & \longleftarrow & \text{using } (3,18) \\ 9 & = & 4a & + & 2b & + & c & \longleftarrow & \text{using } (2,9) \\ 13 & = & 4a & - & 2b & + & c & \longleftarrow & \text{using } (-2,13) \end{array}\right. .$$

The augmented matrix for this system is: $\left[\begin{array}{rrr|r} 9 & 3 & 1 & 18 \\ 4 & 2 & 1 & 9 \\ 4 & -2 & 1 & 13 \end{array}\right]$. The first pivot is the (1,1) entry.

We make this a "1" by performing the following type (I) row operation:

Type (I) operation: $\langle 1\rangle \leftarrow \frac{1}{9}\langle 1\rangle$: $\left[\begin{array}{rrr|r} ① & \frac{1}{3} & \frac{1}{9} & 2 \\ 4 & 2 & 1 & 9 \\ 4 & -2 & 1 & 13 \end{array}\right]$

We now target the (2,1) entry, and make it "0." We do this using the following type (II) row operation:
Type (II) operation: $\langle 2\rangle \leftarrow (-4) \times \langle 1\rangle + \langle 2\rangle$:

Side Calculation

$$\begin{array}{rrrr|r} (-4)\times \text{(row 1)} & -4 & -\frac{4}{3} & -\frac{4}{9} & -8 \\ \text{(row 2)} & 4 & 2 & 1 & 9 \\ \hline \text{(sum)} & 0 & \frac{2}{3} & \frac{5}{9} & 1 \end{array}$$

Resultant Matrix

$$\left[\begin{array}{rrr|r} ① & \frac{1}{3} & \frac{1}{9} & 2 \\ 0 & \frac{2}{3} & \frac{5}{9} & 1 \\ 4 & -2 & 1 & 13 \end{array}\right]$$

Next, we target the (3,1) entry.
Type (II) operation: $\langle 3\rangle \leftarrow (-4) \times \langle 1\rangle + \langle 3\rangle$:

Side Calculation

$$\begin{array}{rrrr|r} (-4)\times \text{(row 1)} & -4 & -\frac{4}{3} & -\frac{4}{9} & -8 \\ \text{(row 3)} & 4 & -2 & 1 & 13 \\ \hline \text{(sum)} & 0 & -\frac{10}{3} & \frac{5}{9} & 5 \end{array}$$

Resultant Matrix

$$\left[\begin{array}{rrr|r} ① & \frac{1}{3} & \frac{1}{9} & 2 \\ 0 & \frac{2}{3} & \frac{5}{9} & 1 \\ 0 & -\frac{10}{3} & \frac{5}{9} & 5 \end{array}\right]$$

Now we move the pivot to the (2,2) entry. We make this entry a "1" using a type (I) row operation.

Type (I) operation: $\langle 2\rangle \leftarrow \frac{3}{2}\langle 2\rangle$: $\left[\begin{array}{rrr|r} 1 & \frac{1}{3} & \frac{1}{9} & 2 \\ 0 & ① & \frac{5}{6} & \frac{3}{2} \\ 0 & -\frac{10}{3} & \frac{5}{9} & 5 \end{array}\right]$

Now we target the (3,2) entry.
Type (II) operation: $\langle 3\rangle \leftarrow (\frac{10}{3}) \times \langle 2\rangle + \langle 3\rangle$:

Side Calculation					Resultant Matrix
$(\frac{10}{3})\times$ (row 2)	0	$\frac{10}{3}$	$\frac{25}{9}$	5	$\left[\begin{array}{ccc\|c} 1 & \frac{1}{3} & \frac{1}{9} & 2 \\ 0 & ① & \frac{5}{6} & \frac{3}{2} \\ 0 & 0 & \frac{10}{3} & 10 \end{array}\right]$
(row 3)	0	$-\frac{10}{3}$	$\frac{5}{9}$	5	
(sum)	0	0	$\frac{10}{3}$	10	

Finally, we move the pivot to the (3,3) entry and perform the following row operation to turn that entry into a "1."

Type (I) operation: $\langle 3\rangle \leftarrow \frac{3}{10}\langle 3\rangle$: $\left[\begin{array}{ccc|c} 1 & \frac{1}{3} & \frac{1}{9} & 2 \\ 0 & 1 & \frac{5}{6} & \frac{3}{2} \\ 0 & 0 & ① & 3 \end{array}\right]$

This matrix corresponds to the following linear system:

$$\begin{cases} a + \frac{1}{3}b + \frac{1}{9}c = 2 \\ \quad b + \frac{5}{6}c = \frac{3}{2} \\ \quad\quad c = 3 \end{cases}.$$

The last equation clearly gives $c = 3$. Plugging $c = 3$ into the second equation yields $b + \frac{5}{6}(3) = \frac{3}{2}$, or, $b = -1$. Substituting these values into the first equation produces $a + \frac{1}{3}(-1) + \frac{1}{9}(3) = 2$, or $a = 2$. Hence the desired quadratic equation is $y = 2x^2 - x + 3$.

(6) First, plug each of the three given points into the equation $x^2 + y^2 + ax + by = c$. This yields the following linear system:

$$\begin{cases} (6^2+8^2) + 6a + 8b = c \longleftarrow \text{using } (6,8) \\ (8^2+4^2) + 8a + 4b = c \longleftarrow \text{using } (8,4) \\ (3^2+9^2) + 3a + 9b = c \longleftarrow \text{using } (3,9) \end{cases}.$$

Rearranging terms leads to the following augmented matrix:

$$\left[\begin{array}{ccc|c} 6 & 8 & -1 & -100 \\ 8 & 4 & -1 & -80 \\ 3 & 9 & -1 & -90 \end{array}\right].$$

The first pivot is the (1,1) entry. We make this a "1" by performing the following type (I) row operation:

Type (I) operation: $\langle 1\rangle \leftarrow \frac{1}{6}\langle 1\rangle$: $\left[\begin{array}{ccc|c} ① & \frac{4}{3} & -\frac{1}{6} & -\frac{50}{3} \\ 8 & 4 & -1 & -80 \\ 3 & 9 & -1 & -90 \end{array}\right]$

We now must target the (2,1) entry, and make it "0." We do this using the following type (II) row operation:

Type (II) operation: $\langle 2\rangle \leftarrow (-8)\times\langle 1\rangle + \langle 2\rangle$:

Side Calculation					Resultant Matrix
$(-8)\times$ (row 1)	-8	$-\frac{32}{3}$	$\frac{4}{3}$	$\frac{400}{3}$	$\left[\begin{array}{ccc\|c} ① & \frac{4}{3} & -\frac{1}{6} & -\frac{50}{3} \\ 0 & -\frac{20}{3} & \frac{1}{3} & \frac{160}{3} \\ 3 & 9 & -1 & -90 \end{array}\right]$
(row 2)	8	4	-1	-80	
(sum)	0	$-\frac{20}{3}$	$\frac{1}{3}$	$\frac{160}{3}$	

Next, we target the (3,1) entry.

Type (II) operation: $\langle 3\rangle \leftarrow (-3) \times \langle 1\rangle + \langle 3\rangle$:

Side Calculation

$(-3)\times$ (row 1)	-3	-4	$\frac{1}{2}$	50
(row 3)	3	9	-1	-90
(sum)	0	5	$-\frac{1}{2}$	-40

Resultant Matrix

$$\left[\begin{array}{ccc|c} \textcircled{1} & \frac{4}{3} & -\frac{1}{6} & -\frac{50}{3} \\ 0 & -\frac{20}{3} & \frac{1}{3} & \frac{160}{3} \\ 0 & 5 & -\frac{1}{2} & -40 \end{array}\right]$$

Now we move the pivot to the (2,2) entry. We make this entry a "1" using a type (I) row operation.

Type (I) operation: $\langle 2\rangle \leftarrow -\frac{3}{20}\langle 2\rangle$: $\left[\begin{array}{ccc|c} 1 & \frac{4}{3} & -\frac{1}{6} & -\frac{50}{3} \\ 0 & \textcircled{1} & -\frac{1}{20} & -8 \\ 0 & 5 & -\frac{1}{2} & -40 \end{array}\right]$

Now we target the (3,2) entry.
Type (II) operation: $\langle 3\rangle \leftarrow (-5) \times \langle 2\rangle + \langle 3\rangle$:

Side Calculation

$(-5)\times$ (row 2)	0	-5	$\frac{1}{4}$	40
(row 3)	0	5	$-\frac{1}{2}$	-40
(sum)	0	0	$-\frac{1}{4}$	0

Resultant Matrix

$$\left[\begin{array}{ccc|c} 1 & \frac{4}{3} & -\frac{1}{6} & -\frac{50}{3} \\ 0 & \textcircled{1} & -\frac{1}{20} & -8 \\ 0 & 0 & -\frac{1}{4} & 0 \end{array}\right]$$

Finally, we move the pivot to the (3,3) entry and perform the following row operation to turn that entry into a "1."

Type (I) operation: $\langle 3\rangle \leftarrow (-4)\langle 3\rangle$: $\left[\begin{array}{ccc|c} 1 & \frac{4}{3} & -\frac{1}{6} & -\frac{50}{3} \\ 0 & 1 & -\frac{1}{20} & -8 \\ 0 & 0 & \textcircled{1} & 0 \end{array}\right]$

This matrix corresponds to the following linear system:

$$\begin{cases} a + \frac{4}{3}b - \frac{1}{6}c = -\frac{50}{3} \\ \quad b - \frac{1}{20}c = -8 \\ \quad\quad c = 0 \end{cases}.$$

The last equation clearly gives $c = 0$. Plugging $c = 0$ into the second equation yields $b - \frac{1}{20}(0) = -8$, or, $b = -8$. Substituting these values into the first equation produces $a + \frac{4}{3}(-8) - \frac{1}{6}(0) = -\frac{50}{3}$, or, $a = -6$. Hence the desired equation of the circle is $x^2 + y^2 - 6x - 8y = 0$, or, $(x-3)^2 + (y-4)^2 = 25$.

(7) (a) First, we compute $\mathbf{AB} = \begin{bmatrix} 26 & 15 & -6 \\ 6 & 4 & 1 \\ 18 & 6 & 15 \\ 10 & 4 & -14 \end{bmatrix}$. To find $R(\mathbf{AB})$ we perform the type (II) row operation $R : \langle 3\rangle \leftarrow (-3) \times \langle 2\rangle + \langle 3\rangle$:

Side Calculation

$(-3)\times$ (row 2)	-18	-12	-3
(row 3)	18	6	15
(sum)	0	-6	12

Resultant Matrix

$$R(\mathbf{AB}) = \begin{bmatrix} 26 & 15 & -6 \\ 6 & 4 & 1 \\ 0 & -6 & 12 \\ 10 & 4 & -14 \end{bmatrix}$$

Next we compute $R(\mathbf{A})$. Using $\mathbf{A} = \begin{bmatrix} 2 & 3 & 4 \\ 0 & 1 & 1 \\ -2 & 1 & 5 \\ 3 & 0 & 1 \end{bmatrix}$ and $R: \langle 3\rangle \leftarrow (-3) \times \langle 2\rangle + \langle 3\rangle$:

Side Calculation				**Resultant Matrix**
$(-3)\times$ (row 2)	0	−3	−3	$R(\mathbf{A}) = \begin{bmatrix} 2 & 3 & 4 \\ 0 & 1 & 1 \\ -2 & -2 & 2 \\ 3 & 0 & 1 \end{bmatrix}$
(row 3)	−2	1	5	
(sum)	−2	−2	2	

Multiplying $R(\mathbf{A})$ by the given matrix $\mathbf{B}$ produces the same result computed above for $R(\mathbf{AB})$.

(8) (a) To save space, we use the notation $\langle \mathbf{C}\rangle_i$ for the ith row of a matrix $\mathbf{C}$. Also, recall the following hint from the textbook, which can rewrite as follows in the notation we have just defined: (Hint: Use the fact from Section 1.5 that $\langle \mathbf{AB}\rangle_k = \langle \mathbf{A}\rangle_k \mathbf{B}$.)

For the Type (I) operation $R: \langle i\rangle \longleftarrow c\langle i\rangle$: Now, $\langle R(\mathbf{AB})\rangle_i = c\langle \mathbf{AB}\rangle_i = c\langle \mathbf{A}\rangle_i\mathbf{B}$ (by the hint) $= \langle R(\mathbf{A})\rangle_i\mathbf{B} = \langle R(\mathbf{A})\mathbf{B}\rangle_i$. But, if $k \neq i$, $\langle R(\mathbf{AB})\rangle_k = \langle \mathbf{AB}\rangle_k = \langle \mathbf{A}\rangle_k\mathbf{B}$ (by the hint) $= \langle R(\mathbf{A})\rangle_k\mathbf{B}$ $= \langle R(\mathbf{A})\mathbf{B}\rangle_k$.

For the Type (II) operation $R: \langle i\rangle \longleftarrow c\langle j\rangle + \langle i\rangle$: Now, $\langle R(\mathbf{AB})\rangle_i = c\langle \mathbf{AB}\rangle_j + \langle \mathbf{AB}\rangle_i =$ $c\langle \mathbf{A}\rangle_j\mathbf{B} + \langle \mathbf{A}\rangle_i\mathbf{B}$ (by the hint) $= (c\langle \mathbf{A}\rangle_j + \langle \mathbf{A}\rangle_i)\mathbf{B} = \langle R(\mathbf{A})\rangle_i\mathbf{B} = \langle R(\mathbf{A})\mathbf{B}\rangle_i$. But, if $k \neq i$, $\langle R(\mathbf{AB})\rangle_k = \langle \mathbf{AB}\rangle_k = \langle \mathbf{A}\rangle_k\mathbf{B}$ (by the hint) $= \langle R(\mathbf{A})\rangle_k\mathbf{B} = \langle R(\mathbf{A})\mathbf{B}\rangle_k$.

For the Type (III) operation $R: \langle i\rangle \longleftrightarrow \langle j\rangle$: Now, $\langle R(\mathbf{AB})\rangle_i = \langle \mathbf{AB}\rangle_j = \langle \mathbf{A}\rangle_j\mathbf{B}$ (by the hint) $= \langle R(\mathbf{A})\rangle_i\mathbf{B} = \langle R(\mathbf{A})\mathbf{B}\rangle_i$. Similarly, $\langle R(\mathbf{AB})\rangle_j = \langle \mathbf{AB}\rangle_i = \langle \mathbf{A}\rangle_i\mathbf{B}$ (by the hint) $= \langle R(\mathbf{A})\rangle_j\mathbf{B}$ $= \langle R(\mathbf{A})\mathbf{B}\rangle_j$. And, if $k \neq i$ and $k \neq j$, $\langle R(\mathbf{AB})\rangle_k = \langle \mathbf{AB}\rangle_k = \langle \mathbf{A}\rangle_k\mathbf{B}$ (by the hint) $= \langle R(\mathbf{A})\rangle_k\mathbf{B}$ $= \langle R(\mathbf{A})\mathbf{B}\rangle_k$.

(11) (a) True. The augmented matrix contains all of the information necessary to solve the system completely.

(b) False. A linear system has either no solutions, one solution, or infinitely many solutions. Exactly three solutions is not one of the possibilities.

(c) False. A linear system is consistent if it has at least one solution. It could have either just one, or infinitely many solutions.

(d) False. Type (I) operations are used to convert nonzero pivot entries to 1. When following the standard algorithm for Gaussian elimination, type (II) row operations are never used for this purpose. Type (II) row operations are only used to convert a nonzero target to a zero.

(e) True. In the standard algorithm for Gaussian elimination, this is the only instance in which a type (III) row operation is used.

(f) True. This is the statement of part (1) of Theorem 2.1 expressed in words rather than symbols.

Section 2.2

(1) The matrix in (e) is in reduced row echelon form, as it satisfies all four conditions. The matrices in (a), (b), (c), (d), and (f) are not in reduced row echelon form.
The matrix in (a) fails condition 2 of the definition, since the first nonzero entry in row 2 is in the 3rd column, while the first nonzero entry in row 3 is only in the second column.
The matrix in (b) fails condition 4 of the definition, as the second row contains all zeroes, but later

rows do not.

The matrix in (c) fails condition 1 of the definition because the first nonzero entry in the second row equals 2, not 1.

The matrix in (d) fails conditions 1, 2, and 3 of the definition. Condition 1 fails because the first nonzero entry in row 3 equals 2, not 1. Condition 2 fails because the first nonzero entries in rows 2 and 3 appear in the same column. Condition 3 fails because the (3,3) entry below the first nonzero entry of the second column is nonzero.

The matrix in (f) fails condition 3 of the definition since the entries in the 5th column above the 1 in the third row do not equal zero.

(2) (a) First convert the pivot (the (1,1) entry) to "1."

$$\langle 1\rangle \leftarrow \left(\tfrac{1}{5}\right)\langle 1\rangle \qquad \left[\begin{array}{ccc|c} \textcircled{1} & 4 & -\frac{18}{5} & -\frac{11}{5} \\ 3 & 12 & -14 & 3 \\ -4 & -16 & 13 & 13 \end{array}\right]$$

Next, target the (2,1) and (3,1) entries.

$$\begin{array}{l} \langle 2\rangle \leftarrow (-3)\times\langle 1\rangle + \langle 2\rangle \\ \langle 3\rangle \leftarrow (4)\times\langle 1\rangle + \langle 3\rangle \end{array} \qquad \left[\begin{array}{ccc|c} \textcircled{1} & 4 & -\frac{18}{5} & -\frac{11}{5} \\ 0 & 0 & -\frac{16}{5} & \frac{48}{5} \\ 0 & 0 & -\frac{7}{5} & \frac{21}{5} \end{array}\right]$$

We try to move the pivot to the second column, but can not, due to the zeroes in the (2,2) and (3,2) entries. Hence, the pivot moves to the (2,3) entry. We set that equal to 1.

$$\langle 2\rangle \leftarrow \left(-\tfrac{5}{16}\right)\langle 2\rangle \qquad \left[\begin{array}{ccc|c} 1 & 4 & -\frac{18}{5} & -\frac{11}{5} \\ 0 & 0 & \textcircled{1} & -3 \\ 0 & 0 & -\frac{7}{5} & \frac{21}{5} \end{array}\right]$$

Finally, we target the (1,3) and (3,3) entries.

$$\begin{array}{l} \langle 1\rangle \leftarrow \left(\tfrac{18}{5}\right)\times\langle 2\rangle + \langle 1\rangle \\ \langle 3\rangle \leftarrow \left(\tfrac{7}{5}\right)\times\langle 2\rangle + \langle 3\rangle \end{array} \qquad \left[\begin{array}{ccc|c} 1 & 4 & 0 & -13 \\ 0 & 0 & \textcircled{1} & -3 \\ 0 & 0 & 0 & 0 \end{array}\right]$$

This is the desired reduced row echelon form matrix.

(b) First convert the pivot (the (1,1) entry) to "1."

$$\langle 1\rangle \leftarrow \left(-\tfrac{1}{2}\right)\langle 1\rangle \qquad \left[\begin{array}{cccc} \textcircled{1} & -\frac{1}{2} & -\frac{1}{2} & -\frac{15}{2} \\ 6 & -1 & -2 & -36 \\ 1 & -1 & -1 & -11 \\ -5 & -5 & -5 & -14 \end{array}\right]$$

Next, target the (2,1), (3,1) and (4,1) entries.

$$\begin{array}{l} \langle 2\rangle \leftarrow (-6)\times\langle 1\rangle + \langle 2\rangle \\ \langle 3\rangle \leftarrow (-1)\times\langle 1\rangle + \langle 3\rangle \\ \langle 4\rangle \leftarrow (5)\times\langle 1\rangle + \langle 4\rangle \end{array} \qquad \left[\begin{array}{cccc} \textcircled{1} & -\frac{1}{2} & -\frac{1}{2} & -\frac{15}{2} \\ 0 & 2 & 1 & 9 \\ 0 & -\frac{1}{2} & -\frac{1}{2} & -\frac{7}{2} \\ 0 & -\frac{15}{2} & -\frac{15}{2} & -\frac{103}{2} \end{array}\right]$$

The pivot now moves to the (2,2) entry.

$$\langle 2\rangle \leftarrow \left(\tfrac{1}{2}\right)\langle 2\rangle \qquad \begin{bmatrix} 1 & -\frac{1}{2} & -\frac{1}{2} & -\frac{15}{2} \\ 0 & \textcircled{1} & \frac{1}{2} & \frac{9}{2} \\ 0 & -\frac{1}{2} & -\frac{1}{2} & -\frac{7}{2} \\ 0 & -\frac{15}{2} & -\frac{15}{2} & -\frac{103}{2} \end{bmatrix}$$

Next, target the (1,2), (3,2) and (4,2) entries.

$$\begin{array}{l} \langle 1\rangle \leftarrow \left(\tfrac{1}{2}\right)\times\langle 2\rangle+\langle 1\rangle \\ \langle 3\rangle \leftarrow \left(\tfrac{1}{2}\right)\times\langle 2\rangle+\langle 3\rangle \\ \langle 4\rangle \leftarrow \left(\tfrac{15}{2}\right)\times\langle 2\rangle+\langle 4\rangle \end{array} \qquad \begin{bmatrix} 1 & 0 & -\frac{1}{4} & -\frac{21}{4} \\ 0 & \textcircled{1} & \frac{1}{2} & \frac{9}{2} \\ 0 & 0 & -\frac{1}{4} & -\frac{5}{4} \\ 0 & 0 & -\frac{15}{4} & -\frac{71}{4} \end{bmatrix}$$

The pivot now moves to the (3,3) entry.

$$\langle 3\rangle \leftarrow (-4)\,\langle 3\rangle \qquad \begin{bmatrix} 1 & 0 & -\frac{1}{4} & -\frac{21}{4} \\ 0 & 1 & \frac{1}{2} & \frac{9}{2} \\ 0 & 0 & \textcircled{1} & 5 \\ 0 & 0 & -\frac{15}{4} & -\frac{71}{4} \end{bmatrix}$$

Next, target the (1,3), (2,3) and (4,3) entries.

$$\begin{array}{l} \langle 1\rangle \leftarrow \left(\tfrac{1}{4}\right)\times\langle 3\rangle+\langle 1\rangle \\ \langle 2\rangle \leftarrow \left(-\tfrac{1}{2}\right)\times\langle 3\rangle+\langle 2\rangle \\ \langle 4\rangle \leftarrow \left(\tfrac{15}{4}\right)\times\langle 3\rangle+\langle 4\rangle \end{array} \qquad \begin{bmatrix} 1 & 0 & 0 & -4 \\ 0 & 1 & 0 & 2 \\ 0 & 0 & \textcircled{1} & 5 \\ 0 & 0 & 0 & 1 \end{bmatrix}$$

The pivot moves to the (4,4) entry. Since this already equals 1, we now only need to target the (1,4), (2,4) and (3,4) entries.

$$\begin{array}{l} \langle 1\rangle \leftarrow (4)\times\langle 4\rangle+\langle 1\rangle \\ \langle 2\rangle \leftarrow (-2)\times\langle 4\rangle+\langle 2\rangle \\ \langle 3\rangle \leftarrow (-5)\times\langle 4\rangle+\langle 3\rangle \end{array} \qquad \begin{bmatrix} 1 & 0 & 0 & 0 \\ 0 & 1 & 0 & 0 \\ 0 & 0 & 1 & 0 \\ 0 & 0 & 0 & \textcircled{1} \end{bmatrix}$$

Thus, $\mathbf{I}_4$ is the desired reduced row echelon form matrix.

(c) First convert the pivot (the (1,1) entry) to "1."

$$\langle 1\rangle \leftarrow \left(-\tfrac{1}{5}\right)\langle 1\rangle \qquad \left[\begin{array}{cccc|c} \textcircled{1} & -2 & \frac{19}{5} & \frac{17}{5} & -4 \\ -3 & 6 & -11 & -11 & 14 \\ -7 & 14 & -26 & -25 & 31 \\ 9 & -18 & 34 & 31 & -37 \end{array}\right]$$

Next, target the (2,1), (3,1) and (4,1) entries.

$$\begin{array}{l} \langle 2\rangle \leftarrow (3)\times\langle 1\rangle+\langle 2\rangle \\ \langle 3\rangle \leftarrow (7)\times\langle 1\rangle+\langle 3\rangle \\ \langle 4\rangle \leftarrow (-9)\times\langle 1\rangle+\langle 4\rangle \end{array} \qquad \left[\begin{array}{cccc|c} \textcircled{1} & -2 & \frac{19}{5} & \frac{17}{5} & -4 \\ 0 & 0 & \frac{2}{5} & -\frac{4}{5} & 2 \\ 0 & 0 & \frac{3}{5} & -\frac{6}{5} & 3 \\ 0 & 0 & -\frac{1}{5} & \frac{2}{5} & -1 \end{array}\right]$$

We try to move the pivot to the second column, but can not, due to the zeroes in the (2,2), (3,2) and (4,2) entries. Hence, the pivot moves to the (2,3) entry. We set that equal to 1.

$\langle 2\rangle \leftarrow (\frac{5}{2})\langle 2\rangle$ $\left[\begin{array}{cccc|c} 1 & -2 & \frac{19}{5} & \frac{17}{5} & -4 \\ 0 & 0 & \textcircled{1} & -2 & 5 \\ 0 & 0 & \frac{3}{5} & -\frac{6}{5} & 3 \\ 0 & 0 & -\frac{1}{5} & \frac{2}{5} & -1 \end{array}\right]$

Next, target the (1,3), (3,3) and (4,3) entries.

$\langle 1\rangle \leftarrow (-\frac{19}{5}) \times \langle 2\rangle + \langle 1\rangle$
$\langle 3\rangle \leftarrow (-\frac{3}{5}) \times \langle 2\rangle + \langle 3\rangle$
$\langle 4\rangle \leftarrow (\frac{1}{5}) \times \langle 2\rangle + \langle 4\rangle$

$\left[\begin{array}{cccc|c} 1 & -2 & 0 & 11 & -23 \\ 0 & 0 & \textcircled{1} & -2 & 5 \\ 0 & 0 & 0 & 0 & 0 \\ 0 & 0 & 0 & 0 & 0 \end{array}\right]$

This is the desired reduced row echelon form matrix.

(e) First convert the pivot (the (1,1) entry) to "1."

$\langle 1\rangle \leftarrow (-\frac{1}{3})\langle 1\rangle$ $\left[\begin{array}{ccccc|c} \textcircled{1} & -2 & \frac{1}{3} & \frac{5}{3} & 0 & \frac{5}{3} \\ -1 & 2 & 3 & -5 & 10 & 5 \end{array}\right]$

Next, target the (2,1) entry.

$\langle 2\rangle \leftarrow (1) \times \langle 1\rangle + \langle 2\rangle$ $\left[\begin{array}{ccccc|c} \textcircled{1} & -2 & \frac{1}{3} & \frac{5}{3} & 0 & \frac{5}{3} \\ 0 & 0 & \frac{10}{3} & -\frac{10}{3} & 10 & \frac{20}{3} \end{array}\right]$

We try to move the pivot to the (2,2) entry, but it equals zero. There are no rows below it to provide a nonzero pivot to be swapped up, and so the pivot moves to the (2,3) entry. We set that equal to 1.

$\langle 2\rangle \leftarrow (\frac{3}{10})\langle 2\rangle$ $\left[\begin{array}{ccccc|c} 1 & -2 & \frac{1}{3} & \frac{5}{3} & 0 & \frac{5}{3} \\ 0 & 0 & \textcircled{1} & -1 & 3 & 2 \end{array}\right]$

Finally, we target the (1,3) entry to obtain the desired reduced row echelon form matrix.

$\langle 1\rangle \leftarrow (-\frac{1}{3}) \times \langle 2\rangle + \langle 1\rangle$ $\left[\begin{array}{ccccc|c} 1 & -2 & 0 & 2 & -1 & 1 \\ 0 & 0 & \textcircled{1} & -1 & 3 & 2 \end{array}\right].$

(3) (a) The final matrix obtained in Exercise 1(a) in Section 2.1 is

$$\left[\begin{array}{ccc|c} 1 & \frac{2}{5} & -\frac{2}{5} & -\frac{14}{5} \\ 0 & 1 & -1 & -2 \\ 0 & 0 & 1 & 5 \end{array}\right].$$

We target the entries above the pivots to obtain the reduced row echelon form matrix. First, we use the (2,2) entry as the pivot and target the (1,2) entry.

$\langle 1\rangle \leftarrow (-\frac{2}{5}) \times \langle 2\rangle + \langle 1\rangle$ $\left[\begin{array}{ccc|c} 1 & 0 & 0 & -2 \\ 0 & \textcircled{1} & -1 & -2 \\ 0 & 0 & 1 & 5 \end{array}\right]$

Next, pivot at the (3,3) entry, and target the (2,3) entry. (Note that the (1,3) entry is already zero.)

$\langle 2\rangle \leftarrow (1) \times \langle 3\rangle + \langle 2\rangle$ $\left[\begin{array}{ccc|c} 1 & 0 & 0 & -2 \\ 0 & 1 & 0 & 3 \\ 0 & 0 & \textcircled{1} & 5 \end{array}\right]$

This matrix corresponds to the linear system

$$\begin{cases} x_1 & & & = & -2 \\ & x_2 & & = & 3 \\ & & x_3 & = & 5 \end{cases}.$$

Hence, the unique solution to the system is $(-2, 3, 5)$.

(e) The final matrix obtained in Exercise 1(e) in Section 2.1 is

$$\left[\begin{array}{ccccc|c} 1 & -2 & -\frac{5}{6} & \frac{8}{3} & -\frac{1}{3} & -\frac{53}{6} \\ 0 & 0 & 1 & -2 & 0 & 5 \\ 0 & 0 & 0 & 0 & 1 & 2 \end{array}\right].$$

We target the entries above the pivots to obtain the reduced row echelon form matrix. First, we use the (2,3) entry as the pivot and target the (1,3) entry.

$$\langle 1\rangle \leftarrow (\tfrac{5}{6}) \times \langle 2\rangle + \langle 1\rangle \qquad \left[\begin{array}{ccccc|c} 1 & -2 & 0 & 1 & -\frac{1}{3} & -\frac{14}{3} \\ 0 & 0 & \textcircled{1} & -2 & 0 & 5 \\ 0 & 0 & 0 & 0 & 1 & 2 \end{array}\right]$$

Next, pivot at the (3,5) entry, and target the (1,5) entry. (Note that the (2,5) entry is already zero.)

$$\langle 1\rangle \leftarrow (\tfrac{1}{3}) \times \langle 3\rangle + \langle 1\rangle \qquad \left[\begin{array}{ccccc|c} 1 & -2 & 0 & 1 & 0 & -4 \\ 0 & 0 & 1 & -2 & 0 & 5 \\ 0 & 0 & 0 & 0 & \textcircled{1} & 2 \end{array}\right]$$

This matrix corresponds to the linear system

$$\begin{cases} x_1 - 2x_2 + x_4 & = -4 \\ x_3 - 2x_4 & = 5 \\ x_5 & = 2 \end{cases}.$$

Columns 2 and 4 are not pivot columns, so x_2 and x_4 are independent variables. Let $x_2 = b$ and $x_4 = d$. Then the first equation gives $x_1 - 2b + d = -4$, or, $x_1 = 2b - d - 4$. The second equation yields $x_3 - 2d = 5$, or, $x_3 = 2d + 5$. The last equation states that $x_5 = 2$. Hence, the complete solution set is $\{(2b - d - 4, b, 2d + 5, d, 2) \mid b, d \in \mathbb{R}\}$.

(g) The final matrix obtained in Exercise 1(g) in Section 2.1 is

$$\left[\begin{array}{ccc|c} 1 & -\frac{1}{2} & -\frac{7}{4} & \frac{5}{4} \\ 0 & 1 & -\frac{1}{4} & -\frac{7}{4} \\ 0 & 0 & 1 & 3 \\ 0 & 0 & 0 & 0 \end{array}\right].$$

We target the entries above the pivots to obtain the reduced row echelon form matrix. First, we use the (2,2) entry as the pivot and target the (1,2) entry.

$$\langle 1\rangle \leftarrow (\tfrac{1}{2}) \times \langle 2\rangle + \langle 1\rangle \qquad \left[\begin{array}{ccc|c} 1 & 0 & -\frac{15}{8} & \frac{3}{8} \\ 0 & \textcircled{1} & -\frac{1}{4} & -\frac{7}{4} \\ 0 & 0 & 1 & 3 \\ 0 & 0 & 0 & 0 \end{array}\right]$$

Next, we use the (3,3) entry as a pivot and target the (1,3) and (2,3) entries.

$\langle 1\rangle \leftarrow (\frac{15}{8}) \times \langle 3\rangle + \langle 1\rangle$
$\langle 2\rangle \leftarrow (\frac{1}{4}) \times \langle 3\rangle + \langle 2\rangle$

$$\left[\begin{array}{ccc|c} 1 & 0 & 0 & 6 \\ 0 & 1 & 0 & -1 \\ 0 & 0 & \textcircled{1} & 3 \\ 0 & 0 & 0 & 0 \end{array}\right]$$

Ignoring the last row, which gives the equation $0 = 0$, this matrix yields the linear system

$$\begin{cases} x_1 & & & = & 6 \\ & x_2 & & = & -1 \\ & & x_3 & = & 3 \end{cases},$$

producing the unique solution $(6, -1, 3)$.

(4) (a) The augmented matrix for the system is $\left[\begin{array}{cccc|c} -2 & -3 & 2 & -13 & 0 \\ -4 & -7 & 4 & -29 & 0 \\ 1 & 2 & -1 & 8 & 0 \end{array}\right]$.

The (1,1) entry is the first pivot. We convert it to 1.

$\langle 1\rangle \leftarrow (-\frac{1}{2})\,\langle 1\rangle$

$$\left[\begin{array}{cccc|c} \textcircled{1} & \frac{3}{2} & -1 & \frac{13}{2} & 0 \\ -4 & -7 & 4 & -29 & 0 \\ 1 & 2 & -1 & 8 & 0 \end{array}\right]$$

Next, we target the (2,1) and (3,1) entries.

$\langle 2\rangle \leftarrow (4) \times \langle 1\rangle + \langle 2\rangle$
$\langle 3\rangle \leftarrow (-1) \times \langle 1\rangle + \langle 3\rangle$

$$\left[\begin{array}{cccc|c} \textcircled{1} & \frac{3}{2} & -1 & \frac{13}{2} & 0 \\ 0 & -1 & 0 & -3 & 0 \\ 0 & \frac{1}{2} & 0 & \frac{3}{2} & 0 \end{array}\right]$$

The (2,2) entry is the second pivot. We convert it to 1.

$\langle 2\rangle \leftarrow (-1)\,\langle 2\rangle$

$$\left[\begin{array}{cccc|c} 1 & \frac{3}{2} & -1 & \frac{13}{2} & 0 \\ 0 & \textcircled{1} & 0 & 3 & 0 \\ 0 & \frac{1}{2} & 0 & \frac{3}{2} & 0 \end{array}\right]$$

Next, we target the (1,2) and (3,2) entries.

$\langle 1\rangle \leftarrow (-\frac{3}{2}) \times \langle 2\rangle + \langle 1\rangle$
$\langle 3\rangle \leftarrow (-\frac{1}{2}) \times \langle 2\rangle + \langle 3\rangle$

$$\left[\begin{array}{cccc|c} 1 & 0 & -1 & 2 & 0 \\ 0 & \textcircled{1} & 0 & 3 & 0 \\ 0 & 0 & 0 & 0 & 0 \end{array}\right]$$

This matrix is in reduced row echelon form. It corresponds to the linear system

$$\begin{cases} x_1 & & - & x_3 & + & 2x_4 & = & 0 \\ & x_2 & & & + & 3x_4 & = & 0 \\ & & & & & 0 & = & 0 \end{cases}.$$

We can ignore the last equation, since it provides us with no information regarding the values of the variables. Columns 3 and 4 of the matrix are nonpivot columns, and so x_3 and x_4 are independent variables. Let $x_3 = c$ and $x_4 = d$. Substituting these into the first equation yields $x_1 - c + 2d = 0$, or, $x_1 = c - 2d$. The second equation gives us $x_2 + 3d = 0$, or, $x_2 = -3d$. Therefore. the complete solution set is $\{(c - 2d, -3d, c, d) \mid c, d \in \mathbb{R}\}$. Letting $c = 1$ and $d = 2$ produces the particular nontrivial solution $(-3, -6, 1, 2)$.

(c) The augmented matrix for the system is

$$\left[\begin{array}{cccccc|c} 7 & 28 & 4 & -2 & 10 & 19 & 0 \\ -9 & -36 & -5 & 3 & -15 & -29 & 0 \\ 3 & 12 & 2 & 0 & 6 & 11 & 0 \\ 6 & 24 & 3 & -3 & 10 & 20 & 0 \end{array}\right].$$

The (1,1) entry is the first pivot. We convert it to 1.

$$\langle 1\rangle \leftarrow \left(\tfrac{1}{7}\right)\langle 1\rangle \qquad \left[\begin{array}{cccccc|c} \textcircled{1} & 4 & \frac{4}{7} & -\frac{2}{7} & \frac{10}{7} & \frac{19}{7} & 0 \\ -9 & -36 & -5 & 3 & -15 & -29 & 0 \\ 3 & 12 & 2 & 0 & 6 & 11 & 0 \\ 6 & 24 & 3 & -3 & 10 & 20 & 0 \end{array}\right]$$

Next, we target the (2,1), (3,1) and (4,1) entries.

$$\begin{array}{l} \langle 2\rangle \leftarrow (9)\times\langle 1\rangle + \langle 2\rangle \\ \langle 3\rangle \leftarrow (-3)\times\langle 1\rangle + \langle 3\rangle \\ \langle 4\rangle \leftarrow (-6)\times\langle 1\rangle + \langle 4\rangle \end{array} \qquad \left[\begin{array}{cccccc|c} \textcircled{1} & 4 & \frac{4}{7} & -\frac{2}{7} & \frac{10}{7} & \frac{19}{7} & 0 \\ 0 & 0 & \frac{1}{7} & \frac{3}{7} & -\frac{15}{7} & -\frac{32}{7} & 0 \\ 0 & 0 & \frac{2}{7} & \frac{6}{7} & \frac{12}{7} & \frac{20}{7} & 0 \\ 0 & 0 & -\frac{3}{7} & -\frac{9}{7} & \frac{10}{7} & \frac{26}{7} & 0 \end{array}\right]$$

We try to move the pivot to the second column, but can not, due to the zeroes in the (2,2), (3,2) and (4,2) entries. Hence, the pivot moves to the (2,3) entry. We convert that to 1.

$$\langle 2\rangle \leftarrow (7)\langle 2\rangle \qquad \left[\begin{array}{cccccc|c} 1 & 4 & \frac{4}{7} & -\frac{2}{7} & \frac{10}{7} & \frac{19}{7} & 0 \\ 0 & 0 & \textcircled{1} & 3 & -15 & -32 & 0 \\ 0 & 0 & \frac{2}{7} & \frac{6}{7} & \frac{12}{7} & \frac{20}{7} & 0 \\ 0 & 0 & -\frac{3}{7} & -\frac{9}{7} & \frac{10}{7} & \frac{26}{7} & 0 \end{array}\right]$$

Next, we target the (1,3), (3,3) and (4,3) entries.

$$\begin{array}{l} \langle 1\rangle \leftarrow \left(-\tfrac{4}{7}\right)\times\langle 2\rangle + \langle 1\rangle \\ \langle 3\rangle \leftarrow \left(-\tfrac{2}{7}\right)\times\langle 2\rangle + \langle 3\rangle \\ \langle 4\rangle \leftarrow \left(\tfrac{3}{7}\right)\times\langle 2\rangle + \langle 4\rangle \end{array} \qquad \left[\begin{array}{cccccc|c} 1 & 4 & 0 & -2 & 10 & 21 & 0 \\ 0 & 0 & \textcircled{1} & 3 & -15 & -32 & 0 \\ 0 & 0 & 0 & 0 & 6 & 12 & 0 \\ 0 & 0 & 0 & 0 & -5 & -10 & 0 \end{array}\right]$$

We try to move the pivot to the third column, but can not, due to the zeroes in the (3,4) and (4,4) entries. Hence, the pivot moves to the (3,5) entry. We convert that to 1.

$$\langle 3\rangle \leftarrow \left(\tfrac{1}{6}\right)\langle 3\rangle \qquad \left[\begin{array}{cccccc|c} 1 & 4 & 0 & -2 & 10 & 21 & 0 \\ 0 & 0 & 1 & 3 & -15 & -32 & 0 \\ 0 & 0 & 0 & 0 & \textcircled{1} & 2 & 0 \\ 0 & 0 & 0 & 0 & -5 & -10 & 0 \end{array}\right]$$

Next, we target the (1,5), (2,5) and (4,5) entries.

$$\begin{array}{l} \langle 1\rangle \leftarrow (-10)\times\langle 3\rangle + \langle 1\rangle \\ \langle 2\rangle \leftarrow (15)\times\langle 3\rangle + \langle 2\rangle \\ \langle 4\rangle \leftarrow (5)\times\langle 3\rangle + \langle 4\rangle \end{array} \qquad \left[\begin{array}{cccccc|c} 1 & 4 & 0 & -2 & 0 & 1 & 0 \\ 0 & 0 & 1 & 3 & 0 & -2 & 0 \\ 0 & 0 & 0 & 0 & \textcircled{1} & 2 & 0 \\ 0 & 0 & 0 & 0 & 0 & 0 & 0 \end{array}\right]$$

This matrix is in reduced row echelon form. It corresponds to the linear system

$$\begin{cases} x_1 + 4x_2 - 2x_4 + x_6 = 0 \\ x_3 + 3x_4 - 2x_6 = 0 \\ x_5 + 2x_6 = 0 \\ 0 = 0 \end{cases}.$$

We can ignore the last equation, since it provides us with no information regarding the values of the variables. Columns 2, 4 and 6 of the matrix are nonpivot columns, and so x_2, x_4 and x_6 are independent variables. Let $x_2 = b$, $x_4 = d$ and $x_6 = f$. Substituting these into the first equation yields $x_1 + 4b - 2d + f = 0$, or, $x_1 = -4b + 2d - f$. The second equation gives us $x_3 + 3d - 2f = 0$, or, $x_3 = -3d + 2f$. The third equation produces $x_5 + 2f = 0$, or, $x_5 = -2f$. Therefore. the complete solution set is $\{(-4b + 2d - f, b, -3d + 2f, d, -2f, f) \mid b, d, f \in \mathbb{R}\}$. Letting $b = 1$, $d = 2$, and $f = 3$ produces the particular nontrivial solution $(-3, 1, 0, 2, -6, 3)$.

(5) (a) The augmented matrix for the system is $\left[\begin{array}{ccc|c} -2 & 1 & 8 & 0 \\ 7 & -2 & -22 & 0 \\ 3 & -1 & -10 & 0 \end{array}\right]$.

The (1,1) entry is the first pivot. We convert it to 1.

$\langle 1\rangle \leftarrow (-\frac{1}{2})\langle 1\rangle$ $\left[\begin{array}{ccc|c} ① & -\frac{1}{2} & -4 & 0 \\ 7 & -2 & -22 & 0 \\ 3 & -1 & -10 & 0 \end{array}\right]$

Next, we target the (2,1) and (3,1) entries.

$\langle 2\rangle \leftarrow (-7) \times \langle 1\rangle + \langle 2\rangle$
$\langle 3\rangle \leftarrow (-3) \times \langle 1\rangle + \langle 3\rangle$ $\left[\begin{array}{ccc|c} ① & -\frac{1}{2} & -4 & 0 \\ 0 & \frac{3}{2} & 6 & 0 \\ 0 & \frac{1}{2} & 2 & 0 \end{array}\right]$

The (2,2) entry is the second pivot. We convert it to 1.

$\langle 2\rangle \leftarrow (\frac{2}{3})\langle 2\rangle$ $\left[\begin{array}{ccc|c} 1 & -\frac{1}{2} & -4 & 0 \\ 0 & ① & 4 & 0 \\ 0 & \frac{1}{2} & 2 & 0 \end{array}\right]$

Next, we target the (1,2) and (3,2) entries.

$\langle 1\rangle \leftarrow (\frac{1}{2}) \times \langle 2\rangle + \langle 1\rangle$
$\langle 3\rangle \leftarrow (-\frac{1}{2}) \times \langle 2\rangle + \langle 3\rangle$ $\left[\begin{array}{ccc|c} 1 & 0 & -2 & 0 \\ 0 & ① & 4 & 0 \\ 0 & 0 & 0 & 0 \end{array}\right]$

This matrix is in reduced row echelon form. It corresponds to the linear system

$$\begin{cases} x_1 - 2x_3 = 0 \\ x_2 + 4x_3 = 0 \\ 0 = 0 \end{cases}.$$

We can ignore the last equation, since it provides us with no information regarding the values of the variables. Column 3 of the matrix is a nonpivot column, and so x_3 is an independent variable. Let $x_3 = c$. Substituting this into the first equation yields $x_1 - 2c = 0$, or, $x_1 = 2c$. The second equation gives us $x_2 + 4c = 0$, or, $x_2 = -4c$. Therefore. the complete solution set is $\{(2c, -4c, c) \mid c \in \mathbb{R}\} = \{c(2, -4, 1) \mid c \in \mathbb{R}\}$.

(c) The augmented matrix for the system is $\left[\begin{array}{cccc|c} 2 & 6 & 13 & 1 & 0 \\ 1 & 4 & 10 & 1 & 0 \\ 2 & 8 & 20 & 1 & 0 \\ 3 & 10 & 21 & 2 & 0 \end{array}\right]$.

The (1,1) entry is the first pivot. We convert it to 1.

$$\langle 1\rangle \leftarrow (\tfrac{1}{2})\,\langle 1\rangle \qquad \left[\begin{array}{cccc|c} ① & 3 & \frac{13}{2} & \frac{1}{2} & 0 \\ 1 & 4 & 10 & 1 & 0 \\ 2 & 8 & 20 & 1 & 0 \\ 3 & 10 & 21 & 2 & 0 \end{array}\right]$$

Next, we target the (2,1), (3,1) and (4,1) entries.

$$\begin{array}{l} \langle 2\rangle \leftarrow (-1)\times\langle 1\rangle + \langle 2\rangle \\ \langle 3\rangle \leftarrow (-2)\times\langle 1\rangle + \langle 3\rangle \\ \langle 4\rangle \leftarrow (-3)\times\langle 1\rangle + \langle 4\rangle \end{array} \qquad \left[\begin{array}{cccc|c} ① & 3 & \frac{13}{2} & \frac{1}{2} & 0 \\ 0 & 1 & \frac{7}{2} & \frac{1}{2} & 0 \\ 0 & 2 & 7 & 0 & 0 \\ 0 & 1 & \frac{3}{2} & \frac{1}{2} & 0 \end{array}\right]$$

The (2,2) entry is the second pivot. It already equals 1, so now we target the (1,2), (3,2) and (4,2) entries.

$$\begin{array}{l} \langle 1\rangle \leftarrow (-3)\times\langle 2\rangle + \langle 1\rangle \\ \langle 3\rangle \leftarrow (-2)\times\langle 2\rangle + \langle 3\rangle \\ \langle 4\rangle \leftarrow (-1)\times\langle 2\rangle + \langle 4\rangle \end{array} \qquad \left[\begin{array}{cccc|c} 1 & 0 & -4 & -1 & 0 \\ 0 & ① & \frac{7}{2} & \frac{1}{2} & 0 \\ 0 & 0 & 0 & -1 & 0 \\ 0 & 0 & -2 & 0 & 0 \end{array}\right]$$

The (3,3) entry equals 0. However, we can switch the 3rd and 4th rows to bring a nonzero number into the (3,3) position.

$$\langle 3\rangle \longleftrightarrow \langle 4\rangle \qquad \left[\begin{array}{cccc|c} 1 & 0 & -4 & -1 & 0 \\ 0 & 1 & \frac{7}{2} & \frac{1}{2} & 0 \\ 0 & 0 & -2 & 0 & 0 \\ 0 & 0 & 0 & -1 & 0 \end{array}\right]$$

We now convert the pivot (the (3,3) entry) to 1.

$$\langle 3\rangle \leftarrow (-\tfrac{1}{2})\,\langle 3\rangle \qquad \left[\begin{array}{cccc|c} 1 & 0 & -4 & -1 & 0 \\ 0 & 1 & \frac{7}{2} & \frac{1}{2} & 0 \\ 0 & 0 & ① & 0 & 0 \\ 0 & 0 & 0 & -1 & 0 \end{array}\right]$$

Next, we target the (1,3) and (2,3) entries. The (4,3) entry already equals 0.

$$\begin{array}{l} \langle 1\rangle \leftarrow (4)\times\langle 3\rangle + \langle 1\rangle \\ \langle 2\rangle \leftarrow (-\tfrac{7}{2})\times\langle 3\rangle + \langle 2\rangle \end{array} \qquad \left[\begin{array}{cccc|c} 1 & 0 & 0 & -1 & 0 \\ 0 & 1 & 0 & \frac{1}{2} & 0 \\ 0 & 0 & ① & 0 & 0 \\ 0 & 0 & 0 & -1 & 0 \end{array}\right]$$

The (4,4) entry is the last pivot. We convert it to 1.

$\langle 4\rangle \leftarrow (-1)\,\langle 4\rangle$ $\quad\left[\begin{array}{cccc|c} 1 & 0 & 0 & -1 & 0 \\ 0 & 1 & 0 & \frac{1}{2} & 0 \\ 0 & 0 & 1 & 0 & 0 \\ 0 & 0 & 0 & \textcircled{1} & 0 \end{array}\right]$

Finally, we target the (1,4) and (2,4) entries. The (3,4) entry is already 0.

$\langle 1\rangle \leftarrow (1) \times \langle 4\rangle + \langle 1\rangle$
$\langle 2\rangle \leftarrow (-\frac{1}{2}) \times \langle 4\rangle + \langle 2\rangle$ $\quad\left[\begin{array}{cccc|c} 1 & 0 & 0 & 0 & 0 \\ 0 & 1 & 0 & 0 & 0 \\ 0 & 0 & 1 & 0 & 0 \\ 0 & 0 & 0 & \textcircled{1} & 0 \end{array}\right]$

This matrix is in reduced row echelon form. It corresponds to the linear system

$$\begin{cases} x_1 & & & & = 0 \\ & x_2 & & & = 0 \\ & & x_3 & & = 0 \\ & & & x_4 & = 0 \end{cases}.$$

Clearly, this system has only the trivial solution. The solution set is $\{(0,0,0,0)\}$.

(6) (a) First we find the system of linear equations corresponding to the given chemical equation by considering each element separately. This produces

$$\begin{cases} 6a & & = & c & & & \leftarrow \text{Carbon equation} \\ 6a & & = & & & 2d & \leftarrow \text{Hydrogen equation} \\ & 2b & = & 2c & + & d & \leftarrow \text{Oxygen equation} \end{cases}.$$

Moving all variables to the left side, and then creating an augmented matrix yields

$$\left[\begin{array}{cccc|c} 6 & 0 & -1 & 0 & 0 \\ 6 & 0 & 0 & -2 & 0 \\ 0 & 2 & -2 & -1 & 0 \end{array}\right].$$

The (1,1) entry is the first pivot. We convert it to 1.

$\langle 1\rangle \leftarrow (\frac{1}{6})\,\langle 1\rangle$ $\quad\left[\begin{array}{cccc|c} \textcircled{1} & 0 & -\frac{1}{6} & 0 & 0 \\ 6 & 0 & 0 & -2 & 0 \\ 0 & 2 & -2 & -1 & 0 \end{array}\right]$

Next, we target the (2,1) entry. (The (3,1) entry is already 0.)

$\langle 2\rangle \leftarrow (-6) \times \langle 1\rangle + \langle 2\rangle$ $\quad\left[\begin{array}{cccc|c} \textcircled{1} & 0 & -\frac{1}{6} & 0 & 0 \\ 0 & 0 & 1 & -2 & 0 \\ 0 & 2 & -2 & -1 & 0 \end{array}\right]$

The pivot moves to the (2,2) entry, which is zero. So, we switch the 2nd and 3rd rows to move a nonzero number into the pivot position.

$\langle 2\rangle \longleftrightarrow \langle 3\rangle$ $\quad\left[\begin{array}{cccc|c} 1 & 0 & -\frac{1}{6} & 0 & 0 \\ 0 & \textcircled{2} & -2 & -1 & 0 \\ 0 & 0 & 1 & -2 & 0 \end{array}\right]$

Now we convert the (2,2) entry to 1.

$\langle 2\rangle \leftarrow (\frac{1}{2})\langle 2\rangle$ $\left[\begin{array}{cccc|c} 1 & 0 & -\frac{1}{6} & 0 & 0 \\ 0 & ① & -1 & -\frac{1}{2} & 0 \\ 0 & 0 & 1 & -2 & 0 \end{array}\right]$

The pivot moves to the (3,3) entry. Since this already equals 1, we target the (1,3) and (2,3) entries.

$\langle 1\rangle \leftarrow (\frac{1}{6}) \times \langle 3\rangle + \langle 1\rangle$
$\langle 2\rangle \leftarrow (1) \times \langle 3\rangle + \langle 2\rangle$ $\left[\begin{array}{cccc|c} 1 & 0 & 0 & -\frac{1}{3} & 0 \\ 0 & 1 & 0 & -\frac{5}{2} & 0 \\ 0 & 0 & ① & -2 & 0 \end{array}\right]$

The matrix is now in reduced row echelon form. It corresponds to the following linear system:

$$\begin{cases} a - \frac{1}{3}d = 0 \\ b - \frac{5}{2}d = 0 \\ c - 2d = 0 \end{cases},$$

or $a = \frac{1}{3}d$, $b = \frac{5}{2}d$, and $c = 2d$. Setting $d = 6$ eliminates all fractions, and provides the smallest solution containing all positive integers. Hence, the desired solution is $a = 2$, $b = 15$, $c = 12$, $d = 6$.

(c) First we find the system of linear equations corresponding to the given chemical equation by considering each element separately. This produces

$$\begin{cases} a = c & \leftarrow \text{Silver equation} \\ a = e & \leftarrow \text{Nitrogen equation} \\ 3a + b = 2d + 3e & \leftarrow \text{Oxygen equation} \\ 2b = e & \leftarrow \text{Hydrogen equation} \end{cases},$$

Moving all variables to the left side, and then creating an augmented matrix yields

$$\left[\begin{array}{ccccc|c} 1 & 0 & -1 & 0 & 0 & 0 \\ 1 & 0 & 0 & 0 & -1 & 0 \\ 3 & 1 & 0 & -2 & -3 & 0 \\ 0 & 2 & 0 & 0 & -1 & 0 \end{array}\right].$$

The (1,1) entry is the first pivot. Since it already equals 1, we target the (2,1) and (3,1) entries. (The (4,1) entry is already 0.)

$\langle 2\rangle \leftarrow (-1) \times \langle 1\rangle + \langle 2\rangle$
$\langle 3\rangle \leftarrow (-3) \times \langle 1\rangle + \langle 3\rangle$ $\left[\begin{array}{ccccc|c} ① & 0 & -1 & 0 & 0 & 0 \\ 0 & 0 & 1 & 0 & -1 & 0 \\ 0 & 1 & 3 & -2 & -3 & 0 \\ 0 & 2 & 0 & 0 & -1 & 0 \end{array}\right]$

The pivot moves to the (2,2) entry. However, it equals zero. So, we switch the 2nd and 3rd rows to move a nonzero number into the pivot position.

$\langle 2\rangle \longleftrightarrow \langle 3\rangle$ $\left[\begin{array}{ccccc|c} 1 & 0 & -1 & 0 & 0 & 0 \\ 0 & ① & 3 & -2 & -3 & 0 \\ 0 & 0 & 1 & 0 & -1 & 0 \\ 0 & 2 & 0 & 0 & -1 & 0 \end{array}\right]$

The (2,2) entry already equals 1, so we target the (4,2) entry. (The (1,2) and (3,2) entries already equal zero, and so do not need to be targeted.)

$\langle 4\rangle \leftarrow (-2)\times\langle 2\rangle + \langle 4\rangle$ $\qquad \left[\begin{array}{ccccc|c} 1 & 0 & -1 & 0 & 0 & 0 \\ 0 & \textcircled{1} & 3 & -2 & -3 & 0 \\ 0 & 0 & 1 & 0 & -1 & 0 \\ 0 & 0 & -6 & 4 & 5 & 0 \end{array}\right]$

The pivot moves to the (3,3) entry. Since this already equals 1, we target the (1,3), (2,3) and (4,3) entries.

$\langle 1\rangle \leftarrow (1)\times\langle 3\rangle + \langle 1\rangle$
$\langle 2\rangle \leftarrow (-3)\times\langle 3\rangle + \langle 2\rangle$
$\langle 4\rangle \leftarrow (6)\times\langle 3\rangle + \langle 4\rangle$ $\qquad \left[\begin{array}{ccccc|c} 1 & 0 & 0 & 0 & -1 & 0 \\ 0 & 1 & 0 & -2 & 0 & 0 \\ 0 & 0 & \textcircled{1} & 0 & -1 & 0 \\ 0 & 0 & 0 & 4 & -1 & 0 \end{array}\right]$

The pivot moves to the (4,4) entry. We convert this to 1.

$\langle 4\rangle \leftarrow (\frac{1}{4})\langle 4\rangle$ $\qquad \left[\begin{array}{ccccc|c} 1 & 0 & 0 & 0 & -1 & 0 \\ 0 & 1 & 0 & -2 & 0 & 0 \\ 0 & 0 & 1 & 0 & -1 & 0 \\ 0 & 0 & 0 & \textcircled{1} & -\frac{1}{4} & 0 \end{array}\right]$

Finally, we target the (2,4) entry.

$\langle 2\rangle \leftarrow (2)\times\langle 4\rangle + \langle 2\rangle$ $\qquad \left[\begin{array}{ccccc|c} 1 & 0 & 0 & 0 & -1 & 0 \\ 0 & 1 & 0 & 0 & -\frac{1}{2} & 0 \\ 0 & 0 & 1 & 0 & -1 & 0 \\ 0 & 0 & 0 & \textcircled{1} & -\frac{1}{4} & 0 \end{array}\right]$

The matrix is now in reduced row echelon form. It corresponds to the following linear system:

$$\left\{\begin{array}{ccccccc} a & & & & - & e & = 0 \\ & b & & & - & \frac{1}{2}e & = 0 \\ & & c & & - & e & = 0 \\ & & & d & - & \frac{1}{4}e & = 0 \end{array}\right.,$$

or $a = e$, $b = \frac{1}{2}e$, $c = e$, and $d = \frac{1}{4}e$. Setting $e = 4$ eliminates all fractions, and provides the smallest solution containing all positive integers. Hence, the desired solution is $a = 4$, $b = 2$, $c = 4$, $d = 1$, $e = 4$.

(7) (a) Combining the fractions on the right side of the given equation over a common denominator produces

$$\frac{A}{(x-1)} + \frac{B}{(x-3)} + \frac{C}{(x+4)} = \frac{A(x-3)(x+4) + B(x-1)(x+4) + C(x-1)(x-3)}{(x-1)(x-3)(x+4)}$$
$$= \frac{(A+B+C)\,x^2 + (A+3B-4C)\,x + (-12A-4B+3C)}{(x-1)(x-3)(x+4)}.$$

Now we set the coefficients of x^2 and x, and the constant term equal to the corresponding coefficients in the numerator of the left side of the equation given in the problem. Doing this yields the following linear system:

$$\left\{\begin{array}{rcrcrcrl} A & + & B & + & C & = & 5 & \leftarrow \text{coefficients of } x^2 \\ A & + & 3B & - & 4C & = & 23 & \leftarrow \text{coefficients of } x \\ -12A & - & 4B & + & 3C & = & -58 & \leftarrow \text{constant terms} \end{array}\right. .$$

The augmented matrix for the system is $\left[\begin{array}{ccc|c} 1 & 1 & 1 & 5 \\ 1 & 3 & -4 & 23 \\ -12 & -4 & 3 & -58 \end{array}\right]$.

The (1,1) entry is the first pivot. Since it already equals 1, we target the (2,1) and (3,1) entries.

$\langle 2\rangle \leftarrow (-1) \times \langle 1\rangle + \langle 2\rangle$
$\langle 3\rangle \leftarrow (12) \times \langle 1\rangle + \langle 3\rangle$
$$\left[\begin{array}{ccc|c} \textcircled{1} & 1 & 1 & 5 \\ 0 & 2 & -5 & 18 \\ 0 & 8 & 15 & 2 \end{array}\right]$$

The (2,2) entry is the second pivot. We convert it to 1.

$\langle 2\rangle \leftarrow (\frac{1}{2})\,\langle 2\rangle$
$$\left[\begin{array}{ccc|c} 1 & 1 & 1 & 5 \\ 0 & \textcircled{1} & -\frac{5}{2} & 9 \\ 0 & 8 & 15 & 2 \end{array}\right]$$

Next, we target the (1,2) and (3,2) entries.

$\langle 1\rangle \leftarrow (-1) \times \langle 2\rangle + \langle 1\rangle$
$\langle 3\rangle \leftarrow (-8) \times \langle 2\rangle + \langle 3\rangle$
$$\left[\begin{array}{ccc|c} 1 & 0 & \frac{7}{2} & -4 \\ 0 & \textcircled{1} & -\frac{5}{2} & 9 \\ 0 & 0 & 35 & -70 \end{array}\right]$$

The (3,3) entry is the last pivot. We convert it to 1.

$\langle 3\rangle \leftarrow (\frac{1}{35})\,\langle 3\rangle$
$$\left[\begin{array}{ccc|c} 1 & 0 & \frac{7}{2} & -4 \\ 0 & 1 & -\frac{5}{2} & 9 \\ 0 & 0 & \textcircled{1} & -2 \end{array}\right]$$

Finally, we target the (1,3) and (2,3) entries.

$\langle 1\rangle \leftarrow (-\frac{7}{2}) \times \langle 3\rangle + \langle 1\rangle$
$\langle 2\rangle \leftarrow (\frac{5}{2}) \times \langle 3\rangle + \langle 2\rangle$
$$\left[\begin{array}{ccc|c} 1 & 0 & 0 & 3 \\ 0 & 1 & 0 & 4 \\ 0 & 0 & \textcircled{1} & -2 \end{array}\right]$$

This matrix is in reduced row echelon form. It corresponds to the linear system

$$\begin{cases} A & & & = & 3 \\ & B & & = & 4 \\ & & C & = & -2 \end{cases},$$

which gives the unique solution to the problem.

(8) We set up the augmented matrix having two columns to the right of the augmentation bar:

$$\left[\begin{array}{ccc|cc} 9 & 2 & 2 & -6 & -12 \\ 3 & 2 & 4 & 0 & -3 \\ 27 & 12 & 22 & 12 & 8 \end{array}\right].$$

The (1,1) entry is the first pivot. We convert it to 1.

$\langle 1\rangle \leftarrow (\frac{1}{9})\,\langle 1\rangle$
$$\left[\begin{array}{ccc|cc} \textcircled{1} & \frac{2}{9} & \frac{2}{9} & -\frac{2}{3} & -\frac{4}{3} \\ 3 & 2 & 4 & 0 & -3 \\ 27 & 12 & 22 & 12 & 8 \end{array}\right]$$

We next target the (2,1) and (3,1) entries.

$$\begin{array}{l} \langle 2\rangle \leftarrow (-3)\times\langle 1\rangle + \langle 2\rangle \\ \langle 3\rangle \leftarrow (-27)\times\langle 1\rangle + \langle 3\rangle \end{array} \quad \left[\begin{array}{ccc|cc} \textcircled{1} & \frac{2}{9} & \frac{2}{9} & -\frac{2}{3} & -\frac{4}{3} \\ 0 & \frac{4}{3} & \frac{10}{3} & 2 & 1 \\ 0 & 6 & 16 & 30 & 44 \end{array}\right]$$

The (2,2) entry is the second pivot. We convert it to 1.

$$\langle 2\rangle \leftarrow \left(\tfrac{3}{4}\right)\langle 2\rangle \quad \left[\begin{array}{ccc|cc} 1 & \frac{2}{9} & \frac{2}{9} & -\frac{2}{3} & -\frac{4}{3} \\ 0 & \textcircled{1} & \frac{5}{2} & \frac{3}{2} & \frac{3}{4} \\ 0 & 6 & 16 & 30 & 44 \end{array}\right]$$

Next, we target the (1,2) and (3,2) entries.

$$\begin{array}{l} \langle 1\rangle \leftarrow (-\tfrac{2}{9})\times\langle 2\rangle + \langle 1\rangle \\ \langle 3\rangle \leftarrow (-6)\times\langle 2\rangle + \langle 3\rangle \end{array} \quad \left[\begin{array}{ccc|cc} 1 & 0 & -\frac{1}{3} & -1 & -\frac{3}{2} \\ 0 & \textcircled{1} & \frac{5}{2} & \frac{3}{2} & \frac{3}{4} \\ 0 & 0 & 1 & 21 & \frac{79}{2} \end{array}\right]$$

The (3,3) entry is the last pivot. Since it already equals 1, we target the (1,3) and (2,3) entries.

$$\begin{array}{l} \langle 1\rangle \leftarrow (\tfrac{1}{3})\times\langle 3\rangle + \langle 1\rangle \\ \langle 2\rangle \leftarrow (-\tfrac{5}{2})\times\langle 3\rangle + \langle 2\rangle \end{array} \quad \left[\begin{array}{ccc|cc} 1 & 0 & 0 & 6 & \frac{35}{3} \\ 0 & 1 & 0 & -51 & -98 \\ 0 & 0 & \textcircled{1} & 21 & \frac{79}{2} \end{array}\right]$$

This matrix is in reduced row echelon form. It gives the unique solution for each of the two original linear systems. In particular, the solution for the system $\mathbf{AX} = \mathbf{B}_1$ is $(6, -51, 21)$, and the solution for the system $\mathbf{AX} = \mathbf{B}_2$ is $(\frac{35}{3}, -98, \frac{79}{2})$.

(11) (b) Any nonhomogeneous system with two equations and two unknowns that has a unique solution will serve as a counterexample. For instance, consider

$$\begin{cases} x + y = 1 \\ x - y = 1 \end{cases}.$$

This system has a unique solution: $(1,0)$. Let (s_1, s_2) and (t_1, t_2) both equal $(1,0)$. Then the sum of solutions is not a solution in this case. Also, if $c \neq 1$, then the scalar multiple of a solution by c is not a solution.

(14) (a) True. The Gaussian elimination method puts the augmented matrix for a system into row echelon form. The statement gives a verbal description of the definition of row echelon form given in the section.

(b) True. The Gaussian elimination method puts the augmented matrix for a system into row echelon form, while the Gauss-Jordan method puts the matrix into reduced row echelon form. The statement describes the difference between those two forms for a matrix.

(c) False. A column is skipped if the pivot entry that would usually be used is zero, and all entries below that position in that column are zero. For example, the reduced row echelon form matrix $\begin{bmatrix} 1 & 2 & 0 \\ 0 & 0 & 1 \end{bmatrix}$ has no pivot in the second column. Rows, however, are never skipped over, although rows of zeroes may occur at the bottom of the matrix.

(d) True. A homogeneous system always has the trivial solution.

(e) False. This statement reverses the roles of the "independent (nonpivot column)" and "dependent (pivot column)" variables. Note, however, that the statement does correctly define the terms "independent" and "dependent."

(f) False. For a counterexample, consider the system

$$\begin{cases} x & & = & 0 \\ & y & = & 0 \\ x & + \; y & = & 0 \end{cases},$$

which clearly has only the trivial solution. The given statement is the reverse of a related true statement, which says: If a homogeneous system has more variables than equations, then the system has a nontrivial solution.

Section 2.3

(1) Recall that a matrix **A** is row equivalent to a matrix **B** if a finite sequence of row operations will convert **A** into **B**. In parts (a) and (c), below, only one row operation is necessary, although longer finite sequences could be used.

(a) A row operation of type (I) converts **A** to **B**: $\langle 2\rangle \longleftarrow -5\langle 2\rangle$.

(c) A row operation of type (II) converts **A** to **B**: $\langle 2\rangle \longleftarrow \langle 3\rangle + \langle 2\rangle$.

(2) (b) We must first compute **B** by putting **A** into reduced row echelon form. Performing the following sequence of row operations converts **A** to $\mathbf{B} = \mathbf{I}_3$:

(I): $\langle 1\rangle \leftarrow \frac{1}{4}\langle 1\rangle$
(II): $\langle 2\rangle \leftarrow 2\langle 1\rangle + \langle 2\rangle$
(II): $\langle 3\rangle \leftarrow -3\langle 1\rangle + \langle 3\rangle$
(III): $\langle 2\rangle \leftrightarrow \langle 3\rangle$
(II): $\langle 1\rangle \leftarrow 5\langle 3\rangle + \langle 1\rangle$

To convert **B** back to **A** we use the inverses of these row operations in the reverse order:

(II): $\langle 1\rangle \leftarrow -5\langle 3\rangle + \langle 1\rangle$	the inverse of (II): $\langle 1\rangle \leftarrow 5\langle 3\rangle + \langle 1\rangle$
(III): $\langle 2\rangle \leftrightarrow \langle 3\rangle$	the inverse of (III): $\langle 2\rangle \leftrightarrow \langle 3\rangle$
(II): $\langle 3\rangle \leftarrow 3\langle 1\rangle + \langle 3\rangle$	the inverse of (II): $\langle 3\rangle \leftarrow -3\langle 1\rangle + \langle 3\rangle$
(II): $\langle 2\rangle \leftarrow -2\langle 1\rangle + \langle 2\rangle$	the inverse of (II): $\langle 2\rangle \leftarrow 2\langle 1\rangle + \langle 2\rangle$
(I): $\langle 1\rangle \leftarrow 4\langle 1\rangle$	the inverse of (I): $\langle 1\rangle \leftarrow \frac{1}{4}\langle 1\rangle$

(3) (a) Both **A** and **B** row reduce to $\mathbf{I}_3$. To convert **A** to $\mathbf{I}_3$, use the following sequence of row operations:

(II): $\langle 3\rangle \longleftarrow 2\langle 2\rangle + \langle 3\rangle$
(I): $\langle 3\rangle \longleftarrow -1\langle 3\rangle$
(II): $\langle 1\rangle \longleftarrow -9\langle 3\rangle + \langle 1\rangle$
(II): $\langle 2\rangle \longleftarrow 3\langle 3\rangle + \langle 2\rangle$

To convert **B** to $\mathbf{I}_3$, use these row operations:

(I): $\langle 1\rangle \longleftarrow -\frac{1}{5}\langle 1\rangle$
(II): $\langle 2\rangle \longleftarrow 2\langle 1\rangle + \langle 2\rangle$
(II): $\langle 3\rangle \longleftarrow 3\langle 1\rangle + \langle 3\rangle$
(I): $\langle 2\rangle \longleftarrow -5\langle 2\rangle$
(II): $\langle 1\rangle \longleftarrow \frac{3}{5}\langle 2\rangle + \langle 1\rangle$
(II): $\langle 3\rangle \longleftarrow \frac{9}{5}\langle 2\rangle + \langle 3\rangle$

(b) First we convert **A** to $\mathbf{I}_3$ using the sequence of row operations we computed in part (a):

(II): $\langle 3\rangle \longleftarrow 2\langle 2\rangle + \langle 3\rangle$

(I): $\langle 3\rangle \longleftarrow -1\,\langle 3\rangle$
(II): $\langle 1\rangle \longleftarrow -9\,\langle 3\rangle + \langle 1\rangle$
(II): $\langle 2\rangle \longleftarrow 3\,\langle 3\rangle + \langle 2\rangle$

Next, we continue by converting $\mathbf{I}_3$ to $\mathbf{B}$. To do this, we use the inverses of the row operations that converted $\mathbf{B}$ to $\mathbf{I}_3$ (from part (a)), in the reverse order:

(II): $\langle 3\rangle \leftarrow -\frac{9}{5}\,\langle 2\rangle + \langle 3\rangle$	the inverse of (II): $\langle 3\rangle \leftarrow \frac{9}{5}\,\langle 2\rangle + \langle 3\rangle$
(II): $\langle 1\rangle \leftarrow -\frac{3}{5}\,\langle 2\rangle + \langle 1\rangle$	the inverse of (II): $\langle 1\rangle \leftarrow \frac{3}{5}\,\langle 2\rangle + \langle 1\rangle$
(I): $\langle 2\rangle \leftarrow -\frac{1}{5}\,\langle 2\rangle$	the inverse of (I): $\langle 2\rangle \leftarrow -5\,\langle 2\rangle$
(II): $\langle 3\rangle \leftarrow -3\,\langle 1\rangle + \langle 3\rangle$	the inverse of (II): $\langle 3\rangle \leftarrow 3\,\langle 1\rangle + \langle 3\rangle$
(II): $\langle 2\rangle \leftarrow -2\,\langle 1\rangle + \langle 2\rangle$	the inverse of (II): $\langle 2\rangle \leftarrow 2\,\langle 1\rangle + \langle 2\rangle$
(I): $\langle 1\rangle \leftarrow -5\,\langle 1\rangle$	the inverse of (I): $\langle 1\rangle \leftarrow -\frac{1}{5}\,\langle 1\rangle$

(5) The rank of a matrix is the number of nonzero rows in its corresponding reduced row echelon form matrix.

(a) The given matrix reduces to $\begin{bmatrix} 1 & 0 & 2 \\ 0 & 1 & -1 \\ 0 & 0 & 0 \end{bmatrix}$, which has two nonzero rows. Hence, its rank is 2.

(c) The given matrix reduces to $\begin{bmatrix} 1 & 0 & 0 \\ 0 & 0 & 1 \\ 0 & 0 & 0 \end{bmatrix}$, which has two nonzero rows. Hence, its rank is 2.

(e) The given matrix reduces to $\begin{bmatrix} 1 & 0 & 0 & 1 \\ 0 & 1 & 0 & -1 \\ 0 & 0 & 1 & \frac{3}{2} \\ 0 & 0 & 0 & 0 \end{bmatrix}$, which has 3 nonzero rows. Hence, its rank is 3.

(6) (a) Corollary 2.6 cannot be used because there are more equations than variables. The augmented matrix for this system is

$$\left[\begin{array}{ccc|c} -2 & 6 & 3 & 0 \\ 5 & -9 & -4 & 0 \\ 4 & -8 & -3 & 0 \\ 6 & -11 & -5 & 0 \end{array}\right] \text{ which reduces to } \left[\begin{array}{ccc|c} 1 & 0 & 0 & 0 \\ 0 & 1 & 0 & 0 \\ 0 & 0 & 1 & 0 \\ 0 & 0 & 0 & 0 \end{array}\right].$$

Since there are three nonzero rows in the reduced row echelon form matrix, the rank = 3. Theorem 2.5 predicts that the system has only the trivial solution. The reduced row echelon form matrix corresponds to the linear system

$$\begin{cases} x_1 & & & = 0 \\ & x_2 & & = 0 \\ & & x_3 & = 0 \\ & & 0 & = 0 \end{cases},$$

which clearly has the solution set $\{(0,0,0)\}$, agreeing with the prediction of Theorem 2.5.

(7) In each case, we give the smallest and largest rank possible, with a particular example of a linear system and the *reduced row echelon form* of its corresponding augmented matrix for each of the two ranks.

(a)

Smallest rank = 1	Largest rank = 4
$\begin{cases} x_1+2x_2+3x_3=4 \\ x_1+2x_2+3x_3=4 \\ x_1+2x_2+3x_3=4 \\ x_1+2x_2+3x_3=4 \end{cases}$	$\begin{cases} x_1 = 0 \\ x_2 = 0 \\ x_3 = 0 \\ 0 = 1 \end{cases}$
$\left[\begin{array}{ccc\|c} 1 & 2 & 3 & 4 \\ 0 & 0 & 0 & 0 \\ 0 & 0 & 0 & 0 \\ 0 & 0 & 0 & 0 \end{array}\right]$	$\left[\begin{array}{ccc\|c} 1 & 0 & 0 & 0 \\ 0 & 1 & 0 & 0 \\ 0 & 0 & 1 & 0 \\ 0 & 0 & 0 & 1 \end{array}\right]$

(c)

Smallest rank = 2	Largest rank = 3
$\begin{cases} x_1+x_2+x_3+x_4=0 \\ x_1+x_2+x_3+x_4=1 \\ x_1+x_2+x_3+x_4=0 \end{cases}$	$\begin{cases} x_1+x_3+x_4=0 \\ x_2+x_3+x_4=0 \\ x_1+x_3+x_4=1 \end{cases}$
$\left[\begin{array}{cccc\|c} 1 & 1 & 1 & 1 & 0 \\ 0 & 0 & 0 & 0 & 1 \\ 0 & 0 & 0 & 0 & 0 \end{array}\right]$	$\left[\begin{array}{cccc\|c} 1 & 0 & 1 & 1 & 0 \\ 0 & 1 & 1 & 1 & 0 \\ 0 & 0 & 0 & 0 & 1 \end{array}\right]$

(8) In each part, to express the vector $\mathbf{x}$ as a linear combination of the vectors $\mathbf{a}_1, \ldots, \mathbf{a}_n$, we need to solve the linear system whose augmented matrix has the vectors $\mathbf{a}_1, \ldots, \mathbf{a}_n$ as columns to the left of the augmentation bar, and has the vector $\mathbf{x}$ as a column to the right of the augmentation bar. A solution of this linear system gives the coefficients for the desired linear combination.

(a) Following the procedure described above, we reduce the matrix

$$\left[\begin{array}{cc|c} 1 & -2 & -3 \\ 4 & 3 & -6 \end{array}\right] \text{ to obtain } \left[\begin{array}{cc|c} 1 & 0 & -\frac{21}{11} \\ 0 & 1 & \frac{6}{11} \end{array}\right].$$

The unique solution to this system provides the coefficients for $\mathbf{x}$ as a linear combination of $\mathbf{a}_1$ and $\mathbf{a}_2$, and so $\mathbf{x} = -\frac{21}{11}\mathbf{a}_1 + \frac{6}{11}\mathbf{a}_2$.

(c) Following the procedure described above, we reduce the matrix

$$\left[\begin{array}{cc|c} 3 & 2 & 2 \\ 6 & 10 & -1 \\ 2 & -4 & 4 \end{array}\right] \text{ to obtain } \left[\begin{array}{cc|c} 1 & 0 & \frac{11}{9} \\ 0 & 1 & -\frac{5}{6} \\ 0 & 0 & -\frac{16}{9} \end{array}\right].$$

The last row of the reduced matrix shows that the system is inconsistent. Therefore, it is not possible to express $\mathbf{x}$ as a linear combination of $\mathbf{a}_1$ and $\mathbf{a}_2$.

(e) Following the procedure described above, we reduce the matrix

$$\left[\begin{array}{ccc|c} 1 & 5 & 4 & 7 \\ -2 & -2 & 0 & 2 \\ 3 & 6 & 3 & 3 \end{array}\right] \text{ to obtain } \left[\begin{array}{ccc|c} 1 & 0 & -1 & -3 \\ 0 & 1 & 1 & 2 \\ 0 & 0 & 0 & 0 \end{array}\right].$$

The linear system has the infinite solution set $\{(c-3, -c+2, c) \mid c \in \mathbb{R}\}$, and hence there are many sets of coefficients b_1, b_2, b_3 for which $\mathbf{x} = b_1\mathbf{a}_1 + b_2\mathbf{a}_2 + b_3\mathbf{a}_3$. Choosing the particular solution when $c = 0$ yields $\mathbf{x} = -3\mathbf{a}_1 + 2\mathbf{a}_2 + 0\mathbf{a}_3$. Other choices for c produce alternate linear combinations.

(g) Following the procedure described above, we reduce the matrix $\left[\begin{array}{ccc|c} 3 & -2 & 6 & 2 \\ 2 & 0 & 1 & 3 \\ -2 & 1 & 2 & -7 \\ 4 & -3 & 8 & 3 \end{array}\right]$ to obtain $\left[\begin{array}{ccc|c} 1 & 0 & 0 & 2 \\ 0 & 1 & 0 & -1 \\ 0 & 0 & 1 & -1 \\ 0 & 0 & 0 & 0 \end{array}\right]$. The unique solution to this system provides the coefficients for $\mathbf{x}$ as a linear combination of $\mathbf{a}_1$, $\mathbf{a}_2$ and $\mathbf{a}_3$, and so $\mathbf{x} = 2\mathbf{a}_1 - \mathbf{a}_2 - \mathbf{a}_3$.

(9) To determine whether a given vector $\mathbf{X}$ is in the row space of matrix $\mathbf{A}$, we row reduce the augmented matrix $[\mathbf{A}^T\,|\mathbf{X}]$. If the corresponding linear system is consistent, the solution set gives coefficients for expressing $\mathbf{X}$ as a linear combination of the rows of $\mathbf{A}$. If the system is inconsistent, then $\mathbf{X}$ is not in the row space of $\mathbf{A}$.

(a) Following the procedure described above, we reduce the matrix $\left[\begin{array}{ccc|c} 3 & 2 & 2 & 7 \\ 6 & 10 & -1 & 1 \\ 2 & -4 & 4 & 18 \end{array}\right]$ to obtain $\left[\begin{array}{ccc|c} 1 & 0 & 0 & 5 \\ 0 & 1 & 0 & -3 \\ 0 & 0 & 1 & -1 \end{array}\right]$. Since the system has the unique solution $(5, -3, -1)$, $\mathbf{X}$ is in the row space of $\mathbf{A}$, and $\mathbf{X} = 5(\text{row }1) - 3(\text{row }2) - 1(\text{row }3)$.

(c) Following the procedure described above, we reduce the matrix $\left[\begin{array}{ccc|c} 4 & -2 & 6 & 2 \\ -1 & 3 & 1 & 2 \\ 2 & 5 & 9 & -3 \end{array}\right]$ to obtain $\left[\begin{array}{ccc|c} 1 & 0 & 2 & 1 \\ 0 & 1 & 1 & 1 \\ 0 & 0 & 0 & -10 \end{array}\right]$. The last row of the reduced matrix shows that the system is inconsistent. Hence, $\mathbf{X}$ is not in the row space of $\mathbf{A}$.

(e) Following the procedure described above, we reduce the matrix $\left[\begin{array}{ccc|c} 2 & 7 & 3 & 1 \\ -4 & -1 & 7 & 11 \\ 1 & -1 & -3 & -4 \\ -3 & 2 & 8 & 11 \end{array}\right]$ to obtain $\left[\begin{array}{ccc|c} 1 & 0 & -2 & -3 \\ 0 & 1 & 1 & 1 \\ 0 & 0 & 0 & 0 \\ 0 & 0 & 0 & 0 \end{array}\right]$.

The linear system has the infinite solution set $\{(2c-3, -c+1, c) \mid c \in \mathbb{R}\}$, and hence there are many sets of coefficients b_1, b_2, b_3 for which $[1, 11, -4, 11] = b_1(\text{row }1) + b_2(\text{row }2) + b_3(\text{row }3)$. Choosing the particular solution when $c = 0$ yields $[1, 11, -4, 11] =$ $-3(\text{row }1) + 1(\text{row }2) + 0(\text{row }3)$. Other choices for c produce alternate linear combinations.

(10) (a) To express the vector $[13, -23, 60]$ as a linear combination of vectors $\mathbf{q}_1$, $\mathbf{q}_2$, and $\mathbf{q}_3$, we need to solve the linear system whose augmented matrix has the vectors $\mathbf{q}_1$, $\mathbf{q}_2$, and $\mathbf{q}_3$ as columns to the left of the augmentation bar, and has the vector $[13, -23, 60]$ as a column to the right of the augmentation bar. Hence, we row reduce

$$\left[\begin{array}{ccc|c} -1 & -10 & 7 & 13 \\ -5 & 3 & -12 & -23 \\ 11 & -8 & 30 & 60 \end{array}\right] \text{ to obtain } \left[\begin{array}{ccc|c} 1 & 0 & 0 & -2 \\ 0 & 1 & 0 & 1 \\ 0 & 0 & 1 & 3 \end{array}\right].$$

The unique solution $(-2, 1, 3)$ for the system corresponding to this matrix gives us the coefficients for the desired linear combination. Hence, $[13, -23, 60] = -2\mathbf{q}_1 + \mathbf{q}_2 + 3\mathbf{q}_3$.

(b) As in part (a), for each $\mathbf{q}_i$, we need to solve a linear system whose augmented matrix of the form $[\mathbf{r}_1\ \mathbf{r}_2\ \mathbf{r}_3 \mid \mathbf{q}_i]$, where the vectors listed represent columns of the matrix. But, using the method for solving several systems simultaneously, we can solve all three problems together by row reducing

$[\mathbf{r}_1\ \mathbf{r}_2\ \mathbf{r}_3 \mid \mathbf{q}_1\ \mathbf{q}_2\ \mathbf{q}_3]$. Hence, we row reduce

$$\left[\begin{array}{rrr|rrr} 3 & 2 & 4 & -1 & -10 & 7 \\ -2 & 1 & -1 & -5 & 3 & -12 \\ 4 & -3 & 2 & 11 & -8 & 30 \end{array}\right] \text{ to get } \left[\begin{array}{rrr|rrr} 1 & 0 & 0 & 3 & 2 & 1 \\ 0 & 1 & 0 & -1 & 2 & -6 \\ 0 & 0 & 1 & -2 & -5 & 4 \end{array}\right].$$

Reading the three solutions $(3,-1,-2)$, $(2,2,-5)$, and $(1,-6,4)$ from the reduced matrix gives the coefficients needed for the desired linear combinations. Hence, $\mathbf{q}_1 = 3\mathbf{r}_1 - \mathbf{r}_2 - 2\mathbf{r}_3$, $\mathbf{q}_2 = 2\mathbf{r}_1 + 2\mathbf{r}_2 - 5\mathbf{r}_3$, and $\mathbf{q}_3 = \mathbf{r}_1 - 6\mathbf{r}_2 + 4\mathbf{r}_3$.

(c) Combining the answers from parts (a) and (b) yields

$$\begin{aligned} [13,-23,60] &= -2\mathbf{q}_1 + \mathbf{q}_2 + 3\mathbf{q}_3 = -2(3\mathbf{r}_1 - \mathbf{r}_2 - 2\mathbf{r}_3) + (2\mathbf{r}_1 + 2\mathbf{r}_2 - 5\mathbf{r}_3) + 3(\mathbf{r}_1 - 6\mathbf{r}_2 + 4\mathbf{r}_3) \\ &= -6\mathbf{r}_1 + 2\mathbf{r}_2 + 4\mathbf{r}_3 + 2\mathbf{r}_1 + 2\mathbf{r}_2 - 5\mathbf{r}_3 + 3\mathbf{r}_1 - 18\mathbf{r}_2 + 12\mathbf{r}_3 \\ &= -\mathbf{r}_1 - 14\mathbf{r}_2 + 11\mathbf{r}_3. \end{aligned}$$

(11) (a) (i) $\mathbf{A}$ row reduces to $\mathbf{B} = \begin{bmatrix} 1 & 0 & -1 & 2 \\ 0 & 1 & 3 & 2 \\ 0 & 0 & 0 & 0 \end{bmatrix}$.

(ii) To check that the ith row, $\mathbf{b}_i$, of $\mathbf{B}$ is in the row space of $\mathbf{A}$, we must solve the system whose augmented matrix is $[\mathbf{A}^T|\mathbf{b}_i]$, where $\mathbf{b}_i$ is a column. But, we can solve this problem for the two nonzero rows of $\mathbf{B}$ simultaneously by row reducing the matrix $[\mathbf{A}^T|\mathbf{b}_1\ \mathbf{b}_2]$. Performing this row reduction produces the matrix

$$\left[\begin{array}{ccc|cc} 1 & 0 & -\frac{3}{8} & -\frac{7}{8} & \frac{1}{4} \\ 0 & 1 & \frac{1}{2} & \frac{1}{2} & 0 \\ 0 & 0 & 0 & 0 & 0 \\ 0 & 0 & 0 & 0 & 0 \end{array}\right].$$

Both systems have infinitely many solutions. Hence, there are many ways to express the two nonzero rows of $\mathbf{B}$ as linear combinations of the rows of $\mathbf{A}$. The two solution sets are $\{(\frac{3}{8}c - \frac{7}{8}, -\frac{1}{2}c + \frac{1}{2}, c) \mid c \in \mathbb{R}\}$ for $\mathbf{b}_1$ and $\{(\frac{3}{8}c + \frac{1}{4}, -\frac{1}{2}c, c) \mid c \in \mathbb{R}\}$ for $\mathbf{b}_2$. If we find particular solutions by choosing $c = 0$, we get

$[1,0,-1,2] = -\frac{7}{8}[0,4,12,8] + \frac{1}{2}[2,7,19,18] + 0[1,2,5,6]$ and

$[0,1,3,2] = \frac{1}{4}[0,4,12,8] + 0[2,7,19,18] + 0[1,2,5,6]$.

(iii) In this part, we perform the same operations as in part (ii), except that we reverse the roles of $\mathbf{A}$ and $\mathbf{B}$. Hence, we row reduce the matrix $[\mathbf{b}_1\ \mathbf{b}_2|\mathbf{A}^T]$, which yields

$$\left[\begin{array}{cc|ccc} 1 & 0 & 0 & 2 & 1 \\ 0 & 1 & 4 & 7 & 2 \\ 0 & 0 & 0 & 0 & 0 \\ 0 & 0 & 0 & 0 & 0 \end{array}\right].$$

In this case, each of the three linear systems has a unique solution. These three solutions are $(0,4)$, $(2,7)$, and $(1,2)$. Each gives the coefficients for expressing a row of $\mathbf{A}$ as a linear combination of the two nonzero rows of $\mathbf{B}$. Hence, we have $[0,4,12,8] = 0[1,0,-1,2] + 4[0,1,3,2]$, $[2,17,19,18] = 2[1,0,-1,2] + 7[0,1,3,2]$, and $[1,2,5,6] = 1[1,0,-1,2] + 2[0,1,3,2]$.

(13) (a) Suppose we are performing row operations on an $m \times n$ matrix $\mathbf{A}$. Throughout this part, we use the notation $\langle \mathbf{B} \rangle_i$ for the ith row of a matrix $\mathbf{B}$.

For the Type (I) operation $R : \langle i \rangle \longleftarrow c\langle i \rangle$: Now R^{-1} is $\langle i \rangle \longleftarrow \frac{1}{c}\langle i \rangle$. Clearly, R and R^{-1} change only the ith row of $\mathbf{A}$. We want to show that $R^{-1}R$ leaves $\langle \mathbf{A} \rangle_i$ unchanged. But $\langle R^{-1}(R(\mathbf{A})) \rangle_i = \frac{1}{c}\langle R(\mathbf{A}) \rangle_i = \frac{1}{c}(c\langle \mathbf{A} \rangle_i) = \langle \mathbf{A} \rangle_i$.
For the Type (II) operation $R : \langle i \rangle \longleftarrow c\langle j \rangle + \langle i \rangle$: Now R^{-1} is $\langle i \rangle \longleftarrow -c\langle j \rangle + \langle i \rangle$. Again, R and R^{-1} change only the ith row of $\mathbf{A}$, and we need to show that $R^{-1}R$ leaves $\langle \mathbf{A} \rangle_i$ unchanged. But $\langle R^{-1}(R(\mathbf{A})) \rangle_i = -c\langle R(\mathbf{A}) \rangle_j + \langle R(\mathbf{A}) \rangle_i = -c\langle \mathbf{A} \rangle_j + \langle R(\mathbf{A}) \rangle_i = -c\langle \mathbf{A} \rangle_j + c\langle \mathbf{A} \rangle_j + \langle \mathbf{A} \rangle_i = \langle \mathbf{A} \rangle_i$.
For the Type (III) operation $R : \langle i \rangle \longleftrightarrow \langle j \rangle$: Now, $R^{-1} = R$. Also, R changes only the ith and jth rows of $\mathbf{A}$, and these get swapped. Obviously, a second application of R swaps them back to where they were, proving that R is indeed its own inverse.

(b) An approach similar to that used for Type (II) operations in the abridged proof of Theorem 2.3 in the text works just as easily for Type (I) and Type (III) operations.

For Type (I) operations: Suppose that the original system has the form

$$\begin{cases} a_{11}x_1 + a_{12}x_2 + a_{13}x_3 + \cdots + a_{1n}x_n = b_1 \\ a_{21}x_1 + a_{22}x_2 + a_{23}x_3 + \cdots + a_{2n}x_n = b_2 \\ \vdots \quad\quad \vdots \quad\quad \vdots \quad\quad \ddots \quad\quad \vdots \quad\quad \vdots \\ a_{m1}x_1 + a_{m2}x_2 + a_{m3}x_3 + \cdots + a_{mn}x_n = b_m \end{cases}$$

and that the row operation used is $R : \langle i \rangle \leftarrow c\,\langle i \rangle$. When R is applied to the corresponding augmented matrix, all rows except the ith row remain unchanged. The new ith equation then has the form $(ca_{i1})x_1+(ca_{i2})x_2+\cdots+(ca_{in})x_n = cb_i$. We must show that any solution $(s_1, s_2, \ldots, s_n)$ of the original system is a solution of the new one. Now, since $(s_1, s_2, \ldots, s_n)$ is a solution of the ith equation in the original system, we have $a_{i1}s_1 + a_{i2}s_2 + \cdots + a_{in}s_n = b_i$. Multiplying this equation by c yields $(ca_{i1})s_1 + (ca_{i2})s_2 + \cdots + (ca_{in})s_n = cb_i$, and so $(s_1, s_2, \ldots, s_n)$ is also a solution of the new ith equation. Now, since none of the other equations in the system have changed, $(s_1, s_2, \ldots, s_n)$ is still a solution for each of them. Therefore, $(s_1, s_2, \ldots, s_n)$ is a solution to the new system formed.

For Type (III) operations: A type (III) row operation merely changes the order of the rows of a matrix. Since the rows of the augmented matrix correspond to equations in the system, performing a type (III) row operation only changes the order in which the equations are written, without making any changes to the actual equations at all. Hence, any solution to the first system of equations must also be a solution to the second system of equations, since none of the actual equations themselves have changed.

There is, however, an alternate approach to this problem that manages to prove the result for all three types of row operations at once: Suppose R is a row operation, and let $\mathbf{X}$ satisfy $\mathbf{AX} = \mathbf{B}$. Multiplying both sides of this matrix equation by the matrix $R(\mathbf{I})$ yields $R(\mathbf{I})\mathbf{AX} = R(\mathbf{I})\mathbf{B}$, implying $R(\mathbf{IA})\mathbf{X} = R(\mathbf{IB})$, by Theorem 2.1. Thus, $R(\mathbf{A})\mathbf{X} = R(\mathbf{B})$, showing that $\mathbf{X}$ is a solution to the new linear system obtained from $\mathbf{AX} = \mathbf{B}$ after the row operation R is performed. You might wonder what motivates the second method of proof used here. The technique is based on the topic of Elementary Matrices, which is covered in Chapter 10 of the textbook. You will have sufficient background to read that section after you finish studying Chapter 2.

(14) The zero vector is a solution to $\mathbf{AX} = \mathbf{O}$, but it is not a solution for $\mathbf{AX} = \mathbf{B}$. Hence these two systems have different solution sets, and thus can not be equivalent systems.

(15) Consider the systems

$$\begin{cases} x + y = 1 \\ x + y = 0 \end{cases} \quad \text{and} \quad \begin{cases} x - y = 1 \\ x - y = 2 \end{cases}.$$

The augmented matrices for these systems are, respectively,

$$\left[\begin{array}{cc|c} 1 & 1 & 1 \\ 1 & 1 & 0 \end{array}\right] \text{ and } \left[\begin{array}{cc|c} 1 & -1 & 1 \\ 1 & -1 & 2 \end{array}\right].$$

The reduced row echelon matrices for these are, respectively,

$$\left[\begin{array}{cc|c} 1 & 1 & 0 \\ 0 & 0 & 1 \end{array}\right] \text{ and } \left[\begin{array}{cc|c} 1 & -1 & 0 \\ 0 & 0 & 1 \end{array}\right],$$

where we have continued the row reduction beyond the augmentation bar. Thus, the original augmented matrices are not row equivalent, since their reduced row echelon forms are different. However, the reduced row echelon form matrices reveal that both systems are inconsistent. Therefore, they both have the same solution set, namely, the empty set. Hence, the systems are equivalent.

(19) (a) As in the abridged proof of Theorem 2.8 in the text, let $\mathbf{a}_1, \ldots, \mathbf{a}_m$ represent the rows of **A**, and let $\mathbf{b}_1, \ldots, \mathbf{b}_m$ represent the rows of **B**.
For the Type (I) operation $R : \langle i \rangle \longleftarrow c\langle i \rangle$: Now $\mathbf{b}_i = 0\mathbf{a}_1 + 0\mathbf{a}_2 + \cdots + c\mathbf{a}_i + 0\mathbf{a}_{i+1} + \cdots + 0\mathbf{a}_m$, and, for $k \neq i$, $\mathbf{b}_k = 0\mathbf{a}_1 + 0\mathbf{a}_2 + \cdots + 1\mathbf{a}_k + 0\mathbf{a}_{k+1} + \cdots + 0\mathbf{a}_m$. Hence, each row of **B** is a linear combination of the rows of **A**, implying it is in the row space of **A**.
For the Type (II) operation $R : \langle i \rangle \longleftarrow c\langle j \rangle + \langle i \rangle$: Now $\mathbf{b}_i = 0\mathbf{a}_1 + 0\mathbf{a}_2 + \cdots + c\mathbf{a}_j + 0\mathbf{a}_{j+1} + \cdots + \mathbf{a}_i + 0\mathbf{a}_{i+1} + \ldots + 0\mathbf{a}_m$, where our notation assumes $i > j$. (An analogous argument works for $i < j$.) And, for $k \neq i$, $\mathbf{b}_k = 0\mathbf{a}_1 + 0\mathbf{a}_2 + \cdots + 1\mathbf{a}_k + 0\mathbf{a}_{k+1} + \cdots + 0\mathbf{a}_m$. Hence, each row of **B** is a linear combination of the rows of **A**, implying it is in the row space of **A**.
For the Type (III) operation $R : \langle i \rangle \longleftrightarrow \langle j \rangle$: Now, $\mathbf{b}_i = 0\mathbf{a}_1 + 0\mathbf{a}_2 + \cdots + 1\mathbf{a}_j + 0\mathbf{a}_{j+1} + \cdots + 0\mathbf{a}_m$, $\mathbf{b}_j = 0\mathbf{a}_1 + 0\mathbf{a}_2 + \cdots + 1\mathbf{a}_i + 0\mathbf{a}_{i+1} + \cdots + 0\mathbf{a}_m$, and, for $k \neq i$, $k \neq j$, $\mathbf{b}_k = 0\mathbf{a}_1 + 0\mathbf{a}_2 + \cdots + 1\mathbf{a}_k + 0\mathbf{a}_{k+1} + \cdots + 0\mathbf{a}_m$. Hence, each row of **B** is a linear combination of the rows of **A**, implying it is in the row space of **A**.

(b) If **x** is in the row space of **B**, then **x** is a linear combination of the rows of **B**. By part (a), each row of **B** is a linear combination of the rows of **A**. Hence, by Lemma 2.7, **x** is a linear combination of the rows of **A** as well. Therefore, **x** is in the row space of **A**. Therefore, since every member of the row space of **B** is also in the row space of **A**, we get that the row space of **B** is contained in the row space of **A**.

(20) Let k be the number of matrices *between* **A** and **B** when performing row operations to get from **A** to **B**. Use a proof by induction on k.
Base Step: If $k = 0$, then there are no intermediary matrices, and Exercise 19 shows that the row space of **B** is contained in the row space of **A**.
Inductive Step: Given the chain

$$\mathbf{A} \rightarrow \mathbf{D}_1 \rightarrow \mathbf{D}_2 \rightarrow \cdots \rightarrow \mathbf{D}_k \rightarrow \mathbf{D}_{k+1} \rightarrow \mathbf{B},$$

we must show that the row space **B** is contained in the row space of **A**. The inductive hypothesis shows that the row space of $\mathbf{D}_{k+1}$ is in the row space of **A**, since there are only k matrices between **A** and $\mathbf{D}_{k+1}$ in the chain. Thus, each row of $\mathbf{D}_{k+1}$ can be expressed as a linear combination of the rows of **A**. But by Exercise 19, each row of **B** can be expressed as a linear combination of the rows of $\mathbf{D}_{k+1}$. Hence, by Lemma 2.7, each row of **B** can be expressed as a linear combination of the rows of **A**, and therefore is in the row space of **A**. By Lemma 2.7 again, the row space of **B** is contained in the row space of **A**.

(22) (a) True. This is the statement of Theorem 2.3.

(b) True. In general, if two matrices are row equivalent, then they are both row equivalent to the same reduced row echelon form matrix. Hence **A** and **B** are both row equivalent to the same matrix **C** in reduced row echelon form. Since the ranks of **A** and **B** are both the number of nonzero rows in their common reduced echelon form matrix **C**, **A** and **B** must have the same rank.

(c) False. The inverse of the type (I) row operation $\langle i\rangle \leftarrow c\,\langle i\rangle$ is the type (I) row operation $\langle i\rangle \leftarrow \frac{1}{c}\,\langle i\rangle$.

(d) False. The statement is true for homogeneous systems, but is false in general. If the system in question is not homogeneous, it could be inconsistent, and thus have no solutions at all, and hence, no nontrivial solutions. For a particular example, consider the system $\begin{cases} x+y+z=1 \\ x+y+z=2 \end{cases}$.
This system has three variables, but the rank of its augmented matrix is ≤ 2, since the matrix has only two rows. However, the system has no solutions because it is impossible for $x+y+z$ to equal both 1 and 2.

(e) False. This statement directly contradicts part (2) of Theorem 2.5.

(f) True. By the definition of row space, **x** is clearly in the row space of **A**. But Theorem 2.8 shows that **A** and **B** have the same row space. Hence **x** is also in the row space of **B**.

Section 2.4

(2) To find the rank of a matrix, find its corresponding reduced row echelon form matrix and count its nonzero rows. An $n \times n$ matrix is nonsingular if and only if its rank equals n.

(a) The given 2×2 matrix row reduces to $\mathbf{I}_2$, which has two nonzero rows. Hence, the original matrix has rank $= 2$. Therefore, the matrix is nonsingular.

(c) The given 3×3 matrix row reduces to $\mathbf{I}_3$, which has three nonzero rows. Hence, the original matrix has rank $= 3$. Therefore, the matrix is nonsingular.

(e) The given 4×4 matrix row reduces to $\begin{bmatrix} 1 & 0 & 0 & 0 \\ 0 & 1 & 0 & 0 \\ 0 & 0 & 1 & -2 \\ 0 & 0 & 0 & 0 \end{bmatrix}$,
which has three nonzero rows. Hence, the rank of the original matrix is 3. But this is less than the number of rows in the matrix, and so the original matrix is singular.

(3) Use the formula for the inverse of a 2×2 matrix given in Theorem 2.13.

(a) First we compute $\delta = (4)(-3) - (2)(9) = -30$. Since $\delta \neq 0$, the given matrix is nonsingular, and its inverse is $\frac{1}{-30}\begin{bmatrix} -3 & -2 \\ -9 & 4 \end{bmatrix} = \begin{bmatrix} \frac{1}{10} & \frac{1}{15} \\ \frac{3}{10} & -\frac{2}{15} \end{bmatrix}$.

(c) First we compute $\delta = (-3)(-8) - (5)(-12) = 84$. Since $\delta \neq 0$, the given matrix is nonsingular, and its inverse is $\frac{1}{84}\begin{bmatrix} -8 & -5 \\ 12 & -3 \end{bmatrix} = \begin{bmatrix} -\frac{2}{21} & -\frac{5}{84} \\ \frac{1}{7} & -\frac{1}{28} \end{bmatrix}$.

(e) First we compute $\delta = (-6)(-8) - (12)(4) = 0$. Since $\delta = 0$, the given matrix is singular; it has no inverse.

(4) To find the inverse of an $n \times n$ matrix $\mathbf{A}$, row reduce $[\mathbf{A}|\mathbf{I}_n]$ to obtain $[\mathbf{I}_n|\mathbf{A}^{-1}]$. If row reduction does not produce $\mathbf{I}_n$ to the left of the augmentation bar, then $\mathbf{A}$ does not have an inverse.

(a) We row reduce $\left[\begin{array}{ccc|ccc} -4 & 7 & 6 & 1 & 0 & 0 \\ 3 & -5 & -4 & 0 & 1 & 0 \\ -2 & 4 & 3 & 0 & 0 & 1 \end{array}\right]$ to obtain $\left[\begin{array}{ccc|ccc} 1 & 0 & 0 & 1 & 3 & 2 \\ 0 & 1 & 0 & -1 & 0 & 2 \\ 0 & 0 & 1 & 2 & 2 & -1 \end{array}\right]$. Therefore,

$\begin{bmatrix} 1 & 3 & 2 \\ -1 & 0 & 2 \\ 2 & 2 & -1 \end{bmatrix}$ is the inverse of the original matrix.

(c) We row reduce $\left[\begin{array}{ccc|ccc} 2 & -2 & 3 & 1 & 0 & 0 \\ 8 & -4 & 9 & 0 & 1 & 0 \\ -4 & 6 & -9 & 0 & 0 & 1 \end{array}\right]$ to obtain $\left[\begin{array}{ccc|ccc} 1 & 0 & 0 & \frac{3}{2} & 0 & \frac{1}{2} \\ 0 & 1 & 0 & -3 & \frac{1}{2} & -\frac{1}{2} \\ 0 & 0 & 1 & -\frac{8}{3} & \frac{1}{3} & -\frac{2}{3} \end{array}\right]$. Therefore,

$\begin{bmatrix} \frac{3}{2} & 0 & \frac{1}{2} \\ -3 & \frac{1}{2} & -\frac{1}{2} \\ -\frac{8}{3} & \frac{1}{3} & -\frac{2}{3} \end{bmatrix}$ is the inverse of the original matrix.

(e) We row reduce $\left[\begin{array}{cccc|cccc} 2 & 0 & -1 & 3 & 1 & 0 & 0 & 0 \\ 1 & -2 & 3 & 1 & 0 & 1 & 0 & 0 \\ 4 & 1 & 0 & -1 & 0 & 0 & 1 & 0 \\ 1 & 3 & -2 & -5 & 0 & 0 & 0 & 1 \end{array}\right]$ to $\left[\begin{array}{cccc|cccc} 1 & 0 & 0 & \frac{8}{15} & \frac{1}{5} & \frac{1}{15} & \frac{2}{15} & 0 \\ 0 & 1 & 0 & -\frac{47}{15} & -\frac{4}{5} & -\frac{4}{15} & \frac{7}{15} & 0 \\ 0 & 0 & 1 & -\frac{29}{15} & -\frac{3}{5} & \frac{2}{15} & \frac{4}{15} & 0 \\ 0 & 0 & 0 & 0 & 1 & 1 & -1 & 1 \end{array}\right]$.

Therefore, since $\mathbf{I}_4$ is not obtained to the left of the augmentation bar, the original matrix does not have an inverse. (Note: If you use a calculator to perform the row reduction above, it might continue to row reduce beyond the augmentation bar.)

(5) (c) To find the inverse of a diagonal matrix, merely find the diagonal matrix whose main diagonal entries are the reciprocals of the main diagonal entries of the original matrix. Hence, when this matrix is multiplied by the original, the product will be a diagonal matrix with all 1's on the main diagonal – that is, the identity matrix. Therefore, the inverse matrix we need is

$$\begin{bmatrix} \frac{1}{a_{11}} & 0 & \cdots & 0 \\ 0 & \frac{1}{a_{22}} & \cdots & 0 \\ \vdots & \vdots & \ddots & \vdots \\ 0 & 0 & \cdots & \frac{1}{a_{nn}} \end{bmatrix}.$$

(6) (a) Using Theorem 2.13, $\delta = (\cos\theta)(\cos\theta) - (\sin\theta)(-\sin\theta) = \cos^2\theta + \sin^2\theta = 1$. Therefore, the general inverse is $\begin{bmatrix} \cos\theta & \sin\theta \\ -\sin\theta & \cos\theta \end{bmatrix}$. Plugging in the given values for θ yields:

θ	Original Matrix	Inverse Matrix
$\frac{\pi}{6}$	$\begin{bmatrix} \frac{\sqrt{3}}{2} & -\frac{1}{2} \\ \frac{1}{2} & \frac{\sqrt{3}}{2} \end{bmatrix}$	$\begin{bmatrix} \frac{\sqrt{3}}{2} & \frac{1}{2} \\ -\frac{1}{2} & \frac{\sqrt{3}}{2} \end{bmatrix}$
$\frac{\pi}{4}$	$\begin{bmatrix} \frac{\sqrt{2}}{2} & -\frac{\sqrt{2}}{2} \\ \frac{\sqrt{2}}{2} & \frac{\sqrt{2}}{2} \end{bmatrix}$	$\begin{bmatrix} \frac{\sqrt{2}}{2} & \frac{\sqrt{2}}{2} \\ -\frac{\sqrt{2}}{2} & \frac{\sqrt{2}}{2} \end{bmatrix}$
$\frac{\pi}{2}$	$\begin{bmatrix} 0 & -1 \\ 1 & 0 \end{bmatrix}$	$\begin{bmatrix} 0 & 1 \\ -1 & 0 \end{bmatrix}$

(b) It is easy to see that if $\mathbf{A}$ and $\mathbf{B}$ are 2×2 matrices with $\mathbf{AB} = \mathbf{C}$, then $\begin{bmatrix} a_{11} & a_{12} & 0 \\ a_{21} & a_{22} & 0 \\ 0 & 0 & 1 \end{bmatrix} \begin{bmatrix} b_{11} & b_{12} & 0 \\ b_{21} & b_{22} & 0 \\ 0 & 0 & 1 \end{bmatrix} = \begin{bmatrix} c_{11} & c_{12} & 0 \\ c_{21} & c_{22} & 0 \\ 0 & 0 & 1 \end{bmatrix}$. (Check it out!) Therefore, to find the inverse of the matrix given in part (b), we merely use the associated 2×2 inverses we found in part (a), glued into the upper corner of a 3×3 matrix, as illustrated. Hence, the general inverse is $\begin{bmatrix} \cos\theta & \sin\theta & 0 \\ -\sin\theta & \cos\theta & 0 \\ 0 & 0 & 1 \end{bmatrix}$. Substituting the given values of θ produces:

θ	Original Matrix	Inverse Matrix
$\frac{\pi}{6}$	$\begin{bmatrix} \frac{\sqrt{3}}{2} & -\frac{1}{2} & 0 \\ \frac{1}{2} & \frac{\sqrt{3}}{2} & 0 \\ 0 & 0 & 1 \end{bmatrix}$	$\begin{bmatrix} \frac{\sqrt{3}}{2} & \frac{1}{2} & 0 \\ -\frac{1}{2} & \frac{\sqrt{3}}{2} & 0 \\ 0 & 0 & 1 \end{bmatrix}$
$\frac{\pi}{4}$	$\begin{bmatrix} \frac{\sqrt{2}}{2} & -\frac{\sqrt{2}}{2} & 0 \\ \frac{\sqrt{2}}{2} & \frac{\sqrt{2}}{2} & 0 \\ 0 & 0 & 1 \end{bmatrix}$	$\begin{bmatrix} \frac{\sqrt{2}}{2} & \frac{\sqrt{2}}{2} & 0 \\ -\frac{\sqrt{2}}{2} & \frac{\sqrt{2}}{2} & 0 \\ 0 & 0 & 1 \end{bmatrix}$
$\frac{\pi}{2}$	$\begin{bmatrix} 0 & -1 & 0 \\ 1 & 0 & 0 \\ 0 & 0 & 1 \end{bmatrix}$	$\begin{bmatrix} 0 & 1 & 0 \\ -1 & 0 & 0 \\ 0 & 0 & 1 \end{bmatrix}$

(7) By Theorem 2.15, if $\mathbf{A}$ is a nonsingular matrix, then the unique solution to $\mathbf{AX} = \mathbf{B}$ is $\mathbf{A}^{-1}\mathbf{B}$.

(a) The given system corresponds to the matrix equation $\begin{bmatrix} 5 & -1 \\ -7 & 2 \end{bmatrix} \begin{bmatrix} x_1 \\ x_2 \end{bmatrix} = \begin{bmatrix} 20 \\ -31 \end{bmatrix}$. We use Theorem 2.13 to compute the inverse of $\begin{bmatrix} 5 & -1 \\ -7 & 2 \end{bmatrix}$. First, $\delta = (5)(2) - (-7)(-1) = 3$. Hence, the inverse $= \frac{1}{3}\begin{bmatrix} 2 & 1 \\ 7 & 5 \end{bmatrix} = \begin{bmatrix} \frac{2}{3} & \frac{1}{3} \\ \frac{7}{3} & \frac{5}{3} \end{bmatrix}$. The unique solution to the system is thus $\begin{bmatrix} \frac{2}{3} & \frac{1}{3} \\ \frac{7}{3} & \frac{5}{3} \end{bmatrix}\begin{bmatrix} 20 \\ -31 \end{bmatrix} = \begin{bmatrix} 3 \\ -5 \end{bmatrix}$. That is, the solution set for the system is $\{(3,-5)\}$.

(c) The system corresponds to the matrix equation $\begin{bmatrix} 0 & -2 & 5 & 1 \\ -7 & -4 & 5 & 22 \\ 5 & 3 & -4 & -16 \\ -3 & -1 & 0 & 9 \end{bmatrix}\begin{bmatrix} x_1 \\ x_2 \\ x_3 \\ x_4 \end{bmatrix} = \begin{bmatrix} 25 \\ -15 \\ 9 \\ -16 \end{bmatrix}$.

We find the inverse of the coefficient matrix by row reducing

$\left[\begin{array}{cccc|cccc} 0 & -2 & 5 & 1 & 1 & 0 & 0 & 0 \\ -7 & -4 & 5 & 22 & 0 & 1 & 0 & 0 \\ 5 & 3 & -4 & -16 & 0 & 0 & 1 & 0 \\ -3 & -1 & 0 & 9 & 0 & 0 & 0 & 1 \end{array}\right]$ to obtain $\left[\begin{array}{cccc|cccc} 1 & 0 & 0 & 0 & 1 & -13 & -15 & 5 \\ 0 & 1 & 0 & 0 & -3 & 3 & 0 & -7 \\ 0 & 0 & 1 & 0 & -1 & 2 & 1 & -3 \\ 0 & 0 & 0 & 1 & 0 & -4 & -5 & 1 \end{array}\right]$. The inverse of the coefficient matrix is the 4×4 matrix to the right of the augmentation bar in the row reduced matrix. Hence, the unique solution to the system is

$$\begin{bmatrix} 1 & -13 & -15 & 5 \\ -3 & 3 & 0 & -7 \\ -1 & 2 & 1 & -3 \\ 0 & -4 & -5 & 1 \end{bmatrix}\begin{bmatrix} 25 \\ -15 \\ 9 \\ -16 \end{bmatrix} = \begin{bmatrix} 5 \\ -8 \\ 2 \\ -1 \end{bmatrix}.$$

That is, the solution set for the linear system is $\{(5,-8,2,-1)\}$.

(8) (a) Through trial and error, we find the involutory matrix $\begin{bmatrix} 0 & 1 \\ 1 & 0 \end{bmatrix}$.

(b) Using the answer to part (a), the comments at the beginning of the answer to Section 2.4, Exercise 6(b) in this manual suggests that we consider the matrix $\begin{bmatrix} 0 & 1 & 0 \\ 1 & 0 & 0 \\ 0 & 0 & 1 \end{bmatrix}$. Squaring this matrix verifies that it is, in fact involutory.

(c) If $\mathbf{A}$ is involutory, then $\mathbf{A}^2 = \mathbf{AA} = \mathbf{I}_n$. Hence, $\mathbf{A}$ itself satisfies the definition of an inverse for $\mathbf{A}$, and so $\mathbf{A}^{-1} = \mathbf{A}$.

(10) (a) Since $\mathbf{A}$ is nonsingular, $\mathbf{A}^{-1}$ exists. Hence $\mathbf{AB} = \mathbf{O}_n \Rightarrow \mathbf{A}^{-1}(\mathbf{AB}) = \mathbf{A}^{-1}\mathbf{O}_n \Rightarrow (\mathbf{A}^{-1}\mathbf{A})\mathbf{B} = \mathbf{O}_n \Rightarrow \mathbf{I}_n\mathbf{B} = \mathbf{O}_n \Rightarrow \mathbf{B} = \mathbf{O}_n$. Thus, $\mathbf{B}$ must be the zero matrix.

(b) No. $\mathbf{AB} = \mathbf{I}_n$ implies $\mathbf{B} = \mathbf{A}^{-1}$, and so $\mathbf{A}$ is nonsingular. We can now apply part (a), substituting $\mathbf{C}$ where $\mathbf{B}$ appears, proving that $\mathbf{AC} = \mathbf{O}_n \Rightarrow \mathbf{C} = \mathbf{O}_n$.

(11) Now $\mathbf{A}^4 = \mathbf{I}_n \Rightarrow \mathbf{A}^3\mathbf{A} = \mathbf{I}_n \Rightarrow \mathbf{A}^{-1} = \mathbf{A}^3$. Also, since $\mathbf{I}_n^k = \mathbf{I}_n$ for every integer k, we see that $\mathbf{A}^{4k} = (\mathbf{A}^4)^k = \mathbf{I}_n^k = \mathbf{I}_n$ for every integer k. Hence, for every integer k, $\mathbf{A}^{4k+3} = \mathbf{A}^{4k}\mathbf{A}^3 = \mathbf{I}_n\mathbf{A}^3 = \mathbf{A}^3 = \mathbf{A}^{-1}$. Thus, all $n \times n$ matrices of the form $\mathbf{A}^{4k+3}$ equal $\mathbf{A}^{-1}$. These are $\ldots, \mathbf{A}^{-9}, \mathbf{A}^{-5}, \mathbf{A}^{-1}, \mathbf{A}^3, \mathbf{A}^7, \mathbf{A}^{11}, \ldots$. Now, could any other powers of $\mathbf{A}$ also equal $\mathbf{A}^{-1}$? Suppose that $\mathbf{A}^m = \mathbf{A}^{-1}$. Then, $\mathbf{A}^{m+1} = \mathbf{A}^m\mathbf{A} = \mathbf{A}^{-1}\mathbf{A} = \mathbf{I}_n$. Now, if $m+1 = 4l$ for some integer l, then $m = 4(l-1)+3$, and so m is already on our list of powers of $\mathbf{A}$ such that $\mathbf{A}^m = \mathbf{A}^{-1}$. On the other hand, if $m+1$ is not a multiple of 4, then, dividing $m+1$ by 4 yields an integer k, with a remainder r of either 1, 2, or 3. Then $m+1 = 4k+r$, and so $\mathbf{A}^{4k+r} = \mathbf{I}_n$ implying $\mathbf{A}^{4k}\mathbf{A}^r = \mathbf{I}_n$. But $\mathbf{A}^{4k} = \mathbf{I}_n$ from above, and so $\mathbf{A}^r = \mathbf{I}_n$. But, since r equals either 1, 2, or 3, this contradicts the given conditions of the problem, and so $m+1$ can not equal $4k+r$. Therefore, the only powers of $\mathbf{A}$ that equal $\mathbf{A}^{-1}$ are of the form $\mathbf{A}^{4k+3}$.

(12) By parts (1) and (3) of Theorem 2.11, $\mathbf{B}^{-1}\mathbf{A}$ is the inverse of $\mathbf{A}^{-1}\mathbf{B}$. (Proof: $(\mathbf{A}^{-1}\mathbf{B})^{-1} = \mathbf{B}^{-1}(\mathbf{A}^{-1})^{-1} = \mathbf{B}^{-1}\mathbf{A}$.) Thus, if $\mathbf{A}^{-1}\mathbf{B}$ is known, simply compute its inverse to find $\mathbf{B}^{-1}\mathbf{A}$.

(14) (a) No step in the row reduction process will alter the column of zeroes, and so the unique reduced row echelon form for the matrix must contain a column of zeroes, and so cannot equal $\mathbf{I}_n$.

(15) (a) Part (1): Since $\mathbf{A}\mathbf{A}^{-1} = \mathbf{I}$, we must have $(\mathbf{A}^{-1})^{-1} = \mathbf{A}$.
Part (2): For $k > 0$, to show $(\mathbf{A}^k)^{-1} = (\mathbf{A}^{-1})^k$, we must show that $\mathbf{A}^k(\mathbf{A}^{-1})^k = \mathbf{I}$. Proceed by induction on k.
Base Step: For $k = 1$, clearly $\mathbf{A}\mathbf{A}^{-1} = \mathbf{I}$.
Inductive Step: Assume $\mathbf{A}^k(\mathbf{A}^{-1})^k = \mathbf{I}$. Prove $\mathbf{A}^{k+1}(\mathbf{A}^{-1})^{k+1} = \mathbf{I}$.
Now, $\mathbf{A}^{k+1}(\mathbf{A}^{-1})^{k+1} = \mathbf{A}\mathbf{A}^k(\mathbf{A}^{-1})^k\mathbf{A}^{-1} = \mathbf{A}\mathbf{I}\mathbf{A}^{-1} = \mathbf{A}\mathbf{A}^{-1} = \mathbf{I}$. This concludes the proof for $k > 0$.
We now show $\mathbf{A}^k(\mathbf{A}^{-1})^k = \mathbf{I}$ for $k \leq 0$.
For $k = 0$, clearly $\mathbf{A}^0(\mathbf{A}^{-1})^0 = \mathbf{I}\,\mathbf{I} = \mathbf{I}$. The case $k = -1$ is covered by part (1) of the theorem.
For $k \leq -2$, $(\mathbf{A}^k)^{-1} = ((\mathbf{A}^{-1})^{-k})^{-1}$ (by definition) $= ((\mathbf{A}^{-k})^{-1})^{-1}$ (by the $k > 0$ case) $= \mathbf{A}^{-k}$ (by part (1)).

(17) (a) Let $p = -s$, $q = -t$. Then $p, q > 0$. Now, $\mathbf{A}^{s+t} = \mathbf{A}^{-(p+q)} = (\mathbf{A}^{-1})^{p+q} = (\mathbf{A}^{-1})^p(\mathbf{A}^{-1})^q$ (by Theorem 1.15) $= \mathbf{A}^{-p}\mathbf{A}^{-q} = \mathbf{A}^s\mathbf{A}^t$.

(21) (a) False. Many $n \times n$ matrices are singular, having no inverse. For example, the 2×2 matrix $\begin{bmatrix} 6 & 3 \\ 8 & 4 \end{bmatrix}$ has no inverse, by Theorem 2.13, because $\delta = (6)(4) - (3)(8) = 0$.

(b) True. This follows directly from Theorem 2.9.

(c) True. $((\mathbf{AB})^T)^{-1} = ((\mathbf{AB})^{-1})^T$ (by part (4) of Theorem 2.11) $= (\mathbf{B}^{-1}\mathbf{A}^{-1})^T$ (by part (3) of Theorem 2.11) $= (\mathbf{A}^{-1})^T(\mathbf{B}^{-1})^T$ (by Theorem 1.16).

(d) False. This statement contradicts Theorem 2.13, which states that $\begin{bmatrix} a & b \\ c & d \end{bmatrix}$ is *nonsingular* if and only if $\delta = ad - bc \neq 0$.

(e) False. This statement contradicts the Inverse Method given in Section 2.4 as well as Theorem 2.14. To correct the statement, change the word "nonsingular" to "singular."

(f) True. This follows directly from combining Theorems 2.14 and 2.15.

Chapter 3

Section 3.1

(1) (a) $\begin{vmatrix} -2 & 5 \\ 3 & 1 \end{vmatrix} = (-2)(1) - (5)(3) = -17$

(c) $\begin{vmatrix} 6 & -12 \\ -4 & 8 \end{vmatrix} = (6)(8) - (-12)(-4) = 0$

(e) To use basketweaving, we form a new array by taking the given matrix, and then adding a second copy of columns 1 and 2 as the new columns 4 and 5. We then form terms using the basketweaving pattern.

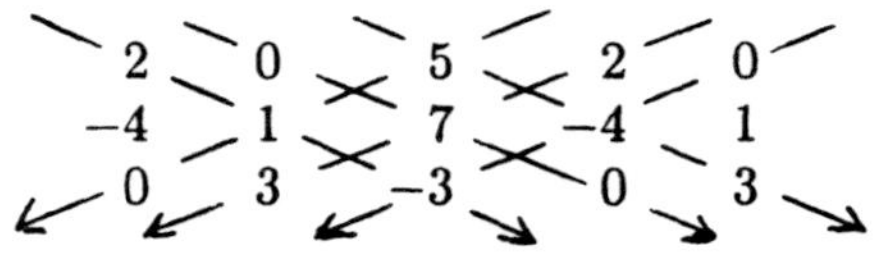

Hence, the determinant equals $(2)(1)(-3) + (0)(7)(0) + (5)(-4)(3) - (5)(1)(0) - (2)(7)(3) - (0)(-4)(-3) = -108$.

(g) To use basketweaving, we form a new array by taking the given matrix, and then adding a second copy of columns 1 and 2 as the new columns 4 and 5. We then form terms using the basketweaving pattern.

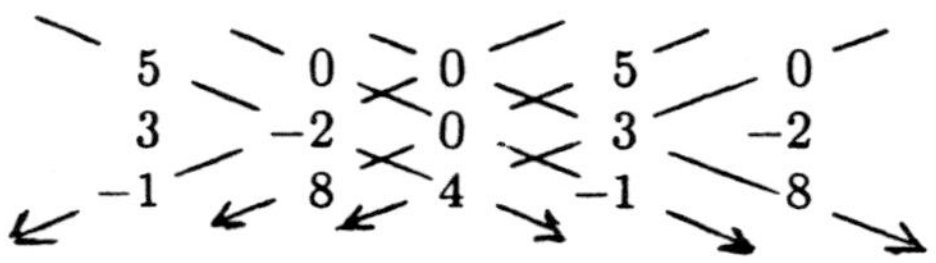

Hence, the determinant equals $(5)(-2)(4) + (0)(0)(-1) + (0)(3)(8) - (0)(-2)(-1) - (5)(0)(8) - (0)(3)(4) = -40$.

(i) To use basketweaving, we form a new array by taking the given matrix, and then adding a second copy of columns 1 and 2 as the new columns 4 and 5. We then form terms using the basketweaving pattern.

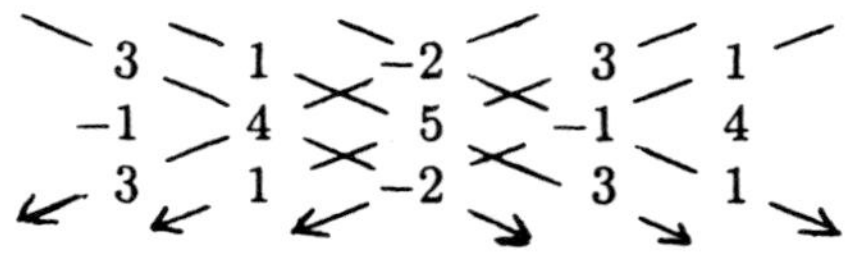

Hence, the determinant equals $(3)(4)(-2) + (1)(5)(3) + (-2)(-1)(1) - (-2)(4)(3) - (3)(5)(1) - (1)(-1)(-2) = 0$.

(j) The determinant of a 1×1 matrix is defined to be the (1,1) entry of the matrix. Therefore, the determinant equals -3.

(2) Recall that the submatrix $\mathbf{A}_{ij}$ is found by deleting the ith row and the jth column from $\mathbf{A}$. The (i,j) minor is the determinant of this submatrix.

(a) $|\mathbf{A}_{21}| = \begin{vmatrix} 4 & 3 \\ -2 & 4 \end{vmatrix} = (4)(4) - (3)(-2) = 22$.

(c) $|\mathbf{C}_{42}| = \begin{vmatrix} -3 & 0 & 5 \\ 2 & -1 & 4 \\ 6 & 4 & 0 \end{vmatrix}$. We will use basketweaving to compute this determinant. To do this, we form a new array by taking $\mathbf{C}_{42}$, and then adding a second copy of columns 1 and 2 as the new columns 4 and 5. We then form terms using the basketweaving pattern.

$$\begin{array}{ccccc} -3 & 0 & 5 & -3 & 0 \\ 2 & -1 & 4 & 2 & -1 \\ 6 & 4 & 0 & 6 & 4 \end{array}$$

Hence, $|\mathbf{C}_{42}| = (-3)(-1)(0) + (0)(4)(6) + (5)(2)(4) - (5)(-1)(6) - (-3)(4)(4) - (0)(2)(0) = 118.$

(3) The cofactor $\mathcal{A}_{ij}$ is defined to be $(-1)^{i+j}|\mathbf{A}_{ij}|$, where $|\mathbf{A}_{ij}|$ is the (i, j) minor – that is, the determinant of the (i, j) submatrix, obtained by deleting the ith row and the jth column from $\mathbf{A}$.

(a) $\mathcal{A}_{22} = (-1)^{2+2}|\mathbf{A}_{22}| = (-1)^4 \begin{vmatrix} 4 & -3 \\ 9 & -7 \end{vmatrix} = (1)\left((4)(-7) - (-3)(9)\right) = -1.$

(c) $\mathcal{C}_{43} = (-1)^{4+3}|\mathbf{C}_{43}| = (-1)^7 \begin{vmatrix} -5 & 2 & 13 \\ -8 & 2 & 22 \\ -6 & -3 & -16 \end{vmatrix} = (-1)(-222) = 222$, where we have computed the 3×3 determinant $|\mathbf{C}_{43}|$ to be -222 using basketweaving as follows:
To compute $|\mathbf{C}_{43}|$, we form a new array by taking $\mathbf{C}_{43}$, and then adding a second copy of columns 1 and 2 as the new columns 4 and 5. We then form terms using the basketweaving pattern.

$$\begin{array}{ccccc} -5 & 2 & 13 & -5 & 2 \\ -8 & 2 & 22 & -8 & 2 \\ -6 & -3 & -16 & -6 & -3 \end{array}$$

Hence, $|\mathbf{C}_{43}| = (-5)(2)(-16) + (2)(22)(-6) + (13)(-8)(-3) - (13)(2)(-6) - (-5)(22)(-3) - (2)(-8)(-16) = -222.$

(d) $\mathcal{D}_{12} = (-1)^{1+2}|\mathbf{D}_{12}| = (-1)^3 \begin{vmatrix} x-4 & x-3 \\ x-1 & x+2 \end{vmatrix} = (-1)\left((x-4)(x+2) - (x-3)(x-1)\right)$
$= (-1)\left((x^2 - 2x - 8) - (x^2 - 4x + 3)\right) = -2x + 11.$

(4) In this problem, we are asked to use only the formal definition (cofactor expansion) to compute determinants. Therefore, we will not use basketweaving for 3×3 determinants, or the simple $a_{11}a_{22} - a_{12}a_{21}$ formula for 2×2 determinants. Of course, the cofactor expansion method still produces the same results obtained in Exercise 1.

(a) Let $\mathbf{A} = \begin{bmatrix} -2 & 5 \\ 3 & 1 \end{bmatrix}$. Then $|\mathbf{A}| = a_{21}\mathcal{A}_{21} + a_{22}\mathcal{A}_{22} = a_{21}(-1)^{2+1}(a_{12}) + a_{22}(-1)^{2+2}(a_{11})$
$= (3)(-1)(5) + (1)(1)(-2) = -17.$

(c) Let $\mathbf{A} = \begin{bmatrix} 6 & -12 \\ -4 & 8 \end{bmatrix}$. Then $|\mathbf{A}| = a_{21}\mathcal{A}_{21} + a_{22}\mathcal{A}_{22} = a_{21}(-1)^{2+1}(a_{12}) + a_{22}(-1)^{2+2}(a_{11})$
$= (-4)(-1)(-12) + (8)(1)(6) = 0.$

(e) Let $\mathbf{A} = \begin{bmatrix} 2 & 0 & 5 \\ -4 & 1 & 7 \\ 0 & 3 & -3 \end{bmatrix}$. Then $|\mathbf{A}| = a_{31}\mathcal{A}_{31} + a_{32}\mathcal{A}_{32} + a_{33}\mathcal{A}_{33}$

$= (0)(-1)^{3+1}\begin{vmatrix} 0 & 5 \\ 1 & 7 \end{vmatrix} + (3)(-1)^{3+2}\begin{vmatrix} 2 & 5 \\ -4 & 7 \end{vmatrix} + (-3)(-1)^{3+3}\begin{vmatrix} 2 & 0 \\ -4 & 1 \end{vmatrix}$. Now, for a 2×2 matrix

$\mathbf{B}$, $|\mathbf{B}| = b_{21}\mathcal{B}_{21} + b_{22}\mathcal{B}_{22} = b_{21}(-1)^{2+1}(b_{12}) + b_{22}(-1)^{2+2}(b_{11})$.

Hence, $\begin{vmatrix} 2 & 5 \\ -4 & 7 \end{vmatrix} = (-4)(-1)(5) + (7)(1)(2) = 34$ and $\begin{vmatrix} 2 & 0 \\ -4 & 1 \end{vmatrix} = (-4)(-1)(0) + (1)(1)(2) = 2$.

It is not necessary to compute $\begin{vmatrix} 0 & 5 \\ 1 & 7 \end{vmatrix}$ because its coefficient in the determinant formula is zero.

Hence, $|\mathbf{A}| = 0 + (3)(-1)(34) + (-3)(1)(2) = -108$.

(g) Let $\mathbf{A} = \begin{bmatrix} 5 & 0 & 0 \\ 3 & -2 & 0 \\ -1 & 8 & 4 \end{bmatrix}$. Then $|\mathbf{A}| = a_{31}\mathcal{A}_{31} + a_{32}\mathcal{A}_{32} + a_{33}\mathcal{A}_{33}$

$= (-1)(-1)^{3+1}\begin{vmatrix} 0 & 0 \\ -2 & 0 \end{vmatrix} + (8)(-1)^{3+2}\begin{vmatrix} 5 & 0 \\ 3 & 0 \end{vmatrix} + (4)(-1)^{3+3}\begin{vmatrix} 5 & 0 \\ 3 & -2 \end{vmatrix}$. Now, for a 2×2 matrix $\mathbf{B}$, $|\mathbf{B}| = b_{21}\mathcal{B}_{21} + b_{22}\mathcal{B}_{22} = b_{21}(-1)^{2+1}(b_{12}) + b_{22}(-1)^{2+2}(b_{11})$.

Hence, $\begin{vmatrix} 0 & 0 \\ -2 & 0 \end{vmatrix} = (-2)(-1)(0) + (0)(1)(0) = 0$, $\begin{vmatrix} 5 & 0 \\ 3 & 0 \end{vmatrix} = (3)(-1)(0) + (0)(1)(5) = 0$

and $\begin{vmatrix} 5 & 0 \\ 3 & -2 \end{vmatrix} = (3)(-1)(0) + (-2)(1)(5) = -10$. Hence, $|\mathbf{A}| = (-1)(1)(0) + (8)(-1)(0) + (4)(1)(-10) = -40$.

(i) Let $\mathbf{A} = \begin{bmatrix} 3 & 1 & -2 \\ -1 & 4 & 5 \\ 3 & 1 & -2 \end{bmatrix}$. Then $|\mathbf{A}| = a_{31}\mathcal{A}_{31} + a_{32}\mathcal{A}_{32} + a_{33}\mathcal{A}_{33}$

$= (3)(-1)^{3+1}\begin{vmatrix} 1 & -2 \\ 4 & 5 \end{vmatrix} + (1)(-1)^{3+2}\begin{vmatrix} 3 & -2 \\ -1 & 5 \end{vmatrix} + (-2)(-1)^{3+3}\begin{vmatrix} 3 & 1 \\ -1 & 4 \end{vmatrix}$. Now, for a 2×2 matrix $\mathbf{B}$, $|\mathbf{B}| = b_{21}\mathcal{B}_{21} + b_{22}\mathcal{B}_{22} = b_{21}(-1)^{2+1}(b_{12}) + b_{22}(-1)^{2+2}(b_{11})$.

Hence, $\begin{vmatrix} 1 & -2 \\ 4 & 5 \end{vmatrix} = (4)(-1)(-2) + (5)(1)(1) = 13$, $\begin{vmatrix} 3 & -2 \\ -1 & 5 \end{vmatrix} = (-1)(-1)(-2) + (5)(1)(3) = 13$, and $\begin{vmatrix} 3 & 1 \\ -1 & 4 \end{vmatrix} = (-1)(-1)(1) + (4)(1)(3) = 13$. Hence, $|\mathbf{A}| = (3)(1)(13) + (1)(-1)(13) + (-2)(1)(13) = 0$.

(j) By definition, the determinant of a 1×1 matrix is defined to be the (1,1) entry of the matrix. Thus, the determinant of $[-3]$ equals -3.

(5) (a) Let $\mathbf{A} = \begin{bmatrix} 5 & 2 & 1 & 0 \\ -1 & 3 & 5 & 2 \\ 4 & 1 & 0 & 2 \\ 0 & 2 & 3 & 0 \end{bmatrix}$. Then $|\mathbf{A}| = a_{41}\mathcal{A}_{41} + a_{42}\mathcal{A}_{42} + a_{43}\mathcal{A}_{43} + a_{44}\mathcal{A}_{44}$

$= (0)(-1)^{4+1}|\mathbf{A}_{41}| + (2)(-1)^{4+2}|\mathbf{A}_{42}| + (3)(-1)^{4+3}|\mathbf{A}_{43}| + (0)(-1)^{4+4}|\mathbf{A}_{44}|$

$= 0 + 2|\mathbf{A}_{42}| - 3|\mathbf{A}_{43}| + 0$. Now, let $\mathbf{B} = \mathbf{A}_{42} = \begin{bmatrix} 5 & 1 & 0 \\ -1 & 5 & 2 \\ 4 & 0 & 2 \end{bmatrix}$. Then $|\mathbf{A}_{42}| = |\mathbf{B}| =$

$b_{31}\mathcal{B}_{31} + b_{32}\mathcal{B}_{32} + b_{33}\mathcal{B}_{33} = (4)(-1)^{3+1}\begin{vmatrix} 1 & 0 \\ 5 & 2 \end{vmatrix} + (0)(-1)^{3+2}\begin{vmatrix} 5 & 0 \\ -1 & 2 \end{vmatrix} + (2)(-1)^{3+3}\begin{vmatrix} 5 & 1 \\ -1 & 5 \end{vmatrix}$

$= (4)(1)\Big((1)(2) - (0)(5)\Big) + 0 + 2(1)\Big((5)(5) - (1)(-1)\Big) = 60$. Also, if $\mathbf{C} = \mathbf{A}_{43} = \begin{bmatrix} 5 & 2 & 0 \\ -1 & 3 & 2 \\ 4 & 1 & 2 \end{bmatrix}$.

Then $|\mathbf{A}_{43}| = |\mathbf{C}| = c_{31}\mathcal{C}_{31} + c_{32}\mathcal{C}_{32} + c_{33}\mathcal{C}_{33}$

$= (4)(-1)^{3+1}\begin{vmatrix} 2 & 0 \\ 3 & 2 \end{vmatrix} + (1)(-1)^{3+2}\begin{vmatrix} 5 & 0 \\ -1 & 2 \end{vmatrix} + (2)(-1)^{3+3}\begin{vmatrix} 5 & 2 \\ -1 & 3 \end{vmatrix}$

$= (4)(1)\Big((2)(2) - (3)(0)\Big) + (1)(-1)\Big((5)(2) - (0)(-1)\Big) + 2(1)\Big((5)(3) - (2)(-1)\Big) = 40$. Hence,

$|\mathbf{A}| = 2|\mathbf{A}_{42}| - 3|\mathbf{A}_{43}| = 2(60) - 3(40) = 0$.

(d) Let $\mathbf{A}$ be the given 5×5 matrix. Then $|\mathbf{A}| = a_{51}\mathcal{A}_{51} + a_{52}\mathcal{A}_{52} + a_{53}\mathcal{A}_{53} + a_{54}\mathcal{A}_{54} + a_{55}\mathcal{A}_{55}$
$= (0)(-1)^{5+1}|\mathbf{A}_{51}| + (3)(-1)^{5+2}|\mathbf{A}_{52}| + (0)(-1)^{5+3}|\mathbf{A}_{53}| + (0)(-1)^{5+4}|\mathbf{A}_{54}| + (2)(-1)^{5+5}|\mathbf{A}_{55}|$
$= (0) + (3)(-1)|\mathbf{A}_{52}| + (0) + (0) + (2)(1)|\mathbf{A}_{55}|$.

Let $\mathbf{B} = \mathbf{A}_{52} = \begin{bmatrix} 0 & 1 & 3 & -2 \\ 2 & 3 & -1 & 0 \\ 3 & 2 & -5 & 1 \\ 1 & -4 & 0 & 0 \end{bmatrix}$. Then, $|\mathbf{A}_{52}| = |\mathbf{B}| = b_{41}\mathcal{B}_{41} + b_{42}\mathcal{B}_{42} + b_{43}\mathcal{B}_{43} + b_{44}\mathcal{B}_{44}$

$= (1)(-1)^{4+1}|\mathbf{B}_{41}| + (-4)(-1)^{4+2}|\mathbf{B}_{42}| + (0)(-1)^{4+3}|\mathbf{B}_{43}| + (0)(-1)^{4+4}|\mathbf{B}_{44}| =$

$(1)(-1)|\mathbf{B}_{41}| + (-4)(1)|\mathbf{B}_{42}| + (0) + (0)$. Now, $\mathbf{B}_{41} = \begin{bmatrix} 1 & 3 & -2 \\ 3 & -1 & 0 \\ 2 & -5 & 1 \end{bmatrix}$. We compute $|\mathbf{B}_{41}|$ using basketweaving.

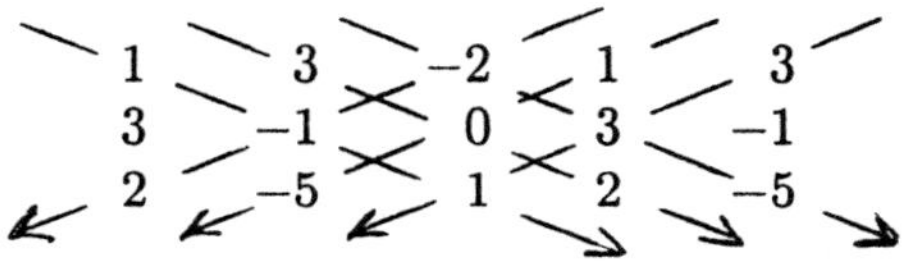

This yields $|\mathbf{B}_{41}| = (1)(-1)(1) + (3)(0)(2) + (-2)(3)(-5) - (-2)(-1)(2) - (1)(0)(-5) - (3)(3)(1) =$
16. Next, $\mathbf{B}_{42} = \begin{bmatrix} 0 & 3 & -2 \\ 2 & -1 & 0 \\ 3 & -5 & 1 \end{bmatrix}$. We compute $|\mathbf{B}_{42}|$ using basketweaving.

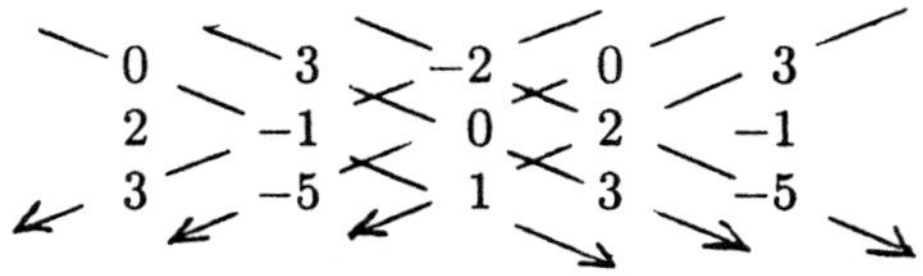

This yields $|\mathbf{B}_{42}| = (0)(-1)(1) + (3)(0)(3) + (-2)(2)(-5) - (-2)(-1)(3) - (0)(0)(-5) - (3)(2)(1) =$
8. Therefore, $|\mathbf{A}_{52}| = (1)(-1)|\mathbf{B}_{41}| + (-4)(1)|\mathbf{B}_{42}| = (1)(-1)(16) + (-4)(1)(8) = -48$. Now,

let $\mathbf{C} = \mathbf{A}_{55} = \begin{bmatrix} 0 & 4 & 1 & 3 \\ 2 & 2 & 3 & -1 \\ 3 & 1 & 2 & -5 \\ 1 & 0 & -4 & 0 \end{bmatrix}$. Then, $|\mathbf{A}_{55}| = |\mathbf{C}| = c_{41}\mathcal{C}_{41} + c_{42}\mathcal{C}_{42} + c_{43}\mathcal{C}_{43} + c_{44}\mathcal{C}_{44}$

$= (1)(-1)^{4+1}|\mathbf{C}_{41}| + (0)(-1)^{4+2}|\mathbf{C}_{42}| + (-4)(-1)^{4+3}|\mathbf{C}_{43}| + (0)(-1)^{4+4}|\mathbf{C}_{44}|$

$= (1)(-1)^{4+1}|\mathbf{C}_{41}| + (0) + (-4)(-1)^{4+3}|\mathbf{C}_{43}| + (0)$. Now, $\mathbf{C}_{41} = \begin{bmatrix} 4 & 1 & 3 \\ 2 & 3 & -1 \\ 1 & 2 & -5 \end{bmatrix}$. We compute $|\mathbf{C}_{41}|$ using basketweaving.

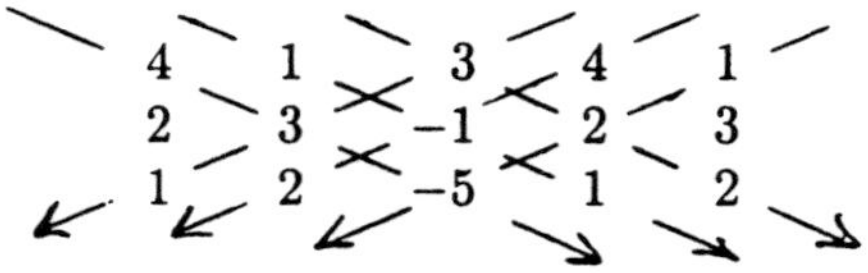

This yields $|\mathbf{C}_{41}| = (4)(3)(-5)+(1)(-1)(1)+(3)(2)(2)-(3)(3)(1)-(4)(-1)(2)-(1)(2)(-5) = -40$.

Similarly, $\mathbf{C}_{43} = \begin{bmatrix} 0 & 4 & 3 \\ 2 & 2 & -1 \\ 3 & 1 & -5 \end{bmatrix}$. We compute $|\mathbf{C}_{43}|$ using basketweaving.

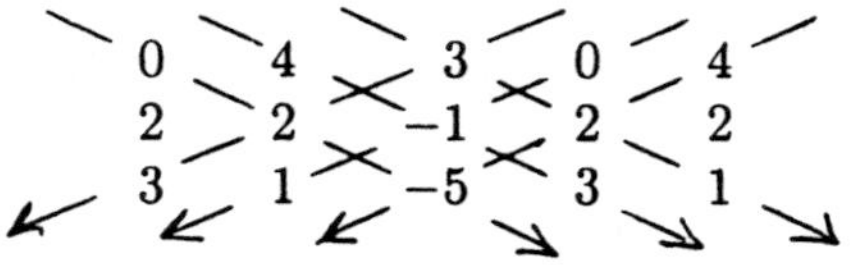

This yields $|\mathbf{C}_{43}| = (0)(2)(-5)+(4)(-1)(3)+(3)(2)(1)-(3)(2)(3)-(0)(-1)(1)-(4)(2)(-5) = 16$.
Therefore, $|\mathbf{A}_{55}| = (1)(-1)^{4+1}|\mathbf{C}_{41}| + (-4)(-1)^{4+3}|\mathbf{C}_{43}| = (1)(-1)(-40) + (-4)(-1)(16) = 104$.
Finally, $|\mathbf{A}| = (0) + (3)(-1)|\mathbf{A}_{52}| + (2)(1)|\mathbf{A}_{55}| = (3)(-1)(-48) + (2)(1)(104) = 352$.

(7) Let $\mathbf{A} = \begin{bmatrix} 1 & 1 \\ 1 & 1 \end{bmatrix}$, and let $\mathbf{B} = \begin{bmatrix} 1 & 0 \\ 0 & 1 \end{bmatrix}$. Then $|\mathbf{A}| = \begin{vmatrix} 1 & 1 \\ 1 & 1 \end{vmatrix} = (1)(1) - (1)(1) = 0$ and $|\mathbf{B}| = \begin{vmatrix} 1 & 0 \\ 0 & 1 \end{vmatrix} = (1)(1) - (0)(0) = 1$. Hence, $|\mathbf{A}| + |\mathbf{B}| = 1$. But, $|\mathbf{A}+\mathbf{B}| = \begin{vmatrix} 2 & 1 \\ 1 & 2 \end{vmatrix} = (2)(2) - (1)(1) = 3$. Hence, $|\mathbf{A}+\mathbf{B}| \neq |\mathbf{A}| + |\mathbf{B}|$.

(9) According to part (1) of Theorem 3.1, the area of the parallelogram is the absolute value of the determinant of the matrix whose rows are the two given vectors. Be careful! If the determinant is negative, we must take the absolute value to find the area.

(a) The area of the parallelogram is the absolute value of the determinant $\begin{vmatrix} 3 & 2 \\ 4 & 5 \end{vmatrix}$. This determinant equals $(3)(5) - (2)(4) = 7$. Hence, the area is 7.

(c) The area of the parallelogram is the absolute value of the determinant $\begin{vmatrix} 5 & -1 \\ -3 & 3 \end{vmatrix}$. This determinant equals $(5)(3) - (-1)(-3) = 12$. Hence, the area is 12.

(10) Let $\mathbf{x} = [x_1, x_2]$ and $\mathbf{y} = [y_1, y_2]$. Consider the hint in the text. Then, the area of the parallelogram is $\|\mathbf{x}\|\,\|\mathbf{h}\|$, where $\mathbf{h} = \mathbf{y} - \mathbf{proj_x y}$. Now, $\mathbf{proj_x y} = \left(\frac{\mathbf{x}\cdot\mathbf{y}}{\|\mathbf{x}\|^2}\right)\mathbf{x} =$
$\frac{1}{\|\mathbf{x}\|^2}[(x_1y_1 + x_2y_2)x_1, (x_1y_1 + x_2y_2)x_2]$. Hence $\mathbf{h} = \mathbf{y} - \mathbf{proj_x y} = \frac{1}{\|\mathbf{x}\|^2}(\|\mathbf{x}\|^2\mathbf{y}) - \left(\frac{\mathbf{x}\cdot\mathbf{y}}{\|\mathbf{x}\|^2}\right)\mathbf{x} =$
$\frac{1}{\|\mathbf{x}\|^2}[(x_1^2 + x_2^2)y_1, (x_1^2 + x_2^2)y_2] - \frac{1}{\|\mathbf{x}\|^2}[(x_1y_1 + x_2y_2)x_1, (x_1y_1 + x_2y_2)x_2]$
$= \frac{1}{\|\mathbf{x}\|^2}[x_1^2y_1 + x_2^2y_1 - x_1^2y_1 - x_1x_2y_2, x_1^2y_2 + x_2^2y_2 - x_1x_2y_1 - x_2^2y_2] =$
$\frac{1}{\|\mathbf{x}\|^2}[x_2^2y_1 - x_1x_2y_2, x_1^2y_2 - x_1x_2y_1] = \frac{1}{\|\mathbf{x}\|^2}[x_2(x_2y_1 - x_1y_2), x_1(x_1y_2 - x_2y_1)] = \frac{x_1y_2 - x_2y_1}{\|\mathbf{x}\|^2}[-x_2, x_1]$. Thus, $\|\mathbf{x}\|\,\|\mathbf{y} - \mathbf{proj_x y}\| = \|\mathbf{x}\|\frac{|x_1y_2 - x_2y_1|}{\|\mathbf{x}\|^2}\sqrt{x_2^2 + x_1^2} = |x_1y_2 - x_2y_1| =$ absolute value of $\begin{vmatrix} x_1 & x_2 \\ y_1 & y_2 \end{vmatrix}$.

(11) According to part (2) of Theorem 3.1, the volume of the parallelepiped is the absolute value of the determinant of the matrix whose rows are the three given vectors. Be careful! If the determinant is negative, we must take the absolute value to find the volume.

(a) The volume of the parallelepiped is the absolute value of the determinant $\begin{vmatrix} -2 & 3 & 1 \\ 4 & 2 & 0 \\ -1 & 3 & 2 \end{vmatrix}$. We use basketweaving to compute this.

$$\begin{array}{ccccc} -2 & 3 & 1 & -2 & 3 \\ 4 & 2 & 0 & 4 & 2 \\ -1 & 3 & 2 & -1 & 3 \end{array}$$

The determinant equals $(-2)(2)(2)+(3)(0)(-1)+(1)(4)(3)-(1)(2)(-1)-(-2)(0)(3)-(3)(4)(2) = -18$. The volume is the absolute value of this result. Hence, the volume equals 18.

(c) The volume of the parallelepiped is the absolute value of the determinant $\begin{vmatrix} -3 & 4 & 0 \\ 6 & -2 & 1 \\ 0 & -3 & 3 \end{vmatrix}$. We use basketweaving to compute this.

$$\begin{array}{ccccc} -3 & 4 & 0 & -3 & 4 \\ 6 & -2 & 1 & 6 & -2 \\ 0 & -3 & 3 & 0 & -3 \end{array}$$

The determinant equals $(-3)(-2)(3) + (4)(1)(0) + (0)(6)(-3) - (0)(-2)(0) - (-3)(1)(-3) - (4)(6)(3) = -63$. The volume is the absolute value of this result. Hence, the volume equals 63.

(12) First, read the hint in the textbook. Then, note that

$$\begin{vmatrix} x_1 & x_2 & x_3 \\ y_1 & y_2 & y_3 \\ z_1 & z_2 & z_3 \end{vmatrix} = x_1 y_2 z_3 + x_2 y_3 z_1 + x_3 y_1 z_2 - x_3 y_2 z_1 - x_1 y_3 z_2 - x_2 y_1 z_3 =$$

$(x_2 y_3 - x_3 y_2) z_1 + (x_3 y_1 - x_1 y_3) z_2 + (x_1 y_2 - x_2 y_1) z_3 = (\mathbf{x} \times \mathbf{y}) \cdot \mathbf{z}$ (from the definition in Exercise 8).

Also note that the formula $\sqrt{(x_2 y_3 - x_3 y_2)^2 + (x_1 y_3 - x_3 y_1)^2 + (x_1 y_2 - x_2 y_1)^2}$ given in the hint for this exercise for the area of the parallelogram (verified below) equals $\|\mathbf{x} \times \mathbf{y}\|$. Hence, the volume of the parallelepiped equals $\|\mathbf{proj}_{(\mathbf{x} \times \mathbf{y})} \mathbf{z}\| \|\mathbf{x} \times \mathbf{y}\| = \left| \frac{\mathbf{z} \cdot (\mathbf{x} \times \mathbf{y})}{\|\mathbf{x} \times \mathbf{y}\|} \right| \|\mathbf{x} \times \mathbf{y}\| = |\mathbf{z} \cdot (\mathbf{x} \times \mathbf{y})| = $ absolute

value of $\begin{vmatrix} x_1 & x_2 & x_3 \\ y_1 & y_2 & y_3 \\ z_1 & z_2 & z_3 \end{vmatrix}$.

Now let us verify the formula $A = \sqrt{(x_2 y_3 - x_3 y_2)^2 + (x_1 y_3 - x_3 y_1)^2 + (x_1 y_2 - x_2 y_1)^2}$ for the area of the parallelogram determined by $\mathbf{x}$ and $\mathbf{y}$. As in the solution for Exercise 10, above, the area of this parallelogram equals $\|\mathbf{x}\| \|\mathbf{y} - \mathbf{proj}_{\mathbf{x}} \mathbf{y}\|$. We must show that $A = \|\mathbf{x}\| \|\mathbf{y} - \mathbf{proj}_{\mathbf{x}} \mathbf{y}\|$. One can verify this by a tedious, brute force, argument. (Algebraically expand and simplify $\|\mathbf{x}\|^2 \|\mathbf{y} - \mathbf{proj}_{\mathbf{x}} \mathbf{y}\|^2$ to get $(x_2 y_3 - x_3 y_2)^2 + (x_1 y_3 - x_3 y_1)^2 + (x_1 y_2 - x_2 y_1)^2$.) An alternate approach is the following:

Now, $A^2 = (x_2 y_3 - x_3 y_2)^2 + (x_1 y_3 - x_3 y_1)^2 + (x_1 y_2 - x_2 y_1)^2$

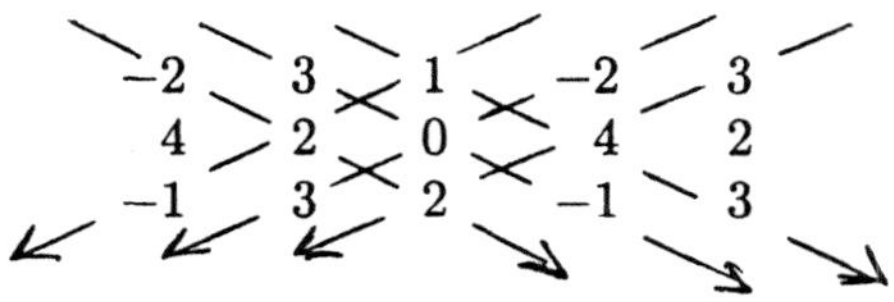

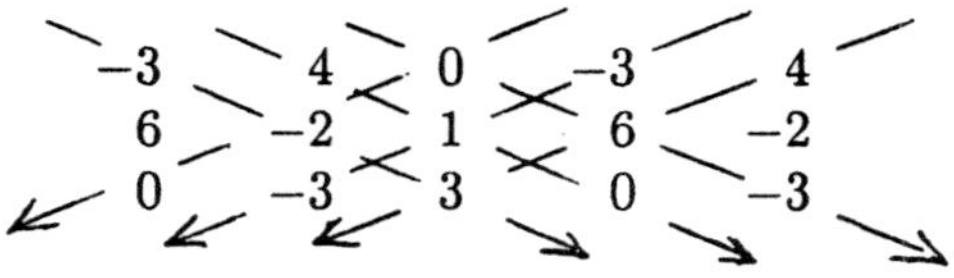

$= x_2^2y_3^2 - 2x_2x_3y_2y_3 + x_3^2y_2^2 + x_1^2y_3^2 - 2x_1x_3y_1y_3 + x_3^2y_1^2 + x_1^2y_2^2 - 2x_1x_2y_1y_2 + x_2^2y_1^2$.
Using some algebraic manipulation, this can be expressed as
$A^2 = (x_1^2+x_2^2+x_3^2)(y_1^2+y_2^2+y_3^2)-(x_1y_1+x_2y_2+x_3y_3)$. (Verify this!) Therefore, $A^2 = \|\mathbf{x}\|^2\,\|\mathbf{y}\|^2-(\mathbf{x}\cdot\mathbf{y})^2$.
Now,

$$\begin{aligned}
\|\mathbf{x}\|^2\|\mathbf{y}-\mathbf{proj_x y}\|^2 &= \|\mathbf{x}\|^2\left\|\mathbf{y}-\frac{\mathbf{x}\cdot\mathbf{y}}{\|\mathbf{x}\|^2}\mathbf{x}\right\|^2 \\
&= \|\mathbf{x}\|^2\left(\mathbf{y}-\frac{(\mathbf{x}\cdot\mathbf{y})\,\mathbf{x}}{\|\mathbf{x}\|^2}\right)\cdot\left(\mathbf{y}-\frac{(\mathbf{x}\cdot\mathbf{y})\,\mathbf{x}}{\|\mathbf{x}\|^2}\right) \\
&= \|\mathbf{x}\|^2\left((\mathbf{y}\cdot\mathbf{y})-2\left(\frac{\mathbf{x}\cdot\mathbf{y}}{\|\mathbf{x}\|^2}\right)(\mathbf{x}\cdot\mathbf{y})+\left(\frac{\mathbf{x}\cdot\mathbf{y}}{\|\mathbf{x}\|^2}\right)^2(\mathbf{x}\cdot\mathbf{x})\right) \\
&= \|\mathbf{x}\|^2\left(\frac{\|\mathbf{x}\|^2(\mathbf{y}\cdot\mathbf{y})}{\|\mathbf{x}\|^2}-2\left(\frac{(\mathbf{x}\cdot\mathbf{y})^2}{\|\mathbf{x}\|^2}\right)+\frac{(\mathbf{x}\cdot\mathbf{y})^2}{\|\mathbf{x}\|^2}\right) \\
&= \|\mathbf{x}\|^2\,\|\mathbf{y}\|^2-(\mathbf{x}\cdot\mathbf{y})^2 = A^2.
\end{aligned}$$

Hence, $A = \|\mathbf{x}\|\,\|\mathbf{y}-\mathbf{proj_x y}\|$.

(15) (a) $\begin{vmatrix} x & 2 \\ 5 & x+3 \end{vmatrix} = 0 \Rightarrow x(x+3)-(2)(5) = 0 \Rightarrow x^2+3x-10 = 0 \Rightarrow (x+5)(x-2) = 0 \Rightarrow x = -5$ or $x = 2$.

(c) First, we compute the given determinant using basketweaving.

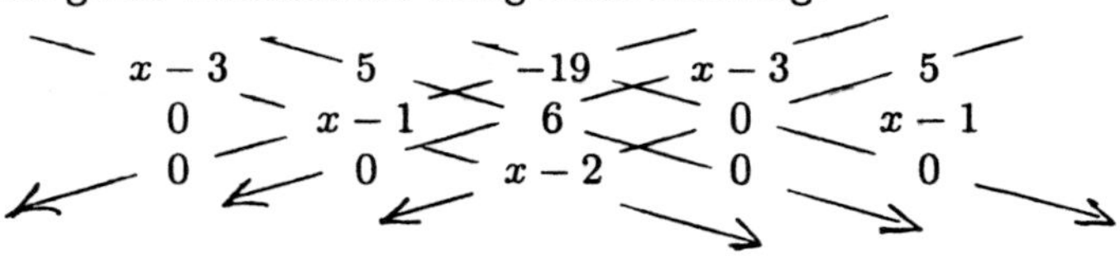

The determinant equals $(x-3)(x-1)(x-2)+(5)(6)(0)+(-19)(0)(0)-(-19)(x-1)(0)$ $-(x-3)(6)(0)-(5)(0)(x-2) = (x-3)(x-1)(x-2)$. Setting this determinant equal to zero yields $x = 3$, $x = 1$, or $x = 2$.

(16) (b) The given matrix is a 3×3 Vandermonde matrix with $a = 2$, $b = 3$, and $c = -2$. Thus, using part (a), the determinant is $(a-b)(b-c)(c-a) = \big(2-3\big)\big(3-(-2)\big)\big((-2)-2\big) = 20$.

(18) (a) False. The basketweaving technique only works to find determinants of 3×3 matrices. For larger matrices, the only method we have learned at this point is the cofactor expansion along the last row, although alternate methods are given in Sections 3.2 and 3.3.

(b) True. Note that $\begin{vmatrix} x_1 & x_2 \\ y_1 & y_2 \end{vmatrix} = x_1y_2 - x_2y_1$. Part (1) of Theorem 3.1 assures us that the absolute value of this determinant gives the area of the desired parallelogram.

(c) False in general. (True for $n = 2$.) An $n\times n$ matrix has n^2 cofactors – one corresponding to each entry in the matrix.

(d) False in general. (True in the special case in which $\mathcal{B}_{23} = 0$.). By definition, $\mathcal{B}_{23} = (-1)^{2+3}|\mathbf{B}_{23}| = -|\mathbf{B}_{23}|$, *not* $|\mathbf{B}_{23}|$.

(e) True. The given formula is the cofactor expansion of **A** along the last row. By definition, this equals the determinant of **A**.

Section 3.2

(1) (a) The row operation used is (II): $\langle 1\rangle \leftarrow -3\langle 2\rangle + \langle 1\rangle$. By Theorem 3.3, performing a type (II) operation does not change the determinant of a matrix. Thus, since $|\mathbf{I}_3| = 1$, the determinant of the given matrix also equals 1.

(c) The row operation used is (I): $\langle 3\rangle \leftarrow -4\langle 3\rangle$. By Theorem 3.3, performing a type (I) operation multiplies the determinant by the constant used in the row operation. Thus, since $|\mathbf{I}_3| = 1$, the determinant of the given matrix equals $(-4)(1) = -4$.

(f) The row operation used is (III): $\langle 1\rangle \leftrightarrow \langle 2\rangle$. By Theorem 3.3, performing a type (III) row operation changes the sign of the determinant. Thus, since $|\mathbf{I}_3| = 1$, the determinant of the given matrix equals -1.

(2) In each part, we use row operations to put the given matrix into upper triangular form. Notice in the solutions, below, that we stop performing row operations as soon as we obtain an upper triangular matrix. We give a chart indicating the row operations used, keeping track of the variable P, as described in Example 5 in Section 3.2 of the textbook. Recall that $P = 1$ at the beginning of the process. We then use the final upper triangular matrix obtained and the value of P to compute the desired determinant.

(a)

Row Operations	Effect	P
(I): $\langle 1\rangle \leftarrow \frac{1}{10}\langle 1\rangle$	Multiply P by 10	10
(II): $\langle 3\rangle \leftarrow 5\langle 1\rangle + \langle 3\rangle$	No change	10
(I): $\langle 2\rangle \leftarrow -\frac{1}{4}\langle 2\rangle$	Multiply P by -4	-40
(II): $\langle 3\rangle \leftarrow (-1)\langle 2\rangle + \langle 3\rangle$	No change	-40

The upper triangular matrix obtained from these operations is $\mathbf{B} = \begin{bmatrix} 1 & \frac{2}{5} & \frac{21}{10} \\ 0 & 1 & -\frac{3}{4} \\ 0 & 0 & -\frac{3}{4} \end{bmatrix}$. By Theorem 3.2, $|\mathbf{B}| = (1)(1)(-\frac{3}{4}) = -\frac{3}{4}$. Hence, as in Example 5 in the textbook, the determinant of the original matrix is $P \times |\mathbf{B}| = (-40)(-\frac{3}{4}) = 30$.

(c)

Row Operations	Effect	P
(II): $\langle 2\rangle \leftarrow 2\langle 1\rangle + \langle 2\rangle$	No change	1
(II): $\langle 3\rangle \leftarrow 3\langle 1\rangle + \langle 3\rangle$	No change	1
(II): $\langle 4\rangle \leftarrow (-2)\langle 1\rangle + \langle 4\rangle$	No change	1
(I): $\langle 2\rangle \leftarrow (-1)\langle 2\rangle$	Multiply P by -1	-1
(II): $\langle 3\rangle \leftarrow \langle 2\rangle + \langle 3\rangle$	No change	-1
(II): $\langle 4\rangle \leftarrow (-1)\langle 2\rangle + \langle 4\rangle$	No change	-1
(III): $\langle 3\rangle \leftrightarrow \langle 4\rangle$	Multiply P by -1	1

The upper triangular matrix obtained from these operations is $\mathbf{B} = \begin{bmatrix} 1 & -1 & 5 & 1 \\ 0 & 1 & -3 & -3 \\ 0 & 0 & 2 & 2 \\ 0 & 0 & 0 & -2 \end{bmatrix}$. By

Theorem 3.2, $|\mathbf{B}| = (1)(1)(2)(-2) = -4$. Hence, as in Example 5 in the textbook, the determinant of the original matrix is $P \times |\mathbf{B}| = 1(-4) = -4$.

(e)

Row Operations	Effect	P
(I): $\langle 1\rangle \leftarrow \frac{1}{5}\langle 1\rangle$	Multiply P by 5	5
(II): $\langle 2\rangle \leftarrow (-\frac{15}{2})\langle 1\rangle + \langle 2\rangle$	No change	5
(II): $\langle 3\rangle \leftarrow \frac{5}{2}\langle 1\rangle + \langle 3\rangle$	No change	5
(II): $\langle 4\rangle \leftarrow (-10)\langle 1\rangle + \langle 4\rangle$	No change	5
(I): $\langle 2\rangle \leftarrow (-\frac{1}{4})\langle 2\rangle$	Multiply P by -4	-20
(II): $\langle 3\rangle \leftarrow (-3)\langle 2\rangle + \langle 3\rangle$	No change	-20
(II): $\langle 4\rangle \leftarrow 9\langle 2\rangle + \langle 4\rangle$	No change	-20
(I): $\langle 3\rangle \leftarrow 4\langle 3\rangle$	Multiply P by $\frac{1}{4}$	-5
(II): $\langle 4\rangle \leftarrow \frac{3}{4}\langle 3\rangle + \langle 4\rangle$	No change	-5

The upper triangular matrix obtained from these operations is $\mathbf{B} = \begin{bmatrix} 1 & \frac{3}{5} & -\frac{8}{5} & \frac{4}{5} \\ 0 & 1 & -\frac{11}{4} & \frac{13}{4} \\ 0 & 0 & 1 & -27 \\ 0 & 0 & 0 & -7 \end{bmatrix}$. By Theorem 3.2, $|\mathbf{B}| = (1)(1)(1)(-7) = -7$. Hence, as in Example 5 in the textbook, the determinant of the original matrix is $P \times |\mathbf{B}| = (-5)(-7) = 35$.

(3) In this problem, first compute the determinant by any convenient method.

(a) $\begin{vmatrix} 5 & 6 \\ -3 & -4 \end{vmatrix} = (5)(-4) - (6)(-3) = -2$. Since the determinant is nonzero, Theorem 3.5 implies that the given matrix is nonsingular.

(c) Using basketweaving on the given matrix produces the value $(-12)(-1)(-8) + (7)(2)(3) + (-27)(4)(2) - (-27)(-1)(3) - (-12)(2)(2) - (7)(4)(-8) = -79$ for the determinant. Since the determinant is nonzero, Theorem 3.5 implies that the given matrix is nonsingular.

(4) (a) The coefficient matrix for the given system is $\mathbf{A} = \begin{bmatrix} -6 & 3 & -22 \\ -7 & 4 & -31 \\ 11 & -6 & 46 \end{bmatrix}$. Using basketweaving yields $|\mathbf{A}| = (-6)(4)(46) + (3)(-31)(11) + (-22)(-7)(-6) - (-22)(4)(11) - (-6)(-31)(-6) - (3)(-7)(46) = -1$. Because $|\mathbf{A}| \neq 0$, Corollary 3.6 shows that $\text{rank}(\mathbf{A}) = 3$. Hence, by Theorem 2.5, the system has only the trivial solution.

(6) Perform the following type (III) row operations on $\mathbf{A}$:

$\begin{cases} \text{(III)}: \langle 1\rangle \leftrightarrow \langle 6\rangle \\ \text{(III)}: \langle 2\rangle \leftrightarrow \langle 5\rangle \\ \text{(III)}: \langle 3\rangle \leftrightarrow \langle 4\rangle \end{cases}$. This produces the matrix $\mathbf{B} = \begin{bmatrix} a_{61} & a_{62} & a_{63} & a_{64} & a_{65} & a_{66} \\ 0 & a_{52} & a_{53} & a_{54} & a_{55} & a_{56} \\ 0 & 0 & a_{43} & a_{44} & a_{45} & a_{46} \\ 0 & 0 & 0 & a_{34} & a_{35} & a_{36} \\ 0 & 0 & 0 & 0 & a_{25} & a_{26} \\ 0 & 0 & 0 & 0 & 0 & a_{16} \end{bmatrix}$.

Since $\mathbf{B}$ is upper triangular, Theorem 3.2 gives $|\mathbf{B}| = a_{61}a_{52}a_{43}a_{34}a_{25}a_{16}$. But, applying a type (III) row operation to a matrix changes the sign of its determinant. Since we performed three type (III) row operations, $|\mathbf{B}| = -(-(-|\mathbf{A}|))$, or $|\mathbf{A}| = -|\mathbf{B}| = -a_{61}a_{52}a_{43}a_{34}a_{25}a_{16}$.

(16) (a) False in general, although Theorem 3.2 shows that it is true for upper triangular matrices. For a counterexample to the general statement, consider the matrix $\mathbf{A} = \begin{bmatrix} 1 & 1 \\ 1 & 1 \end{bmatrix}$ whose determinant is $(1)(1) - (1)(1) = 0$. However, the product of the main diagonal entries of $\mathbf{A}$ is $(1)(1) = 1$.

(b) True. Part (3) of Theorem 3.3 shows that performing a type (III) row operation changes the sign of the determinant. Therefore, performing two type (III) row operations in succession changes the sign twice. Hence, there is no overall effect on the determinant.

(c) False in general. If $\mathbf{A}$ is a 4×4 matrix, Corollary 3.4 shows that $|3\mathbf{A}| = 3^4|\mathbf{A}| = 81|\mathbf{A}|$. This equals $3|\mathbf{A}|$ only in the exceptional case in which $|\mathbf{A}| = 0$. The matrix $\mathbf{A} = \mathbf{I}_4$ provides a specific counterexample to the original statement.

(d) False. If $\mathbf{A}$ is a matrix having (row i) = (row j), then $|\mathbf{A}| = 0 \neq 1$. To prove this, note that performing the type (III) row operation $\langle i \rangle \leftrightarrow \langle j \rangle$ on $\mathbf{A}$ results in $\mathbf{A}$. Thus, by part (3) of Theorem 3.3, $|\mathbf{A}| = -|\mathbf{A}|$. Hence, $|\mathbf{A}| = 0$. For a specific counterexample to the original statement, consider $\mathbf{O}_n$, for any $n \geq 2$.

(e) False. This statement is in direct contradiction with Theorem 3.5.

(f) True. This statement is the contrapositive of Corollary 3.6.

Section 3.3

(1) (a) $|\mathbf{A}| = a_{31}\mathcal{A}_{31} + a_{32}\mathcal{A}_{32} + a_{33}\mathcal{A}_{33} + a_{34}\mathcal{A}_{34}$
$= a_{31}(-1)^{3+1}|\mathbf{A}_{31}| + a_{32}(-1)^{3+2}|\mathbf{A}_{32}| + a_{33}(-1)^{3+3}|\mathbf{A}_{33}| + a_{34}(-1)^{3+4}|\mathbf{A}_{34}|$
$= a_{31}|\mathbf{A}_{31}| - a_{32}|\mathbf{A}_{32}| + a_{33}|\mathbf{A}_{33}| - a_{34}|\mathbf{A}_{34}|$.

(c) $|\mathbf{A}| = a_{14}\mathcal{A}_{14} + a_{24}\mathcal{A}_{24} + a_{34}\mathcal{A}_{34} + a_{44}\mathcal{A}_{44}$
$= a_{14}(-1)^{1+4}|\mathbf{A}_{14}| + a_{24}(-1)^{2+4}|\mathbf{A}_{24}| + a_{34}(-1)^{3+4}|\mathbf{A}_{34}| + a_{44}(-1)^{4+4}|\mathbf{A}_{44}|$
$= -a_{14}|\mathbf{A}_{14}| + a_{24}|\mathbf{A}_{24}| - a_{34}|\mathbf{A}_{34}| + a_{44}|\mathbf{A}_{44}|$.

(2) (a) Let $\mathbf{A}$ be the given matrix. Then $|\mathbf{A}| = a_{21}\mathcal{A}_{21} + a_{22}\mathcal{A}_{22} + a_{23}\mathcal{A}_{23} = a_{21}(-1)^{2+1}|\mathbf{A}_{21}| + a_{22}(-1)^{2+2}|\mathbf{A}_{22}| + a_{23}(-1)^{2+3}|\mathbf{A}_{23}| = -(0)\begin{vmatrix} -1 & 4 \\ -2 & -3 \end{vmatrix} + (3)\begin{vmatrix} 2 & 4 \\ 5 & -3 \end{vmatrix} - (-2)\begin{vmatrix} 2 & -1 \\ 5 & -2 \end{vmatrix} =$
$0 + (3)\big((2)(-3) - (4)(5)\big) - (-2)\big((2)(-2) - (-1)(5)\big) = -76$.

(c) Let $\mathbf{C}$ be the given matrix. Then $|\mathbf{C}| = c_{11}\mathcal{C}_{11} + c_{21}\mathcal{C}_{21} + c_{31}\mathcal{C}_{31} = c_{11}(-1)^{1+1}|\mathbf{C}_{11}|$
$+ c_{21}(-1)^{2+1}|\mathbf{C}_{21}| + c_{31}(-1)^{3+1}|\mathbf{C}_{31}| = (4)\begin{vmatrix} -1 & -2 \\ 3 & 2 \end{vmatrix} - (5)\begin{vmatrix} -2 & 3 \\ 3 & 2 \end{vmatrix} + (3)\begin{vmatrix} -2 & 3 \\ -1 & -2 \end{vmatrix} =$
$(4)\big((-1)(2) - (-2)(3)\big) - (5)\big((-2)(2) - (3)(3)\big) + (3)\big((-2)(-2) - (3)(-1)\big) = 102$.

(3) (a) Let $\mathbf{A}$ represent the given matrix. First, we compute each cofactor:

$$\mathcal{A}_{11} = (-1)^{1+1}\begin{vmatrix} 0 & -3 \\ -2 & -33 \end{vmatrix} = -6, \quad \mathcal{A}_{12} = (-1)^{1+2}\begin{vmatrix} 2 & -3 \\ 20 & -33 \end{vmatrix} = 6,$$

$$\mathcal{A}_{21} = (-1)^{2+1}\begin{vmatrix} -1 & -21 \\ -2 & -33 \end{vmatrix} = 9, \quad \mathcal{A}_{22} = (-1)^{2+2}\begin{vmatrix} 14 & -21 \\ 20 & -33 \end{vmatrix} = -42,$$

$$\mathcal{A}_{31} = (-1)^{3+1}\begin{vmatrix} -1 & -21 \\ 0 & -3 \end{vmatrix} = 3, \quad \mathcal{A}_{32} = (-1)^{3+2}\begin{vmatrix} 14 & -21 \\ 2 & -3 \end{vmatrix} = 0,$$

$$\mathcal{A}_{13} = (-1)^{1+3}\begin{vmatrix} 2 & 0 \\ 20 & -2 \end{vmatrix} = -4, \quad \mathcal{A}_{23} = (-1)^{2+3}\begin{vmatrix} 14 & -1 \\ 20 & -2 \end{vmatrix} = 8,$$

$$\mathcal{A}_{33} = (-1)^{3+3}\begin{vmatrix} 14 & -1 \\ 2 & 0 \end{vmatrix} = 2.$$

The adjoint matrix $\mathcal{A}$ is the 3×3 matrix whose (i, j) entry is $\mathcal{A}_{ji}$. Hence, $\mathcal{A} = \begin{bmatrix} -6 & 9 & 3 \\ 6 & -42 & 0 \\ -4 & 8 & 2 \end{bmatrix}$.

Next, to find the determinant, we perform a cofactor expansion along the second column, since $a_{22} = 0$. Using the cofactors we have already computed, $|\mathbf{A}| = a_{12}\mathcal{A}_{12} + a_{22}\mathcal{A}_{22} + a_{32}\mathcal{A}_{32} = (-1)(6) + (0)(-42) + (-2)(0) = -6$. Finally, by Corollary 3.12,

$$\mathbf{A}^{-1} = \frac{1}{|\mathbf{A}|}\mathcal{A} = \frac{1}{(-6)}\begin{bmatrix} -6 & 9 & 3 \\ 6 & -42 & 0 \\ -4 & 8 & 2 \end{bmatrix} = \begin{bmatrix} 1 & -\frac{3}{2} & -\frac{1}{2} \\ -1 & 7 & 0 \\ \frac{2}{3} & -\frac{4}{3} & -\frac{1}{3} \end{bmatrix}.$$

(c) Let $\mathbf{A}$ represent the given matrix. First, we compute each cofactor. Each of the 3×3 determinants can be calculated using basketweaving.

$$\mathcal{A}_{11} = (-1)^{1+1}\begin{vmatrix} -4 & 1 & 4 \\ 11 & -2 & -8 \\ 10 & -2 & -7 \end{vmatrix} = -3, \quad \mathcal{A}_{21} = (-1)^{2+1}\begin{vmatrix} 1 & 0 & -1 \\ 11 & -2 & -8 \\ 10 & -2 & -7 \end{vmatrix} = 0,$$

$$\mathcal{A}_{31} = (-1)^{3+1}\begin{vmatrix} 1 & 0 & -1 \\ -4 & 1 & 4 \\ 10 & -2 & -7 \end{vmatrix} = 3, \quad \mathcal{A}_{41} = (-1)^{4+1}\begin{vmatrix} 1 & 0 & -1 \\ -4 & 1 & 4 \\ 11 & -2 & -8 \end{vmatrix} = -3,$$

$$\mathcal{A}_{12} = (-1)^{1+2}\begin{vmatrix} 7 & 1 & 4 \\ -14 & -2 & -8 \\ -12 & -2 & -7 \end{vmatrix} = 0, \quad \mathcal{A}_{22} = (-1)^{2+2}\begin{vmatrix} -2 & 0 & -1 \\ -14 & -2 & -8 \\ -12 & -2 & -7 \end{vmatrix} = 0,$$

$$\mathcal{A}_{32} = (-1)^{3+2}\begin{vmatrix} -2 & 0 & -1 \\ 7 & 1 & 4 \\ -12 & -2 & -7 \end{vmatrix} = 0, \quad \mathcal{A}_{42} = (-1)^{4+2}\begin{vmatrix} -2 & 0 & -1 \\ 7 & 1 & 4 \\ -14 & -2 & -8 \end{vmatrix} = 0,$$

$$\mathcal{A}_{13} = (-1)^{1+3}\begin{vmatrix} 7 & -4 & 4 \\ -14 & 11 & -8 \\ -12 & 10 & -7 \end{vmatrix} = -3, \quad \mathcal{A}_{23} = (-1)^{2+3}\begin{vmatrix} -2 & 1 & -1 \\ -14 & 11 & -8 \\ -12 & 10 & -7 \end{vmatrix} = 0,$$

$$\mathcal{A}_{33} = (-1)^{3+3}\begin{vmatrix} -2 & 1 & -1 \\ 7 & -4 & 4 \\ -12 & 10 & -7 \end{vmatrix} = 3, \quad \mathcal{A}_{43} = (-1)^{4+3}\begin{vmatrix} -2 & 1 & -1 \\ 7 & -4 & 4 \\ -14 & 11 & -8 \end{vmatrix} = -3,$$

$$\mathcal{A}_{14} = (-1)^{1+4}\begin{vmatrix} 7 & -4 & 1 \\ -14 & 11 & -2 \\ -12 & 10 & -2 \end{vmatrix} = 6, \qquad \mathcal{A}_{24} = (-1)^{2+4}\begin{vmatrix} -2 & 1 & 0 \\ -14 & 11 & -2 \\ -12 & 10 & -2 \end{vmatrix} = 0,$$

$$\mathcal{A}_{34} = (-1)^{3+4}\begin{vmatrix} -2 & 1 & 0 \\ 7 & -4 & 1 \\ -12 & 10 & -2 \end{vmatrix} = -6, \qquad \mathcal{A}_{44} = (-1)^{4+4}\begin{vmatrix} -2 & 1 & 0 \\ 7 & -4 & 1 \\ -14 & 11 & -2 \end{vmatrix} = 6.$$

The adjoint matrix $\mathcal{A}$ is the 4×4 matrix whose (i,j) entry is $\mathcal{A}_{ji}$. So, $\mathcal{A} = \begin{bmatrix} -3 & 0 & 3 & -3 \\ 0 & 0 & 0 & 0 \\ -3 & 0 & 3 & -3 \\ 6 & 0 & -6 & 6 \end{bmatrix}$.

Next, we use a cofactor expansion along the first row to compute the determinant of $\mathbf{A}$, because $a_{13} = 0$. Using the cofactors we calculated above:

$$|\mathbf{A}| = a_{11}\mathcal{A}_{11} + a_{12}\mathcal{A}_{12} + a_{13}\mathcal{A}_{13} + a_{14}\mathcal{A}_{14} = (-2)(-3) + (1)(0) + (0)(-3) + (-1)(6) = 0.$$

Finally, since $|\mathbf{A}| = 0$, $\mathbf{A}$ has no inverse, by Theorem 3.5.

(e) Let $\mathbf{A}$ represent the given matrix. First, we compute each cofactor:

$$\mathcal{A}_{11} = (-1)^{1+1}\begin{vmatrix} -3 & 2 \\ 0 & -1 \end{vmatrix} = 3, \qquad \mathcal{A}_{12} = (-1)^{1+2}\begin{vmatrix} 0 & 2 \\ 0 & -1 \end{vmatrix} = 0,$$

$$\mathcal{A}_{21} = (-1)^{2+1}\begin{vmatrix} -1 & 0 \\ 0 & -1 \end{vmatrix} = -1, \qquad \mathcal{A}_{22} = (-1)^{2+2}\begin{vmatrix} 3 & 0 \\ 0 & -1 \end{vmatrix} = -3,$$

$$\mathcal{A}_{31} = (-1)^{3+1}\begin{vmatrix} -1 & 0 \\ -3 & 2 \end{vmatrix} = -2, \qquad \mathcal{A}_{32} = (-1)^{3+2}\begin{vmatrix} 3 & 0 \\ 0 & 2 \end{vmatrix} = -6,$$

$$\mathcal{A}_{13} = (-1)^{1+3}\begin{vmatrix} 0 & -3 \\ 0 & 0 \end{vmatrix} = 0, \qquad \mathcal{A}_{23} = (-1)^{2+3}\begin{vmatrix} 3 & -1 \\ 0 & 0 \end{vmatrix} = 0,$$

$$\mathcal{A}_{33} = (-1)^{3+3}\begin{vmatrix} 3 & -1 \\ 0 & -3 \end{vmatrix} = -9.$$

The adjoint matrix $\mathcal{A}$ is the 3×3 matrix whose (i,j) entry is $\mathcal{A}_{ji}$. Hence, $\mathcal{A} = \begin{bmatrix} 3 & -1 & -2 \\ 0 & -3 & -6 \\ 0 & 0 & -9 \end{bmatrix}$.

Next, since $\mathbf{A}$ is upper triangular, $|\mathbf{A}| = (3)(-3)(-1) = 9$. Finally, by Corollary 3.12,

$$\mathbf{A}^{-1} = \frac{1}{|\mathbf{A}|}\mathcal{A} = \frac{1}{9}\begin{bmatrix} 3 & -1 & -2 \\ 0 & -3 & -6 \\ 0 & 0 & -9 \end{bmatrix} = \begin{bmatrix} \frac{1}{3} & -\frac{1}{9} & -\frac{2}{9} \\ 0 & -\frac{1}{3} & -\frac{2}{3} \\ 0 & 0 & -1 \end{bmatrix}.$$

(4) (a) Let $\mathbf{A}$ be the coefficient matrix $\begin{bmatrix} 3 & -1 & -1 \\ 2 & -1 & -2 \\ -9 & 1 & 0 \end{bmatrix}$. The matrices $\mathbf{A}_1$, $\mathbf{A}_2$, and $\mathbf{A}_3$ are formed by removing, respectively, the first, second, and third columns, from $\mathbf{A}$, and replacing them with $[-8, 3, 39]$ (as a column vector). Hence, $\mathbf{A}_1 = \begin{bmatrix} -8 & -1 & -1 \\ 3 & -1 & -2 \\ 39 & 1 & 0 \end{bmatrix}$, $\mathbf{A}_2 = \begin{bmatrix} 3 & -8 & -1 \\ 2 & 3 & -2 \\ -9 & 39 & 0 \end{bmatrix}$,

and $\mathbf{A}_3 = \begin{bmatrix} 3 & -1 & -8 \\ 2 & -1 & 3 \\ -9 & 1 & 39 \end{bmatrix}$. Basketweaving, or any other convenient method, can be used to compute the determinant of each of these matrices, yielding $|\mathbf{A}| = -5$, $|\mathbf{A}_1| = 20$, $|\mathbf{A}_2| = -15$, and $|\mathbf{A}_3| = 35$. Therefore, Cramer's Rule states that the unique solution to the system is given by $x_1 = \frac{|\mathbf{A}_1|}{|\mathbf{A}|} = \frac{20}{(-5)} = -4$, $x_2 = \frac{|\mathbf{A}_2|}{|\mathbf{A}|} = \frac{(-15)}{(-5)} = 3$, and $x_3 = \frac{|\mathbf{A}_3|}{|\mathbf{A}|} = \frac{35}{(-5)} = -7$. The full solution set is $\{(-4, 3, -7)\}$.

(d) Let $\mathbf{A}$ be the coefficient matrix $\begin{bmatrix} -5 & 2 & -2 & 1 \\ 2 & -1 & 2 & -2 \\ 5 & -2 & 3 & -1 \\ -6 & 2 & -2 & 1 \end{bmatrix}$. The matrices $\mathbf{A}_1$, $\mathbf{A}_2$, $\mathbf{A}_3$ and $\mathbf{A}_4$ are formed by removing, respectively, the first, second, third, and fourth columns, from $\mathbf{A}$, and replacing them with $[-10, -9, 7, -14]$ (as a column vector). Hence,

$$\mathbf{A}_1 = \begin{bmatrix} -10 & 2 & -2 & 1 \\ -9 & -1 & 2 & -2 \\ 7 & -2 & 3 & -1 \\ -14 & 2 & -2 & 1 \end{bmatrix}, \mathbf{A}_2 = \begin{bmatrix} -5 & -10 & -2 & 1 \\ 2 & -9 & 2 & -2 \\ 5 & 7 & 3 & -1 \\ -6 & -14 & -2 & 1 \end{bmatrix}, \mathbf{A}_3 = \begin{bmatrix} -5 & 2 & -10 & 1 \\ 2 & -1 & -9 & -2 \\ 5 & -2 & 7 & -1 \\ -6 & 2 & -14 & 1 \end{bmatrix}$$

and $\mathbf{A}_4 = \begin{bmatrix} -5 & 2 & -2 & -10 \\ 2 & -1 & 2 & -9 \\ 5 & -2 & 3 & 7 \\ -6 & 2 & -2 & -14 \end{bmatrix}$. Row reduction or cofactor expansion can be used to compute the determinant of each of these matrices, yielding $|\mathbf{A}| = 3$, $|\mathbf{A}_1| = 12$, $|\mathbf{A}_2| = -3$, $|\mathbf{A}_3| = -9$ and $|\mathbf{A}_4| = 18$. Therefore, Cramer's Rule states that the unique solution to the system is given by $x_1 = \frac{|\mathbf{A}_1|}{|\mathbf{A}|} = \frac{12}{3} = 4$, $x_2 = \frac{|\mathbf{A}_2|}{|\mathbf{A}|} = \frac{(-3)}{3} = -1$, $x_3 = \frac{|\mathbf{A}_3|}{|\mathbf{A}|} = \frac{(-9)}{3} = -3$, and $x_4 = \frac{|\mathbf{A}_4|}{|\mathbf{A}|} = \frac{18}{3} = 6$. The full solution set is $\{(4, -1, -3, 6)\}$.

(8) (b) Try $n = 2$. All 2×2 skew-symmetric matrices are of the form $\begin{bmatrix} 0 & -a \\ a & 0 \end{bmatrix}$. Letting $a = 1$ yields $\mathbf{A} = \begin{bmatrix} 0 & -1 \\ 1 & 0 \end{bmatrix}$. Note that $|\mathbf{A}| = 1 \neq 0$.

(9) (b) Use trial and error, rearranging the rows of $\mathbf{I}_3$. For example, $\mathbf{A} = \begin{bmatrix} 0 & 1 & 0 \\ 1 & 0 & 0 \\ 0 & 0 & 1 \end{bmatrix}$ works, since $\mathbf{A}\mathbf{A}^T = \mathbf{A}\mathbf{A}$ (because $\mathbf{A}$ is symmetric) $= \mathbf{I}_3$. (Note: all of the four other matrices obtained by rearranging the rows of $\mathbf{I}_3$ are also orthogonal matrices!)

(13) (b) Simply choose any nonsingular 2×2 matrix $\mathbf{P}$ and compute $\mathbf{B} = \mathbf{P}\mathbf{A}\mathbf{P}^{-1}$. For example, using $\mathbf{P} = \begin{bmatrix} 1 & -1 \\ -1 & 2 \end{bmatrix}$, we get $\mathbf{P}^{-1} = \begin{bmatrix} 2 & 1 \\ 1 & 1 \end{bmatrix}$ (see Theorem 2.13), yielding

$\mathbf{B} = \mathbf{P}\mathbf{A}\mathbf{P}^{-1} = \begin{bmatrix} 1 & -1 \\ -1 & 2 \end{bmatrix}\begin{bmatrix} 1 & 2 \\ 3 & 4 \end{bmatrix}\begin{bmatrix} 2 & 1 \\ 1 & 1 \end{bmatrix} = \begin{bmatrix} -6 & -4 \\ 16 & 11 \end{bmatrix}$. Similarly, letting $\mathbf{P} = \begin{bmatrix} 3 & 5 \\ 1 & 2 \end{bmatrix}$ produces $\mathbf{B} = \begin{bmatrix} 3 & 5 \\ 1 & 2 \end{bmatrix}\begin{bmatrix} 1 & 2 \\ 3 & 4 \end{bmatrix}\begin{bmatrix} 2 & -5 \\ -1 & 3 \end{bmatrix} = \begin{bmatrix} 10 & -12 \\ 4 & -5 \end{bmatrix}$.

(14) By Corollary 3.12, $\mathbf{A}^{-1} = \frac{1}{|\mathbf{A}|}\mathcal{A}$, and $\mathbf{B}^{-1} = \frac{1}{|\mathbf{B}|}\mathcal{B}$. Hence, by part (3) of Theorem 2.11, $(\mathbf{AB})^{-1} = \mathbf{B}^{-1}\mathbf{A}^{-1} = \left(\frac{1}{|\mathbf{B}|}\mathcal{B}\right)\left(\frac{1}{|\mathbf{A}|}\mathcal{A}\right) = \mathcal{BA}/(|\mathbf{A}|\,|\mathbf{B}|)$.

(18) (b) If we try a skew-symmetric 2×2 matrix as a possible example, we find that the adjoint *is* skew-symmetric. So we consider a skew-symmetric 3×3 matrix, such as $\mathbf{A} = \begin{bmatrix} 0 & 1 & 1 \\ -1 & 0 & 1 \\ -1 & -1 & 0 \end{bmatrix}$. Then each of the nine minors $|\mathbf{A}_{ij}|$ are easily seen to equal 1. Hence, the (i,j) entry of $\mathcal{A}$ is $\mathcal{A}_{ji} = (-1)^{j+i}|\mathbf{A}_{ji}| = (-1)^{j+i}(1) = (-1)^{j+i}$. Therefore, $\mathcal{A} = \begin{bmatrix} 1 & -1 & 1 \\ -1 & 1 & -1 \\ 1 & -1 & 1 \end{bmatrix}$, which is not skew-symmetric.

(22) (a) True. Corollary 3.8 shows that $|\mathbf{A}^{-1}| = \frac{1}{|\mathbf{A}|}$. But Theorem 3.9 asserts that $|\mathbf{A}| = |\mathbf{A}^T|$. Combining these two equations produces the desired result.

(b) True. According to Theorem 3.10, both cofactor expansions produce $|\mathbf{A}|$ as an answer. (The size of the square matrix is not important.)

(c) False. Type (III) column operations applied to square matrices change the sign of the determinant, just as type (III) row operations do. Hence, $|\mathbf{B}| = -|\mathbf{A}|$. The equation $|\mathbf{B}| = |\mathbf{A}|$ can only be true in the special case in which $|\mathbf{A}| = 0$. For a particular counterexample to $|\mathbf{B}| = |\mathbf{A}|$, consider $\mathbf{A} = \mathbf{I}_2$, and use the column operation $\langle \text{col. } 1\rangle \leftrightarrow \langle \text{col. } 2\rangle$. Then $\mathbf{B} = \begin{bmatrix} 0 & 1 \\ 1 & 0 \end{bmatrix}$. Note that $|\mathbf{B}| = -1$, but $|\mathbf{A}| = 1$.

(d) True. The (i,j) entry of the adjoint is defined to be $\mathcal{A}_{ji} = (-1)^{j+i}|\mathbf{A}_{ji}|$, which clearly equals $(-1)^{i+j}|\mathbf{A}_{ji}|$.

(e) False. Theorem 3.11 states that $\mathbf{A}\mathcal{A} = |\mathbf{A}|\,\mathbf{I}$. Hence, the equation $\mathbf{A}\mathcal{A} = \mathbf{I}$ can only be true when $|\mathbf{A}| = 1$. For a specific counterexample, consider $\mathbf{A} = 2\mathbf{I}_2$. A very short computation shows that $\mathcal{A} = 2\mathbf{I}_2$ as well. Hence, $\mathbf{A}\mathcal{A} = (2\mathbf{I}_2)(2\mathbf{I}_2) = 4\mathbf{I}_2 \neq \mathbf{I}_2$.

(f) True. This follows directly from Cramer's Rule. The coefficient matrix $\mathbf{A}$ for the given system is upper triangular, and so $|\mathbf{A}|$ is easily found to be $(4)(-3)(1) = -12$ (see Theorem 3.2). Notice that $\mathbf{A}_2 = \begin{bmatrix} 4 & -6 & -1 \\ 0 & 5 & 4 \\ 0 & 3 & 1 \end{bmatrix}$. Cramer's Rule then states that $x_2 = \frac{|\mathbf{A}_2|}{|\mathbf{A}|} = -\frac{1}{12}|\mathbf{A}_2|$.

(23) (a) Let R be the type (III) row operation $\langle k\rangle \longleftrightarrow \langle k-1\rangle$, let $\mathbf{A}$ be an $n \times n$ matrix and let $\mathbf{B} = R(\mathbf{A})$. Then, by part (3) of Theorem 3.3, $|\mathbf{B}| = (-1)|\mathbf{A}|$. Next, notice that the submatrix $\mathbf{A}_{(k-1)j} = \mathbf{B}_{kj}$ because the $(k-1)$st row of $\mathbf{A}$ becomes the kth row of $\mathbf{B}$, implying the same row is being eliminated from both matrices, and since all other rows maintain their original relative positions (notice the same column is being eliminated in both cases). Hence, $a_{(k-1)1}\mathcal{A}_{(k-1)1} + a_{(k-1)2}\mathcal{A}_{(k-1)2} + \cdots + a_{(k-1)n}\mathcal{A}_{(k-1)n} = b_{k1}(-1)^{(k-1)+1}|\mathbf{A}_{(k-1)1}| + b_{k2}(-1)^{(k-1)+2}|\mathbf{A}_{(k-1)2}| + \cdots + b_{kn}(-1)^{(k-1)+n}|\mathbf{A}_{(k-1)n}|$ (because $a_{(k-1)j} = b_{kj}$ for $1 \leq j \leq n$) $= b_{k1}(-1)^{k}|\mathbf{B}_{k1}| + b_{k2}(-1)^{k+1}|\mathbf{B}_{k2}| + \cdots + b_{kn}(-1)^{k+n-1}|\mathbf{B}_{kn}| = (-1)(b_{k1}(-1)^{k+1}|\mathbf{B}_{k1}| + b_{k2}(-1)^{k+2}|\mathbf{B}_{k2}| + \cdots + b_{kk}(-1)^{k+n}|\mathbf{B}_{kn}|) = (-1)|\mathbf{B}|$ (by applying Theorem 3.10 along the kth row of $\mathbf{B}$) $= (-1)|R(\mathbf{A})| = (-1)(-1)|\mathbf{A}| = |\mathbf{A}|$, finishing the proof.

(b) The definition of the determinant is the Base Step for an induction proof on the row number, counting down from n to 1. Part (a) is the Inductive Step.

(24) Suppose row i of $\mathbf{A}$ equals row j of $\mathbf{A}$. Let R be the type (III) row operation $\langle i\rangle \longleftrightarrow \langle j\rangle$: Then, clearly $\mathbf{A} = R(\mathbf{A})$. Thus, by part (3) of Theorem 3.3, $|\mathbf{A}| = -|\mathbf{A}|$, or $2|\mathbf{A}| = 0$, implying $|\mathbf{A}| = 0$.

(25) Let $\mathbf{A}$, i, j, and $\mathbf{B}$ be as given in the exercise and its hint. Then by Exercise 24, $|\mathbf{B}| = 0$, since its ith and jth rows are the same. Also, since every row of $\mathbf{A}$ equals the corresponding row of $\mathbf{B}$, with the exception of the jth row, the submatrices $\mathbf{A}_{jk}$ and $\mathbf{B}_{jk}$ are equal for $1 \le k \le n$. Hence, $\mathcal{A}_{jk} = \mathcal{B}_{jk}$ for $1 \le k \le n$. Now, computing the determinant of $\mathbf{B}$ using a cofactor expansion along the jth row (allowed by Exercise 23) yields $0 = |\mathbf{B}| = b_{j1}\mathcal{B}_{j1} + b_{j2}\mathcal{B}_{j2} + \cdots + b_{jn}\mathcal{B}_{jn} = a_{i1}\mathcal{B}_{j1} + a_{i2}\mathcal{B}_{j2} + \cdots + a_{in}\mathcal{B}_{jn}$ (because the jth row of $\mathbf{B}$ equals the ith row of $\mathbf{A}$) $= a_{i1}\mathcal{A}_{j1} + a_{i2}\mathcal{A}_{j2} + \cdots + a_{in}\mathcal{A}_{jn}$, completing the proof.

(26) Using the fact that the general (k,m) entry of $\mathcal{A}$ is $\mathcal{A}_{mk}$, we see that the (i,j) entry of $\mathbf{A}\mathcal{A}$ equals $a_{i1}\mathcal{A}_{j1} + a_{i2}\mathcal{A}_{j2} + \cdots + a_{in}\mathcal{A}_{jn}$. If $i = j$, Exercise 23 implies that this sum equals $|\mathbf{A}|$, while if $i \ne j$, Exercise 25 implies that the sum equals 0. Hence, $\mathbf{A}\mathcal{A}$ equals $|\mathbf{A}|$ on the main diagonal and 0 off the main diagonal, yielding $(|\mathbf{A}|)\mathbf{I}_n$.

(27) Suppose $\mathbf{A}$ is nonsingular. Then $|\mathbf{A}| \ne 0$ by Theorem 3.5. Thus, by Exercise 26, $\mathbf{A}\left(\frac{1}{|\mathbf{A}|}\mathcal{A}\right) = \mathbf{I}_n$. Therefore, by Theorem 2.9, $\left(\frac{1}{|\mathbf{A}|}\mathcal{A}\right)\mathbf{A} = \mathbf{I}_n$, implying $\mathcal{A}\mathbf{A} = (|\mathbf{A}|)\mathbf{I}_n$.

(28) Using the fact that the general (k,m) entry of $\mathcal{A}$ is $\mathcal{A}_{mk}$, we see that the (j,j) entry of $\mathcal{A}\mathbf{A}$ equals $a_{1j}\mathcal{A}_{1j} + a_{2j}\mathcal{A}_{2j} + \cdots + a_{nj}\mathcal{A}_{nj}$. But all main diagonal entries of $\mathcal{A}\mathbf{A}$ equal $|\mathbf{A}|$ by Exercise 27.

(29) Let $\mathbf{A}$ be a singular $n \times n$ matrix. Then $|\mathbf{A}| = 0$ by Theorem 3.5. By part (4) of Theorem 2.11, $\mathbf{A}^T$ is also singular (or else $\mathbf{A} = (\mathbf{A}^T)^T$ is nonsingular). Hence, $|\mathbf{A}^T| = 0$ by Theorem 3.5, and so $|\mathbf{A}| = |\mathbf{A}^T|$ in this case.

(30) Case 1: Assume $1 \le k < j$ and $1 \le i < m$. Then the (i,k) entry of $(\mathbf{A}_{jm})^T = (k,i)$ entry of $\mathbf{A}_{jm}$ $= (k,i)$ entry of $\mathbf{A} = (i,k)$ entry of $\mathbf{A}^T = (i,k)$ entry of $(\mathbf{A}^T)_{mj}$. Case 2: Assume $j \le k < n$ and $1 \le i < m$. Then the (i,k) entry of $(\mathbf{A}_{jm})^T = (k,i)$ entry of $\mathbf{A}_{jm} = (k+1,i)$ entry of $\mathbf{A} = (i,k+1)$ entry of $\mathbf{A}^T = (i,k)$ entry of $(\mathbf{A}^T)_{mj}$. Case 3: Assume $1 \le k < j$ and $m < i \le n$. Then the (i,k) entry of $(\mathbf{A}_{jm})^T = (k,i)$ entry of $\mathbf{A}_{jm} = (k,i+1)$ entry of $\mathbf{A} = (i+1,k)$ entry of $\mathbf{A}^T = (i,k)$ entry of $(\mathbf{A}^T)_{mj}$. Case 4: Assume $j \le k < n$ and $m < i \le n$. Then the (i,k) entry of $(\mathbf{A}_{jm})^T = (k,i)$ entry of $\mathbf{A}_{jm} = (k+1,i+1)$ entry of $\mathbf{A} = (i+1,k+1)$ entry of $\mathbf{A}^T = (i,k)$ entry of $(\mathbf{A}^T)_{mj}$. Hence, the corresponding entries of $(\mathbf{A}_{jm})^T$ and $(\mathbf{A}^T)_{mj}$ are all equal, proving that the matrices themselves are equal.

(31) Suppose $\mathbf{A}$ is a nonsingular $n \times n$ matrix. Then, by part (4) of Theorem 2.11, $\mathbf{A}^T$ is also nonsingular. We prove $|\mathbf{A}| = |\mathbf{A}^T|$ by induction on n.
Base Step: Assume $n = 1$. Then $\mathbf{A} = [a_{11}] = \mathbf{A}^T$, and so their determinants must be equal.
Inductive Step: Assume that $|\mathbf{B}| = |\mathbf{B}^T|$ for any $(n-1) \times (n-1)$ nonsingular matrix $\mathbf{B}$, and prove that $|\mathbf{A}| = |\mathbf{A}^T|$ for any $n \times n$ nonsingular matrix $\mathbf{A}$. Now, first note that, by Exercise 30, $|(\mathbf{A}^T)_{ni}| = |(\mathbf{A}_{in})^T|$. But, $|(\mathbf{A}_{in})^T| = |\mathbf{A}_{in}|$, either by the inductive hypothesis, if $\mathbf{A}_{in}$ is nonsingular, or by Exercise 29, if $\mathbf{A}_{in}$ is singular. Hence, $|(\mathbf{A}^T)_{ni}| = |\mathbf{A}_{in}|$. Let $\mathbf{D} = \mathbf{A}^T$. So, $|\mathbf{D}_{ni}| = |\mathbf{A}_{in}|$. Then, by Exercise 28, $|\mathbf{A}| = a_{1n}\mathcal{A}_{1n} + a_{2n}\mathcal{A}_{2n} + \cdots + a_{nn}\mathcal{A}_{nn} = d_{n1}(-1)^{1+n}|\mathbf{A}_{1n}| + d_{n2}(-1)^{2+n}|\mathbf{A}_{2n}| + \cdots + d_{nn}(-1)^{n+n}|\mathbf{A}_{nn}| = d_{n1}(-1)^{1+n}|\mathbf{D}_{n1}| + d_{n2}(-1)^{2+n}|\mathbf{D}_{n2}| + \cdots + d_{nn}(-1)^{n+n}|\mathbf{D}_{nn}| = |\mathbf{D}|$, since this is the cofactor expansion for $|\mathbf{D}|$ along the last row of $\mathbf{D}$. This completes the proof.

(32) Let $\mathbf{A}$ be an $n \times n$ singular matrix. Let $\mathbf{D} = \mathbf{A}^T$. $\mathbf{D}$ is also singular by part (4) of Theorem 2.11 (or else $\mathbf{A} = \mathbf{D}^T$ is nonsingular). Now, Exercise 30 shows that $|\mathbf{D}_{jk}| = |(\mathbf{A}_{kj})^T|$, which equals $|\mathbf{A}_{kj}|$, by Exercise 29 if $\mathbf{A}_{kj}$ is singular, or by Exercise 31 if $\mathbf{A}_{kj}$ is nonsingular. So, $|\mathbf{A}| = |\mathbf{D}|$ (by Exercise 29) $= d_{j1}\mathcal{D}_{j1} + d_{j2}\mathcal{D}_{j2} + \cdots + d_{jn}\mathcal{D}_{jn}$ (by Exercise 23) $= a_{1j}(-1)^{j+1}|\mathbf{D}_{j1}| + a_{2j}(-1)^{j+2}|\mathbf{D}_{j2}| + \cdots + a_{nj}(-1)^{j+n}|\mathbf{D}_{jn}| = a_{1j}(-1)^{1+j}|\mathbf{A}_{1j}| + a_{2j}(-1)^{2+j}|\mathbf{A}_{2j}| + \cdots + a_{nj}(-1)^{n+j}|\mathbf{A}_{nj}|$ $= a_{1j}\mathcal{A}_{1j} + a_{2j}\mathcal{A}_{2j} + \cdots + a_{nj}\mathcal{A}_{nj}$, and we are finished.

(33) If $\mathbf{A}$ has two identical columns, then $\mathbf{A}^T$ has two identical rows. Hence $|\mathbf{A}^T| = 0$ by Exercise 24. Thus, $|\mathbf{A}| = 0$, by Theorem 3.9 (which is proven in Exercises 29 and 31).

(34) Let $\mathbf{B}$ be the $n \times n$ matrix, all of whose entries are equal to the corresponding entries in $\mathbf{A}$, except that the jth column of $\mathbf{B}$ equals the ith column of $\mathbf{A}$ (as does the ith column of $\mathbf{B}$). Then $|\mathbf{B}| = 0$ by Exercise 33. Also, since every column of $\mathbf{A}$ equals the corresponding column of $\mathbf{B}$, with the exception of the jth column, the submatrices $\mathbf{A}_{kj}$ and $\mathbf{B}_{kj}$ are equal for $1 \le k \le n$. Hence, $\mathcal{A}_{kj} = \mathcal{B}_{kj}$ for $1 \le k \le n$. Now, computing the determinant of $\mathbf{B}$ using a cofactor expansion along the jth column (allowed by part (2) of Theorem 3.10, proven in Exercises 28 and 32) yields $0 = |\mathbf{B}| = b_{1j}\mathcal{B}_{1j} + b_{2j}\mathcal{B}_{2j} + \cdots + b_{nj}\mathcal{B}_{nj}$ $= a_{1i}\mathcal{B}_{1j} + a_{2i}\mathcal{B}_{2j} + \cdots + a_{ni}\mathcal{B}_{nj}$ (because the jth column of $\mathbf{B}$ equals the ith column of $\mathbf{A}$) $=$ $a_{1i}\mathcal{A}_{1j} + a_{2i}\mathcal{A}_{2j} + \cdots + a_{ni}\mathcal{A}_{nj}$, completing the proof.

(35) Using the fact that the general (k, m) entry of $\mathcal{A}$ is $\mathcal{A}_{mk}$, we see that the (i, j) entry of $\mathbf{A}\mathcal{A}$ equals $a_{1j}\mathcal{A}_{1i} + a_{2j}\mathcal{A}_{2i} + \cdots + a_{nj}\mathcal{A}_{ni}$. If $i = j$, Exercise 32 implies that this sum equals $|\mathbf{A}|$, while if $i \ne j$, Exercise 34 implies that the sum equals 0. Hence, $\mathbf{A}\mathcal{A}$ equals $|\mathbf{A}|$ on the main diagonal and 0 off the main diagonal, and so $\mathbf{A}\mathcal{A} = (|\mathbf{A}|)\mathbf{I}_n$. (Note that, since $\mathbf{A}$ is singular, $|\mathbf{A}| = 0$, and so, in fact, $\mathcal{A}\mathbf{A} = \mathbf{O}_n$.)

(36) (a) By Exercise 27, $\frac{1}{|\mathbf{A}|}\mathcal{A}\mathbf{A} = \mathbf{I}_n$. Hence, $\mathbf{AX} = \mathbf{B} \Rightarrow \frac{1}{|\mathbf{A}|}\mathcal{A}\mathbf{AX} = \frac{1}{|\mathbf{A}|}\mathcal{A}\mathbf{B} \Rightarrow \mathbf{I}_n\mathbf{X} = \frac{1}{|\mathbf{A}|}\mathcal{A}\mathbf{B} \Rightarrow$ $\mathbf{X} = \frac{1}{|\mathbf{A}|}\mathcal{A}\mathbf{B}$.

(b) Using part (a) and the fact that the general (k, m)th entry of $\mathcal{A}$ is $\mathcal{A}_{mk}$, we see that the kth entry of $\mathbf{X}$ equals the kth entry of $\frac{1}{|\mathbf{A}|}\mathcal{A}\mathbf{B} = \frac{1}{|\mathbf{A}|}(b_1\mathcal{A}_{1k} + \cdots + b_n\mathcal{A}_{nk})$.

(c) The matrix $\mathbf{A}_k$ is defined to equal $\mathbf{A}$ in every entry, except in the kth column, which equals the column vector $\mathbf{B}$. Thus, since this kth column is ignored when computing $(\mathbf{A}_k)_{ik}$, the (i, k) submatrix of $\mathbf{A}_k$, we see that $(\mathbf{A}_k)_{ik} = \mathbf{A}_{ik}$ for each $1 \le i \le n$. Hence, computing $|\mathbf{A}_k|$ by performing a cofactor expansion down the kth column of $\mathbf{A}_k$ (by part (2) of Theorem 3.10) yields $|\mathbf{A}_k| = b_1(-1)^{1+k}|(\mathbf{A}_k)_{1k}| + b_2(-1)^{2+k}|(\mathbf{A}_k)_{2k}| + \cdots + b_n(-1)^{n+k}|(\mathbf{A}_k)_{nk}| = b_1(-1)^{1+k}|\mathbf{A}_{1k}| + b_2(-1)^{2+k}|\mathbf{A}_{2k}| + \cdots + b_n(-1)^{n+k}|\mathbf{A}_{nk}| = b_1\mathcal{A}_{1k} + b_2\mathcal{A}_{2k} + \cdots + b_n\mathcal{A}_{nk}$.

(d) Theorem 3.13 claims that the kth entry of the solution $\mathbf{X}$ of $\mathbf{AX} = \mathbf{B}$ is $|\mathbf{A}_k|/|\mathbf{A}|$. Part (b) gives the expression $\frac{1}{|\mathbf{A}|}(b_1\mathcal{A}_{1k} + \cdots + b_n\mathcal{A}_{nk})$ for this kth entry, and part (c) replaces $(b_1\mathcal{A}_{1k} + \cdots + b_n\mathcal{A}_{nk})$ with $|\mathbf{A}_k|$.

Section 3.4

(1) In each part, let $\mathbf{A}$ represent the given matrix.

(a) $p_{\mathbf{A}}(x) = |x\mathbf{I}_2 - \mathbf{A}| = \begin{vmatrix} x-3 & -1 \\ 2 & x-4 \end{vmatrix} = (x-3)(x-4) - (2)(-1) = x^2 - 7x + 14$.

(c) $p_{\mathbf{A}}(x) = |x\mathbf{I}_3 - \mathbf{A}| = \begin{vmatrix} x-2 & -1 & 1 \\ 6 & x-6 & 0 \\ -3 & 0 & x \end{vmatrix}$. At this point, we could use basketweaving, but instead,

we will use a cofactor expansion on the last column. Hence,

$$p_{\mathbf{A}}(x) = (1)(-1)^{1+3}\begin{vmatrix} 6 & x-6 \\ -3 & 0 \end{vmatrix} + (0) + x(-1)^{3+3}\begin{vmatrix} x-2 & -1 \\ 6 & x-6 \end{vmatrix}$$
$$= (1)(0+3(x-6)) + x\left((x-2)(x-6)+6\right) = x^3 - 8x^2 + 21x - 18.$$

(e) $p_{\mathbf{A}}(x) = |x\mathbf{I}_4 - \mathbf{A}| = \begin{vmatrix} x & 1 & 0 & -1 \\ 5 & x-2 & 1 & -2 \\ 0 & -1 & x-1 & 0 \\ -4 & 1 & -3 & x \end{vmatrix}$. Following the hint, we do a cofactor expansion along the third row: $p_{\mathbf{A}}(x) = (0) + (-1)(-1)^{3+2}\begin{vmatrix} x & 0 & -1 \\ 5 & 1 & -2 \\ -4 & -3 & x \end{vmatrix}$ $+ (x-1)(-1)^{3+3}\begin{vmatrix} x & 1 & -1 \\ 5 & x-2 & -2 \\ -4 & 1 & x \end{vmatrix} + (0)$. Using basketweaving produces $p_{\mathbf{A}}(x) =$
$(1)\left((x)(1)(x) + (0)(-2)(-4) + (-1)(5)(-3) - (-1)(1)(-4) - (x)(-2)(-3) - (0)(5)(x)\right) +$
$(x-1)\left((x)(x-2)(x) + (1)(-2)(-4) + (-1)(5)(1) - (-1)(x-2)(-4) - (x)(-2)(1) - (1)(5)(x)\right)$,
which, after algebraic simplification, equals $x^4 - 3x^3 - 4x^2 + 12x$.

(2) In each part, let $\mathbf{A}$ represent the given matrix. Then, to find the eigenspace E_λ, we form the matrix $[(\lambda\mathbf{I}_n - \mathbf{A})|\mathbf{0}]$ and row reduce to solve the associated linear system.

(a) Since $\lambda = 2$, $[(\lambda\mathbf{I}_2 - \mathbf{A})|\mathbf{0}] = [(2\mathbf{I}_2 - \mathbf{A})|\mathbf{0}] = \left[\begin{array}{cc|c} 1 & -1 & 0 \\ 2 & -2 & 0 \end{array}\right]$. This row reduces to $\left[\begin{array}{cc|c} 1 & -1 & 0 \\ 0 & 0 & 0 \end{array}\right]$, which corresponds to the system $\begin{cases} x_1 - x_2 = 0 \\ 0 = 0 \end{cases}$. The independent variable is x_2. Let $x_2 = b$. The first equation then yields $x_1 = b$ as well. Hence, E_2, the solution set for this system, is $\{b[1,1] \mid b \in \mathbb{R}\}$.

(c) Since $\lambda = -1$, $[(\lambda\mathbf{I}_3 - \mathbf{A})|\mathbf{0}] = [((-1)\mathbf{I}_3 - \mathbf{A})|\mathbf{0}] = \left[\begin{array}{ccc|c} 4 & -2 & 0 & 0 \\ 8 & -4 & 0 & 0 \\ -4 & 2 & 0 & 0 \end{array}\right]$. This reduces to $\left[\begin{array}{ccc|c} 1 & -\frac{1}{2} & 0 & 0 \\ 0 & 0 & 0 & 0 \\ 0 & 0 & 0 & 0 \end{array}\right]$, which corresponds to the system $\begin{cases} x_1 - \frac{1}{2}x_2 = 0 \\ 0 = 0 \\ 0 = 0 \end{cases}$. The second and third columns are nonpivot columns, and so x_2 and x_3 are independent variables. Let $x_2 = b$ and $x_3 = c$. The first equation then yields $x_1 = \frac{1}{2}b$. We can eliminate fractions by letting $b = 2t$, giving $x_1 = t$ and $x_2 = 2t$. Hence, E_{-1}, the solution set for this system, is $\{[t, 2t, c] \mid c, t \in \mathbb{R}\} = \{t[1,2,0] + c[0,0,1] \mid c, t \in \mathbb{R}\}$.

(3) In each part, let $\mathbf{A}$ represent the given matrix. We will follow the first three steps of the Diagonalization Method, as described in Section 3.4 of the textbook.

(a) $p_{\mathbf{A}}(x) = |x\mathbf{I}_2 - \mathbf{A}| = \begin{vmatrix} x-1 & -3 \\ 0 & x-1 \end{vmatrix} = (x-1)^2$. The eigenvalues of $\mathbf{A}$ are the roots of $p_{\mathbf{A}}(x)$. Clearly, $\lambda = 1$ is the only root of $(x-1)^2$. Since the linear factor $(x-1)$ appears to the 2nd power in $p_{\mathbf{A}}(x)$, this eigenvalue has algebraic multiplicity 2. Next, we solve for the eigenspace E_1 by solving the homogeneous system whose augmented matrix is $[(1\mathbf{I}_2 - \mathbf{A})|\mathbf{0}]$. Now $[(1\mathbf{I}_2 - \mathbf{A})|\mathbf{0}] =$ $\left[\begin{array}{cc|c} 0 & -3 & 0 \\ 0 & 0 & 0 \end{array}\right]$, which reduces to $\left[\begin{array}{cc|c} 0 & 1 & 0 \\ 0 & 0 & 0 \end{array}\right]$. The associated linear system is $\left\{\begin{array}{rcl} x_2 & = & 0 \\ 0 & = & 0 \end{array}\right.$. Since column 1 is not a pivot column, x_1 is an independent variable. Let $x_1 = a$. Then the solution set is $E_1 = \{[a, 0] \mid a \in \mathbb{R}\} = \{a[1, 0] \mid a \in \mathbb{R}\}$.

(c) $p_{\mathbf{A}}(x) = |x\mathbf{I}_3 - \mathbf{A}| = \left|\begin{array}{ccc} x-1 & 0 & -1 \\ 0 & x-2 & 3 \\ 0 & 0 & x+5 \end{array}\right| = (x-1)(x-2)(x+5)$, where we have used the fact that $|x\mathbf{I}_3 - \mathbf{A}|$ is upper triangular to calculate the determinant quickly, using Theorem 3.2. The eigenvalues of $\mathbf{A}$ are the roots of $p_{\mathbf{A}}(x)$. These are clearly $\lambda_1 = 1$, $\lambda_2 = 2$, and $\lambda_3 = -5$. Since each of the linear factors $(x-1)$, $(x-2)$, and $(x+5)$ appears in $p_{\mathbf{A}}(x)$ raised to the 1st power, each of these eigenvalues has algebraic multiplicity 1. Next, we solve for each eigenspace.

For $\lambda_1 = 1$, we solve the homogeneous system whose augmented matrix is $[(1\mathbf{I}_3 - \mathbf{A})|\mathbf{0}]$. Now $[(1\mathbf{I}_3 - \mathbf{A})|\mathbf{0}] = \left[\begin{array}{ccc|c} 0 & 0 & -1 & 0 \\ 0 & -1 & 3 & 0 \\ 0 & 0 & 6 & 0 \end{array}\right]$, which reduces to $\left[\begin{array}{ccc|c} 0 & 1 & 0 & 0 \\ 0 & 0 & 1 & 0 \\ 0 & 0 & 0 & 0 \end{array}\right]$. The associated linear system is $\left\{\begin{array}{rcl} x_2 \quad & = & 0 \\ x_3 & = & 0 \\ 0 & = & 0 \end{array}\right.$. Since column 1 is not a pivot column, x_1 is an independent variable. Let $x_1 = a$. Then the solution set is $E_1 = \{[a, 0, 0] \mid a \in \mathbb{R}\} = \{a[1, 0, 0] \mid a \in \mathbb{R}\}$.

For $\lambda_2 = 2$, we solve the homogeneous system whose augmented matrix is $[(2\mathbf{I}_3 - \mathbf{A})|\mathbf{0}]$. Now $[(2\mathbf{I}_3 - \mathbf{A})|\mathbf{0}] = \left[\begin{array}{ccc|c} 1 & 0 & -1 & 0 \\ 0 & 0 & 3 & 0 \\ 0 & 0 & 7 & 0 \end{array}\right]$, which reduces to $\left[\begin{array}{ccc|c} 1 & 0 & 0 & 0 \\ 0 & 0 & 1 & 0 \\ 0 & 0 & 0 & 0 \end{array}\right]$. The associated linear system is $\left\{\begin{array}{rcl} x_1 \quad & = & 0 \\ x_3 & = & 0 \\ 0 & = & 0 \end{array}\right.$. Since column 2 is not a pivot column, x_2 is an independent variable. Let $x_2 = b$. Then the solution set is $E_2 = \{[0, b, 0] \mid b \in \mathbb{R}\} = \{b[0, 1, 0] \mid a \in \mathbb{R}\}$.

For $\lambda_3 = -5$, we solve the homogeneous system whose augmented matrix is $[((-5)\mathbf{I}_3 - \mathbf{A})|\mathbf{0}]$. Now $[((-5)\mathbf{I}_3 - \mathbf{A})|\mathbf{0}] = \left[\begin{array}{ccc|c} -6 & 0 & -1 & 0 \\ 0 & -7 & 3 & 0 \\ 0 & 0 & 0 & 0 \end{array}\right]$, which reduces to $\left[\begin{array}{ccc|c} 1 & 0 & \frac{1}{6} & 0 \\ 0 & 1 & -\frac{3}{7} & 0 \\ 0 & 0 & 0 & 0 \end{array}\right]$. The associated linear system is $\left\{\begin{array}{rcl} x_1 \qquad + \frac{1}{6}x_3 & = & 0 \\ x_2 - \frac{3}{7}x_3 & = & 0 \\ 0 & = & 0 \end{array}\right.$. Since column 3 is not a pivot column, x_3 is an independent variable. Let $x_3 = c$. Then $x_1 = -\frac{1}{6}x_3 = -\frac{1}{6}c$ and $x_2 = \frac{3}{7}x_3 = \frac{3}{7}c$. Hence, the solution set is $E_{-5} = \left\{\left[-\frac{1}{6}c, \frac{3}{7}c, c\right] \mid c \in \mathbb{R}\right\} = \left\{c\left[-\frac{1}{6}, \frac{3}{7}, 1\right] \mid c \in \mathbb{R}\right\}$.

(e) $p_{\mathbf{A}}(x) = |x\mathbf{I}_3 - \mathbf{A}| = \begin{vmatrix} x-4 & 0 & 2 \\ -6 & x-2 & 6 \\ -4 & 0 & x+2 \end{vmatrix}$. Cofactor expansion along the second column yields $p_{\mathbf{A}}(x) = (0) + (x-2)(-1)^{2+2}\begin{vmatrix} x-4 & 2 \\ -4 & x+2 \end{vmatrix} + (0) = (x-2)\Big((x-4)(x+2) - (-4)(2)\Big) = (x-2)(x^2-2x) = x(x-2)^2$. The eigenvalues of $\mathbf{A}$ are the roots of $p_{\mathbf{A}}(x)$. These are clearly $\lambda_1 = 0$, and $\lambda_2 = 2$. Since the linear factor $(x-0) = x$ appears in $p_{\mathbf{A}}(x)$ raised to the 1st power, $\lambda_1 = 0$ has algebraic multiplicity 1. Similarly, since the linear factor $(x-2)$ appears to the 2nd power in $p_{\mathbf{A}}(x)$, $\lambda_2 = 2$ has algebraic multiplicity 2. Next, we solve for each eigenspace.

For $\lambda_1 = 0$, we solve the homogeneous system whose augmented matrix is $[(0\mathbf{I}_3 - \mathbf{A})|\mathbf{0}] = [-\mathbf{A}|\mathbf{0}]$. Now $[-\mathbf{A}|\mathbf{0}] = \left[\begin{array}{ccc|c} -4 & 0 & 2 & 0 \\ -6 & -2 & 6 & 0 \\ -4 & 0 & 2 & 0 \end{array}\right]$, which reduces to $\left[\begin{array}{ccc|c} 1 & 0 & -\frac{1}{2} & 0 \\ 0 & 1 & -\frac{3}{2} & 0 \\ 0 & 0 & 0 & 0 \end{array}\right]$. The associated linear system is $\begin{cases} x_1 \quad - \frac{1}{2}x_3 = 0 \\ \quad x_2 - \frac{3}{2}x_3 = 0 \\ \quad\quad 0 = 0 \end{cases}$. Since column 3 is not a pivot column, x_3 is an independent variable. Let $x_3 = 2c$, where the coefficient "2" is used to eliminate fractions. Then $x_1 = \frac{1}{2}x_3 = c$, and $x_2 = \frac{3}{2}x_3 = 3c$. Hence, the solution set is $E_0 = \{[c, 3c, 2c] \mid c \in \mathbb{R}\} = \{c[1, 3, 2] \mid c \in \mathbb{R}\}$.

For $\lambda_2 = 2$, we solve the homogeneous system whose augmented matrix is $[(2\mathbf{I}_3 - \mathbf{A})|\mathbf{0}]$. Now $[(2\mathbf{I}_3 - \mathbf{A})|\mathbf{0}] = \left[\begin{array}{ccc|c} -2 & 0 & 2 & 0 \\ -6 & 0 & 6 & 0 \\ -4 & 0 & 4 & 0 \end{array}\right]$, which reduces to $\left[\begin{array}{ccc|c} 1 & 0 & -1 & 0 \\ 0 & 0 & 0 & 0 \\ 0 & 0 & 0 & 0 \end{array}\right]$. The associated linear system is $\begin{cases} x_1 - x_3 = 0 \\ \quad 0 = 0 \\ \quad 0 = 0 \end{cases}$. Since columns 2 and 3 are not pivot columns, x_2 and x_3 are independent variables. Let $x_2 = b$ and $x_3 = c$. Then $x_1 = x_3 = c$, and the solution set is $E_2 = \{[c, b, c] \mid b, c \in \mathbb{R}\} = \{b[0, 1, 0] + c[1, 0, 1] \mid b, c \in \mathbb{R}\}$.

(h) $p_{\mathbf{A}}(x) = |x\mathbf{I}_4 - \mathbf{A}| = \begin{vmatrix} x-3 & 1 & -4 & 1 \\ 0 & x-3 & 3 & -3 \\ 6 & -2 & x+8 & -2 \\ 6 & 4 & 2 & x+4 \end{vmatrix}$. We use a cofactor expansion along the first column. This gives $p_{\mathbf{A}}(x) = (x-3)(-1)^{1+1}\begin{vmatrix} x-3 & 3 & -3 \\ -2 & x+8 & -2 \\ 4 & 2 & x+4 \end{vmatrix} + (0)$

$+(6)(-1)^{3+1}\begin{vmatrix} 1 & -4 & 1 \\ x-3 & 3 & -3 \\ 4 & 2 & x+4 \end{vmatrix} + (6)(-1)^{4+1}\begin{vmatrix} 1 & -4 & 1 \\ x-3 & 3 & -3 \\ -2 & x+8 & -2 \end{vmatrix}$. Using basketweaving to find each of the 3×3 determinants, we get $p_{\mathbf{A}}(x) = (x-3)\Big((x-3)(x+8)(x+4) + (3)(-2)(4)$
$+(-3)(-2)(2) - (-3)(x+8)(4) - (x-3)(-2)(2) - (3)(-2)(x+4)\Big) + 6\Big((1)(3)(x+4)$

$+(-4)(-3)(4) + (1)(x-3)(2) - (1)(3)(4) - (1)(-3)(2) - (-4)(x-3)(x+4)\Big) - 6\Big((1)(3)(-2)$
$+(-4)(-3)(-2) + (1)(x-3)(x+8) - (1)(3)(-2) - (1)(-3)(x+8) - (-4)(x-3)(-2)\Big)$
$= (x-3)(x^3+9x^2+18x) + 6(4x^2+9x) - 6(x^2) = x^4 + 6x^3 + 9x^2$
$= x^2(x^2+6x+9) = x^2(x+3)^2$. The eigenvalues of $\mathbf{A}$ are the roots of $p_{\mathbf{A}}(x)$. Hence, there are two eigenvalues, $\lambda_1 = 0$ and $\lambda_2 = -3$. Since each of the linear factors $(x-0) = x$ and $(x+3)$ appears in $p_{\mathbf{A}}(x)$ raised to the 2nd power, each of these eigenvalues has algebraic multiplicity 2. Next, we solve for each eigenspace.

For $\lambda_1 = 0$, we solve the homogeneous system whose augmented matrix is $[(0\mathbf{I}_4 - \mathbf{A})|\mathbf{0}] = [-\mathbf{A}|\mathbf{0}]$. Now $[-\mathbf{A}|\mathbf{0}] = \left[\begin{array}{cccc|c} -3 & 1 & -4 & 1 & 0 \\ 0 & -3 & 3 & -3 & 0 \\ 6 & -2 & 8 & -2 & 0 \\ 6 & 4 & 2 & 4 & 0 \end{array}\right]$, reducing to $\left[\begin{array}{cccc|c} 1 & 0 & 1 & 0 & 0 \\ 0 & 1 & -1 & 1 & 0 \\ 0 & 0 & 0 & 0 & 0 \\ 0 & 0 & 0 & 0 & 0 \end{array}\right]$. The associated linear system is $\begin{cases} x_1 \quad\quad + x_3 \quad\quad = 0 \\ \quad\quad x_2 - x_3 + x_4 = 0 \\ \quad\quad\quad\quad\quad\quad 0 = 0 \\ \quad\quad\quad\quad\quad\quad 0 = 0 \end{cases}$. Since columns 3 and 4 are not pivot columns, x_3 and x_4 are independent variables. Let $x_3 = c$ and $x_4 = d$. Thus, $x_1 = -x_3 = -c$ and $x_2 = x_3 - x_4 = c - d$. Hence, the solution set is $E_0 = \{[-c, c-d, c, d] \mid c, d \in \mathbb{R}\} = \{c[-1,1,1,0] + d[0,-1,0,1] \mid c, d \in \mathbb{R}\}$.

For $\lambda_2 = -3$, we solve the homogeneous system whose augmented matrix is $[((-3)\mathbf{I}_4 - \mathbf{A})|\mathbf{0}]$. Now $[((-3)\mathbf{I}_4 - \mathbf{A})|\mathbf{0}] = \left[\begin{array}{cccc|c} -6 & 1 & -4 & 1 & 0 \\ 0 & -6 & 3 & -3 & 0 \\ 6 & -2 & 5 & -2 & 0 \\ 6 & 4 & 2 & 1 & 0 \end{array}\right]$, reducing to $\left[\begin{array}{cccc|c} 1 & 0 & 0 & \frac{1}{2} & 0 \\ 0 & 1 & 0 & 0 & 0 \\ 0 & 0 & 1 & -1 & 0 \\ 0 & 0 & 0 & 0 & 0 \end{array}\right]$. The associated linear system is $\begin{cases} x_1 \quad\quad\quad + \frac{1}{2}x_4 = 0 \\ \quad x_2 \quad\quad\quad\quad = 0 \\ \quad\quad x_3 - \quad x_4 = 0 \\ \quad\quad\quad\quad\quad 0 = 0 \end{cases}$. Since column 4 is not a pivot column, x_4 is an independent variable. Let $x_4 = 2d$, where we have included the factor "2" to eliminate fractions. Thus, $x_1 = -\frac{1}{2}x_4 = -d$, $x_2 = 0$, and $x_3 = x_4 = 2d$. Hence, the solution set is $E_{-3} = \{[-d, 0, 2d, 2d] \mid d \in \mathbb{R}\} = \{d[-1,0,2,2] \mid d \in \mathbb{R}\}$.

(4) (a) Step 1: $p_{\mathbf{A}}(x) = |x\mathbf{I}_2 - \mathbf{A}| = \begin{vmatrix} x-19 & 48 \\ -8 & x+21 \end{vmatrix} = (x-19)(x+21) - (48)(-8) = x^2 + 2x - 15 = (x-3)(x+5)$.

Step 2: The eigenvalues of $\mathbf{A}$ are the roots of $p_{\mathbf{A}}(x)$. Hence, there are two eigenvalues, $\lambda_1 = 3$ and $\lambda_2 = -5$.

Step 3: Now we solve for eigenvectors.

For $\lambda_1 = 3$, we solve the homogeneous system whose augmented matrix is $[(3\mathbf{I}_2 - \mathbf{A})|\mathbf{0}]$. Now $[(3\mathbf{I}_2 - \mathbf{A})|\mathbf{0}] = \left[\begin{array}{cc|c} -16 & 48 & 0 \\ -8 & 24 & 0 \end{array}\right]$, which reduces to $\left[\begin{array}{cc|c} 1 & -3 & 0 \\ 0 & 0 & 0 \end{array}\right]$. The associated linear system is $\left\{ \begin{array}{rcrcl} x_1 & - & 3x_2 & = & 0 \\ & & 0 & = & 0 \end{array} \right.$. Since column 2 is not a pivot column, x_2 is an independent variable. Let $x_2 = 1$. Then $x_1 = 3x_2 = 3$. This yields the eigenvector $[3, 1]$.

For $\lambda_2 = -5$, we solve the homogeneous system whose augmented matrix is $[((-5)\mathbf{I}_2 - \mathbf{A})|\mathbf{0}]$. Now $[((-5)\mathbf{I}_2 - \mathbf{A})|\mathbf{0}] = \left[\begin{array}{rr|r} -24 & 48 & 0 \\ -8 & 16 & 0 \end{array} \right]$, which reduces to $\left[\begin{array}{rr|r} 1 & -2 & 0 \\ 0 & 0 & 0 \end{array} \right]$. The associated linear system is $\left\{ \begin{array}{rcrcl} x_1 & - & 2x_2 & = & 0 \\ & & 0 & = & 0 \end{array} \right.$. Since column 2 is not a pivot column, x_2 is an independent variable. Let $x_2 = 1$. Then $x_1 = 2x_2 = 2$. This yields the eigenvector $[2, 1]$.

Step 4: Since $n = 2$, and since we have found 2 eigenvectors, $\mathbf{A}$ is diagonalizable.

Step 5: We form the 2×2 matrix whose columns are the two vectors we have found: $\mathbf{P} = \left[\begin{array}{cc} 3 & 2 \\ 1 & 1 \end{array} \right]$.

Step 6: The matrix whose diagonal entries are the eigenvalues of $\mathbf{A}$ (in the correct order) is $\mathbf{D} = \left[\begin{array}{rr} 3 & 0 \\ 0 & -5 \end{array} \right]$. Also, using Theorem 2.13, we get $\mathbf{P}^{-1} = \left[\begin{array}{rr} 1 & -2 \\ -1 & 3 \end{array} \right]$. It is easy to verify for these matrices that $\mathbf{D} = \mathbf{P}^{-1}\mathbf{AP}$.

(c) Step 1: $p_{\mathbf{A}}(x) = |x\mathbf{I}_2 - \mathbf{A}| = \left| \begin{array}{cc} x - 13 & 34 \\ -5 & x + 13 \end{array} \right| = (x - 13)(x + 13) - (34)(-5) = x^2 + 1$.

Step 2: The eigenvalues of $\mathbf{A}$ are the roots of $p_{\mathbf{A}}(x)$. But $p_{\mathbf{A}}(x) = x^2 + 1$ has no real roots. Therefore, $\mathbf{A}$ has no eigenvalues.

Step 3: Because $\mathbf{A}$ has no eigenvalues, there are no associated eigenvectors.

Step 4: Since there are no eigenvectors for $\mathbf{A}$, $\mathbf{A}$ is not diagonalizable.

(d) Step 1: $p_{\mathbf{A}}(x) = |x\mathbf{I}_3 - \mathbf{A}| = \left| \begin{array}{ccc} x + 13 & 3 & -18 \\ 20 & x + 4 & -26 \\ 14 & 3 & x - 19 \end{array} \right|$. Using basketweaving gives $p_{\mathbf{A}}(x) = (x + 13)(x + 4)(x - 19) + (3)(-26)(14) + (-18)(20)(3) - (-18)(x + 4)(14) - (x + 13)(-26)(3) - (3)(20)(x - 19) = x^3 - 2x^2 - x + 2 = (x^2 - 1)(x - 2) = (x - 1)(x + 1)(x - 2)$.

Step 2: The eigenvalues of $\mathbf{A}$ are the roots of $p_{\mathbf{A}}(x)$. Hence, there are three eigenvalues, $\lambda_1 = 1$, $\lambda_2 = -1$, and $\lambda_3 = 2$.

Step 3: Now we solve for eigenvectors.

For $\lambda_1 = 1$, we solve the homogeneous system whose augmented matrix is $[(1\mathbf{I}_3 - \mathbf{A})|\mathbf{0}]$. Now $[(1\mathbf{I}_3 - \mathbf{A})|\mathbf{0}] = \left[\begin{array}{rrr|r} 14 & 3 & -18 & 0 \\ 20 & 5 & -26 & 0 \\ 14 & 3 & -18 & 0 \end{array} \right]$, which reduces to $\left[\begin{array}{rrr|r} 1 & 0 & -\frac{6}{5} & 0 \\ 0 & 1 & -\frac{2}{5} & 0 \\ 0 & 0 & 0 & 0 \end{array} \right]$. The associated linear system is $\left\{ \begin{array}{rcrcrcl} x_1 & & & - & \frac{6}{5}x_3 & = & 0 \\ & & x_2 & - & \frac{2}{5}x_3 & = & 0 \\ & & & & 0 & = & 0 \end{array} \right.$. Since column 3 is not a pivot column, x_3 is an independent variable. We choose $x_3 = 5$ to eliminate fractions. Then $x_1 = \frac{6}{5}x_3 = 6$ and $x_2 = \frac{2}{5}x_3 = 2$. This yields the eigenvector $[6, 2, 5]$.

For $\lambda_2 = -1$, we solve the homogeneous system whose augmented matrix is $[((-1)\mathbf{I}_3 - \mathbf{A})|\mathbf{0}]$. Now

$[((-1)\mathbf{I}_3 - \mathbf{A})|\mathbf{0}] = \left[\begin{array}{ccc|c} 12 & 3 & -18 & 0 \\ 20 & 3 & -26 & 0 \\ 14 & 3 & -20 & 0 \end{array}\right]$, which reduces to $\left[\begin{array}{ccc|c} 1 & 0 & -1 & 0 \\ 0 & 1 & -2 & 0 \\ 0 & 0 & 0 & 0 \end{array}\right]$. The associated linear system is $\left\{\begin{array}{rcrcl} x_1 & & - & x_3 & = 0 \\ & x_2 & - & 2x_3 & = 0 \\ & & & 0 & = 0 \end{array}\right.$. Since column 3 is not a pivot column, x_3 is an independent variable. Let $x_3 = 1$. Then $x_1 = x_3 = 1$ and $x_2 = 2x_3 = 2$. This yields the eigenvector $[1, 2, 1]$.

For $\lambda_3 = 2$, we solve the homogeneous system whose augmented matrix is $[(2\mathbf{I}_3 - \mathbf{A})|\mathbf{0}]$. Now $[(2\mathbf{I}_3 - \mathbf{A})|\mathbf{0}] = \left[\begin{array}{ccc|c} 15 & 3 & -18 & 0 \\ 20 & 6 & -26 & 0 \\ 14 & 3 & -17 & 0 \end{array}\right]$, which reduces to $\left[\begin{array}{ccc|c} 1 & 0 & -1 & 0 \\ 0 & 1 & -1 & 0 \\ 0 & 0 & 0 & 0 \end{array}\right]$. The associated linear system is $\left\{\begin{array}{rcrcl} x_1 & & - & x_3 & = 0 \\ & x_2 & - & x_3 & = 0 \\ & & & 0 & = 0 \end{array}\right.$. Since column 3 is not a pivot column, x_3 is an independent variable. Let $x_3 = 1$. Then $x_1 = x_3 = 1$ and $x_2 = x_3 = 1$. This yields the eigenvector $[1, 1, 1]$.

Step 4: Since $n = 3$, and we have found 3 eigenvectors, $\mathbf{A}$ is diagonalizable.

Step 5: We form the 3×3 matrix whose columns are the 3 vectors we have found: $\mathbf{P} = \begin{bmatrix} 6 & 1 & 1 \\ 2 & 2 & 1 \\ 5 & 1 & 1 \end{bmatrix}$.

Step 6: The matrix whose diagonal entries are the eigenvalues of $\mathbf{A}$ (in the correct order) is $\mathbf{D} = \begin{bmatrix} 1 & 0 & 0 \\ 0 & -1 & 0 \\ 0 & 0 & 2 \end{bmatrix}$. Also, by row reducing $[\mathbf{P}|\mathbf{I}_3]$, we get $\mathbf{P}^{-1} = \begin{bmatrix} 1 & 0 & -1 \\ 3 & 1 & -4 \\ -8 & -1 & 10 \end{bmatrix}$. It is easy to verify for these matrices that $\mathbf{D} = \mathbf{P}^{-1}\mathbf{A}\mathbf{P}$.

(f) Step 1: $p_{\mathbf{A}}(x) = |x\mathbf{I}_3 - \mathbf{A}| = \begin{vmatrix} x-5 & 8 & 12 \\ 2 & x-3 & -4 \\ -4 & 6 & x+9 \end{vmatrix}$. Using basketweaving gives us $p_{\mathbf{A}}(x) = (x-5)(x-3)(x+9) + (8)(-4)(-4) + (12)(2)(6) - (12)(x-3)(-4) - (x-5)(-4)(6) - (8)(2)(x+9)$
$= x^3 + x^2 - x - 1 = (x^2 - 1)(x+1) = (x-1)(x+1)^2$.

Step 2: The eigenvalues of $\mathbf{A}$ are the roots of $p_{\mathbf{A}}(x)$. Hence, there are two eigenvalues, $\lambda_1 = 1$, and $\lambda_2 = -1$.

Step 3: Now we solve for eigenvectors.

For $\lambda_1 = 1$, we solve the homogeneous system whose augmented matrix is $[(1\mathbf{I}_3 - \mathbf{A})|\mathbf{0}]$. Now $[(1\mathbf{I}_3 - \mathbf{A})|\mathbf{0}] = \left[\begin{array}{ccc|c} -4 & 8 & 12 & 0 \\ 2 & -2 & -4 & 0 \\ -4 & 6 & 10 & 0 \end{array}\right]$, which reduces to $\left[\begin{array}{ccc|c} 1 & 0 & -1 & 0 \\ 0 & 1 & 1 & 0 \\ 0 & 0 & 0 & 0 \end{array}\right]$. The associated linear system is $\left\{\begin{array}{rcrcl} x_1 & & - & x_3 & = 0 \\ & x_2 & + & x_3 & = 0 \\ & & & 0 & = 0 \end{array}\right.$. Since column 3 is not a pivot column, x_3 is an independent variable. Let $x_3 = 1$. Then $x_1 = x_3 = 1$ and $x_2 = -x_3 = -1$. This yields the eigenvector $[1, -1, 1]$.

For $\lambda_2 = -1$, we solve the homogeneous system whose augmented matrix is $[((-1)\mathbf{I}_3 - \mathbf{A})|\mathbf{0}]$. Now $[((-1)\mathbf{I}_3 - \mathbf{A})|\mathbf{0}] = \left[\begin{array}{ccc|c} -6 & 8 & 12 & 0 \\ 2 & -4 & -4 & 0 \\ -4 & 6 & 8 & 0 \end{array}\right]$, reducing to $\left[\begin{array}{ccc|c} 1 & 0 & -2 & 0 \\ 0 & 1 & 0 & 0 \\ 0 & 0 & 0 & 0 \end{array}\right]$. The associated linear system is $\begin{cases} x_1 \quad - 2x_3 = 0 \\ x_2 = 0 \\ 0 = 0 \end{cases}$. Since column 3 is not a pivot column, x_3 is an independent variable. Let $x_3 = 1$. Then $x_1 = 2x_3 = 2$ and $x_2 = 0$. This yields the eigenvector $[2, 0, 1]$.

Step 4: Since $n = 3$, and the process has produced only 2 eigenvectors, $\mathbf{A}$ is not diagonalizable.

(g) Step 1: $p_{\mathbf{A}}(x) = |x\mathbf{I}_3 - \mathbf{A}| = \begin{vmatrix} x-2 & 0 & 0 \\ 3 & x-4 & -1 \\ -3 & 2 & x-1 \end{vmatrix}$. Using a cofactor expansion along the first row gives us $p_{\mathbf{A}}(x) = (x-2)(-1)^{1+1}\begin{vmatrix} x-4 & -1 \\ 2 & x-1 \end{vmatrix} = (x-2)\Big((x-4)(x-1) - (-1)(2)\Big) = (x-2)(x^2 - 5x + 6) = (x-2)^2(x-3)$.

Step 2: The eigenvalues of $\mathbf{A}$ are the roots of $p_{\mathbf{A}}(x)$. Hence, there are two eigenvalues, $\lambda_1 = 2$, and $\lambda_2 = 3$.

Step 3: Now we solve for eigenvectors.

For $\lambda_1 = 2$, we solve the homogeneous system whose augmented matrix is $[(2\mathbf{I}_3 - \mathbf{A})|\mathbf{0}]$. Now $[(2\mathbf{I}_3 - \mathbf{A})|\mathbf{0}] = \left[\begin{array}{ccc|c} 0 & 0 & 0 & 0 \\ 3 & -2 & -1 & 0 \\ -3 & 2 & 1 & 0 \end{array}\right]$, which reduces to $\left[\begin{array}{ccc|c} 1 & -\frac{2}{3} & -\frac{1}{3} & 0 \\ 0 & 0 & 0 & 0 \\ 0 & 0 & 0 & 0 \end{array}\right]$. The associated linear system is $\begin{cases} x_1 - \frac{2}{3}x_2 - \frac{1}{3}x_3 = 0 \\ 0 = 0 \\ 0 = 0 \end{cases}$. Since columns 2 and 3 are not pivot columns, x_2 and x_3 are independent variables. First, we choose $x_2 = 3$ (to eliminate fractions) and $x_3 = 0$. Then $x_1 = \frac{2}{3}x_2 + \frac{1}{3}x_3 = 2$, producing the eigenvector $[2, 3, 0]$. Second, we choose $x_2 = 0$ and $x_3 = 3$ (to eliminate fractions). This gives $x_1 = \frac{2}{3}x_2 + \frac{1}{3}x_3 = 1$, yielding the eigenvector $[1, 0, 3]$.

For $\lambda_2 = 3$, we solve the homogeneous system whose augmented matrix is $[(3\mathbf{I}_3 - \mathbf{A})|\mathbf{0}]$. Now $[(3\mathbf{I}_3 - \mathbf{A})|\mathbf{0}] = \left[\begin{array}{ccc|c} 1 & 0 & 0 & 0 \\ 3 & -1 & -1 & 0 \\ -3 & 2 & 2 & 0 \end{array}\right]$, which reduces to $\left[\begin{array}{ccc|c} 1 & 0 & 0 & 0 \\ 0 & 1 & 1 & 0 \\ 0 & 0 & 0 & 0 \end{array}\right]$. The associated linear system is $\begin{cases} x_1 = 0 \\ x_2 + x_3 = 0 \\ 0 = 0 \end{cases}$. Since column 3 is not a pivot column, x_3 is an independent variable. Let $x_3 = 1$. Then $x_1 = 0$ and $x_2 = -x_3 = -1$. This yields the eigenvector $[0, -1, 1]$.

Step 4: Since $n = 3$, and we have found 3 eigenvectors, $\mathbf{A}$ is diagonalizable.

Step 5: We form the 3×3 matrix whose columns are the 3 vectors we have found:

$$\mathbf{P} = \begin{bmatrix} 2 & 1 & 0 \\ 3 & 0 & -1 \\ 0 & 3 & 1 \end{bmatrix}.$$

Step 6: The matrix whose diagonal entries are the eigenvalues of $\mathbf{A}$ (in the correct order) is $\mathbf{D} = \begin{bmatrix} 2 & 0 & 0 \\ 0 & 2 & 0 \\ 0 & 0 & 3 \end{bmatrix}$. Also, by row reducing $[\mathbf{P}|\mathbf{I}_3]$, we get $\mathbf{P}^{-1} = \begin{bmatrix} 1 & -\frac{1}{3} & -\frac{1}{3} \\ -1 & \frac{2}{3} & \frac{2}{3} \\ 3 & -2 & -1 \end{bmatrix}$. It is easy to verify for these matrices that $\mathbf{D} = \mathbf{P}^{-1}\mathbf{AP}$.

(i) Step 1: $p_{\mathbf{A}}(x) = |x\mathbf{I}_4 - \mathbf{A}| = \begin{vmatrix} x-3 & -1 & 6 & 2 \\ -4 & x & 6 & 4 \\ -2 & 0 & x+3 & 2 \\ 0 & -1 & 2 & x-1 \end{vmatrix}$. Using a cofactor expansion along the last row gives us $p_{\mathbf{A}}(x) = (0) + (-1)(-1)^{4+2}\begin{vmatrix} x-3 & 6 & 2 \\ -4 & 6 & 4 \\ -2 & x+3 & 2 \end{vmatrix} + (2)(-1)^{4+3}\begin{vmatrix} x-3 & -1 & 2 \\ -4 & x & 4 \\ -2 & 0 & 2 \end{vmatrix}$

$+(x-1)(-1)^{4+4}\begin{vmatrix} x-3 & -1 & 6 \\ -4 & x & 6 \\ -2 & 0 & x+3 \end{vmatrix}$. Using basketweaving on each 3×3 determinant produces $p_{\mathbf{A}}(x) = -\Big((x-3)(6)(2) + (6)(4)(-2) + (2)(-4)(x+3) - (2)(6)(-2) - (x-3)(4)(x+3)$
$-(6)(-4)(2)\Big) -2\Big((x-3)(x)(2) + (-1)(4)(-2) + (2)(-4)(0) - (2)(x(-2) - (x-3)(4)(0)$
$-(-1)(-4)(2)\Big) +(x-1)\Big((x-3)(x)(x+3) + (-1)(6)(-2) + (6)(-4)(0) - (6)(x)(-2)$
$-(x-3)(6)(0) - (-1)(-4)(x+3)\Big) = -(-4x^2 + 4x) - 2(2x^2 - 2x) + (x-1)(x^3 - x)$
$= 4x(x-1) - 4x(x-1) + (x-1)x(x^2-1) = (x-1)^2(x+1)x.$

Step 2: The eigenvalues of $\mathbf{A}$ are the roots of $p_{\mathbf{A}}(x)$. Hence, there are three eigenvalues, $\lambda_1 = 1$, $\lambda_2 = -1$ and $\lambda_3 = 0$.

Step 3: Now we solve for eigenvectors.

For $\lambda_1 = 1$, we solve the homogeneous system whose augmented matrix is $[(1\mathbf{I}_3 - \mathbf{A})|\mathbf{0}]$. Now $[(1\mathbf{I}_3 - \mathbf{A})|\mathbf{0}] = \left[\begin{array}{cccc|c} -2 & -1 & 6 & 2 & 0 \\ -4 & 1 & 6 & 4 & 0 \\ -2 & 0 & 4 & 2 & 0 \\ 0 & -1 & 2 & 0 & 0 \end{array}\right]$, reducing to $\left[\begin{array}{cccc|c} 1 & 0 & -2 & -1 & 0 \\ 0 & 1 & -2 & 0 & 0 \\ 0 & 0 & 0 & 0 & 0 \\ 0 & 0 & 0 & 0 & 0 \end{array}\right]$. The associated linear system is $\begin{cases} x_1 \quad - 2x_3 - x_4 = 0 \\ \quad x_2 - 2x_3 = 0 \\ \quad\quad 0 = 0 \\ \quad\quad 0 = 0 \end{cases}$. Since columns 3 and 4 are not pivot columns, x_3 and x_4 are independent variables. First, we let $x_3 = 1$ and $x_4 = 0$. Then $x_1 = 2x_3 + x_4 = 2$ and $x_2 = 2x_3 = 2$. This yields the eigenvector $[2, 2, 1, 0]$. Second, we let $x_3 = 0$

and $x_4 = 1$. Then $x_1 = 2x_3 + x_4 = 1$ and $x_2 = 2x_3 = 0$, producing the eigenvector $[1,0,0,1]$.

For $\lambda_2 = -1$, we solve the homogeneous system whose augmented matrix is $[((-1)\mathbf{I}_3 - \mathbf{A})|\mathbf{0}]$.

Now $[((-1)\mathbf{I}_3 - \mathbf{A})|\mathbf{0}] = \left[\begin{array}{cccc|c} -4 & -1 & 6 & 2 & 0 \\ -4 & -1 & 6 & 4 & 0 \\ -2 & 0 & 2 & 2 & 0 \\ 0 & -1 & 2 & -2 & 0 \end{array}\right]$, reducing to $\left[\begin{array}{cccc|c} 1 & 0 & -1 & 0 & 0 \\ 0 & 1 & -2 & 0 & 0 \\ 0 & 0 & 0 & 1 & 0 \\ 0 & 0 & 0 & 0 & 0 \end{array}\right]$. The associated linear system is $\begin{cases} x_1 \quad - \quad x_3 = 0 \\ x_2 - 2x_3 = 0 \\ x_4 = 0 \\ 0 = 0 \end{cases}$. Since column 3 is not a pivot column, x_3 is an independent variable. Let $x_3 = 1$. Then $x_1 = x_3 = 1$, $x_2 = 2x_3 = 2$, and $x_4 = 0$. This yields the eigenvector $[1,2,1,0]$.

For $\lambda_3 = 0$, we solve the homogeneous system whose augmented matrix is $[(0\mathbf{I}_3 - \mathbf{A})|\mathbf{0}] = [-\mathbf{A}|\mathbf{0}]$.

Now $[-\mathbf{A}|\mathbf{0}] = \left[\begin{array}{cccc|c} -3 & -1 & 6 & 2 & 0 \\ -4 & 0 & 6 & 4 & 0 \\ -2 & 0 & 3 & 2 & 0 \\ 0 & -1 & 2 & -1 & 0 \end{array}\right]$, reducing to $\left[\begin{array}{cccc|c} 1 & 0 & 0 & -1 & 0 \\ 0 & 1 & 0 & 1 & 0 \\ 0 & 0 & 1 & 0 & 0 \\ 0 & 0 & 0 & 0 & 0 \end{array}\right]$. The associated linear system is $\begin{cases} x_1 \quad - x_4 = 0 \\ x_2 + x_4 = 0 \\ x_3 = 0 \\ 0 = 0 \end{cases}$. Since column 4 is not a pivot column, x_4 is an independent variable. Let $x_4 = 1$. Then $x_1 = x_4 = 1$, $x_2 = -x_4 = -1$, and $x_3 = 0$. This yields the eigenvector $[1,-1,0,1]$.

Step 4: Since $n = 4$, and we have found 4 eigenvectors, $\mathbf{A}$ is diagonalizable.

Step 5: We form the 4×4 matrix whose columns are the four vectors we have found:

$$\mathbf{P} = \begin{bmatrix} 2 & 1 & 1 & 1 \\ 2 & 0 & 2 & -1 \\ 1 & 0 & 1 & 0 \\ 0 & 1 & 0 & 1 \end{bmatrix}.$$

Step 6: The matrix whose diagonal entries are the eigenvalues of $\mathbf{A}$ (in the correct order) is $\mathbf{D} = \begin{bmatrix} 1 & 0 & 0 & 0 \\ 0 & 1 & 0 & 0 \\ 0 & 0 & -1 & 0 \\ 0 & 0 & 0 & 0 \end{bmatrix}$. Also, by row reducing $[\mathbf{P}|\mathbf{I}_4]$, we get $\mathbf{P}^{-1} = \begin{bmatrix} 1 & 0 & -1 & -1 \\ 0 & 1 & -2 & 1 \\ -1 & 0 & 2 & 1 \\ 0 & -1 & 2 & 0 \end{bmatrix}$. It is easy to verify for the given matrices that $\mathbf{D} = \mathbf{P}^{-1}\mathbf{A}\mathbf{P}$.

(5) In each part, we use the Diagonalization Method to find a matrix $\mathbf{P}$ and a diagonal matrix $\mathbf{D}$ such that $\mathbf{A} = \mathbf{P}\mathbf{D}\mathbf{P}^{-1}$. Then, $\mathbf{A}^n = \mathbf{P}\mathbf{D}^n\mathbf{P}^{-1}$. For a diagonal matrix $\mathbf{D}$, $\mathbf{D}^n$ is easily computed as the diagonal matrix whose (i,i) entry is d_{ii}^n.

(a) First, we diagonalize $\mathbf{A}$. Step 1: $p_{\mathbf{A}}(x) = |x\mathbf{I}_2 - \mathbf{A}| = \begin{vmatrix} x-4 & 6 \\ -3 & x+5 \end{vmatrix}$
$= (x-4)(x+5) - (6)(-3) = x^2 + x - 2 = (x-1)(x+2)$.

Step 2: The eigenvalues of $\mathbf{A}$ are the roots of $p_{\mathbf{A}}(x)$. Hence, there are two eigenvalues, $\lambda_1 = 1$ and $\lambda_2 = -2$.

Step 3: Now we solve for eigenvectors.

For $\lambda_1 = 1$, we solve the homogeneous system whose augmented matrix is $[(1\mathbf{I}_2 - \mathbf{A})|\mathbf{0}]$. Now $[(1\mathbf{I}_2 - \mathbf{A})|\mathbf{0}] = \left[\begin{array}{cc|c} -3 & 6 & 0 \\ -3 & 6 & 0 \end{array}\right]$, which reduces to $\left[\begin{array}{cc|c} 1 & -2 & 0 \\ 0 & 0 & 0 \end{array}\right]$. The associated linear system is $\left\{\begin{array}{rcl} x_1 - 2x_2 & = & 0 \\ 0 & = & 0 \end{array}\right.$. Since column 2 is not a pivot column, x_2 is an independent variable.

Let $x_2 = 1$. Then $x_1 = 2x_2 = 2$. This yields the eigenvector $[2, 1]$.

For $\lambda_2 = -2$, we solve the homogeneous system whose augmented matrix is $[((-2)\mathbf{I}_2 - \mathbf{A})|\mathbf{0}]$. Now $[((-2)\mathbf{I}_2 - \mathbf{A})|\mathbf{0}] = \left[\begin{array}{cc|c} -6 & 6 & 0 \\ -3 & 3 & 0 \end{array}\right]$, which reduces to $\left[\begin{array}{cc|c} 1 & -1 & 0 \\ 0 & 0 & 0 \end{array}\right]$. The associated linear system is $\left\{\begin{array}{rcl} x_1 - x_2 & = & 0 \\ 0 & = & 0 \end{array}\right.$. Since column 2 is not a pivot column, x_2 is an independent variable. Let $x_2 = 1$. Then $x_1 = x_2 = 1$. This yields the eigenvector $[1, 1]$.

Step 4: Since $n = 2$, and we have found 2 eigenvectors, $\mathbf{A}$ is diagonalizable.

Step 5: We form the 2×2 matrix whose columns are the two vectors we have found: $\mathbf{P} = \begin{bmatrix} 2 & 1 \\ 1 & 1 \end{bmatrix}$.

Step 6: The matrix whose diagonal entries are the eigenvalues of $\mathbf{A}$ (in the correct order) is $\mathbf{D} = \begin{bmatrix} 1 & 0 \\ 0 & -2 \end{bmatrix}$. Also, using Theorem 2.13, we get $\mathbf{P}^{-1} = \begin{bmatrix} 1 & -1 \\ -1 & 2 \end{bmatrix}$. Thus,

$$\mathbf{A}^{15} = \mathbf{P}\mathbf{D}^{15}\mathbf{P}^{-1} = \begin{bmatrix} 2 & 1 \\ 1 & 1 \end{bmatrix} \begin{bmatrix} 1^{15} & 0 \\ 0 & (-2)^{15} \end{bmatrix} \begin{bmatrix} 1 & -1 \\ -1 & 2 \end{bmatrix} = \begin{bmatrix} 32770 & -65538 \\ 32769 & -65537 \end{bmatrix}.$$

(c) First, we diagonalize $\mathbf{A}$. Step 1: $p_{\mathbf{A}}(x) = |x\mathbf{I}_3 - \mathbf{A}| = \begin{vmatrix} x-11 & 6 & 12 \\ -13 & x+6 & 16 \\ -5 & 3 & x+5 \end{vmatrix}$. Using basketweaving gives $p_{\mathbf{A}}(x) = (x-11)(x+6)(x+5) + (6)(16)(-5) + (12)(-13)(3) - (12)(x+6)(-5) - (x-11)(16)(3) - (6)(-13)(x+5) = x^3 - x = x(x^2-1) = x(x-1)(x+1)$.

Step 2: The eigenvalues of $\mathbf{A}$ are the roots of $p_{\mathbf{A}}(x)$. Hence, there are three eigenvalues, $\lambda_1 = 0$, $\lambda_2 = 1$ and $\lambda_3 = -1$.

Normally, at this point, we solve for eigenvectors. However, as we will see, we do not need to compute the actual matrix $\mathbf{P}$ in this case. Note that we get (at least) one eigenvector from each eigenvalue, so we obtain 3 eigenvectors. Hence, $\mathbf{A}$ is diagonalizable, and $\mathbf{A} = \mathbf{P}\mathbf{D}\mathbf{P}^{-1}$ for some matrix $\mathbf{P}$, where $\mathbf{D} = \begin{bmatrix} 0 & 0 & 0 \\ 0 & 1 & 0 \\ 0 & 0 & -1 \end{bmatrix}$ is the diagonal matrix whose diagonal entries are the eigenvalues of $\mathbf{A}$. Now $\mathbf{D}^{49} = \begin{bmatrix} 0 & 0 & 0 \\ 0 & 1^{49} & 0 \\ 0 & 0 & (-1)^{49} \end{bmatrix} = \mathbf{D}$. Thus, $\mathbf{A}^{49} = \mathbf{P}\mathbf{D}^{49}\mathbf{P}^{-1} = \mathbf{P}\mathbf{D}\mathbf{P}^{-1} = \mathbf{A}$.

(e) First, we diagonalize $\mathbf{A}$. Step 1: $p_{\mathbf{A}}(x) = |x\mathbf{I}_3 - \mathbf{A}| = \begin{vmatrix} x-7 & -9 & 12 \\ -10 & x-16 & 22 \\ -8 & -12 & x+16 \end{vmatrix}$. Using basketweaving gives $p_{\mathbf{A}}(x) = (x-7)(x-16)(x+16) + (-9)(22)(-8) + (12)(-10)(-12)$ $-(12)(x-16)(-8) - (x-7)(22)(-12) - (-9)(-10)(x+16) = x^3 - 7x^2 + 14x - 8$ $= (x-1)(x-2)(x-4)$.

Step 2: The eigenvalues of $\mathbf{A}$ are the roots of $p_{\mathbf{A}}(x)$. Hence, there are three eigenvalues, $\lambda_1 = 1$, $\lambda_2 = 2$ and $\lambda_3 = 4$.

Step 3: Now we solve for eigenvectors.

For $\lambda_1 = 1$, we solve the homogeneous system whose augmented matrix is $[(1\mathbf{I}_3 - \mathbf{A})|\mathbf{0}]$. Now $[(1\mathbf{I}_3 - \mathbf{A})|\mathbf{0}] = \left[\begin{array}{ccc|c} -6 & -9 & 12 & 0 \\ -10 & -15 & 22 & 0 \\ -8 & -12 & 17 & 0 \end{array}\right]$, which reduces to $\left[\begin{array}{ccc|c} 1 & \frac{3}{2} & 0 & 0 \\ 0 & 0 & 1 & 0 \\ 0 & 0 & 0 & 0 \end{array}\right]$. The associated linear system is $\begin{cases} x_1 + \frac{3}{2}x_2 = 0 \\ x_3 = 0 \\ 0 = 0 \end{cases}$. Since column 2 is not a pivot column, x_2 is an independent variable. We choose $x_2 = 2$ to eliminate fractions. Then $x_1 = -\frac{3}{2}x_2 = -3$ and $x_3 = 0$, producing the eigenvector $[-3, 2, 0]$.

For $\lambda_2 = 2$, we solve the homogeneous system whose augmented matrix is $[(2\mathbf{I}_3 - \mathbf{A})|\mathbf{0}]$. Now $[(2\mathbf{I}_3 - \mathbf{A})|\mathbf{0}] = \left[\begin{array}{ccc|c} -5 & -9 & 12 & 0 \\ -10 & -14 & 22 & 0 \\ -8 & -12 & 18 & 0 \end{array}\right]$, which reduces to $\left[\begin{array}{ccc|c} 1 & 0 & -\frac{3}{2} & 0 \\ 0 & 1 & -\frac{1}{2} & 0 \\ 0 & 0 & 0 & 0 \end{array}\right]$. The associated linear system is $\begin{cases} x_1 - \frac{3}{2}x_3 = 0 \\ x_2 - \frac{1}{2}x_3 = 0 \\ 0 = 0 \end{cases}$. Since column 3 is not a pivot column, x_3 is an independent variable. We choose $x_3 = 2$ to eliminate fractions. Then $x_1 = \frac{3}{2}x_3 = 3$ and $x_2 = \frac{1}{2}x_3 = 1$, yielding the eigenvector $[3, 1, 2]$.

For $\lambda_3 = 4$, we solve the homogeneous system whose augmented matrix is $[(4\mathbf{I}_3 - \mathbf{A})|\mathbf{0}]$. Now $[(4\mathbf{I}_3 - \mathbf{A})|\mathbf{0}] = \left[\begin{array}{ccc|c} -3 & -9 & 12 & 0 \\ -10 & -12 & 22 & 0 \\ -8 & -12 & 20 & 0 \end{array}\right]$, which reduces to $\left[\begin{array}{ccc|c} 1 & 0 & -1 & 0 \\ 0 & 1 & -1 & 0 \\ 0 & 0 & 0 & 0 \end{array}\right]$. The associated linear system is $\begin{cases} x_1 - x_3 = 0 \\ x_2 - x_3 = 0 \\ 0 = 0 \end{cases}$. Since column 3 is not a pivot column, x_3 is an independent variable. Let $x_3 = 1$. Then $x_1 = x_3 = 1$ and $x_2 = x_3 = 1$. This yields the eigenvector $[1, 1, 1]$.

Step 4: Since $n = 3$, and we have found 3 eigenvectors, $\mathbf{A}$ is diagonalizable.

Step 5: We form the 3×3 matrix whose columns are the 3 eigenvectors we have found:

$$\mathbf{P} = \begin{bmatrix} -3 & 3 & 1 \\ 2 & 1 & 1 \\ 0 & 2 & 1 \end{bmatrix}.$$

Step 6: The matrix whose diagonal entries are the eigenvalues of $\mathbf{A}$ (in the correct order) is $\mathbf{D} = \begin{bmatrix} 1 & 0 & 0 \\ 0 & 2 & 0 \\ 0 & 0 & 4 \end{bmatrix}$. Also, by row reducing $[\mathbf{P}|\mathbf{I}_3]$, we get $\mathbf{P}^{-1} = \begin{bmatrix} -1 & -1 & 2 \\ -2 & -3 & 5 \\ 4 & 6 & -9 \end{bmatrix}$. Thus,

$$\mathbf{A}^{10} = \mathbf{P}\mathbf{D}^{10}\mathbf{P}^{-1} = \begin{bmatrix} -3 & 3 & 1 \\ 2 & 1 & 1 \\ 0 & 2 & 1 \end{bmatrix} \begin{bmatrix} 1^{10} & 0 & 0 \\ 0 & 2^{10} & 0 \\ 0 & 0 & 4^{10} \end{bmatrix} \begin{bmatrix} -1 & -1 & 2 \\ -2 & -3 & 5 \\ 4 & 6 & -9 \end{bmatrix}$$

$$= \begin{bmatrix} 4188163 & 6282243 & -9421830 \\ 4192254 & 6288382 & -9432060 \\ 4190208 & 6285312 & -9426944 \end{bmatrix}.$$

(7) (b) If $\mathbf{A}$ has all eigenvalues nonnegative, then $\mathbf{A}$ has a square root. Suppose $\mathbf{A} = \mathbf{PDP}^{-1}$ for some diagonal matrix $\mathbf{D}$. Let $\mathbf{C}$ be the diagonal matrix with $c_{ii} = \sqrt{d_{ii}}$. Thus, $\mathbf{C}^2 = \mathbf{D}$.
Let $\mathbf{B} = \mathbf{PCP}^{-1}$. Then, $\mathbf{B}^2 = (\mathbf{PCP}^{-1})^2 = (\mathbf{PCP}^{-1})(\mathbf{PCP}^{-1}) = \mathbf{PC}(\mathbf{P}^{-1}\mathbf{P})\mathbf{CP}^{-1} = \mathbf{PCI}_n\mathbf{CP}^{-1} = \mathbf{PC}^2\mathbf{P}^{-1} = \mathbf{PDP}^{-1} = \mathbf{A}$.

(8) Let $\mathbf{B} = \begin{bmatrix} 15 & -14 & -14 \\ -13 & 16 & 17 \\ 20 & -22 & -23 \end{bmatrix}$. We will use the Diagonalization Method to find a matrix $\mathbf{P}$ and a diagonal matrix $\mathbf{D}$ such that $\mathbf{B} = \mathbf{PDP}^{-1}$. If $\mathbf{C}$ is the diagonal matrix with $c_{ii} = \sqrt[3]{d_{ii}}$, then $\mathbf{C}^3 = \mathbf{D}$. So, if $\mathbf{A} = \mathbf{PCP}^{-1}$, then $\mathbf{A}^3 = (\mathbf{PCP}^{-1})^3 = (\mathbf{PCP}^{-1})(\mathbf{PCP}^{-1})(\mathbf{PCP}^{-1}) = \mathbf{PC}(\mathbf{P}^{-1}\mathbf{P})\mathbf{C}(\mathbf{P}^{-1}\mathbf{P})\mathbf{CP}^{-1}$ $= \mathbf{PCI}_n\mathbf{CI}_n\mathbf{CP}^{-1} = \mathbf{PC}^3\mathbf{P}^{-1} = \mathbf{PDP}^{-1} = \mathbf{B}$. Hence, to solve this problem, we first need to find $\mathbf{P}$ and $\mathbf{D}$. To do this, we diagonalize $\mathbf{B}$.

Step 1: $p_{\mathbf{B}}(x) = |x\mathbf{I}_3 - \mathbf{B}| = \begin{vmatrix} x-15 & 14 & 14 \\ 13 & x-16 & -17 \\ -20 & 22 & x+23 \end{vmatrix}$.

Using basketweaving, we obtain $p_{\mathbf{B}}(x) = (x-15)(x-16)(x+23) + (14)(-17)(-20) + (14)(13)(22)$ $-(14)(x-16)(-20) - (x-15)(-17)(22) - (14)(13)(x+23) = x^3 - 8x^2 - x + 8 = (x^2-1)(x-8) =$ $(x-1)(x+1)(x-8)$.

Step 2: The eigenvalues of $\mathbf{B}$ are the roots of $p_{\mathbf{B}}(x)$. Hence, there are three eigenvalues, $\lambda_1 = 1$, $\lambda_2 = -1$ and $\lambda_3 = 8$.

Step 3: Now we solve for eigenvectors.

For $\lambda_1 = 1$, we solve the homogeneous system whose augmented matrix is $[(1\mathbf{I}_3 - \mathbf{B})|\mathbf{0}]$. Now $[(1\mathbf{I}_3 - \mathbf{B})|\mathbf{0}] = \left[\begin{array}{ccc|c} -14 & 14 & 14 & 0 \\ 13 & -15 & -17 & 0 \\ -20 & 22 & 24 & 0 \end{array}\right]$, which reduces to $\left[\begin{array}{ccc|c} 1 & 0 & 1 & 0 \\ 0 & 1 & 2 & 0 \\ 0 & 0 & 0 & 0 \end{array}\right]$. The associated linear system is $\begin{cases} x_1 \quad + x_3 = 0 \\ \quad x_2 + 2x_3 = 0 \\ \qquad 0 = 0 \end{cases}$. Since column 3 is not a pivot column, x_3 is an independent variable. Let $x_3 = 1$. Then $x_1 = -x_3 = -1$ and $x_2 = -2x_3 = -2$. This yields the eigenvector $[-1, -2, 1]$.

For $\lambda_2 = -1$, we solve the homogeneous system whose augmented matrix is $[((-1)\mathbf{I}_3 - \mathbf{B})|\mathbf{0}]$. Now $[((-1)\mathbf{I}_3 - \mathbf{B})|\mathbf{0}] = \left[\begin{array}{ccc|c} -16 & 14 & 14 & 0 \\ 13 & -17 & -17 & 0 \\ -20 & 22 & 22 & 0 \end{array}\right]$, which reduces to $\left[\begin{array}{ccc|c} 1 & 0 & 0 & 0 \\ 0 & 1 & 1 & 0 \\ 0 & 0 & 0 & 0 \end{array}\right]$. The associated linear system is $\begin{cases} x_1 & & = 0 \\ & x_2 + x_3 & = 0 \\ & 0 & = 0 \end{cases}$. Since column 3 is not a pivot column, x_3 is an independent variable. Let $x_3 = 1$. Then $x_1 = 0$ and $x_2 = -x_3 = -1$, producing the eigenvector $[0, -1, 1]$.

For $\lambda_3 = 8$, we solve the homogeneous system whose augmented matrix is $[(8\mathbf{I}_3 - \mathbf{B})|\mathbf{0}]$. Now $[(8\mathbf{I}_3 - \mathbf{B})|\mathbf{0}] = \left[\begin{array}{ccc|c} -7 & 14 & 14 & 0 \\ 13 & -8 & -17 & 0 \\ -20 & 22 & 31 & 0 \end{array}\right]$, which reduces to $\left[\begin{array}{ccc|c} 1 & 0 & -1 & 0 \\ 0 & 1 & \frac{1}{2} & 0 \\ 0 & 0 & 0 & 0 \end{array}\right]$. The associated linear system is $\begin{cases} x_1 - x_3 & = 0 \\ x_2 + \frac{1}{2}x_3 & = 0 \\ 0 & = 0 \end{cases}$. Since column 3 is not a pivot column, x_3 is an independent variable. We choose $x_3 = 2$ to eliminate fractions. Then $x_1 = x_3 = 2$ and $x_2 = -\frac{1}{2}x_3 = -1$, yielding the eigenvector $[2, -1, 2]$.

Step 4: Since $n = 3$, and we have found 3 eigenvectors, $\mathbf{B}$ is diagonalizable.

Step 5: We form the 3×3 matrix whose columns are the 3 vectors we have found: $\mathbf{P} = \begin{bmatrix} -1 & 0 & 2 \\ -2 & -1 & -1 \\ 1 & 1 & 2 \end{bmatrix}$.

Step 6: The matrix whose diagonal entries are the eigenvalues of $\mathbf{B}$ (in the correct order) is $\mathbf{D} = \begin{bmatrix} 1 & 0 & 0 \\ 0 & -1 & 0 \\ 0 & 0 & 8 \end{bmatrix}$. Also, by row reducing $[\mathbf{P}|\mathbf{I}_3]$, we get $\mathbf{P}^{-1} = \begin{bmatrix} 1 & -2 & -2 \\ -3 & 4 & 5 \\ 1 & -1 & -1 \end{bmatrix}$.

Next, we compute the diagonal matrix $\mathbf{C}$ whose main diagonal entries are the cube roots of the eigenvalues of $\mathbf{B}$. That is, $\mathbf{C} = \begin{bmatrix} 1 & 0 & 0 \\ 0 & -1 & 0 \\ 0 & 0 & 2 \end{bmatrix}$. Then

$\mathbf{A} = \mathbf{PCP}^{-1} = \begin{bmatrix} -1 & 0 & 2 \\ -2 & -1 & -1 \\ 1 & 1 & 2 \end{bmatrix} \begin{bmatrix} 1 & 0 & 0 \\ 0 & -1 & 0 \\ 0 & 0 & 2 \end{bmatrix} \begin{bmatrix} 1 & -2 & -2 \\ -3 & 4 & 5 \\ 1 & -1 & -1 \end{bmatrix} = \begin{bmatrix} 3 & -2 & -2 \\ -7 & 10 & 11 \\ 8 & -10 & -11 \end{bmatrix}$. Direct calculation of $\mathbf{A}^3$ verifies that $\mathbf{A}^3 = \mathbf{B}$.

(10) (b) Consider the matrix $\mathbf{A} = \begin{bmatrix} 0 & -1 \\ 1 & 0 \end{bmatrix}$. Now $\mathbf{A}$ has no eigenvalues, because $p_{\mathbf{A}}(x) = |x\mathbf{I}_2 - \mathbf{A}| = \begin{vmatrix} x & 1 \\ -1 & x \end{vmatrix} = x^2+1$, which has no real roots. However, $\mathbf{A}^4 = \mathbf{I}_2$ has 1 as an eigenvalue. Thus, with $k = 4$, $\lambda^4 = 1$ is an eigenvalue for $\mathbf{A}^4$, but neither solution for $\lambda = \pm\sqrt[4]{1} = \pm 1$ is an eigenvalue for $\mathbf{A}$.

(24) (a) True. This is the definition of what it means for 5 to be an eigenvalue of $\mathbf{A}$.

(b) False. The eigenvalues of $\mathbf{A}$ are the solutions of $p_{\mathbf{A}}(x) = |x\mathbf{I}_n - \mathbf{A}| = 0$. The statement given in the text is missing the determinant symbol, and also has a zero matrix on the right side of the equation.

(c) True. This is part of Theorem 3.14.

(d) True. This fact is stated in Step 6 of the Diagonalization Method.

(e) False. If every column of $\mathbf{P}$ is a multiple of the same eigenvector, then $|\mathbf{P}| = 0$, and $\mathbf{P}$ is singular. Consider $\mathbf{A} = \begin{bmatrix} 1 & 1 \\ 0 & 2 \end{bmatrix}$, which has eigenvalues $\lambda_1 = 1$ and $\lambda_2 = 2$. (Do you see why?) The Diagonalization Method produces the eigenvectors $[1,0]$ for λ_1 and $[1,1]$ for λ_2. But note that $[2,0]$ is also an eigenvector for λ_1. If we use the eigenvectors $[1,0]$ and $[2,0]$ as columns for $\mathbf{P}$, instead of $[1,0]$ and $[1,1]$ as we should, we get $\mathbf{P} = \begin{bmatrix} 1 & 2 \\ 0 & 0 \end{bmatrix}$. Note that the columns of $\mathbf{P}$ are eigenvectors for $\mathbf{A}$, but $\mathbf{P}$ is singular ($|\mathbf{P}| = 0$).

(f) True. The given characteristic polynomial indicates that the algebraic multiplicity of the eigenvalue (-1) is 1. We stated in Section 3.4 (without proof) that the number of eigenvectors produced for a given eigenvalue in Step 3 of the Diagonalization Method is always *less than or equal to* the algebraic multiplicity of the eigenvalue. (This is proven in Chapter 5.) Hence, no more than one eigenvector for (-1) can be produced in the Diagonalization Method. (Note that since the Diagonalization Method always produces *at least* one eigenvector for each eigenvalue, in this case the method must produce *exactly* one eigenvector for the eigenvalue (-1).)

(g) True. Since Step 3 of the Diagonalization Method always produces *at least* one eigenvector for each eigenvalue of $\mathbf{A}$, the method must yield (at least) three eigenvectors. Step 4 then assures us that $\mathbf{A}$ is diagonalizable.

(h) False. In the discussion of using eigenvalues and eigenvectors to compute large powers of a matrix (in Section 3.4), we noted that, in general, $\mathbf{A} = \mathbf{PDP}^{-1}$ implies, $\mathbf{A}^n = \mathbf{PD}^n\mathbf{P}^{-1}$ (where $\mathbf{D}$ is a diagonal matrix). Neither $\mathbf{P}$ nor $\mathbf{P}^{-1}$ should be raised to the nth power. For a specific counterexample to the equation $\mathbf{A}^n = \mathbf{P}^n\mathbf{D}^n(\mathbf{P}^{-1})^n$ given in the problem, consider $\mathbf{A} = \begin{bmatrix} 19 & -48 \\ 8 & -21 \end{bmatrix}$, from Exercise 4(a). In the solution to that exercise (see above), we found that $\mathbf{D} = \mathbf{P}^{-1}\mathbf{AP}$, and so $\mathbf{A} = \mathbf{PDP}^{-1}$, for the matrices $\mathbf{P} = \begin{bmatrix} 3 & 2 \\ 1 & 1 \end{bmatrix}$ and $\mathbf{D} = \begin{bmatrix} 3 & 0 \\ 0 & -5 \end{bmatrix}$. Now, a short computation shows that $\mathbf{A}^2 = \begin{bmatrix} -23 & 96 \\ -16 & 57 \end{bmatrix}$, but $\mathbf{P}^2\mathbf{D}^2(\mathbf{P}^{-1})^2 = \begin{bmatrix} -503 & 1408 \\ -192 & 537 \end{bmatrix}$ instead. Note, however, that $\mathbf{PD}^2\mathbf{P}^{-1} = \mathbf{A}^2$.

Chapter 4

Section 4.1

(5) The set is not closed under addition. For example,

$\begin{bmatrix} 1 & 1 \\ 1 & 1 \end{bmatrix} + \begin{bmatrix} 1 & 2 \\ 2 & 4 \end{bmatrix} = \begin{bmatrix} 2 & 3 \\ 3 & 5 \end{bmatrix}$. Both $\begin{bmatrix} 1 & 1 \\ 1 & 1 \end{bmatrix}$ and $\begin{bmatrix} 1 & 2 \\ 2 & 4 \end{bmatrix}$ are singular (their determinants are zero), but $\begin{bmatrix} 2 & 3 \\ 3 & 5 \end{bmatrix}$ is nonsingular (its determinant is $1 \neq 0$).

(8) Properties (2), (3), and (6) are not satisfied. Property (4) makes no sense without Property (3). The following is a counterexample for Property (2): $3 \oplus (4 \oplus 5) = 3 \oplus 18 = 42$, but $(3 \oplus 4) \oplus 5 = 14 \oplus 5 = 38$.
While this single counterexample for just one of the properties is enough to prove that the set is not a vector space, we will also illustrate why properties (3) and (6) do not hold.
For Property (3), suppose $z \in \mathbb{R}$ is the zero vector. Then, for all $x \in \mathbb{R}$, $x \oplus z = x \Rightarrow 2(x+z) = x \Rightarrow z = -\frac{x}{2}$. But then *different* values of x would produce *different* values of z. For if $x = 2$, $z = -1$, and if $x = 4$, $z = -2$. However, the value of z must be unique.
For a counterexample to Property (6), note that $(1+2)(3) = (3)(3) = 9$, but $(1)(3) \oplus (2)(3) = 3 \oplus 6 = 18$.

(20) (a) False. The two closure properties and the eight remaining properties must be true for all elements in $\mathbb{R}^n$ in order for it to be a vector space under some "addition" and "scalar multiplication." For example, Example 10 in Section 4.1 illustrates operations on $\mathbb{R}^2$ for which $\mathbb{R}^2$ is not a vector space. Also, Exercise 8, whose answer appears above, provides a counterexample in $\mathbb{R}$.

(b) False. This set is not closed under addition. Both $x^7 + x$ and $-x^7$ are in this set, but their sum is x, which does not have degree 7. (Also note that the zero polynomial is not in the set.)

(c) True. This is the vector space $\mathcal{P}_7$ described in Example 4 of Section 4.1.

(d) True. This statement is logically equivalent to part (4) of Theorem 4.1. Note that part (4) of Theorem 4.1 is stated in the form "If A then B or C," with A = "$c\mathbf{x} = \mathbf{0}$," B = "$c = 0$," and C = "$\mathbf{x} = \mathbf{0}$" (where we use c and $\mathbf{x}$ instead of a and $\mathbf{v}$). Now, in Section 1.3 we found that "If A then B or C" is logically equivalent to "If A is true and B is false, then C." Here, B is given to be false (since $c \neq 0$), and so A implies C. That is, when $c \neq 0$, "$c\mathbf{x} = \mathbf{0}$" implies "$\mathbf{x} = \mathbf{0}$."

(e) False. Scalar multiplication by the zero scalar *always* results in the zero *vector*, not a scalar. This is part (2) of Theorem 4.1.

(f) True. This is the statement of part (3) of Theorem 4.1.

(g) True. Suppose $\mathcal{V}$ represents the given set of functions. Then $\mathcal{V}$ is closed under addition because if $\mathbf{f}, \mathbf{g} \in \mathcal{V}$, then $\mathbf{h} = \mathbf{f} + \mathbf{g}$ is a real valued function, and $\mathbf{h}(1) = \mathbf{f}(1) + \mathbf{g}(1) = 0 + 0 = 0$. Similarly, $\mathcal{V}$ is closed under scalar multiplication because if $\mathbf{f} \in \mathcal{V}$ and $c \in \mathbb{R}$, then $\mathbf{h} = c\mathbf{f}$ is a real valued function, and $\mathbf{h}(1) = c\mathbf{f}(1) = c0 = 0$. The proofs of Properties (1), (2), (5), (6), (7) and (8) are exactly identical to those in Example 6 of Section 4.1 of the textbook. For Property (3), $\mathbf{z}(x) = 0$ is in $\mathcal{V}$ because $\mathbf{z}(1) = 0$. The same argument for Property (3) in Example 6 then shows that $\mathbf{z}$ works as the identity element for addition. Similarly, for Property (4): if $\mathbf{f} \in \mathcal{V}$, then the function $-\mathbf{f}$ defined by $[-\mathbf{f}](x) = -(\mathbf{f}(x))$ is in $\mathcal{V}$ because $[-\mathbf{f}](1) = -(\mathbf{f}(1)) = -(0) = 0$. Thus, Example 6 finishes the proof that Property (4) holds.

Section 4.2

(1) In each part, we will refer to the given set as $\mathcal{V}$. If $\mathcal{V}$ is a subspace of $\mathbb{R}^2$, we will prove it using Theorem 4.2 by showing that it is nonempty and that it is closed under both vector addition and scalar multiplication. If $\mathcal{V}$ is not a subspace, we will show that it lacks at least one of these properties. Or, since every subspace of $\mathbb{R}^2$ contains the zero vector $[0,0]$, we can also show that $\mathcal{V}$ is not a subspace by proving that it does not contain the zero vector. In that case, it is not necessary to also prove that one of the closure properties fails.

(a) This set $\mathcal{V}$ is not a subspace of $\mathbb{R}^2$ because the zero vector, $[0,0]$, is not a unit vector, and so is not in $\mathcal{V}$. (Note: It can also be shown that $\mathcal{V}$ is not closed under either operation.)

(c) This set $\mathcal{V}$ is a subspace of $\mathbb{R}^2$. $\mathcal{V}$ is nonempty since $[0,0] \in \mathcal{V}$ because it is of the form $[a, 2a]$ with $a = 0$. $\mathcal{V}$ is closed under addition because $[a, 2a] + [b, 2b] = [(a+b), 2(a+b)] = [A, 2A] \in \mathcal{V}$, where $A = (a+b)$. Finally, $\mathcal{V}$ is closed under scalar multiplication since $c[a, 2a] = [ca, 2(ca)] = [B, 2B] \in \mathcal{V}$, where $B = ca$.

(e) This set $\mathcal{V}$ is not a subspace of $\mathbb{R}^2$ since the zero vector, $[0,0]$, does not equal $[1,2]$, and so is not in $\mathcal{V}$. (Note: It can also be shown that $\mathcal{V}$ is not closed under either operation.)

(g) This set $\mathcal{V}$ is not a subspace of $\mathbb{R}^2$ because it is not closed under addition. The counterexample $[1,1] + [1,-1] = [2,0] \notin \mathcal{V}$ (since $|2| \neq |0|$) proves this.

(j) This set $\mathcal{V}$ is not a subspace of $\mathbb{R}^2$ because it is not closed under addition. To prove this, consider the following counterexample: $[1,1] + [2,4] = [3,5] \notin \mathcal{V}$, since $5 \neq 3^2$. (Note: It can also be shown that $\mathcal{V}$ is not closed under scalar multiplication.)

(l) This set $\mathcal{V}$ is not a subspace of $\mathbb{R}^2$ because it is not closed under scalar multiplication. To prove this, consider the following counterexample: $2[0.75, 0] = [1.5, 0] \notin \mathcal{V}$. (Note: It can also be shown that $\mathcal{V}$ is not closed under addition.)

(2) In each part, we will refer to the given set as $\mathcal{V}$. If $\mathcal{V}$ is a subspace of $\mathcal{M}_{22}$, we will prove it using Theorem 4.2 by showing that it is nonempty and that it is closed under both matrix addition and scalar multiplication. If $\mathcal{V}$ is not a subspace, we will show that it lacks at least one of these properties. Or, since every subspace of $\mathcal{M}_{22}$ contains the zero matrix $\mathbf{O}_{22}$, we can also show that $\mathcal{V}$ is not a subspace by proving that it does not contain $\mathbf{O}_{22}$. In that case, it is not necessary to also prove that one of the closure properties fails.

(a) This set $\mathcal{V}$ is a subspace of $\mathcal{M}_{22}$. $\mathcal{V}$ is nonempty since setting $a = b = 0$ shows that $\mathbf{O}_{22} \in \mathcal{V}$. $\mathcal{V}$ is closed under matrix addition since

$$\begin{bmatrix} a & -a \\ b & 0 \end{bmatrix} + \begin{bmatrix} c & -c \\ d & 0 \end{bmatrix} = \begin{bmatrix} (a+c) & -(a+c) \\ (b+d) & 0 \end{bmatrix} = \begin{bmatrix} A & -A \\ B & 0 \end{bmatrix} \in \mathcal{V}, \text{ where } A = (a+c) \text{ and}$$

$B = (b+d)$. Similarly, $\mathcal{V}$ is closed under scalar multiplication because

$$c\begin{bmatrix} a & -a \\ b & 0 \end{bmatrix} = \begin{bmatrix} (ca) & -(ca) \\ (cb) & 0 \end{bmatrix} = \begin{bmatrix} C & -C \\ D & 0 \end{bmatrix} \in \mathcal{V}, \text{ where } C = (ca) \text{ and } D = (cb).$$

(c) This set $\mathcal{V}$ is a subspace of $\mathcal{M}_{22}$. $\mathcal{V}$ is clearly nonempty since $\mathbf{O}_{22}$ is symmetric, and hence in $\mathcal{V}$. $\mathcal{V}$ is closed under matrix addition since $\begin{bmatrix} a & b \\ b & c \end{bmatrix} + \begin{bmatrix} d & e \\ e & f \end{bmatrix} = \begin{bmatrix} (a+d) & (b+e) \\ (b+e) & (c+f) \end{bmatrix}$, which is in $\mathcal{V}$ since it is symmetric. Similarly, $\mathcal{V}$ is closed under scalar multiplication because

$$c\begin{bmatrix} a & b \\ b & d \end{bmatrix} = \begin{bmatrix} (ca) & (cb) \\ (cb) & (cd) \end{bmatrix} \text{ which is in } \mathcal{V}, \text{ since it is symmetric.}$$

(e) This set $\mathcal{V}$ is a subspace of $\mathcal{M}_{22}$. Since the sum of the entries of $\mathbf{O}_{22}$ is zero, $\mathbf{O}_{22} \in \mathcal{V}$. Thus $\mathcal{V}$ is nonempty. Note that a typical element of $\mathcal{V}$ has the form $\begin{bmatrix} a & b \\ c & d \end{bmatrix}$, where $d = -(a+b+c)$.

Now, $\mathcal{V}$ is closed under addition because $\begin{bmatrix} a & b \\ c & -(a+b+c) \end{bmatrix} + \begin{bmatrix} d & e \\ f & -(d+e+f) \end{bmatrix}$

$= \begin{bmatrix} a+d & b+e \\ c+f & -(a+b+c)-(d+e+f) \end{bmatrix} = \begin{bmatrix} a+d & b+e \\ c+f & -((a+d)+(b+e)+(c+f)) \end{bmatrix}$, which is in $\mathcal{V}$, since its entries sum to zero. Finally, $\mathcal{V}$ is closed under scalar multiplication since

$k\begin{bmatrix} a & b \\ c & -(a+b+c) \end{bmatrix} = \begin{bmatrix} ka & kb \\ kc & k(-(a+b+c)) \end{bmatrix} = \begin{bmatrix} ka & kb \\ kc & -((ka)+(kb)+(kc)) \end{bmatrix}$,

which is, again, in $\mathcal{V}$, since its entries sum to zero.

(g) This set $\mathcal{V}$ is a subspace of $\mathcal{M}_{22}$. Let $\mathbf{B} = \begin{bmatrix} 1 & 3 \\ -2 & -6 \end{bmatrix}$. Because $\mathbf{O}_{22}\mathbf{B} = \mathbf{O}_{22}$, $\mathbf{O}_{22} \in \mathcal{V}$, and so $\mathcal{V}$ is nonempty. Next we show that $\mathcal{V}$ is closed under addition. If $\mathbf{A}, \mathbf{C} \in \mathcal{V}$, then $(\mathbf{A}+\mathbf{C})\mathbf{B} = \mathbf{AB} + \mathbf{CB} = \mathbf{O}_{22} + \mathbf{O}_{22} = \mathbf{O}_{22}$. Hence, $(\mathbf{A}+\mathbf{C}) \in \mathcal{V}$. Finally, we show that $\mathcal{V}$ is closed under scalar multiplication. If $\mathbf{A} \in \mathcal{V}$, then $(c\mathbf{A})\mathbf{B} = c(\mathbf{AB}) = c\mathbf{O}_{22} = \mathbf{O}_{22}$, and so $(c\mathbf{A}) \in \mathcal{V}$.

(h) This set $\mathcal{V}$ is not a subspace of $\mathcal{M}_{22}$ since it is not closed under addition. This is proven by the following counterexample: $\begin{bmatrix} 1 & 1 \\ 0 & 0 \end{bmatrix} + \begin{bmatrix} 0 & 0 \\ 1 & 1 \end{bmatrix} = \begin{bmatrix} 1 & 1 \\ 1 & 1 \end{bmatrix} \notin \mathcal{V}$ (because the product $(1)(1)(1)(1)$ of its entries is not zero).

(3) In each part, we will refer to the given set as $\mathcal{V}$. If $\mathcal{V}$ is a subspace of $\mathcal{P}_5$, we will prove it using Theorem 4.2 by showing that it is nonempty and that it is closed under both polynomial addition and scalar multiplication. If $\mathcal{V}$ is not a subspace, we will show that it lacks at least one of these properties. Or, since every subspace of $\mathcal{P}_5$ contains $\mathbf{z}$, the zero polynomial, we can also show that $\mathcal{V}$ is not a subspace by proving that it does not contain $\mathbf{z}$. In that case, it is not necessary to also prove that one of the closure properties fails.

(a) This set $\mathcal{V}$ is a subspace of $\mathcal{P}_5$. Clearly $\mathbf{z} \in \mathcal{V}$ because both the fifth degree term and the first degree term of $\mathbf{z}$ are zero. Thus $\mathcal{V}$ is nonempty. Next we show that $\mathcal{V}$ is closed under addition. $(ax^5 + bx^4 + cx^3 + dx^2 + ax + e) + (rx^5 + sx^4 + tx^3 + ux^2 + rx + w) =$ $(a+r)x^5 + (b+s)x^4 + (c+t)x^3 + (d+u)x^2 + (a+r)x + (e+w)$, which is in $\mathcal{V}$ since the coefficient of x^5 equals the coefficient of x. Finally, we show that $\mathcal{V}$ is closed under scalar multiplication. $t(ax^5 + bx^4 + cx^3 + dx^2 + ax + e) = (ta)x^5 + (tb)x^4 + (tc)x^3 + (td)x^2 + (ta)x + (te)$, which is in $\mathcal{V}$, since the coefficient of x^5 equals the coefficient of x.

(b) This set $\mathcal{V}$ is a subspace of $\mathcal{P}_5$. Now $\mathbf{z}(3) = 0$, so $\mathbf{z} \in \mathcal{V}$. Hence $\mathcal{V}$ is nonempty. Suppose $\mathbf{p}, \mathbf{q} \in \mathcal{V}$. Then $\mathcal{V}$ is closed under addition because $(\mathbf{p}+\mathbf{q})(3) = \mathbf{p}(3) + \mathbf{q}(3) = 0 + 0 = 0$, proving $(\mathbf{p}+\mathbf{q}) \in \mathcal{V}$. Also, $\mathcal{V}$ is closed under scalar multiplication because $(c\mathbf{p})(3) = c\mathbf{p}(3) = c(0) = 0$, which shows that $(c\mathbf{p}) \in \mathcal{V}$.

(e) This set $\mathcal{V}$ is not a subspace of $\mathcal{P}_5$. Note that $\mathbf{z} \notin \mathcal{V}$, since the degree of $\mathbf{z}$ is 0. (Note: It can also be shown that $\mathcal{V}$ is not closed under either operation.)

(g) This set $\mathcal{V}$ is a subspace of $\mathcal{P}_5$. Now $\mathbf{z}'(4) = 0$, so $\mathbf{z} \in \mathcal{V}$. Hence $\mathcal{V}$ is nonempty. Suppose $\mathbf{p}, \mathbf{q} \in \mathcal{V}$. Then $\mathcal{V}$ is closed under addition because $(\mathbf{p}+\mathbf{q})'(4) = \mathbf{p}'(4) + \mathbf{q}'(4) = 0 + 0 = 0$, proving

$(\mathbf{p}+\mathbf{q}) \in \mathcal{V}$. Also, $\mathcal{V}$ is closed under scalar multiplication because $(c\mathbf{p})'(4) = c\mathbf{p}'(4) = c(0) = 0$, which shows that $(c\mathbf{p}) \in \mathcal{V}$.

(12) (e) No; if $|\mathbf{A}| \neq 0$ and $c = 0$, then $|c\mathbf{A}| = |\mathbf{O}_n| = 0$. (For a specific counterexample, consider $\mathbf{A} = \mathbf{I}_n$.)

(15) If λ is not an eigenvalue for $\mathbf{A}$, then, by the definition of an eigenvalue, there are no nonzero vectors $\mathbf{X}$ such that $\mathbf{AX} = \lambda\mathbf{X}$. Thus, no nonzero vector can be in S. However, $\mathbf{A0} = \mathbf{0} = \lambda\mathbf{0}$, hence $\mathbf{0} \in S$. Therefore, $S = \{\mathbf{0}\}$, the trivial subspace of $\mathbb{R}^n$.

(22) (a) False. By Theorem 4.2, a nonempty subset $\mathcal{W}$ of a vector space $\mathcal{V}$ must be closed under addition and scalar multiplication in order to be a subspace of $\mathcal{V}$. For example, in Example 5 in the textbook, the first quadrant in $\mathbb{R}^2$ is a nonempty subset of $\mathbb{R}^2$ that is not a subspace of $\mathbb{R}^2$.

(b) True. Every vector space has itself as a subspace. The trivial subspace $\{\mathbf{0}\}$ is also a subspace of every vector space. (Note that if $\mathcal{V} = \{\mathbf{0}\}$, these two subspaces are equal.)

(c) False. The plane $\mathcal{W}$ must pass through the origin, or else $\mathbf{0} \notin \mathcal{W}$, and so $\mathcal{W}$ is not a subspace. For example, the plane $x + y + z = 1$ is not a subspace for this reason.

(d) True. Let $\mathcal{V}$ be the set of lower triangular 5×5 matrices. To show that $\mathcal{V}$ is a subspace of $\mathcal{M}_{55}$, we must show that it is nonempty, and that it is closed under both addition and scalar multiplication. Since $\mathbf{O}_5$ is lower triangular, $\mathbf{O}_5 \in \mathcal{V}$, and so $\mathcal{V}$ is nonempty. Next, let $\mathbf{A}, \mathbf{B} \in \mathcal{V}$. Then $a_{ij} = b_{ij} = 0$ for all $i < j$. If $\mathbf{C} = \mathbf{A} + \mathbf{B}$, then for $i < j$, $c_{ij} = a_{ij} + b_{ij} = 0 + 0 = 0$, and so $\mathbf{C} \in \mathcal{V}$. Hence, $\mathcal{V}$ is closed under addition. Similarly, if $\mathbf{D} = r\mathbf{A}$, then for $i < j$, $d_{ij} = r(a_{ij}) = r(0) = 0$, proving $\mathbf{D} \in \mathcal{V}$. Therefore $\mathcal{V}$ is also closed under scalar multiplication.

(e) True. Let $\mathcal{V}$ be the given subset of $\mathbb{R}^4$. To show that $\mathcal{V}$ is a subspace of $\mathbb{R}^4$, we must show that it is nonempty, and that it is closed under both matrix addition and scalar multiplication. Letting $a = b = 0$ shows that $[0,0,0,0] \in \mathcal{V}$, hence $\mathcal{V}$ is nonempty. $\mathcal{V}$ is closed under addition because $[0,a,b,0] + [0,c,d,0] = [0,(a+c),(b+d),0] = [0,A,B,0] \in \mathcal{V}$, where $A = (a+c)$ and $B = (b+d)$. Finally, $\mathcal{V}$ is closed under scalar multiplication since $c[0,a,b,0] = [0,ca,cb,0] = [0,C,D,0] \in \mathcal{V}$, where $C = ca$ and $D = cb$.

(f) False. While containing the zero vector is a *necessary* condition for a subset of a vector space to be a subspace, it is *not* a *sufficient* condition. For example, the subset $\mathcal{W} = \{[0,0],\ [1,0]\}$ of $\mathbb{R}^2$ contains the zero vector but is not a subspace of $\mathbb{R}^2$. It is not closed under scalar multiplication because $2[1,0] = [2,0] \notin \mathcal{W}$. (Note that $\mathcal{W}$ also fails to be closed under addition.)

(g) True. This is the statement of Theorem 4.3.

(h) True. Theorem 4.4 asserts that the eigenspace corresponding to any eigenvalue of any $n \times n$ matrix is a subspace of $\mathbb{R}^n$.

Section 4.3

(1) In each part we follow the three steps of the Simplified Span Method.

(a) Step 1: Form the matrix $\mathbf{A} = \begin{bmatrix} 1 & 1 & 0 \\ 2 & -3 & -5 \end{bmatrix}$.

Step 2: $\mathbf{A}$ row reduces to $\mathbf{C} = \begin{bmatrix} 1 & 0 & -1 \\ 0 & 1 & 1 \end{bmatrix}$.

Step 3: $\text{span}(S) = \{a[1,0,-1] + b[0,1,1] \mid a,b \in \mathbb{R}\} = \{[a,b,-a+b] \mid a,b \in \mathbb{R}\}$.

(c) Step 1: Form the matrix $\mathbf{A} = \begin{bmatrix} 1 & -1 & 1 \\ 2 & -3 & 3 \\ 0 & 1 & -1 \end{bmatrix}$.

Step 2: $\mathbf{A}$ row reduces to $\mathbf{C} = \begin{bmatrix} 1 & 0 & 0 \\ 0 & 1 & -1 \\ 0 & 0 & 0 \end{bmatrix}$.

Step 3: Using only the nonzero rows of $\mathbf{C}$ gives
$\text{span}(S) = \{a[1,0,0] + b[0,1,-1] \,|\, a,b \in \mathbb{R}\} = \{[a,b,-b] \,|\, a,b \in \mathbb{R}\}$.

(e) Step 1: Form the matrix $\mathbf{A} = \begin{bmatrix} 1 & 3 & 0 & 1 \\ 0 & 0 & 1 & 1 \\ 0 & 1 & 0 & 1 \\ 1 & 5 & 1 & 4 \end{bmatrix}$.

Step 2: $\mathbf{A}$ row reduces to $\mathbf{C} = \begin{bmatrix} 1 & 0 & 0 & -2 \\ 0 & 1 & 0 & 1 \\ 0 & 0 & 1 & 1 \\ 0 & 0 & 0 & 0 \end{bmatrix}$.

Step 3: Using only the nonzero rows of $\mathbf{C}$ gives $\text{span}(S) =$
$\{a[1,0,0,-2] + b[0,1,0,1] + c[0,0,1,1] \,|\, a,b,c \in \mathbb{R}\} = \{[a,b,c,-2a+b+c] \,|\, a,b,c \in \mathbb{R}\}$.

(2) (a) First we convert the polynomials in S into vectors in $\mathbb{R}^4$:
$(x^3-1) \rightarrow [1,0,0,-1]$, $(x^2-x) \rightarrow [0,1,-1,0]$, and $(x-1) \rightarrow [0,0,1,-1]$. Now, we use the Simplified Span Method on these vectors. To do this, we row reduce the matrix

$$\mathbf{A} = \begin{bmatrix} 1 & 0 & 0 & -1 \\ 0 & 1 & -1 & 0 \\ 0 & 0 & 1 & -1 \end{bmatrix} \text{ to obtain } \mathbf{C} = \begin{bmatrix} 1 & 0 & 0 & -1 \\ 0 & 1 & 0 & -1 \\ 0 & 0 & 1 & -1 \end{bmatrix}.$$

We must convert the nonzero rows of $\mathbf{C}$ to polynomial form:
$[1,0,0,-1] \rightarrow (x^3-1)$, $[0,1,0,-1] \rightarrow (x^2-1)$, and $[0,0,1,-1] \rightarrow (x-1)$. Hence, span($S$) is the set of linear combinations of these 3 polynomials. That is, $\text{span}(S) =$
$\{a(x^3-1) + b(x^2-1) + c(x-1) \mid a,b,c \in \mathbb{R}\} = \{ax^3 + bx^2 + cx - (a+b+c) \mid a,b,c \in \mathbb{R}\}$.

(c) First we convert the polynomials in S into vectors in $\mathbb{R}^4$:
$(x^3-x+5) \rightarrow [1,0,-1,5]$, $(3x^3-3x+10) \rightarrow [3,0,-3,10]$, $(5x^3-5x-6) \rightarrow [5,0,-5,-6]$ and $(6x-6x^3-13) \rightarrow [-6,0,6,-13]$. Now we use the Simplified Span Method on these vectors. To do this, we row reduce the matrix

$$\mathbf{A} = \begin{bmatrix} 1 & 0 & -1 & 5 \\ 3 & 0 & -3 & 10 \\ 5 & 0 & -5 & -6 \\ -6 & 0 & 6 & -13 \end{bmatrix} \text{ to obtain } \mathbf{C} = \begin{bmatrix} 1 & 0 & -1 & 0 \\ 0 & 0 & 0 & 1 \\ 0 & 0 & 0 & 0 \\ 0 & 0 & 0 & 0 \end{bmatrix}.$$

We must convert the nonzero rows of $\mathbf{C}$ to polynomial form:
$[1,0,-1,0] \rightarrow (x^3-x)$, and $[0,0,0,1] \rightarrow (1)$. Hence, span($S$) is the set of linear combinations of these 2 polynomials. That is, $\text{span}(S) = \{a(x^3-x) + b(1) \,|\, a,b \in \mathbb{R}\} = \{ax^3 - ax + b \,|\, a,b \in \mathbb{R}\}$.

(3) (a) First we convert the matrices in S into vectors in $\mathbb{R}^4$:

$\begin{bmatrix} -1 & 1 \\ 0 & 0 \end{bmatrix} \rightarrow [-1,1,0,0]$, $\begin{bmatrix} 0 & 0 \\ 1 & -1 \end{bmatrix} \rightarrow [0,0,1,-1]$, and $\begin{bmatrix} -1 & 0 \\ 0 & 1 \end{bmatrix} \rightarrow [-1,0,0,1]$. Now we use the Simplified Span Method on these vectors. To do this, we row reduce the matrix

$$\mathbf{A} = \begin{bmatrix} -1 & 1 & 0 & 0 \\ 0 & 0 & 1 & -1 \\ -1 & 0 & 0 & 1 \end{bmatrix} \text{ to obtain } \mathbf{C} = \begin{bmatrix} 1 & 0 & 0 & -1 \\ 0 & 1 & 0 & -1 \\ 0 & 0 & 1 & -1 \end{bmatrix}.$$

We must convert the nonzero rows of $\mathbf{C}$ to matrix form:

$[1,0,0,-1] \to \begin{bmatrix} 1 & 0 \\ 0 & -1 \end{bmatrix}$, $[0,1,0,-1] \to \begin{bmatrix} 0 & 1 \\ 0 & -1 \end{bmatrix}$, and $[0,0,1,-1] \to \begin{bmatrix} 0 & 0 \\ 1 & -1 \end{bmatrix}$. Hence, span($S$) is the set of linear combinations of these 3 matrices. That is, span(S) =

$$\left\{ a\begin{bmatrix} 1 & 0 \\ 0 & -1 \end{bmatrix} + b\begin{bmatrix} 0 & 1 \\ 0 & -1 \end{bmatrix} + c\begin{bmatrix} 0 & 0 \\ 1 & -1 \end{bmatrix} \middle| \, a,b,c \in \mathbb{R} \right\} = \left\{ \begin{bmatrix} a & b \\ c & -a-b-c \end{bmatrix} \middle| \, a,b,c \in \mathbb{R} \right\}.$$

(c) First we convert the matrices in S into vectors in $\mathbb{R}^4$:

$\begin{bmatrix} 1 & -1 \\ 3 & 0 \end{bmatrix} \to [1,-1,3,0]$, $\begin{bmatrix} 2 & -1 \\ 8 & -1 \end{bmatrix} \to [2,-1,8,-1]$, $\begin{bmatrix} -1 & 4 \\ 4 & -1 \end{bmatrix} \to [-1,4,4,-1]$,

and $\begin{bmatrix} 3 & -4 \\ 5 & 6 \end{bmatrix} \to [3,-4,5,6]$. Now we use the Simplified Span Method on these vectors. To do this, we row reduce the matrix

$$\mathbf{A} = \begin{bmatrix} 1 & -1 & 3 & 0 \\ 2 & -1 & 8 & -1 \\ -1 & 4 & 4 & -1 \\ 3 & -4 & 5 & 6 \end{bmatrix} \text{ to obtain } \mathbf{C} = \begin{bmatrix} 1 & 0 & 0 & 0 \\ 0 & 1 & 0 & 0 \\ 0 & 0 & 1 & 0 \\ 0 & 0 & 0 & 1 \end{bmatrix}.$$

We must convert the nonzero rows of $\mathbf{C}$ to matrix form: $[1,0,0,0] \to \begin{bmatrix} 1 & 0 \\ 0 & 0 \end{bmatrix}$,

$[0,1,0,0] \to \begin{bmatrix} 0 & 1 \\ 0 & 0 \end{bmatrix}$, $[0,0,1,0] \to \begin{bmatrix} 0 & 0 \\ 1 & 0 \end{bmatrix}$, and $[0,0,0,1] \to \begin{bmatrix} 0 & 0 \\ 0 & 1 \end{bmatrix}$. Hence, span($S$) is the set of linear combinations of these 4 matrices.

That is, $\text{span}(S) = \left\{ a\begin{bmatrix} 1 & 0 \\ 0 & 0 \end{bmatrix} + b\begin{bmatrix} 0 & 1 \\ 0 & 0 \end{bmatrix} + c\begin{bmatrix} 0 & 0 \\ 1 & 0 \end{bmatrix} + d\begin{bmatrix} 0 & 0 \\ 0 & 1 \end{bmatrix} \middle| \, a,b,c,d \in \mathbb{R} \right\}$

$= \left\{ \begin{bmatrix} a & b \\ c & d \end{bmatrix} \middle| \, a,b,c,d \in \mathbb{R} \right\} = \mathcal{M}_{22}$.

(4) (a) $\mathcal{W} = \{[a+b, a+c, b+c, c\} \mid a,b,c \in \mathbb{R}\} = \{a[1,1,0,0] + b[1,0,1,0] + c[0,1,1,1] \mid a,b,c \in \mathbb{R}\}$

$= \text{the row space of } \mathbf{A}, \text{ where } \mathbf{A} = \begin{bmatrix} 1 & 1 & 0 & 0 \\ 1 & 0 & 1 & 0 \\ 0 & 1 & 1 & 1 \end{bmatrix}$.

(b) $\mathbf{A}$ row reduces to $\mathbf{B} = \begin{bmatrix} 1 & 0 & 0 & -\frac{1}{2} \\ 0 & 1 & 0 & \frac{1}{2} \\ 0 & 0 & 1 & \frac{1}{2} \end{bmatrix}$.

(c) From the Simplified Span Method, we know that $\mathcal{W}$ also equals the row space of $\mathbf{B}$, which equals $\{a[1,0,0,-\frac{1}{2}] + b[0,1,0,\frac{1}{2}] + c[0,0,1,\frac{1}{2}] \mid a,b,c \in \mathbb{R}\} = \{[a,b,c,-\frac{1}{2}a + \frac{1}{2}b + \frac{1}{2}c] \mid a,b,c \in \mathbb{R}\}$.

(11) Suppose $a(x^3 - 2x^2 + x - 3) + b(2x^3 - 3x^2 + 2x + 5) + c(4x^2 + x - 3) + d(4x^3 - 7x^2 + 4x - 1) = 3x^3 - 8x^2 + 2x + 16$. Equating the coefficients of the x^3 term, the x^2 term, etc., from each side of

this equation produces the following linear system:

$$\begin{cases} a + 2b + 4d = 3 & \leftarrow x^3 \text{ terms} \\ -2a - 3b + 4c - 7d = -8 & \leftarrow x^2 \text{ terms} \\ a + 2b + c + 4d = 2 & \leftarrow x \text{ terms} \\ -3a + 5b - 3c - d = 16 & \leftarrow \text{constant terms} \end{cases}$$

Using either Gaussian elimination or the Gauss-Jordan method on this system produces the solution set $\{[-2d-1, -d+2, -1, d] \mid d \in \mathbb{R}\}$. Choosing $d = 0$ produces the particular solution $[-1, 2, -1, 0]$. This vector provides coefficients for the desired linear combination. That is, $3x^3 - 8x^2 + 2x + 16 = -1(x^3 - 2x^2 + x - 3) + 2(2x^3 - 3x^2 + 2x + 5) - 1(4x^2 + x - 3) + 0(4x^3 - 7x^2 + 4x - 1)$. Other particular solutions to the system will yield alternate linear combinations that produce $3x^3 - 8x^2 + 2x + 16$. Finally, note that the augmented matrix that was row reduced in this solution has the coefficients of each polynomial as its *columns*.

(14) (a) Theorem 1.13 shows that every 2×2 matrix $\mathbf{A}$ can be expressed as the sum of a symmetric matrix $\mathbf{S}$ and a skew-symmetric matrix $\mathbf{V}$. Hence, every matrix in $\mathcal{M}_{22}$ is a linear combination of a matrix $\mathbf{S}$ in S_1 and a matrix $\mathbf{V}$ in S_2. Therefore, $\mathcal{M}_{22} \subseteq \text{span}(S_1 \cup S_2)$. But since $\text{span}(S_1 \cup S_2) \subseteq \mathcal{M}_{22}$, we have $\text{span}(S_1 \cup S_2) = \mathcal{M}_{22}$.

(16) (a) Row reduce $[(1\mathbf{I}_3 - \mathbf{A})|\mathbf{0}] = \left[\begin{array}{ccc|c} 10 & 15 & -8 & 0 \\ 10 & 15 & -8 & 0 \\ 30 & 45 & -24 & 0 \end{array}\right]$ to obtain $\left[\begin{array}{ccc|c} 1 & \frac{3}{2} & -\frac{4}{5} & 0 \\ 0 & 0 & 0 & 0 \\ 0 & 0 & 0 & 0 \end{array}\right]$.

The solution set to the associated linear system is $\{[-\frac{3}{2}b + \frac{4}{5}c, b, c] \mid b, c \in \mathbb{R}\}$. Choosing $b = 2$ (to eliminate fractions) and $c = 0$ produces the eigenvector $[-3, 2, 0]$. Choosing $b = 0$ and $c = 5$ (to eliminate fractions) yields the eigenvector $[4, 0, 5]$. Hence, the desired set of two eigenvectors is $\{[-3, 2, 0], [4, 0, 5]\}$.

(21) First, $S_1 \subseteq S_2 \subseteq \text{span}(S_2)$, by Theorem 4.5, part (1). Then, since $\text{span}(S_2)$ is a subspace of $\mathcal{V}$ containing S_1 (by Theorem 4.5, part (2)), $\text{span}(S_1) \subseteq \text{span}(S_2)$ by Theorem 4.5, part (3).

(24) (b) Suppose $S_1 = \{[1,0,0], [0,1,0]\}$ and $S_2 = \{[0,1,0], [0,0,1]\}$. Then $S_1 \cap S_2 = \{[0,1,0]\}$, and so $\text{span}(S_1 \cap S_2) = \{a[0,1,0] \mid a \in \mathbb{R}\} = \{[0,a,0] \mid a \in \mathbb{R}\}$.
Next, $\text{span}(S_1) = \{b[1,0,0] + c[0,1,0] \mid b, c \in \mathbb{R}\} = \{[b, c, 0] \mid b, c \in \mathbb{R}\}$.
Similarly, $\text{span}(S_2) = \{d[0,1,0] + e[0,0,1] \mid d, e \in \mathbb{R}\} = \{[0, d, e] \mid d, e \in \mathbb{R}\}$. We want to show that $\text{span}(S_1 \cap S_2) = \text{span}(S_1) \cap \text{span}(S_2)$. This is done by proving the two set inclusions $\text{span}(S_1 \cap S_2) \subseteq \text{span}(S_1) \cap \text{span}(S_2)$ and $\text{span}(S_1 \cap S_2) \supseteq \text{span}(S_1) \cap \text{span}(S_2)$. The first of these is already proven for any sets S_1 and S_2 in part (a). To prove the second, we must show that if $\mathbf{x} \in \text{span}(S_1) \cap \text{span}(S_2)$, then $\mathbf{x} \in \text{span}(S_1 \cap S_2)$. So, suppose $\mathbf{x} \in \text{span}(S_1) \cap \text{span}(S_2)$. Thus, $\mathbf{x}$ is in both $\text{span}(S_1)$ and $\text{span}(S_2)$, and so must have the correct form for both sets. But $\mathbf{x} = [x_1, x_2, x_3] \in \text{span}(S_1)$ implies that $x_3 = 0$. Similarly, $\mathbf{x} \in \text{span}(S_2)$ implies that $x_1 = 0$. Hence, $\mathbf{x}$ must have the form $[0, x_2, 0]$, implying $\mathbf{x} \in \text{span}(S_1 \cap S_2)$, since it has the proper form.

(c) Suppose $S_1 = \{[1,0,0], [0,1,0]\}$ and $S_2 = \{[1,0,0], [1,1,0]\}$. Then $S_1 \cap S_2 = \{[1,0,0]\}$, and so $\text{span}(S_1 \cap S_2) = \{a[1,0,0] \mid a \in \mathbb{R}\} = \{[a,0,0] \mid a \in \mathbb{R}\}$. Now $\mathbf{x} = [0,1,0] \in S_1$, and so $\mathbf{x} \in \text{span}(S_1)$, by part (1) of Theorem 4.5. Also, $\mathbf{x} = [0,1,0] = (-1)[1,0,0] + (1)[1,1,0]$ shows that $\mathbf{x} \in \text{span}(S_2)$. Hence, $\mathbf{x} \in \text{span}(S_1) \cap \text{span}(S_2)$. But, since $\mathbf{x}$ has "1" as its second coordinate, it is *not* in $\text{span}(S_1 \cap S_2)$. Therefore, $\text{span}(S_1 \cap S_2) \neq \text{span}(S_1) \cap \text{span}(S_2)$.

(25) (c) Suppose $S_1 = \{x^5\}$ and $S_2 = \{x^4\}$. Then $S_1 \cup S_2 = \{x^5, x^4\}$. Now $\text{span}(S_1) = \{ax^5 \mid a \in \mathbb{R}\}$ and $\text{span}(S_2) = \{bx^4 \mid b \in \mathbb{R}\}$. Consider $\mathbf{p} = x^5 + x^4$. Then $\mathbf{p}$ is clearly in neither $\text{span}(S_1)$ nor

span(S_2), since it does not fit the description of the polynomials in either set. Hence, $\mathbf{p} \notin \text{span}(S_1) \cup \text{span}(S_2)$. However, $\mathbf{p} \in \text{span}(S_1 \cup S_2)$ because it is a linear combination of x^5 and x^4. Therefore, $\text{span}(S_1 \cup S_2) \neq \text{span}(S_1) \cup \text{span}(S_2)$.

(28) (a) False. The set S is permitted to be infinite. The restriction is that span(S) contains only the finite linear combinations from the set S. That is, in each separate linear combination, only a finite number of the vectors in S are used. (See Example 3 of Section 4.3 in the textbook for an example of the span of an infinite set.)

(b) True. Span(S) is defined to be the set of all such finite linear combinations.

(c) False. Theorem 4.5 shows that span(S) is the smallest *subspace* of $\mathcal{V}$ containing S, not the smallest *set* in $\mathcal{V}$ containing S. The smallest set containing S is clearly just S itself (no matter what S is). So, for example, if $S = \{[1,0]\} \subseteq \mathbb{R}^2$, then $\text{span}(S) = \{a[1,0] \mid a \in \mathbb{R}\}$. Hence span($S$) contains vectors such as $[2,0]$ and $[-1,0]$, and so is bigger than S itself.

(d) False. Part (3) of Theorem 4.5 shows that the set inclusion given in this problem should be reversed. That is, $\text{span}(S) \subseteq \mathcal{W}$. For a specific counterexample to the statement in the problem, consider $\mathcal{V} = \mathcal{W} = \mathbb{R}^2$ and $S = \{[1,0]\}$. Then $\mathcal{W}$ contains S, but $\text{span}(S) = \{a[1,0] \mid a \in \mathbb{R}\}$ clearly does not contain all of $\mathcal{W}$, since $[0,1] \in \mathcal{W}$, but $[0,1] \notin \text{span}(S)$.

(e) False. The row space of an $m \times n$ matrix is a subspace of $\mathbb{R}^n$, since it consists of the set of linear combinations of the *rows* of the matrix, each of which has n entries. Therefore, the row space of a 4×5 matrix is a subspace of $\mathbb{R}^5$, not $\mathbb{R}^4$.

(f) True. This statement summarizes the Simplified Span Method.

(g) False. The eigenspace E_λ is the solution set of the homogeneous system $(\lambda\mathbf{I}_n - \mathbf{A})\mathbf{X} = \mathbf{0}$, which is generally different from the row space of $(\lambda\mathbf{I}_n - \mathbf{A})$. For a particular example in which these two sets are different, consider the matrix $\mathbf{A} = \begin{bmatrix} 1 & 0 \\ 0 & 2 \end{bmatrix}$ and the eigenvalue $\lambda = 1$. Then $\lambda\mathbf{I}_2 - \mathbf{A} = \mathbf{I}_2 - \mathbf{A} = \begin{bmatrix} 0 & 0 \\ 0 & -1 \end{bmatrix}$. The row space of $(1\mathbf{I}_2 - \mathbf{A})$ is $\{a[0,-1] \mid a \in \mathbb{R}\}$, which is the set of all 2-vectors having a zero in the first coordinate. However, E_1 is the solution set to the homogeneous system $\begin{cases} 0 = 0 \\ -x_2 = 0 \end{cases}$, which is $\{a[1,0] \mid a \in \mathbb{R}\}$, the set of all 2-vectors having a zero in the second coordinate. Hence, $[1,0] \in E_1$, but $[1,0]$ is not in the row space of $(1\mathbf{I}_2 - \mathbf{A})$. Therefore, E_1 and the row space of $1\mathbf{I}_2 - \mathbf{A}$ are different sets.

Section 4.4

(1) In each part, let S represent the given set.

(a) Because S contains only 1 element, S is *not* linearly dependent (from the definition) since that element is not the zero vector. Therefore, S is linearly independent, by the definition of linear independence.

(b) Because S contains 2 elements, S is linearly dependent if and only if either vector in S is a linear combination of the other vector in S. But since neither vector in S is a scalar multiple of the other, S is linearly independent instead.

(c) Because S contains 2 elements, S is linearly dependent if and only if either vector in S is a linear combination of the other vector in S. Since $[-2,-4,10] = (-2)[1,2,-5]$, S is linearly dependent.

(d) S is linearly dependent because S contains the zero vector $[0,0,0]$, as explained in Example 5 in Section 4.4 of the textbook.

(e) S is a subset of $\mathbb{R}^3$ that contains 4 distinct vectors. Hence, S is linearly dependent by Corollary 4.8.

(2) In each part, let S represent the given set of vectors.

(a) We follow the Independence Test Method.

Step 1: Form the matrix $\mathbf{A} = \begin{bmatrix} 1 & 3 & -2 \\ 9 & 4 & 5 \\ -2 & 5 & -7 \end{bmatrix}$, whose columns are the vectors in S.

Step 2: $\mathbf{A}$ row reduces to $\mathbf{B} = \begin{bmatrix} 1 & 0 & 1 \\ 0 & 1 & -1 \\ 0 & 0 & 0 \end{bmatrix}$.

Step 3: There is no pivot in column 3, so S is linearly dependent.

(b) We follow the Independence Test Method.

Step 1: Form the matrix $\mathbf{A} = \begin{bmatrix} 2 & 4 & -2 \\ -1 & -1 & 0 \\ 3 & 6 & 2 \end{bmatrix}$, whose columns are the vectors in S.

Step 2: $\mathbf{A}$ row reduces to $\mathbf{B} = \mathbf{I}_3$.

Step 3: Since every column in $\mathbf{I}_3$ is a pivot column, S is linearly independent.

(e) We follow the Independence Test Method.

Step 1: Form the matrix $\mathbf{A} = \begin{bmatrix} 2 & 4 & 1 \\ 5 & 3 & -1 \\ -1 & 1 & 1 \\ 6 & 4 & -1 \end{bmatrix}$, whose columns are the vectors in S.

Step 2: $\mathbf{A}$ row reduces to $\mathbf{B} = \begin{bmatrix} 1 & 0 & -\frac{1}{2} \\ 0 & 1 & \frac{1}{2} \\ 0 & 0 & 0 \\ 0 & 0 & 0 \end{bmatrix}$.

Step 3: There is no pivot in column 3, so S is linearly dependent.

(3) In each part, let S represent the given set of polynomials.

(a) First, we convert the polynomials in S into vectors in $\mathbb{R}^3$:
$x^2 + x + 1 \rightarrow [1,1,1]$, $x^2 - 1 \rightarrow [1,0,-1]$, and $x^2 + 1 \rightarrow [1,0,1]$. Next, we use the Independence Test Method on this set of 3-vectors.

Step 1: Form the matrix $\mathbf{A} = \begin{bmatrix} 1 & 1 & 1 \\ 1 & 0 & 0 \\ 1 & -1 & 1 \end{bmatrix}$, whose columns are the vectors converted from the polynomials in S.

Step 2: $\mathbf{A}$ row reduces to $\mathbf{B} = \mathbf{I}_3$.

Step 3: Since every column in $\mathbf{I}_3$ is a pivot column, S is linearly independent.

(c) First, we convert the polynomials in S into vectors in $\mathbb{R}^2$: $2x - 6 \to [2, -6]$, $7x + 2 \to [7, 2]$, and $12x - 7 \to [12, -7]$. Next, we use the Independence Test Method on this set of 2-vectors.

Step 1: Form the matrix $\mathbf{A} = \begin{bmatrix} 2 & 7 & 12 \\ -6 & 2 & -7 \end{bmatrix}$, whose columns are the polynomials in S converted to vectors.

Step 2: $\mathbf{A}$ row reduces to $\mathbf{B} = \begin{bmatrix} 1 & 0 & \frac{73}{46} \\ 0 & 1 & \frac{29}{23} \end{bmatrix}$.

Step 3: There is no pivot in column 3, so S is linearly dependent. (Note that, in this problem, once we converted the polynomials to vectors in $\mathbb{R}^2$, we could easily see from Corollary 4.8 that the vectors are linearly dependent, confirming the result we obtained using the Independence Test Method.)

(4) In each part, let S represent the given set of polynomials.

(a) First, we convert the polynomials in S into vectors in $\mathbb{R}^3$: $x^2 - 1 \to [1, 0, -1]$, $x^2 + 1 \to [1, 0, 1]$, and $x^2 + x \to [1, 1, 0]$. Next, we use the Independence Test Method on this set of 3-vectors.

Step 1: Form the matrix $\mathbf{A} = \begin{bmatrix} 1 & 1 & 1 \\ 0 & 0 & 1 \\ -1 & 1 & 0 \end{bmatrix}$, whose columns are the polynomials in S converted into vectors.

Step 2: $\mathbf{A}$ row reduces to $\mathbf{B} = \mathbf{I}_3$.

Step 3: Since every column in $\mathbf{I}_3$ is a pivot column, S is linearly independent.

(c) First, we convert the polynomials in S into vectors in $\mathbb{R}^3$: $4x^2 + 2 \to [4, 0, 2]$, $x^2 + x - 1 \to [1, 1, -1]$, $x \to [0, 1, 0]$, and $x^2 - 5x - 3 \to [1, -5, -3]$. But this set of four vectors in $\mathbb{R}^3$ is linearly dependent by Corollary 4.8. Hence, S is linearly dependent.

(e) No vector in S can be expressed as a linear combination of the others since each polynomial in S has a distinct degree, and no other polynomial in S has any term of that degree. Therefore, S does not satisfy the definition of a linearly dependent set, and so S is linearly independent.

(7) It is useful to have a simplified description of the vectors in span(S). Using the Simplified Span Method, we row reduce $\begin{bmatrix} 1 & 1 & 0 \\ -2 & 0 & 1 \end{bmatrix}$ to obtain $\begin{bmatrix} 1 & 0 & -\frac{1}{2} \\ 0 & 1 & \frac{1}{2} \end{bmatrix}$. Hence, $\text{span}(S) = \{[a, b, (-\frac{1}{2}a + \frac{1}{2}b)] \mid a, b \in \mathbb{R}\}$.

(b) We want to choose a vector $\mathbf{v}$ outside of span(S), for otherwise, $S \cup \{\mathbf{v}\}$ would be linearly dependent, by the alternate characterization of linear dependence. Let $\mathbf{v} = [0, 1, 0]$, which, using the simplified form of span(S), is easily seen not to be in span(S). Then $S \cup \{\mathbf{v}\} = \{[1, 1, 0], [-2, 0, 1], [0, 1, 0]\}$. We now use the Independence Test Method.

Step 1: Form the matrix $\mathbf{A} = \begin{bmatrix} 1 & -2 & 0 \\ 1 & 0 & 1 \\ 0 & 1 & 0 \end{bmatrix}$, whose columns are the vectors in $S \cup \{\mathbf{v}\}$.

Step 2: $\mathbf{A}$ row reduces to $\mathbf{B} = \mathbf{I}_3$.

Step 3: Since every column in $\mathbf{I}_3$ is a pivot column, $S \cup \{\mathbf{v}\}$ is linearly independent.

(c) Many other choices for $\mathbf{v}$ will work. In fact any vector in $\mathbb{R}^3$ that is outside of span(S) can be chosen here. In particular, consider $\mathbf{v} = [0, 0, 1]$, which does not have the proper form to be in span(S). You can verify that $S \cup \{\mathbf{v}\} = \{[1,1,0],[-2,0,1],[0,0,1]\}$ is linearly independent using the Independence Test Method in a manner similar to that used in part (b).

(d) Any vector $\mathbf{u} \in \text{span}(S)$ other than $[1,1,0]$ or $[-2,0,1]$ will provide the example we need. In particular, consider $\mathbf{u} = [1,1,0]+[-2,0,1] = [-1,1,1]$. Then $S \cup \{\mathbf{u}\} = \{[1,1,0],[-2,0,1],[-1,1,1]\}$. You can verify that $S \cup \{\mathbf{u}\}$ is linearly dependent by using the Independence Test Method. (The underlying principle behind why $S \cup \{\mathbf{u}\}$ is linearly dependent is that $\mathbf{u} \in \text{span}(S) = \text{span}((S \cup \{\mathbf{u}\}) - \{\mathbf{u}\})$. Therefore, we can assert that $S \cup \{\mathbf{u}\}$ is linearly dependent by the alternate characterization of linear dependence.)

(11) In each part, there are many different correct answers. Only one possibility is given here.

(a) Let $S = \{\mathbf{e}_1, \mathbf{e}_2, \mathbf{e}_3, \mathbf{e}_4\}$. We will use Theorem 4.7 to verify that S is linearly independent. Now $a_1\mathbf{e}_1 + a_2\mathbf{e}_2 + a_3\mathbf{e}_3 + a_4\mathbf{e}_4 = [a_1, a_2, a_3, a_4]$, which clearly equals the zero vector $[0,0,0,0]$ if and only if $a_1 = a_2 = a_3 = a_4 = 0$. Therefore, S is linearly independent.

(c) Let $S = \{1, x, x^2, x^3\}$. We will use Theorem 4.7 to verify that S is linearly independent. Now $a_1(1) + a_2x + a_3x^2 + a_4x^3$ clearly equals the zero polynomial if and only if $a_1 = a_2 = a_3 = a_4 = 0$. Therefore, S is linearly independent.

(e) Let $S = \left\{ \begin{bmatrix} 1 & 0 & 0 \\ 0 & 0 & 0 \\ 0 & 0 & 0 \end{bmatrix}, \begin{bmatrix} 0 & 1 & 0 \\ 1 & 0 & 0 \\ 0 & 0 & 0 \end{bmatrix}, \begin{bmatrix} 0 & 0 & 1 \\ 0 & 0 & 0 \\ 1 & 0 & 0 \end{bmatrix}, \begin{bmatrix} 0 & 0 & 0 \\ 0 & 1 & 0 \\ 0 & 0 & 0 \end{bmatrix} \right\}$.

Notice that each matrix in S is symmetric. We will use Theorem 4.7 to verify that S is linearly independent. Now

$$a_1 \begin{bmatrix} 1 & 0 & 0 \\ 0 & 0 & 0 \\ 0 & 0 & 0 \end{bmatrix} + a_2 \begin{bmatrix} 0 & 1 & 0 \\ 1 & 0 & 0 \\ 0 & 0 & 0 \end{bmatrix} + a_3 \begin{bmatrix} 0 & 0 & 1 \\ 0 & 0 & 0 \\ 1 & 0 & 0 \end{bmatrix} + a_4 \begin{bmatrix} 0 & 0 & 0 \\ 0 & 1 & 0 \\ 0 & 0 & 0 \end{bmatrix} = \begin{bmatrix} a_1 & a_2 & a_3 \\ a_2 & a_4 & 0 \\ a_3 & 0 & 0 \end{bmatrix},$$

which clearly equals the zero matrix $\mathbf{O}_3$ if and only if $a_1 = a_2 = a_3 = a_4 = 0$. Therefore, S is linearly independent.

(13) (b) Let S be the given set of three vectors. We prove that $\mathbf{v} = [0,0,-6,0]$ is redundant. To do this, we need to show that $\text{span}(S - \{\mathbf{v}\}) = \text{span}(S)$. We will do this by applying the Simplified Span Method to both $S - \{\mathbf{v}\}$ and S. If the method leads to the same reduced row echelon form (except, perhaps, with an extra row of zeroes) with or without $\mathbf{v}$ as one of the rows, then the two spans are equal.

For $\text{span}(S - \{\mathbf{v}\})$:

$$\begin{bmatrix} 1 & 1 & 0 & 0 \\ 1 & 1 & 1 & 0 \end{bmatrix} \text{ row reduces to } \begin{bmatrix} 1 & 1 & 0 & 0 \\ 0 & 0 & 1 & 0 \end{bmatrix}.$$

For $\text{span}(S)$:

$$\begin{bmatrix} 1 & 1 & 0 & 0 \\ 1 & 1 & 1 & 0 \\ 0 & 0 & -6 & 0 \end{bmatrix} \text{ row reduces to } \begin{bmatrix} 1 & 1 & 0 & 0 \\ 0 & 0 & 1 & 0 \\ 0 & 0 & 0 & 0 \end{bmatrix}.$$

Since the reduced row echelon form matrices are the same, except for the extra row of zeroes, the two spans are equal, and $\mathbf{v} = [0,0,-6,0]$ is a redundant vector.

For an alternate approach to proving that $\text{span}(S - \{\mathbf{v}\}) = \text{span}(S)$, first note that $S - \{\mathbf{v}\} \subseteq S$, and so $\text{span}(S - \{\mathbf{v}\}) \subseteq \text{span}(S)$ by Theorem 4.6. Next, $[1,1,0,0]$ and $[1,1,1,0]$ are clearly in $\text{span}(S - \{\mathbf{v}\})$ by part (1) of Theorem 4.5. But $\mathbf{v} \in \text{span}(S - \{\mathbf{v}\})$ as well, because $[0,0,-6,0] =$

$(6)[1,1,0,0] + (-6)[1,1,1,0]$. Hence, S is a subset of the subspace span$(S - \{\mathbf{v}\})$. Therefore, by part (3) of Theorem 4.5, span$(S) \subseteq$ span$(S - \{\mathbf{v}\})$. Thus, since we have proven subset inclusion in both directions, span$(S - \{\mathbf{v}\}) =$ span(S).

(16) Let $\mathbf{A}$ be the $n \times k$ matrix having the vectors in $S = \{\mathbf{v}_1, \ldots, \mathbf{v}_k\}$ as columns. Then the system $\mathbf{AX} = \mathbf{O}$ has nontrivial solutions by Corollary 2.6. But, if $\mathbf{X} = [x_1, \ldots, x_k]$, then $\mathbf{AX} = x_1\mathbf{v}_1 + \cdots + x_k\mathbf{v}_k$, and so the existence of a nontrivial solution to $\mathbf{AX} = \mathbf{O}$ implies that S is linearly dependent (by Theorem 4.7).

(19) (b) The converse to the statement is: If $S = \{\mathbf{v}_1, \ldots, \mathbf{v}_k\}$ is a linearly independent subset of $\mathbb{R}^m$, then $T = \{\mathbf{A}\mathbf{v}_1, \ldots, \mathbf{A}\mathbf{v}_k\}$ is a linearly independent subset of $\mathbb{R}^n$. For a specific counterexample, consider $S = \{\mathbf{e}_1\}$, a linearly independent subset of $\mathbb{R}^m$. Let $\mathbf{A}$ be the zero matrix $\mathbf{O}_{nm}$. Then $T = \{\mathbf{0}\}$, a linearly dependent subset of $\mathbb{R}^n$.

(25) We prove the contrapositive of Theorem 4.9 as given in the exercise. We may assume that S is an infinite set, since the finite case is already covered by Theorem 4.7. Now, suppose that S is linearly dependent. Then there is a vector $\mathbf{v}$ in S such that $\mathbf{v} \in$ span$(S - \{\mathbf{v}\})$. Therefore, there are vectors $\mathbf{v}_1, \ldots, \mathbf{v}_n \in S - \{\mathbf{v}\}$ and real numbers $a_1, \ldots, a_n$ such that $\mathbf{v} = a_1\mathbf{v}_1 + \cdots + a_n\mathbf{v}_n$. Hence $\mathbf{0} = (-1)\mathbf{v} + a_1\mathbf{v}_1 + \cdots + a_n\mathbf{v}_n$. Note that the coefficient of $\mathbf{v}$ is nonzero.

Conversely, assume $\{\mathbf{v}_1, \ldots, \mathbf{v}_n\} \subseteq S$ with coefficients $a_1, \ldots, a_n$ such that $\mathbf{0} = a_1\mathbf{v}_1 + \cdots + a_n\mathbf{v}_n$, with $a_i \neq 0$ for some i. Then, $\mathbf{v}_i = (-\frac{a_1}{a_i})\mathbf{v}_1 + \cdots + (-\frac{a_{i-1}}{a_i})\mathbf{v}_{i-1} + (-\frac{a_{i+1}}{a_i})\mathbf{v}_{i+1} + \cdots + (-\frac{a_n}{a_i})\mathbf{v}_n$, which shows that $\mathbf{v}_i \in$ span$(S - \{\mathbf{v}_i\})$. Hence S is linearly dependent.

(28) (a) False. $[-8, 12, -4] = (-4)[2, -3, 1]$. Thus, one vector is a scalar multiple (hence a linear combination) of the other vector. Therefore, the set is linearly dependent.

(b) True. This follows directly from the definition of linear dependence.

(c) True. This is the definition of linear dependence for a one-element set.

(d) False. This statement directly contradicts the alternate characterization for linear independence.

(e) True. This follows directly from Theorem 4.7.

(f) True. This follows directly from Corollary 4.8.

(g) False. In order to determine linear independence, we need to row reduce the matrix whose *columns* are the vectors in S, not the matrix whose *rows* are the vectors in S. (The matrix $\mathbf{A}$ whose rows are the vectors in S is used, instead, in the Simplified Span Method to find a simplified form for span(S).) For a specific counterexample to the given statement, consider $S = \{\mathbf{i}, \mathbf{j}, \mathbf{0}\}$ in $\mathbb{R}^2$. Then $\mathbf{A} = \begin{bmatrix} 1 & 0 \\ 0 & 1 \\ 0 & 0 \end{bmatrix}$, which row reduces to $\mathbf{A}$ itself. Note that there is a pivot in both columns of $\mathbf{A}$.

However, S is linearly dependent since it contains the zero vector (see Example 5 in Section 4.4 of the textbook).

(h) True. This follows directly from Theorem 4.10.

(i) True. This follows directly from the definition of linear dependence, since the statement indicates that $\mathbf{v}_3$ is a linear combination of $\mathbf{v}_1$ and $\mathbf{v}_2$.

Section 4.5

(4) (a) The set S is not a basis because S can not span $\mathbb{R}^4$. For if S spans $\mathbb{R}^4$, then by part(1) of Corollary 4.14, $4 = \dim(\mathbb{R}^4) \leq |S| = 3$, a contradiction.

(c) The set S is a basis. To prove this, we will show that S spans $\mathbb{R}^4$, and then apply Corollary 4.14. We show that S spans $\mathbb{R}^4$ using the Simplified Span Method. The matrix $\mathbf{A} = \begin{bmatrix} 7 & 1 & 2 & 0 \\ 8 & 0 & 1 & -1 \\ 1 & 0 & 0 & -2 \\ 3 & 0 & 1 & -1 \end{bmatrix}$ whose rows are the vectors in S reduces to $\mathbf{I}_4$. Hence, $\text{span}(S) = \text{span}(\{\mathbf{e}_1, \mathbf{e}_2, \mathbf{e}_3, \mathbf{e}_4\}) = \mathbb{R}^4$. Now, $|S| = 4 = \dim(\mathbb{R}^4)$, and so part (1) of Corollary 4.14 implies that S is a basis for $\mathbb{R}^4$.

(e) The set S is not a basis since S is not linearly independent. For if S is linearly independent, then by part (2) of Corollary 4.14 $5 = |S| \leq \dim(\mathbb{R}^4) = 4$, a contradiction.

(5) (b) From part (a), the given set B is a maximal linearly independent subset of S. Theorem 4.15 then implies that B is a basis for $\text{span}(S)$. Hence, $\dim(\text{span}(S)) = |B| = 2$, by the definition of dimension.

(c) No. $\text{Span}(S) \neq \mathbb{R}^4$ because $\dim(\text{span}(S)) = 2 \neq 4 = \dim(\mathbb{R}^4)$. If $\text{span}(S)$ and $\mathbb{R}^4$ were the same vector space, they would have the same dimension.

(11) (b) Part (a) shows that the given set B is a basis for $\mathcal{V}$. Since $|B| = 5$, the definition of dimension yields $\dim(\mathcal{V}) = 5$.

(c) In part (a), we used the strategy of creating a basis for $\mathcal{V}$ by taking the polynomials 1, x, x^2, x^3, and x^4 and multiplying them by $(x-2)$ so that the results would have $x = 2$ as a root. So, in this part we multiply 1, x, x^2, and x^3 by $(x-2)(x-3)$ so that both $x = 2$ and $x = 3$ are roots of the resulting polynomials. This produces the set
$D = \{(x-2)(x-3), x(x-2)(x-3), x^2(x-2)(x-3), x^3(x-2)(x-3)\}$. We need to show that D is a basis for $\mathcal{W}$.
First, we show that D is linearly independent. We use Theorem 4.7. Suppose
$a_1(x-2)(x-3) + a_2x(x-2)(x-3) + a_3x^2(x-2)(x-3) + a_4x^3(x-2)(x-3) = 0$. Then, factoring out $(x-2)(x-3)$ yields $(x-2)(x-3)(a_1 + a_2x + a_3x^2 + a_4x^3) = 0$. But since $(x-2)(x-3)$ is only zero if $x = 2$ or $x = 3$, $a_1 + a_2x + a_3x^2 + a_4x^3 = 0$ for all other values of x. Since this is a polynomial, it can only equal zero for that many values of x if $a_1 = a_2 = a_3 = a_4 = 0$. Hence, D is linearly independent.
To show that D spans $\mathcal{W}$, let $\mathbf{p} \in \mathcal{W}$. Then $\mathbf{p}(2) = \mathbf{p}(3) = 0$. Therefore, $(x-2)$ and $(x-3)$ are factors of $\mathbf{p}$. Thus, since $\mathbf{p}$ has degree at most 5, dividing $\mathbf{p}$ by $(x-2)(x-3)$ produces a polynomial of the form $ax^3 + bx^2 + cx + d$. That is, $\mathbf{p}(x) = (x-2)(x-3)(ax^3 + bx^2 + cx + d)$. Hence, $\mathbf{p}(x) = ax^3(x-2)(x-3) + bx^2(x-2)(x-3) + cx(x-2)(x-3) + d(x-2)(x-3)$, which shows that $\mathbf{p} \in \text{span}(D)$. Therefore, $\mathcal{W} \subseteq \text{span}(D)$. Since D is clearly a subset of $\mathcal{W}$, part (3) of Theorem 4.5 shows that $\text{span}(D) \subseteq \mathcal{W}$. Thus $\text{span}(D) = \mathcal{W}$. This completes the proof that D is a basis for $\mathcal{W}$.

(d) Part (c) shows that the set D defined there is a basis for $\mathcal{W}$. Since $|D| = 4$, the definition of dimension yields $\dim(\mathcal{W}) = 4$.

(12) (a) Let $\mathcal{V} = \mathbb{R}^3$, and let $S = \{[1,0,0], [2,0,0], [3,0,0]\}$. Now $\dim(\mathcal{V}) = 3 = |S|$. However, S does not span $\mathbb{R}^3$. Every element of $\text{span}(S)$ is of the form $a[1,0,0] + b[2,0,0] + c[3,0,0]$, which must have zeroes in the second and third coordinates. Therefore, vectors in $\mathbb{R}^3$ such as $[0,1,0]$ and $[0,0,1]$ are not in $\text{span}(S)$ because they are not of the correct form.

(b) Let $\mathcal{V} = \mathbb{R}^3$, and let $T = \{[1,0,0], [2,0,0], [3,0,0]\}$. Now $\dim(\mathcal{V}) = 3 = |T|$, but T is not linearly independent. This is clear, since $[3,0,0] = [1,0,0] + [2,0,0]$, showing that one vector in T is a linear combination of the other vectors in T.

(18) (b) First, span(B) is a subspace of $\mathcal{V}$, and so by part (3) of Theorem 4.5, span(S) $\subseteq$ span(B). Thus, since span(S) $= \mathcal{V}$, $\mathcal{V} \subseteq$ span(B). Also, $B \subseteq \mathcal{V}$, and so span(B) $\subseteq \mathcal{V}$, again by part (3) of Theorem 4.5. Hence, span(B) $= \mathcal{V}$.

(c) Now $B \subset C \subseteq S$, but $B \neq C$, since $\mathbf{w} \notin B$. Therefore, the definition for "B is a maximal linearly independent subset of S" given in the textbook implies that C is linearly dependent.

(19) (b) Let S be a spanning set for a vector space $\mathcal{V}$. If S is linearly dependent, then there is a subset C of S such that $C \neq S$ and C spans $\mathcal{V}$.

(c) The statement in part (b) is essentially half of the "if and only if" statement in Exercise 12 in Section 4.4. We rewrite the appropriate half of the proof of that exercise.
Since S is linearly dependent, then S is nonempty and $a_1\mathbf{v}_1 + \cdots + a_n\mathbf{v}_n = \mathbf{0}$ for some $\{\mathbf{v}_1, \ldots, \mathbf{v}_n\} \subseteq S$ and $a_1, \ldots, a_n \in \mathbb{R}$, with some $a_i \neq 0$. Then
$\mathbf{v}_i = -\frac{a_1}{a_i}\mathbf{v}_1 - \cdots - \frac{a_{i-1}}{a_i}\mathbf{v}_{i-1} - \frac{a_{i+1}}{a_i}\mathbf{v}_{i+1} - \cdots - \frac{a_n}{a_i}\mathbf{v}_n$. Now, if $\mathbf{w} \in$ span(S), then $\mathbf{w} = b\mathbf{v}_i + c_1\mathbf{w}_1 + \cdots + c_k\mathbf{w}_k$, for some $\{\mathbf{w}_1, \ldots, \mathbf{w}_k\} \subseteq S - \{\mathbf{v}_i\}$ and $b, c_1, \ldots, c_k \in \mathbb{R}$, where b could be zero. Hence, $\mathbf{w} = -\frac{ba_1}{a_i}\mathbf{v}_1 - \cdots - \frac{ba_{i-1}}{a_i}\mathbf{v}_{i-1} - \frac{ba_{i+1}}{a_i}\mathbf{v}_{i+1} - \cdots - \frac{ba_n}{a_i}\mathbf{v}_n + c_1\mathbf{w}_1 + \cdots + c_k\mathbf{w}_k \in$ span($S - \{\mathbf{v}_i\}$). So, span(S) $\subseteq$ span($S - \{\mathbf{v}_i\}$). Also, span($S - \{\mathbf{v}_i\}$) $\subseteq$ span(S) by Theorem 4.6. Thus span($S - \{\mathbf{v}_i\}$) $=$ span(S) $= \mathcal{V}$. Letting $C = S - \{\mathbf{v}_i\}$ completes the proof.

(22) (c) First, T is linearly independent by the definition of A. Next, suppose $C \subseteq \mathcal{W}$ with $T \subset C$ and $T \neq C$. We must show that C is linearly dependent. We prove C is linearly dependent by contradiction. So, suppose C is linearly independent instead. Then $|C| \in A$, by the definition of A. (Note that $|C|$ is finite, by an argument similar to that given in part (b).) But $T \subset C$ and $T \neq C$ implies that $|C| > |T| = n$. This contradicts the fact that n is the largest element of A. Therefore, C must be linearly dependent, proving that T is a *maximal* linearly independent subset of $\mathcal{W}$.

(d) In part (c) we proved that T is a maximal linearly independent subset of $\mathcal{W}$. But $\mathcal{W}$ is a spanning set for itself since $\mathcal{W}$ is a subspace of $\mathcal{V}$. Hence, Theorem 4.15 shows that T is a basis for $\mathcal{W}$.

(25) (a) True. This is the definition of a basis for a vector space.

(b) False. As we saw in Example 7 in Section 4.5 of the textbook, $\dim(\mathcal{P}_n) = n + 1$. Hence, $\dim(\mathcal{P}_4) = 5$, implying that every basis for $\mathcal{P}_4$ has five elements, not four.

(c) False. As we saw in Example 8 in Section 4.5 of the textbook, $\dim(\mathcal{M}_{mn}) = m \cdot n$. Hence, $\dim(\mathcal{M}_{43}) = 4 \cdot 3 = 12$, not 7.

(d) False. Part (1) of Corollary 4.14 shows, instead, that $|S| \geq n$. For a specific counterexample to the statement in the problem, consider $S = \{[1, 0], [0, 1], [1, 1]\}$, a subset of $\mathbb{R}^2$. Now S spans $\mathbb{R}^2$ because every 2-vector can be expressed as a linear combination of vectors in S as follows: $[a, b] = a[1, 0] + b[0, 1]$. However, $|S| = 3 > 2 = \dim(\mathbb{R}^2)$.

(e) False. Corollary 4.14 shows, instead, that $|T| \leq n$. For a specific counterexample to the equation $|T| = n$ in the given statement, consider $T = \{[1, 0]\}$, a subset of $\mathbb{R}^2$. Then T is linearly independent by definition, since $[1, 0] \neq [0, 0]$. But, $|T| = 1 < 2 = \dim(\mathbb{R}^2)$.

(f) True. This is proven using both parts of Corollary 4.14 as follows: $|T| \leq \dim(\mathcal{W})$ (by part (2)) $\leq |S|$ (by part (1)).

(g) False. Recall that $\mathcal{V}$ is a subspace of itself. Letting $\mathcal{W} = \mathcal{V}$ gives an case in which $\dim(\mathcal{W}) = \dim(\mathcal{V})$, contradicting the strict inequality given in the problem.

(h) False. Consider the infinite dimensional vector space $\mathcal{P}$ from Exercise 16 in Section 4.5. $\mathcal{P}_5$ is a finite dimensional subspace of $\mathcal{P}$. (Note: $\dim(\mathcal{P}_5) = 6$.)

(i) False. We can only conclude that both S and T are bases for $\mathcal{V}$ (Theorems 4.15 and 4.16). If $\mathcal{V}$ is finite dimensional, that would imply that $|S| = |T|$, but not necessarily that $S = T$. Any two different bases for the same vector space can be used to provide a counterexample. For a particular counterexample, let $S = \{[1,0],[0,1]\}$ and let $T = \{[2,0],[0,2]\}$. Now S is a spanning set for $\mathbb{R}^2$ because $[a,b] = a[1,0] + b[0,1]$. Also, any proper subset of S has fewer that 2 elements, and so by part (1) of Corollary 4.14, can not span $\mathbb{R}^2$. Hence S is a minimal spanning set for $\mathbb{R}^2$. Also, T is a linearly independent subset of $\mathbb{R}^2$ because $a[2,0] + b[0,2] = [2a, 2b] = [0,0]$ implies $2a = 2b = 0$, yielding $a = b = 0$. But any larger set of vectors containing T will have more than 2 elements, and so can not be linearly independent, by part (2) of Corollary 4.14. Thus, T is a maximal linearly independent subset of $\mathbb{R}^2$. Note that $S \neq T$.

(j) True. Since $\mathbf{A}$ is nonsingular, $\mathbf{A}$ row reduces to $\mathbf{I}_4$. The Simplified Span Method then shows that the span of the rows of $\mathbf{A}$ equals the span of the rows of $\mathbf{I}_4$, which equals $\mathbb{R}^4$. But, since $\mathbf{A}$ has exactly 4 rows, the size of this spanning set equals $\dim(\mathbb{R}^4) = 4$. Part (1) of Corollary 4.14 then proves that the rows of $\mathbf{A}$ form a basis for $\mathbb{R}^4$.

Section 4.6

(1) (a) In accordance with the Simplified Span Method, we row reduce the matrix whose rows are the vectors in the given set S. Hence, we row reduce

$$\begin{bmatrix} 1 & 2 & 3 & -1 & 0 \\ 3 & 6 & 8 & -2 & 0 \\ -1 & -1 & -3 & 1 & 1 \\ -2 & -3 & -5 & 1 & 1 \end{bmatrix} \text{ to obtain } \begin{bmatrix} 1 & 0 & 0 & 2 & -2 \\ 0 & 1 & 0 & 0 & 1 \\ 0 & 0 & 1 & -1 & 0 \\ 0 & 0 & 0 & 0 & 0 \end{bmatrix}.$$

The set of nonzero rows of the row reduced matrix is a basis B for span(S). That is, $B = \{[1,0,0,2,-2],[0,1,0,0,1],[0,0,1,-1,0]\}$.

(d) In accordance with the Simplified Span Method, we row reduce the matrix whose rows are the vectors in the given set S. Hence, we row reduce

$$\begin{bmatrix} 1 & 1 & 1 & 1 & 1 \\ 1 & 2 & 3 & 4 & 5 \\ 0 & 1 & 2 & 3 & 4 \\ 0 & 0 & 4 & 0 & -1 \end{bmatrix} \text{ to obtain } \begin{bmatrix} 1 & 0 & 0 & -2 & -\frac{13}{4} \\ 0 & 1 & 0 & 3 & \frac{9}{2} \\ 0 & 0 & 1 & 0 & -\frac{1}{4} \\ 0 & 0 & 0 & 0 & 0 \end{bmatrix}.$$

The set of nonzero rows of the row reduced matrix is basis B for span(S). That is, $B = \{[1,0,0,-2,-\frac{13}{4}],[0,1,0,3,\frac{9}{2}],[0,0,1,0,-\frac{1}{4}]\}$.

(2) First, we convert the polynomials in S into vectors in $\mathbb{R}^4$: $(x^3 - 3x^2 + 2) \to [1,-3,0,2]$, $(2x^3 - 7x^2 + x - 3) \to [2,-7,1,-3]$, and $(4x^3 - 13x^2 + x + 5) \to [4,-13,1,5]$. Next, using the Simplified Span Method, we row reduce the matrix whose rows are the converted vectors, above. Hence, we row reduce

$$\begin{bmatrix} 1 & -3 & 0 & 2 \\ 2 & -7 & 1 & -3 \\ 4 & -13 & 1 & 5 \end{bmatrix} \text{ to obtain } \begin{bmatrix} 1 & 0 & -3 & 0 \\ 0 & 1 & -1 & 0 \\ 0 & 0 & 0 & 1 \end{bmatrix}.$$

A basis B for span(S) is found by converting the nonzero rows of the reduced matrix into polynomials in $\mathcal{P}_3$. This produces $B = \{x^3 - 3x, x^2 - x, 1\}$.

(3) First, we convert the matrices in S into vectors in $\mathbb{R}^6$. This is done by changing each 3×2 matrix $\mathbf{A}$ into its corresponding 6-vector $[a_{11}, a_{12}, a_{21}, a_{22}, a_{31}, a_{32}]$. Doing this for each of the 4 matrices in

S yields the vectors $[1, 4, 0, -1, 2, 2]$, $[2, 5, 1, -1, 4, 9]$, $[1, 7, -1, -2, 2, -3]$, and $[3, 6, 2, -1, 6, 12]$. Next, using the Simplified Span Method, we row reduce the matrix whose rows are these vectors, above. Hence, we row reduce

$$\begin{bmatrix} 1 & 4 & 0 & -1 & 2 & 2 \\ 2 & 5 & 1 & -1 & 4 & 9 \\ 1 & 7 & -1 & -2 & 2 & -3 \\ 3 & 6 & 2 & -1 & 6 & 12 \end{bmatrix} \text{ to obtain } \begin{bmatrix} 1 & 0 & \frac{4}{3} & \frac{1}{3} & 2 & 0 \\ 0 & 1 & -\frac{1}{3} & -\frac{1}{3} & 0 & 0 \\ 0 & 0 & 0 & 0 & 0 & 1 \\ 0 & 0 & 0 & 0 & 0 & 0 \end{bmatrix}.$$

A basis B for span(S) is found by converting the nonzero rows of the row reduced matrix back into 3×2 matrices. Hence, $B = \left\{ \begin{bmatrix} 1 & 0 \\ \frac{4}{3} & \frac{1}{3} \\ 2 & 0 \end{bmatrix}, \begin{bmatrix} 0 & 1 \\ -\frac{1}{3} & -\frac{1}{3} \\ 0 & 0 \end{bmatrix}, \begin{bmatrix} 0 & 0 \\ 0 & 0 \\ 0 & 1 \end{bmatrix} \right\}$.

(4) In parts (a), (c) and (e), we use the Independence Test Method, as described in Section 4.6. That is, we create a matrix whose columns are the vectors in the given set S. We row reduce this matrix and note which columns are pivot columns. The vectors in the corresponding columns of the original matrix constitute a basis for span(S) which is a subset of S. In part (h), we must use a different procedure because the set S is infinite.

(a) Creating the matrix whose columns are the vectors in S, we get

$$\begin{bmatrix} 3 & 0 & 6 \\ 1 & 0 & 2 \\ -2 & 0 & -3 \end{bmatrix}, \text{ which row reduces to } \begin{bmatrix} 1 & 0 & 0 \\ 0 & 0 & 1 \\ 0 & 0 & 0 \end{bmatrix}.$$

Since columns 1 and 3 are pivot columns, we choose the 1st and 3rd columns of the original matrix for our basis for span(S). Hence, $B = \{[3, 1, -2], [6, 2, -3]\}$.

(c) Creating the matrix whose columns are the vectors in S, we get

$$\begin{bmatrix} 1 & 2 & 3 & 0 & -2 \\ 3 & 1 & -6 & 1 & 1 \\ -2 & 4 & 18 & -1 & -6 \end{bmatrix}, \text{ row reducing to } \begin{bmatrix} 1 & 0 & -3 & 0 & 0 \\ 0 & 1 & 3 & 0 & -1 \\ 0 & 0 & 0 & 1 & 2 \end{bmatrix}.$$

Since columns 1, 2 and 4 are pivot columns, we choose the 1st, 2nd and 4th columns of the original matrix for our basis for span(S). Hence, $B = \{[1, 3, -2], [2, 1, 4], [0, 1, -1]\}$.

(e) Creating the matrix whose columns are the vectors in S, we get

$$\begin{bmatrix} 3 & 1 & 3 & -1 \\ -2 & 2 & -2 & -10 \\ 2 & -1 & 7 & 6 \end{bmatrix}, \text{ which row reduces to } \begin{bmatrix} 1 & 0 & 0 & 1 \\ 0 & 1 & 0 & -4 \\ 0 & 0 & 1 & 0 \end{bmatrix}.$$

Since columns 1, 2 and 3 are pivot columns, we choose the 1st, 2nd and 3rd columns of the original matrix for our basis for span(S). Hence, $B = \{[3, -2, 2], [1, 2, -1], [3, -2, 7]\}$.

(h) Because the given set S is infinite, we will use the Inspection Method. We notice that the 2nd coordinates of the vectors in S are described in terms of their 1st and 3rd coordinates, that is, $x_2 = -3x_1 + x_3$. This inspires us to consider the 1st and 3rd coordinates as "independent variables" and solve for the second coordinate, creating vectors in a manner similar to Step 3 of the Diagonalization Method of Section 3.4. Thus, we let $x_1 = 1$ and $x_3 = 0$ to obtain $x_2 = (-3)(1) + 0 = -3$. This gives us the vector $[1, -3, 0]$. Next, letting $x_1 = 0$ and $x_3 = 1$ yields $x_2 = (-3)(0) + 1 = 1$, producing $[0, 1, 1]$. Now the set $B = \{[1, -3, 0], [0, 1, 1]\}$ is clearly linearly independent, since neither vector is a scalar multiple of the other. We claim that B is a maximal linearly independent subset of S, and hence is a basis for span(S) by Theorem 4.15. To do this, we need to show that every vector in S is a linear combination of the vectors in B, because then, adding another vector from S to B would make B linearly dependent. So, let $\mathbf{v} = [v_1, v_2, v_3] \in S$. Then, $v_2 = -3v_1 + v_3$. Hence $\mathbf{v} = [v_1, -3v_1 + v_3, v_3] = v_1[1, -3, 0] + v_3[0, 1, 1]$, and so $\mathbf{v}$ is a linear combination of the vectors in B.

(5) (a) First, we convert the polynomials in S into vectors in $\mathbb{R}^4$: $(x^3 - 8x^2 + 1) \to [1, -8, 0, 1]$, $(3x^3 - 2x^2 + x) \to [3, -2, 1, 0]$, $(4x^3 + 2x - 10) \to [4, 0, 2, -10]$, $(x^3 - 20x^2 - x + 12) \to [1, -20, -1, 12]$, $(x^3 + 24x^2 + 2x - 13) \to [1, 24, 2, -13]$, and $(x^3 + 14x^2 - 7x + 18) \to [1, 14, -7, 18]$. Next, we use the Independence Test Method. That is, we create a matrix whose columns are the vectors above. We row reduce this matrix and note which columns are pivot columns. The polynomials for the vectors in the corresponding columns of the original matrix constitute a basis for span(S) which is a subset of S. Thus, we row reduce the matrix

$$\begin{bmatrix} 1 & 3 & 4 & 1 & 1 & 1 \\ -8 & -2 & 0 & -20 & 24 & 14 \\ 0 & 1 & 2 & -1 & 2 & -7 \\ 1 & 0 & -10 & 12 & -13 & 18 \end{bmatrix} \text{ to } \begin{bmatrix} 1 & 0 & 0 & 0 & -3 & 205 \\ 0 & 1 & 0 & 0 & 0 & 83 \\ 0 & 0 & 1 & 0 & 1 & -\frac{181}{2} \\ 0 & 0 & 0 & 1 & 0 & -91 \end{bmatrix}.$$

Since columns 1, 2, 3 and 4 are pivot columns, we choose the polynomials corresponding to the 1st, 2nd, 3rd and 4th columns of the original matrix for our basis for span(S). Hence, $B = \{x^3 - 8x^2 + 1,\ 3x^3 - 2x^2 + x,\ 4x^3 + 2x - 10,\ x^3 - 20x^2 - x + 12\}$.

(c) Because the set S is infinite, we use the Inspection Method. We begin by choosing the nonzero polynomial x^3 in S. Then we must choose more elements from S that are independent from those previously chosen. We choose x^2 and x, each of which is clearly not in the span of the polynomials chosen before them. This gives us $B = \{x^3, x^2, x\}$. We must show that B is a maximal linearly independent subset of S, which will make it a basis for span(S), by Theorem 4.15. To do this, we must show that every polynomial in S can be expressed as a linear combination of the polynomials in B. For then, adding any polynomial from S to B would make B linearly dependent. Now every polynomial $\mathbf{p}$ in S is of the form $\mathbf{p}(x) = ax^3 + bx^2 + cx$, since it is in $\mathcal{P}_3$ and has zero constant term. But this form clearly expresses $\mathbf{p}$ as a linear combination of the polynomials in B.

(e) Because the set S is infinite, we use the Inspection Method. We begin by choosing the nonzero polynomial $x^3 + x^2$ in S. Then we must choose more elements from S that are independent from those previously chosen. We choose x and 1, each of which is clearly not in the span of the polynomials chosen before them. This gives us $B = \{x^3 + x^2, x, 1\}$. We must show that B is a maximal linearly independent subset of S, which will make it a basis for span(S), by Theorem 4.15. To do this, we must show that every polynomial in S can be expressed as a linear combination of the polynomials in B. For then, adding any polynomial from S to B would make B linearly dependent. Now every polynomial $\mathbf{p}$ in S is of the form $\mathbf{p}(x) = ax^3 + ax^2 + bx + c$, since it is in $\mathcal{P}_3$ and has the coefficients of its x^2 and x^3 terms equal. But then $\mathbf{p}(x) = a(x^3 + x^2) + bx + c(1)$, a linear combination of the polynomials in B.

(6) (a) Recall that the standard basis for $\mathcal{M}_{33}$ is $B = \{\Psi_{ij} \mid 1 \le i \le 3,\ 1 \le j \le 3\}$, where Ψ_{ij} is the 3×3 matrix having 1 as its (i, j) entry and zeroes elsewhere. (See Example 3 in Section 4.5.) Note that B is a subset of the given set S. B is linearly independent and spans $\mathcal{M}_{33}$. Now span(S) $\supseteq$ span(B) $= \mathcal{M}_{33}$, and so, since span(S) can not be bigger than $\mathcal{M}_{33}$, span(S) $= \mathcal{M}_{33}$. Hence, B is a basis for span(S).

(c) Because the given set S is infinite, we will use the method described in Section 4.6 for shrinking an infinite spanning set to a basis. The first step is to guess at such a basis. To help us make this guess, we first observe that

$$\begin{bmatrix} a & b & c \\ b & d & e \\ c & e & f \end{bmatrix} = a\begin{bmatrix} 1 & 0 & 0 \\ 0 & 0 & 0 \\ 0 & 0 & 0 \end{bmatrix} + b\begin{bmatrix} 0 & 1 & 0 \\ 1 & 0 & 0 \\ 0 & 0 & 0 \end{bmatrix} + c\begin{bmatrix} 0 & 0 & 1 \\ 0 & 0 & 0 \\ 1 & 0 & 0 \end{bmatrix} + d\begin{bmatrix} 0 & 0 & 0 \\ 0 & 1 & 0 \\ 0 & 0 & 0 \end{bmatrix}$$

$+e\begin{bmatrix} 0 & 0 & 0 \\ 0 & 0 & 1 \\ 0 & 1 & 0 \end{bmatrix} + f\begin{bmatrix} 0 & 0 & 0 \\ 0 & 0 & 0 \\ 0 & 0 & 1 \end{bmatrix}$. Thus, we choose $B =$

$$\left\{\begin{bmatrix} 1 & 0 & 0 \\ 0 & 0 & 0 \\ 0 & 0 & 0 \end{bmatrix}, \begin{bmatrix} 0 & 1 & 0 \\ 1 & 0 & 0 \\ 0 & 0 & 0 \end{bmatrix}, \begin{bmatrix} 0 & 0 & 1 \\ 0 & 0 & 0 \\ 1 & 0 & 0 \end{bmatrix}, \begin{bmatrix} 0 & 0 & 0 \\ 0 & 1 & 0 \\ 0 & 0 & 0 \end{bmatrix}, \begin{bmatrix} 0 & 0 & 0 \\ 0 & 0 & 1 \\ 0 & 1 & 0 \end{bmatrix}, \begin{bmatrix} 0 & 0 & 0 \\ 0 & 0 & 0 \\ 0 & 0 & 1 \end{bmatrix}\right\}.$$

Note that every matrix in B is symmetric. Also note that we have already shown that every matrix in S is in span(B), since we expressed every symmetric matrix as a linear combination of matrices in B.

Finally, we need to check that B is linearly independent. But this is also clear, because none of the matrices in B can be expressed as a linear combination of the others, since each contains a 1 in at least one entry in which all of the other matrices in B have a 0.

Now, no further matrices in S can be added to B and maintain linear independence since every matrix in S is already in span(B). Thus, B is a maximal linearly independent subset of S, and so is a basis for span(S).

(7) (a) We use the Enlarging Method.

Step 1: We let $A = \{\mathbf{e}_1, \mathbf{e}_2, \mathbf{e}_3, \mathbf{e}_4, \mathbf{e}_5\}$ be the standard basis for $\mathbb{R}^5$.

Step 2: Form the spanning set $S = \{[1,-3,0,1,4], [2,2,1,-3,1], \mathbf{e}_1, \mathbf{e}_2, \mathbf{e}_3, \mathbf{e}_4, \mathbf{e}_5\}$.

Step 3: We apply the Independence Test Method. To do this, we create the matrix whose columns are the vectors in S. We row reduce this matrix and note which columns are pivot columns. The vectors in the corresponding columns of the original matrix constitute a basis B for span(S) $= \mathbb{R}^5$. Since the vectors in T are listed first, this basis will contain T. Hence, we row reduce

$$\begin{bmatrix} 1 & 2 & 1 & 0 & 0 & 0 & 0 \\ -3 & 2 & 0 & 1 & 0 & 0 & 0 \\ 0 & 1 & 0 & 0 & 1 & 0 & 0 \\ 1 & -3 & 0 & 0 & 0 & 1 & 0 \\ 4 & 1 & 0 & 0 & 0 & 0 & 1 \end{bmatrix} \text{ to } \begin{bmatrix} 1 & 0 & 0 & 0 & 0 & \frac{1}{13} & \frac{3}{13} \\ 0 & 1 & 0 & 0 & 0 & -\frac{4}{13} & \frac{1}{13} \\ 0 & 0 & 1 & 0 & 0 & \frac{7}{13} & -\frac{5}{13} \\ 0 & 0 & 0 & 1 & 0 & \frac{11}{13} & \frac{7}{13} \\ 0 & 0 & 0 & 0 & 1 & \frac{4}{13} & -\frac{1}{13} \end{bmatrix}.$$

Since there are pivots in the first 5 columns, the first 5 columns of the original matrix form the desired basis. That is, $B = \{[1,-3,0,1,4], [2,2,1,-3,1], [1,0,0,0,0], [0,1,0,0,0], [0,0,1,0,0]\}$.

(c) We use the Enlarging Method.

Step 1: We let $A = \{\mathbf{e}_1, \mathbf{e}_2, \mathbf{e}_3, \mathbf{e}_4, \mathbf{e}_5\}$ be the standard basis for $\mathbb{R}^5$.

Step 2: Form the spanning set $S = \{[1,0,-1,0,0], [0,1,-1,1,0], [2,3,-8,-1,0], \mathbf{e}_1, \mathbf{e}_2, \mathbf{e}_3, \mathbf{e}_4, \mathbf{e}_5\}$.

Step 3: We apply the Independence Test Method. To do this, we create the matrix whose columns are the vectors in S. We row reduce this matrix and note which columns are pivot columns. The vectors in the corresponding columns of the original matrix constitute a basis B for span(S) $= \mathbb{R}^5$. Since the vectors in T are listed first, this basis will contain T. Hence, we row reduce

$$\begin{bmatrix} 1 & 0 & 2 & 1 & 0 & 0 & 0 & 0 \\ 0 & 1 & 3 & 0 & 1 & 0 & 0 & 0 \\ -1 & -1 & -8 & 0 & 0 & 1 & 0 & 0 \\ 0 & 1 & -1 & 0 & 0 & 0 & 1 & 0 \\ 0 & 0 & 0 & 0 & 0 & 0 & 0 & 1 \end{bmatrix} \text{ to obtain } \begin{bmatrix} 1 & 0 & 0 & 0 & -\frac{9}{4} & -1 & \frac{5}{4} & 0 \\ 0 & 1 & 0 & 0 & \frac{1}{4} & 0 & \frac{3}{4} & 0 \\ 0 & 0 & 1 & 0 & \frac{1}{4} & 0 & -\frac{1}{4} & 0 \\ 0 & 0 & 0 & 1 & \frac{7}{4} & 1 & -\frac{3}{4} & 0 \\ 0 & 0 & 0 & 0 & 0 & 0 & 0 & 1 \end{bmatrix}.$$

Since there are pivots in the first 4 columns and in column 8, the first 4 columns and column 8 of the original matrix form the desired basis. That is,
$B = \{[1,0,-1,0,0], [0,1,-1,1,0], [2,3,-8,-1,0], [1,0,0,0,0], [0,0,0,0,1]\}$.

(8) (a) First we convert the polynomials in T to vectors in $\mathbb{R}^5$:
$(x^3 - x^2) \to [0,1,-1,0,0]$, and $(x^4 - 3x^3 + 5x^2 - x) \to [1,-3,5,-1,0]$. Now we use the Enlarging Method on this set of vectors to find a basis for $\mathbb{R}^5$.
Step 1: We let $A = \{\mathbf{e}_1, \mathbf{e}_2, \mathbf{e}_3, \mathbf{e}_4, \mathbf{e}_5\}$ be the standard basis for $\mathbb{R}^5$.
Step 2: Form the spanning set $S = \{[0,1,-1,0,0], [1,-3,5,-1,0], \mathbf{e}_1, \mathbf{e}_2, \mathbf{e}_3, \mathbf{e}_4, \mathbf{e}_5\}$.
Step 3: We apply the Independence Test Method. To do this, we create the matrix whose columns are the vectors in S. We row reduce this matrix and note which columns are pivot columns. The vectors in the corresponding columns of the original matrix constitute a basis B for span$(S) = \mathbb{R}^5$. Since the vectors corresponding to the polynomials in T are listed first, this basis will contain T. Hence, we row reduce

$$\begin{bmatrix} 0 & 1 & 1 & 0 & 0 & 0 & 0 \\ 1 & -3 & 0 & 1 & 0 & 0 & 0 \\ -1 & 5 & 0 & 0 & 1 & 0 & 0 \\ 0 & -1 & 0 & 0 & 0 & 1 & 0 \\ 0 & 0 & 0 & 0 & 0 & 0 & 1 \end{bmatrix} \text{ to } \begin{bmatrix} 1 & 0 & 0 & 0 & -1 & -5 & 0 \\ 0 & 1 & 0 & 0 & 0 & -1 & 0 \\ 0 & 0 & 1 & 0 & 0 & 1 & 0 \\ 0 & 0 & 0 & 1 & 1 & 2 & 0 \\ 0 & 0 & 0 & 0 & 0 & 0 & 1 \end{bmatrix}.$$

Since there are pivots in columns 1, 2, 3, 4 and 7, columns 1, 2, 3, 4, and 7 of the original matrix form the basis $\{[0,1,-1,0,0], [1,-3,5,-1,0], \mathbf{e}_1, \mathbf{e}_2, \mathbf{e}_5\}$ for $\mathbb{R}^5$. Converting these 5-vectors back to polynomials in $\mathcal{P}_4$ yields the basis $B = \{x^3 - x^2, x^4 - 3x^3 + 5x^2 - x, x^4, x^3, 1\}$.

(c) First we convert the polynomials in T to vectors in $\mathbb{R}^5$:
$(x^4 - x^3 + x^2 - x + 1) \to [1,-1,1,-1,1]$, $(x^3 - x^2 + x - 1) \to [0,1,-1,1,-1]$, and $(x^2 - x + 1) \to [0,0,1,-1,1]$. Now we use the Enlarging Method on this set of vectors to find a basis for $\mathbb{R}^5$.
Step 1: We let $A = \{\mathbf{e}_1, \mathbf{e}_2, \mathbf{e}_3, \mathbf{e}_4, \mathbf{e}_5\}$ be the standard basis for $\mathbb{R}^5$.
Step 2: Form the spanning set $S = \{[1,-1,1,-1,1], [0,1,-1,1,-1], [0,0,1,-1,1], \mathbf{e}_1, \mathbf{e}_2, \mathbf{e}_3, \mathbf{e}_4, \mathbf{e}_5\}$.
Step 3: We apply the Independence Test Method. To do this, we create the matrix whose columns are the vectors in S. We row reduce this matrix and note which columns are pivot columns. The vectors in the corresponding columns of the original matrix constitute a basis B for span$(S) = \mathbb{R}^5$. Since the vectors corresponding to the polynomials in T are listed first, this basis will contain T. Hence, we row reduce

$$\begin{bmatrix} 1 & 0 & 0 & 1 & 0 & 0 & 0 & 0 \\ -1 & 1 & 0 & 0 & 1 & 0 & 0 & 0 \\ 1 & -1 & 1 & 0 & 0 & 1 & 0 & 0 \\ -1 & 1 & -1 & 0 & 0 & 0 & 1 & 0 \\ 1 & -1 & 1 & 0 & 0 & 0 & 0 & 1 \end{bmatrix} \text{ to obtain } \begin{bmatrix} 1 & 0 & 0 & 1 & 0 & 0 & 0 & 0 \\ 0 & 1 & 0 & 1 & 1 & 0 & 0 & 0 \\ 0 & 0 & 1 & 0 & 1 & 0 & 0 & 1 \\ 0 & 0 & 0 & 0 & 0 & 1 & 0 & -1 \\ 0 & 0 & 0 & 0 & 0 & 0 & 1 & 1 \end{bmatrix}.$$

Since there are pivots in columns 1, 2, 3, 6, and 7, columns 1, 2, 3, 6 and 7 of the original matrix form the basis $\{[1,-1,1,-1,1], [0,1,-1,1,-1], [0,0,1,-1,1], \mathbf{e}_3, \mathbf{e}_4\}$ for $\mathbb{R}^5$. Converting these 5-

vectors back to polynomials in $\mathcal{P}_4$ yields the basis
$B = \{x^4 - x^3 + x^2 - x + 1, x^3 - x^2 + x - 1, x^2 - x + 1, x^2, x\}$.

(9) (a) First we convert the matrices in T to vectors in $\mathbb{R}^6$. This is done by changing each 3×2 matrix $\mathbf{A}$ into the 6-vector $[a_{11}, a_{12}, a_{21}, a_{22}, a_{31}, a_{32}]$. Doing this for both of the matrices in T yields the vectors $[1, -1, -1, 1, 0, 0]$ and $[0, 0, 1, -1, -1, 1]$. Now we use the Enlarging Method on this set of vectors to find a basis for $\mathbb{R}^6$.
Step 1: We let $A = \{\mathbf{e}_1, \mathbf{e}_2, \mathbf{e}_3, \mathbf{e}_4, \mathbf{e}_5, \mathbf{e}_6\}$ be the standard basis for $\mathbb{R}^6$.
Step 2: Form the spanning set $S = \{[1, -1, -1, 1, 0, 0], [0, 0, 1, -1, -1, 1], \mathbf{e}_1, \mathbf{e}_2, \mathbf{e}_3, \mathbf{e}_4, \mathbf{e}_5, \mathbf{e}_6\}$.
Step 3: We apply the Independence Test Method. To do this, we create the matrix whose columns are the vectors in S. We row reduce this matrix and note which columns are pivot columns. The vectors in the corresponding columns of the original matrix constitute a basis B for span$(S) = \mathbb{R}^6$. Since the vectors corresponding to the matrices in T are listed first, this basis will contain T. Hence, we row reduce

$$\begin{bmatrix} 1 & 0 & 1 & 0 & 0 & 0 & 0 & 0 \\ -1 & 0 & 0 & 1 & 0 & 0 & 0 & 0 \\ -1 & 1 & 0 & 0 & 1 & 0 & 0 & 0 \\ 1 & -1 & 0 & 0 & 0 & 1 & 0 & 0 \\ 0 & -1 & 0 & 0 & 0 & 0 & 1 & 0 \\ 0 & 1 & 0 & 0 & 0 & 0 & 0 & 1 \end{bmatrix} \text{ to obtain } \begin{bmatrix} 1 & 0 & 0 & 0 & 0 & 1 & 0 & 1 \\ 0 & 1 & 0 & 0 & 0 & 0 & 0 & 1 \\ 0 & 0 & 1 & 0 & 0 & -1 & 0 & -1 \\ 0 & 0 & 0 & 1 & 0 & 1 & 0 & 1 \\ 0 & 0 & 0 & 0 & 1 & 1 & 0 & 0 \\ 0 & 0 & 0 & 0 & 0 & 0 & 1 & 1 \end{bmatrix}.$$

Since there are pivots in columns 1, 2, 3, 4, 5 and 7, columns 1, 2, 3, 4, 5 and 7 of the original matrix form the basis $\{[1, -1, -1, 1, 0, 0], [0, 0, 1, -1, -1, 1], \mathbf{e}_1, \mathbf{e}_2, \mathbf{e}_3, \mathbf{e}_5\}$ for $\mathbb{R}^6$. Converting these 6-vectors back to matrices in $\mathcal{M}_{32}$ yields the basis

$$B = \left\{ \begin{bmatrix} 1 & -1 \\ -1 & 1 \\ 0 & 0 \end{bmatrix}, \begin{bmatrix} 0 & 0 \\ 1 & -1 \\ -1 & 1 \end{bmatrix}, \begin{bmatrix} 1 & 0 \\ 0 & 0 \\ 0 & 0 \end{bmatrix}, \begin{bmatrix} 0 & 1 \\ 0 & 0 \\ 0 & 0 \end{bmatrix}, \begin{bmatrix} 0 & 0 \\ 1 & 0 \\ 0 & 0 \end{bmatrix}, \begin{bmatrix} 0 & 0 \\ 0 & 0 \\ 1 & 0 \end{bmatrix} \right\}.$$

(c) First we convert the matrices in T to vectors in $\mathbb{R}^6$. This is done by changing each 3×2 matrix $\mathbf{A}$ into the 6-vector $[a_{11}, a_{12}, a_{21}, a_{22}, a_{31}, a_{32}]$. Doing this for the four matrices in T yields the vectors $[3, 0, -1, 7, 0, 1]$, $[-1, 0, 1, 3, 0, -2]$, $[2, 0, 3, 1, 0, -1]$ and $[6, 0, 0, 1, 0, -1]$. Now we use the Enlarging Method on this set of vectors to find a basis for $\mathbb{R}^6$.
Step 1: We let $A = \{\mathbf{e}_1, \mathbf{e}_2, \mathbf{e}_3, \mathbf{e}_4, \mathbf{e}_5, \mathbf{e}_6\}$ be the standard basis for $\mathbb{R}^6$.
Step 2: Form the spanning set
$S = \{[3, 0, -1, 7, 0, 1], [-1, 0, 1, 3, 0, -2], [2, 0, 3, 1, 0, -1], [6, 0, 0, 1, 0, -1], \mathbf{e}_1, \mathbf{e}_2, \mathbf{e}_3, \mathbf{e}_4, \mathbf{e}_5, \mathbf{e}_6\}$.
Step 3: We apply the Independence Test Method. To do this, we create the matrix whose columns are the vectors in S. We row reduce this matrix and note which columns are pivot columns. The vectors in the corresponding columns of the original matrix constitute a basis B for span$(S) = \mathbb{R}^6$. Since the vectors corresponding to the matrices in T are listed first, this basis will contain T. Hence, we row reduce

$$\begin{bmatrix} 3 & -1 & 2 & 6 & 1 & 0 & 0 & 0 & 0 & 0 \\ 0 & 0 & 0 & 0 & 0 & 1 & 0 & 0 & 0 & 0 \\ -1 & 1 & 3 & 0 & 0 & 0 & 1 & 0 & 0 & 0 \\ 7 & 3 & 1 & 1 & 0 & 0 & 0 & 1 & 0 & 0 \\ 0 & 0 & 0 & 0 & 0 & 0 & 0 & 0 & 1 & 0 \\ 1 & -2 & -1 & -1 & 0 & 0 & 0 & 0 & 0 & 1 \end{bmatrix}$$

to obtain $\begin{bmatrix} 1 & 0 & 0 & 0 & \frac{1}{101} & 0 & \frac{4}{303} & \frac{35}{303} & 0 & \frac{53}{303} \\ 0 & 1 & 0 & 0 & -\frac{8}{101} & 0 & -\frac{32}{303} & \frac{23}{303} & 0 & -\frac{121}{303} \\ 0 & 0 & 1 & 0 & \frac{3}{101} & 0 & \frac{113}{303} & \frac{4}{303} & 0 & \frac{58}{303} \\ 0 & 0 & 0 & 1 & \frac{14}{101} & 0 & -\frac{15}{101} & -\frac{5}{101} & 0 & -\frac{22}{101} \\ 0 & 0 & 0 & 0 & 0 & 1 & 0 & 0 & 0 & 0 \\ 0 & 0 & 0 & 0 & 0 & 0 & 0 & 0 & 1 & 0 \end{bmatrix}$.

Since there are pivots in columns 1, 2, 3, 4, 6 and 9, columns 1, 2, 3, 4, 6 and 9 of the original matrix form the basis $\{[3,0,-1,7,0,1], [-1,0,1,3,0,-2], [2,0,3,1,0,-1], [6,0,0,1,0,-1], \mathbf{e}_2, \mathbf{e}_5\}$ for $\mathbb{R}^6$. Converting these 6-vectors back to matrices in $\mathcal{M}_{32}$ yields the basis

$$B = \left\{ \begin{bmatrix} 3 & 0 \\ -1 & 7 \\ 0 & 1 \end{bmatrix}, \begin{bmatrix} -1 & 0 \\ 1 & 3 \\ 0 & -2 \end{bmatrix}, \begin{bmatrix} 2 & 0 \\ 3 & 1 \\ 0 & -1 \end{bmatrix}, \begin{bmatrix} 6 & 0 \\ 0 & 1 \\ 0 & -1 \end{bmatrix}, \begin{bmatrix} 0 & 1 \\ 0 & 0 \\ 0 & 0 \end{bmatrix}, \begin{bmatrix} 0 & 0 \\ 0 & 0 \\ 1 & 0 \end{bmatrix} \right\}.$$

(10) To find a basis for $\mathcal{U}_4$, we first guess at a basis B, and then verify that B is, indeed, a basis for $\mathcal{U}_4$. Now, recall that the standard basis for $\mathcal{M}_{44}$ is $\{\mathbf{\Psi}_{ij} \mid 1 \le i \le 4,\ 1 \le j \le 4\}$, where $\mathbf{\Psi}_{ij}$ is the 4×4 matrix having 1 as its (i,j) entry and zeroes elsewhere. (See Example 3 in Section 4.5.) However, not all of these matrices are in $\mathcal{U}_4$, since some have nonzero entries below the main diagonal. Let B be the set of these matrices that are upper triangular. That is, $B = \{\mathbf{\Psi}_{ij} \mid 1 \le i \le j \le 4\}$. There are ten such matrices. We claim that B is a basis for $\mathcal{U}_4$. First we show that B is a linearly independent set. Suppose $\sum_{i \le j} d_{ij}\mathbf{\Psi}_{ij} = \mathbf{O}_4$. Since the (i,j) entry of $\sum_{i \le j} d_{ij}\mathbf{\Psi}_{ij}$ is d_{ij}, each $d_{ij} = 0$. Therefore, B is linearly independent, by Theorem 4.7. Next, we prove that B spans $\mathcal{U}_4$. Let $\mathbf{Q} \in \mathcal{U}_4$. Then, since each nonzero entry of q_{ij} of $\mathbf{Q}$ is on or above the main diagonal of $\mathbf{Q}$, we see that $\mathbf{Q} = \sum_{i \le j} q_{ij}\mathbf{\Psi}_{ij}$, which shows that $\mathbf{Q} \in \text{span}(B)$. This completes the proof.

(11) (b) To find a basis for $\mathcal{V}$, we first guess at a basis B, and then verify that B is, indeed, a basis for $\mathcal{V}$. Now, recall that the standard basis for $\mathcal{M}_{33}$ is $\{\mathbf{\Psi}_{ij} \mid 1 \le i \le 3,\ 1 \le j \le 3\}$, where $\mathbf{\Psi}_{ij}$ is the 3×3 matrix having 1 as its (i,j) entry and zeroes elsewhere. (See Example 3 in Section 4.5.) However, not all of these matrices are in $\mathcal{V}$, since some have nonzero trace. However, the six matrices $\mathbf{\Psi}_{ij}$ with $i \ne j$ are all in $\mathcal{V}$. But these 6 are not sufficient to span $\mathcal{V}$, since all have zero entries on the main diagonal. Thus, to get our basis, we want to add more matrices to these, linearly independent of the first 6, and of each other. However, we can not add more than two matrices, or else the basis would have 9 or more elements, contradicting Theorem 4.17, since $\mathcal{V} \ne \mathcal{M}_{33}$. Consider

$\mathbf{C}_1 = \begin{bmatrix} 1 & 0 & 0 \\ 0 & 0 & 0 \\ 0 & 0 & -1 \end{bmatrix}$ and $\mathbf{C}_2 = \begin{bmatrix} 0 & 0 & 0 \\ 0 & 1 & 0 \\ 0 & 0 & -1 \end{bmatrix}$, which are clearly in $\mathcal{V}$. We claim that $B = \{\mathbf{\Psi}_{ij} \mid 1 \le i \le 3,\ 1 \le j \le 3,\ i \ne j\} \cup \{\mathbf{C}_1, \mathbf{C}_2\}$ is a basis for $\mathcal{V}$.

First we show that B is a linearly independent set. Suppose $\sum_{i \ne j} d_{ij}\mathbf{\Psi}_{ij} + c_1\mathbf{C}_1 + c_2\mathbf{C}_2 = \mathbf{O}_3$.

Then, $\begin{bmatrix} c_1 & d_{12} & d_{13} \\ d_{21} & c_2 & d_{23} \\ d_{31} & d_{32} & -c_1 - c_2 \end{bmatrix} = \begin{bmatrix} 0 & 0 & 0 \\ 0 & 0 & 0 \\ 0 & 0 & 0 \end{bmatrix}$. Hence, each $d_{ij} = 0$ for $i \ne j$ and $c_1 = c_2 = 0$.

Therefore, B is linearly independent, by Theorem 4.7.

Next, we prove that B spans $\mathcal{V}$. Let $\mathbf{Q} \in \mathcal{V}$. Then, $q_{33} = -q_{11} - q_{22}$, and so $\mathbf{Q}$ has the form

$\begin{bmatrix} q_{11} & q_{12} & q_{13} \\ q_{21} & q_{22} & q_{23} \\ q_{31} & q_{32} & -q_{11}-q_{22} \end{bmatrix}$. Thus $\mathbf{Q}$ can be expressed as $\mathbf{Q} = \sum_{i \neq j} q_{ij}\mathbf{\Psi}_{ij} + q_{11}\mathbf{C}_1 + q_{22}\mathbf{C}_2$. Hence, $\mathbf{Q} \in \text{span}(B)$. This completes the proof that B is a basis. Therefore, $\dim(\mathcal{V}) = |B| = 8$.

(d) To find a basis for $\mathcal{V}$, we first guess at a basis B, and then verify that B is, indeed, a basis for $\mathcal{V}$. Consider that the definition of $\mathcal{V}$ involves three "independent variables" a, b, and c. Hence, use the polynomials obtained by first setting $a = 1$, $b = 0$, $c = 0$, then $a = 0$, $b = 1$, $c = 0$, and $a = 0$, $b = 0$, $c = 1$. This yields $B = \{x^6 + x^4 + x^2 - x + 3, -x^5 + x^2 - 2, -x^3 + x^2 + x + 16\}$. Clearly, B spans $\mathcal{V}$, since $ax^6 - bx^5 + ax^4 - cx^3 + (a+b+c)x^2 - (a-c)x + (3a - 2b + 16c)$ $= a(x^6 + x^4 + x^2 - x + 3) + b(-x^5 + x^2 - 2) + c(-x^3 + x^2 + x + 16)$. To show B is also linearly independent, set $a(x^6 + x^4 + x^2 - x + 3) + b(-x^5 + x^2 - 2) + c(-x^3 + x^2 + x + 16) = 0$. Examining the coefficient of x^6 yields $a = 0$. Similarly, looking at the coefficients of x^5 and x^3 produces $b = 0$ and $c = 0$. Hence, Theorem 4.7 shows that B is linearly independent. This completes the proof that B is a basis. Therefore, $\dim(\mathcal{V}) = |B| = 3$.

(12) (b) Let $\mathbf{C}_{ij} = \mathbf{\Psi}_{ij} - \mathbf{\Psi}_{ij}^T$, where $\mathbf{\Psi}_{ij}$ is the $n \times n$ matrix having 1 as its (i,j) entry and zeroes elsewhere. That is, $\mathbf{C}_{ij}$ is the $n \times n$ matrix whose (i,j) entry is 1, (j,i) entry is -1, and all other entries are zero. A basis for the skew-symmetric $n \times n$ matrices is $B = \{\mathbf{C}_{ij} \,|\, 1 \leq i < j \leq n\}$. To show this is a basis, we first prove that B is linearly independent. But each $\mathbf{C}_{ij}$, for $i < j$, has its (i,j) entry equal to 1, while every other matrix in B has its (i,j) entry equal to zero. Hence, no $\mathbf{C}_{ij}$ is a linear combination of the other matrices in B. Therefore, B is linearly independent. Also, if $\mathbf{Q}$ is skew-symmetric, then $q_{ji} = -q_{ij}$ for all $i < j$. That is, entries q_{ij} above the main diagonal are paired with their additive inverse q_{ji} below the main diagonal. Thus, $\mathbf{Q} = \sum_{i<j}(q_{ij}\mathbf{\Psi}_{ij} + q_{ji}\mathbf{\Psi}_{ji}) = \sum_{i<j}(q_{ij}\mathbf{\Psi}_{ij} - q_{ij}\mathbf{\Psi}_{ji}) = \sum_{i<j} q_{ij}(\mathbf{\Psi}_{ij} - \mathbf{\Psi}_{ji}) = \sum_{i<j} q_{ij}\mathbf{C}_{ij}$, which shows that B spans the skew-symmetric matrices. This completes the proof that B is a basis. Hence, dim(skew-symmetric matrices) $= |B| = (n^2 - n)/2$. (The number $(n^2 - n)/2$ is computed as follows: The number of non-diagonal entries in an $n \times n$ matrix is $n^2 - n$. Half of these, $(n^2 - n)/2$, are above the main diagonal. Since B contains one matrix corresponding to each of these entries, $|B| = (n^2 - n)/2$.)

(14) Let $\mathcal{V}$ and S be as given in the statement of the theorem.
Let $A = \{k \,|\, \text{a set } T \text{ exists with } T \subseteq S,\ |T| = k, \text{ and } T \text{ linearly independent}\}$.
Step 1: The empty set is linearly independent and is a subset of S. Hence $0 \in A$, and so A is nonempty.
Step 2: Suppose $k \in A$. Then, every linearly independent subset T of S is also a linearly independent subset of $\mathcal{V}$. Hence, using part (2) of Corollary 4.14 on T shows that $k = |T| \leq \dim(\mathcal{V})$.
Step 3: Suppose n is the largest number in A, which must exist, since A is nonempty and all its elements are $\leq \dim(\mathcal{V})$. Let B be a linearly independent subset of S with $|B| = n$, which exists because $n \in A$. We want to show that B is a maximal linearly independent subset of S. Now, B is given as linearly independent. Suppose $C \subseteq S$ with $B \subset C$ and $B \neq C$. We must show that C is linearly dependent. We prove this by contradiction. Thus, we suppose that C is linearly independent instead. Then $|C| \in A$, by the definition of A. (Note that $|C|$ is finite, by an argument similar to that given in Step 2.) But $B \subset C$ and $B \neq C$ implies that $|C| > |B| = n$. This contradicts the fact that n is the largest element of A. Hence C is linearly dependent, and so B is a maximal linearly independent subset of S.
Step 4: The set B from Step 3 is a basis for $\mathcal{V}$, by Theorem 4.15. This completes the proof.

(15) (b) No; consider the subspace $\mathcal{W}$ of $\mathbb{R}^3$ given by $\mathcal{W} = \{[a,0,0] \,|\, a \in \mathbb{R}\}$. No subset of the basis $B = \{[1,1,0], [1,-1,0], [0,0,1]\}$ for $\mathbb{R}^3$ is a basis for $\mathcal{W}$, since none of the vectors in B are in $\mathcal{W}$. (To verify that B is a basis for $\mathbb{R}^3$, note that no vector in B is a linear combination of the other

vectors in B. Hence, B is a linearly independent subset of $\mathbb{R}^3$ with $|B| = 3 = \dim(\mathbb{R}^3)$. Thus, B is a basis for $\mathbb{R}^3$ by part (2) of Corollary 4.14.)

(c) Yes; consider $\mathcal{Y} = \text{span}(B')$. B' is linearly independent because it is a subset of the linearly independent set B. Also, B' clearly spans $\mathcal{Y}$, so B' is a basis for $\mathcal{Y}$. Finally, $B' \subseteq B \subseteq \mathcal{V}$, and so $\mathcal{Y} = \text{span}(B') \subseteq \mathcal{V}$, by part (3) of Theorem 4.5. Of course, $\mathcal{Y}$ is a subspace of $\mathcal{V}$ by part (2) of Theorem 4.5.

(16) (b) In $\mathbb{R}^3$, consider $\mathcal{W} = \{[a, b, 0] \,|\, a, b \in \mathbb{R}\}$. We could let $\mathcal{W}_1' = \{[0, 0, c] \,|\, c \in \mathbb{R}\}$ or $\mathcal{W}_2' = \{[0, c, c] \,|\, c \in \mathbb{R}\}$. We need to show that $\mathcal{W}_1'$ and $\mathcal{W}_2'$ both satisfy the given conditions.

For $\mathcal{W}_1'$: Suppose $\mathbf{v} = [v_1, v_2, v_3] \in \mathbb{R}^3$. Let $\mathbf{w} = [v_1, v_2, 0]$ and $\mathbf{w}' = [0, 0, v_3]$. Then $\mathbf{w} \in \mathcal{W}$ and $\mathbf{w}' \in \mathcal{W}_1'$, and $\mathbf{v} = \mathbf{w} + \mathbf{w}'$, showing existence for the decomposition. To prove uniqueness, suppose $\mathbf{u} = [u_1, u_2, 0] \in \mathcal{W}$ and $\mathbf{u}' = [0, 0, u_3'] \in \mathcal{W}_1'$, and $\mathbf{v} = \mathbf{u} + \mathbf{u}'$. Then $[v_1, v_2, v_3] = [u_1, u_2, u_3'] \Rightarrow$ $v_1 = u_1$, $v_2 = u_2$, and $v_3 = u_3' \Rightarrow \mathbf{u} = \mathbf{w}$, and $\mathbf{u}' = \mathbf{w}'$.

For $\mathcal{W}_2'$: Suppose $\mathbf{v} = [v_1, v_2, v_3] \in \mathbb{R}^3$. Let $\mathbf{w} = [v_1, v_2 - v_3, 0]$ and $\mathbf{w}' = [0, v_3, v_3]$. Then $\mathbf{w} \in \mathcal{W}$ and $\mathbf{w}' \in \mathcal{W}_2'$, and $\mathbf{v} = \mathbf{w} + \mathbf{w}'$, showing existence.
To prove uniqueness, suppose $\mathbf{u} = [u_1, u_2, 0] \in \mathcal{W}$ and $\mathbf{u}' = [0, u_3', u_3'] \in \mathcal{W}_2'$, with $\mathbf{v} = \mathbf{u} + \mathbf{u}'$. Then $[v_1, v_2, v_3] = [u_1, u_2 + u_3', u_3']$. Hence, $v_3 = u_3'$, so $\mathbf{u}' = \mathbf{w}'$. Now, $\mathbf{w} + \mathbf{w}' = \mathbf{v} = \mathbf{u} + \mathbf{u}'$. Subtracting $\mathbf{w}' = \mathbf{u}'$ from both sides of this equation yields $\mathbf{u} = \mathbf{w}$, finishing the proof.

(20) (a) True. This is the statement of Theorem 4.18.

(b) True. This is the statement of Theorem 4.19.

(c) False. The Simplified Span Method produces a basis B for span(S), but B is not necessarily a subset of S. Instead, B contains vectors of a simpler form that those in S. For example, applying the Simplified Span Method to the spanning set $S = \{[1, 1], [1, -2]\}$ for $\mathbb{R}^2$ produces the standard basis $B = \{\mathbf{i}, \mathbf{j}\}$ for $\mathbb{R}^2$. (Try it.) Notice that B is not a subset of S.

(d) True. The Independence Test Method uses row reduction to select a maximal linearly independent subset of S, which, by Theorem 4.15, is a basis for span(S).

(e) True. The Inspection Method, by repeatedly selecting vectors of S linearly independent from those already chosen, produces a maximal linearly independent subset of S, which, by Theorem 4.15, is a basis for span(S).

(f) False. The basis B created by the Enlarging Method satisfies $B \supseteq T$ rather than $B \subseteq T$. (This can be seen in Example 8 in Section 4.6 of the textbook.)

(g) False. The statement has the two methods reversed. The Simplified Span Method actually places the vectors of the given spanning set S as rows in a matrix, while the Independence Test Method places the vectors in S as columns.

Section 4.7

(1) (a) We use the Coordinatization Method.
Steps 1 and 2: Form the augmented matrix $[\mathbf{A}|\mathbf{v}]$, where the columns of $\mathbf{A}$ are the vectors in the basis B, and then find its corresponding reduced row echelon form:

$$[\mathbf{A}|\mathbf{v}] = \left[\begin{array}{ccc|c} 1 & 5 & 0 & 2 \\ -4 & -7 & -4 & -1 \\ 1 & 2 & 1 & 0 \end{array}\right] \longrightarrow \left[\begin{array}{ccc|c} 1 & 0 & 0 & 7 \\ 0 & 1 & 0 & -1 \\ 0 & 0 & 1 & -5 \end{array}\right]$$

Step 3: The row reduced matrix contains no rows with all zero entries on the left and a nonzero

entry on the right, so we proceed to Step 4.
Step 4: $[\mathbf{v}]_B = [7, -1, -5]$, the last column of the row reduced matrix.

(c) We use the Coordinatization Method.
Steps 1 and 2: Form the augmented matrix $[\mathbf{A}|\mathbf{v}]$, where the columns of $\mathbf{A}$ are the vectors in the basis B, and then find its corresponding reduced row echelon form:

$$[\mathbf{A}|\mathbf{v}] = \left[\begin{array}{rrr|r} 2 & 4 & 1 & 7 \\ 3 & 3 & 2 & -4 \\ 1 & 3 & 1 & 5 \\ -2 & 1 & -1 & 13 \\ 2 & -1 & 1 & -13 \end{array}\right] \longrightarrow \left[\begin{array}{rrr|r} 1 & 0 & 0 & -2 \\ 0 & 1 & 0 & 4 \\ 0 & 0 & 1 & -5 \\ 0 & 0 & 0 & 0 \\ 0 & 0 & 0 & 0 \end{array}\right].$$

Step 3: The row reduced matrix contains no rows with all zero entries on the left and a nonzero entry on the right, so we proceed to Step 4.
Step 4: We eliminate the last two rows of the row reduced matrix, since they contain all zeroes. Then $[\mathbf{v}]_B = [-2, 4, -5]$, the remaining entries of the last column.

(e) We convert the given polynomials to vectors in $\mathbb{R}^3$ by changing $ax^2 + bx + c$ to $[a, b, c]$. Then we use the Coordinatization Method.
Steps 1 and 2: Form the augmented matrix $[\mathbf{A}|\mathbf{v}]$, where the columns of $\mathbf{A}$ are the vectors obtained by converting the polynomials in the basis B. Then find its corresponding reduced row echelon form:

$$[\mathbf{A}|\mathbf{v}] = \left[\begin{array}{rrr|r} 3 & 1 & 2 & 13 \\ -1 & 2 & 3 & -5 \\ 2 & -3 & -1 & 20 \end{array}\right] \longrightarrow \left[\begin{array}{rrr|r} 1 & 0 & 0 & 4 \\ 0 & 1 & 0 & -5 \\ 0 & 0 & 1 & 3 \end{array}\right].$$

Step 3: The row reduced matrix contains no rows with all zero entries on the left and a nonzero entry on the right, so we proceed to Step 4.
Step 4: $[\mathbf{v}]_B = [4, -5, 3]$, the entries of the last column of the row reduced matrix. (Note that, even though we began the problem with polynomials for vectors, the answer to the problem, the coordinatization of the polynomial $\mathbf{v}$, is a vector in $\mathbb{R}^3$.)

(g) We convert the given polynomials to vectors in $\mathbb{R}^4$ by changing $ax^3 + bx^2 + cx + d$ to $[a, b, c, d]$. Then we use the Coordinatization Method.
Steps 1 and 2: Form the augmented matrix $[\mathbf{A}|\mathbf{v}]$, where the columns of $\mathbf{A}$ are the vectors obtained by converting the polynomials in the basis B. Then find its corresponding reduced row echelon form:

$$[\mathbf{A}|\mathbf{v}] = \left[\begin{array}{rrr|r} 2 & 1 & -3 & 8 \\ -1 & 2 & -1 & 11 \\ 3 & -1 & 1 & -9 \\ -1 & 3 & 1 & 11 \end{array}\right] \longrightarrow \left[\begin{array}{rrr|r} 1 & 0 & 0 & -1 \\ 0 & 1 & 0 & 4 \\ 0 & 0 & 1 & -2 \\ 0 & 0 & 0 & 0 \end{array}\right].$$

Step 3: The row reduced matrix contains no rows with all zero entries on the left and a nonzero entry on the right, so we proceed to Step 4.
Step 4: We eliminate the last row of the row reduced matrix, since it contains all zeroes. Then, $[\mathbf{v}]_B = [-1, 4, -2]$, the remaining entries of the last column of the row reduced matrix. (Note that, even though we began the problem with polynomials for vectors, the answer to the problem, the coordinatization of the polynomial $\mathbf{v}$, is a vector in $\mathbb{R}^3$.)

(h) We convert the given matrices in $\mathcal{M}_{22}$ to vectors in $\mathbb{R}^4$ by changing $\left[\begin{array}{cc} a & b \\ c & d \end{array}\right]$ to $[a, b, c, d]$. Then we use the Coordinatization Method.

Steps 1 and 2: Form the augmented matrix $[\mathbf{A}|\mathbf{v}]$, where the columns of $\mathbf{A}$ are the vectors obtained by converting the matrices in the basis B. Then find its corresponding reduced row echelon form:

$$[\mathbf{A}|\mathbf{v}] = \left[\begin{array}{ccc|c} 1 & 2 & 1 & -3 \\ -2 & -1 & -1 & -2 \\ 0 & 1 & 3 & 0 \\ 1 & 0 & 1 & 3 \end{array}\right] \longrightarrow \left[\begin{array}{ccc|c} 1 & 0 & 0 & 2 \\ 0 & 1 & 0 & -3 \\ 0 & 0 & 1 & 1 \\ 0 & 0 & 0 & 0 \end{array}\right].$$

Step 3: The row reduced matrix contains no rows with all zero entries on the left and a nonzero entry on the right, so we proceed to Step 4.

Step 4: We eliminate the last row of the row reduced matrix, since it contains all zeroes. Then, $[\mathbf{v}]_B = [2, -3, 1]$, the remaining entries of the last column of the row reduced matrix. (Note that, even though we began the problem with matrices for vectors, the answer to the problem, the coordinatization of the matrix $\mathbf{v}$, is a vector in $\mathbb{R}^3$.)

(j) We convert the given matrices in $\mathcal{M}_{32}$ to vectors in $\mathbb{R}^6$ by changing $\begin{bmatrix} a & b & c \\ d & e & f \end{bmatrix}$ to $[a, b, c, d, e, f]$.

Then we use the Coordinatization Method.

Steps 1 and 2: Form the augmented matrix $[\mathbf{A}|\mathbf{v}]$, where the columns of $\mathbf{A}$ are the vectors obtained by converting the matrices in the basis B. Then find its corresponding reduced row echelon form:

$$[\mathbf{A}|\mathbf{v}] = \left[\begin{array}{cc|c} 1 & -3 & 11 \\ 3 & 1 & 13 \\ -1 & 7 & -19 \\ 2 & 1 & 8 \\ 1 & 2 & 1 \\ 4 & 5 & 10 \end{array}\right] \longrightarrow \left[\begin{array}{cc|c} 1 & 0 & 5 \\ 0 & 1 & -2 \\ 0 & 0 & 0 \\ 0 & 0 & 0 \\ 0 & 0 & 0 \\ 0 & 0 & 0 \end{array}\right].$$

Step 3: The row reduced matrix contains no rows with all zero entries on the left and a nonzero entry on the right, so we proceed to Step 4.

Step 4: We eliminate the last four rows of the row reduced matrix, since it contains all zeroes. Then, $[\mathbf{v}]_B = [5, -2]$, the remaining entries of the last column of the row reduced matrix. (Note that, even though we began the problem with matrices for vectors, the answer to the problem, the coordinatization of the matrix $\mathbf{v}$, is a vector in $\mathbb{R}^2$.)

(2) (a) We use the Transition Matrix Method. To do this, we form the augmented matrix $[\mathbf{C}|\mathbf{B}]$, in which the columns of $\mathbf{B}$ are the vectors in the basis B, and the columns of $\mathbf{C}$ are the vectors in the basis C. Then, we find the reduced row echelon form for this matrix:

$$\left[\begin{array}{ccc|ccc} 1 & 1 & 1 & 1 & 0 & 0 \\ 5 & 6 & 3 & 0 & 1 & 0 \\ 1 & -6 & 14 & 0 & 0 & 1 \end{array}\right] \longrightarrow \left[\begin{array}{ccc|ccc} 1 & 0 & 0 & -102 & 20 & 3 \\ 0 & 1 & 0 & 67 & -13 & -2 \\ 0 & 0 & 1 & 36 & -7 & -1 \end{array}\right].$$

The transition matrix is the matrix on the right side of the row reduced matrix. That is,

$$\mathbf{P} = \begin{bmatrix} -102 & 20 & 3 \\ 67 & -13 & -2 \\ 36 & -7 & -1 \end{bmatrix}.$$

(c) We convert the given polynomials to vectors in $\mathbb{R}^3$ by changing $ax^2 + bx + c$ to $[a, b, c]$. Then we use the Transition Matrix Method. To do this, we form the augmented matrix $[\mathbf{C}|\mathbf{B}]$, in which the columns of $\mathbf{B}$ are the vectors converted from the basis B, and the columns of $\mathbf{C}$ are the vectors converted from the basis C. Then, we find the reduced row echelon form for this matrix:

$$\left[\begin{array}{ccc|ccc} 1 & 3 & 10 & 2 & 8 & 1 \\ 3 & 4 & 17 & 3 & 1 & 0 \\ 1 & 1 & 5 & -1 & 1 & 6 \end{array}\right] \longrightarrow \left[\begin{array}{ccc|ccc} 1 & 0 & 0 & 20 & -30 & -69 \\ 0 & 1 & 0 & 24 & -24 & -80 \\ 0 & 0 & 1 & -9 & 11 & 31 \end{array}\right].$$

The transition matrix is the matrix on the right side of the row reduced matrix. That is,

$$\mathbf{P} = \begin{bmatrix} 20 & -30 & -69 \\ 24 & -24 & -80 \\ -9 & 11 & 31 \end{bmatrix}.$$

(d) We convert the given matrices in $\mathcal{M}_{22}$ to vectors in $\mathbb{R}^4$ by changing $\begin{bmatrix} a & b \\ c & d \end{bmatrix}$ to $[a, b, c, d]$. Then we use the Transition Matrix Method. To do this, we form the augmented matrix $[\mathbf{C}|\mathbf{B}]$, in which the columns of $\mathbf{B}$ are the vectors converted from the basis B, and the columns of $\mathbf{C}$ are the vectors converted from the basis C. Then, we find the reduced row echelon form for this matrix:

$$\left[\begin{array}{cccc|cccc} -1 & 1 & 3 & 1 & 1 & 2 & 3 & 0 \\ 1 & 0 & -4 & -1 & 3 & 1 & 1 & 2 \\ 3 & 0 & -7 & -2 & 5 & 0 & 1 & -4 \\ -1 & 1 & 4 & 1 & 1 & 4 & 0 & 1 \end{array}\right] \longrightarrow \left[\begin{array}{cccc|cccc} 1 & 0 & 0 & 0 & -1 & -4 & 2 & -9 \\ 0 & 1 & 0 & 0 & 4 & 5 & 1 & 3 \\ 0 & 0 & 1 & 0 & 0 & 2 & -3 & 1 \\ 0 & 0 & 0 & 1 & -4 & -13 & 13 & -15 \end{array}\right].$$

The transition matrix is the matrix on the right side of the row reduced matrix. That is,

$$\mathbf{P} = \begin{bmatrix} -1 & -4 & 2 & -9 \\ 4 & 5 & 1 & 3 \\ 0 & 2 & -3 & 1 \\ -4 & -13 & 13 & -15 \end{bmatrix}.$$

(f) We convert the given polynomials to vectors in $\mathbb{R}^5$ by changing $ax^4 + bx^3 + cx^2 + dx + e$ to $[a, b, c, d, e]$. Then we use the Transition Matrix Method. To do this, we form the augmented matrix $[\mathbf{C}|\mathbf{B}]$, in which the columns of $\mathbf{B}$ are the vectors converted from the basis B, and the columns of $\mathbf{C}$ are the vectors converted from the basis C. Then, we find the reduced row echelon form for this matrix:

$$\left[\begin{array}{ccc|ccc} 1 & 2 & 2 & 6 & 1 & 0 \\ 3 & 7 & 5 & 20 & 5 & 5 \\ 0 & 4 & -3 & 7 & 7 & 17 \\ 4 & 3 & 8 & 19 & -1 & -10 \\ -2 & 1 & -7 & -4 & 6 & 19 \end{array}\right] \longrightarrow \left[\begin{array}{ccc|ccc} 1 & 0 & 0 & 6 & 1 & 2 \\ 0 & 1 & 0 & 1 & 1 & 2 \\ 0 & 0 & 1 & -1 & -1 & -3 \\ 0 & 0 & 0 & 0 & 0 & 0 \\ 0 & 0 & 0 & 0 & 0 & 0 \end{array}\right].$$

Now, we eliminate the last two rows of the row reduced matrix, since it contains all zeroes. Then, the transition matrix consists of the remaining entries in the matrix on the right side of the row reduced matrix. That is, $\mathbf{P} = \begin{bmatrix} 6 & 1 & 2 \\ 1 & 1 & 2 \\ -1 & -1 & -3 \end{bmatrix}$.

(4) (a) In each subpart, we use the Transition Matrix Method. In order to find the transition matrix from basis X to basis Y, we form the augmented matrix $[\mathbf{Y}|\mathbf{X}]$, in which the columns of $\mathbf{X}$ are the vectors in the basis X, and the columns of $\mathbf{Y}$ are the vectors in the basis Y. Then, in each case, we convert this matrix to reduced row echelon form. Since, in this problem, there are no rows of all zeroes in any of the row reduced matrices obtained, the matrices to the right of the augmentation bar are the desired transition matrices.

(i) To find the transition matrix from B to C, we row reduce

$\left[\begin{array}{cc|cc} 3 & 2 & 3 & 7 \\ 7 & 5 & 1 & 2 \end{array}\right]$, obtaining $\left[\begin{array}{cc|cc} 1 & 0 & 13 & 31 \\ 0 & 1 & -18 & -43 \end{array}\right]$. Hence, $\mathbf{P} = \left[\begin{array}{cc} 13 & 31 \\ -18 & -43 \end{array}\right]$.

(ii) To find the transition matrix from C to D, we row reduce

$\left[\begin{array}{cc|cc} 5 & 2 & 3 & 2 \\ 2 & 1 & 7 & 5 \end{array}\right]$, obtaining $\left[\begin{array}{cc|cc} 1 & 0 & -11 & -8 \\ 0 & 1 & 29 & 21 \end{array}\right]$. Hence, $\mathbf{Q} = \left[\begin{array}{cc} -11 & -8 \\ 29 & 21 \end{array}\right]$.

(iii) To find the transition matrix from B to D, we row reduce

$\left[\begin{array}{cc|cc} 5 & 2 & 3 & 7 \\ 2 & 1 & 1 & 2 \end{array}\right]$, obtaining $\left[\begin{array}{cc|cc} 1 & 0 & 1 & 3 \\ 0 & 1 & -1 & -4 \end{array}\right]$. Hence $\mathbf{T} = \left[\begin{array}{cc} 1 & 3 \\ -1 & -4 \end{array}\right]$. A straightforward matrix multiplication verifies that $\mathbf{T} = \mathbf{QP}$.

(c) First, we convert the polynomials in each of the given bases to vectors in $\mathbb{R}^3$ by changing ax^2+bx+c to $[a,b,c]$. Then, in each subpart, we use the Transition Matrix Method. Given bases of converted vectors X and Y, in order to find the transition matrix from X to Y, we form the augmented matrix $[\mathbf{Y}|\mathbf{X}]$, in which the columns of $\mathbf{X}$ are the vectors in the basis X, and the columns of $\mathbf{Y}$ are the vectors in the basis Y. Then, in each case, we convert this matrix to reduced row echelon form. Since, in this problem, there are no rows of all zeroes in any of the row reduced matrices obtained, the matrices to the right of the augmentation bar are the desired transition matrices.
(i) To find the transition matrix from B to C, we row reduce

$\left[\begin{array}{ccc|ccc} 1 & 2 & 1 & 1 & 3 & 3 \\ 4 & 1 & 0 & 2 & 7 & 9 \\ 1 & 0 & 0 & 2 & 8 & 13 \end{array}\right]$ to $\left[\begin{array}{c|ccc} & 2 & 8 & 13 \\ \mathbf{I}_3 & -6 & -25 & -43 \\ & 11 & 45 & 76 \end{array}\right]$. Hence, $\mathbf{P} = \left[\begin{array}{ccc} 2 & 8 & 13 \\ -6 & -25 & -43 \\ 11 & 45 & 76 \end{array}\right]$.

(ii) To find the transition matrix from C to D, we row reduce

$\left[\begin{array}{ccc|ccc} 7 & 1 & 1 & 1 & 2 & 1 \\ -3 & 7 & -2 & 4 & 1 & 0 \\ 2 & -3 & 1 & 1 & 0 & 0 \end{array}\right]$ to $\left[\begin{array}{c|ccc} & -24 & -2 & 1 \\ \mathbf{I}_3 & 30 & 3 & -1 \\ & 139 & 13 & -5 \end{array}\right]$. Hence, $\mathbf{Q} = \left[\begin{array}{ccc} -24 & -2 & 1 \\ 30 & 3 & -1 \\ 139 & 13 & -5 \end{array}\right]$.

(iii) To find the transition matrix from B to D, we row reduce

$\left[\begin{array}{ccc|ccc} 7 & 1 & 1 & 1 & 3 & 3 \\ -3 & 7 & -2 & 2 & 7 & 9 \\ 2 & -3 & 1 & 2 & 8 & 13 \end{array}\right]$ to $\left[\begin{array}{c|ccc} & -25 & -97 & -150 \\ \mathbf{I}_3 & 31 & 120 & 185 \\ & 145 & 562 & 868 \end{array}\right]$.

Hence $\mathbf{T} = \left[\begin{array}{ccc} -25 & -97 & -150 \\ 31 & 120 & 185 \\ 145 & 562 & 868 \end{array}\right]$. A straightforward matrix multiplication verifies that $\mathbf{T} = \mathbf{QP}$.

(5) (a) (i) In accordance with the Simplified Span Method, we row reduce

$\left[\begin{array}{ccccc} 1 & -4 & 1 & 2 & 1 \\ 6 & -24 & 5 & 8 & 3 \\ 3 & -12 & 3 & 6 & 2 \end{array}\right]$ to $\left[\begin{array}{ccccc} 1 & -4 & 0 & -2 & 0 \\ 0 & 0 & 1 & 4 & 0 \\ 0 & 0 & 0 & 0 & 1 \end{array}\right]$.

Hence, $C = ([1,-4,0,-2,0],[0,0,1,4,0],[0,0,0,0,1])$.

(ii) We use the Transition Matrix Method. To do this, we form the augmented matrix $[\mathbf{C}|\mathbf{B}]$, in which the *columns* of $\mathbf{B}$ are the vectors in the basis B, and the *columns* of $\mathbf{C}$ are the vectors in the basis C. Then, we convert this matrix to reduced row echelon form:

$$\left[\begin{array}{ccc|ccc} 1 & 0 & 0 & 1 & 6 & 3 \\ -4 & 0 & 0 & -4 & -24 & -12 \\ 0 & 1 & 0 & 1 & 5 & 3 \\ -2 & 4 & 0 & 2 & 8 & 6 \\ 0 & 0 & 1 & 1 & 3 & 2 \end{array}\right] \longrightarrow \left[\begin{array}{ccc|ccc} 1 & 0 & 0 & 1 & 6 & 3 \\ 0 & 1 & 0 & 1 & 5 & 3 \\ 0 & 0 & 1 & 1 & 3 & 2 \\ 0 & 0 & 0 & 0 & 0 & 0 \\ 0 & 0 & 0 & 0 & 0 & 0 \end{array}\right].$$

Now, we eliminate the last two rows of the row reduced matrix, since it contains all zeroes. Then, the transition matrix consists of the remaining entries in the matrix on the right side of the row reduced matrix. That is, $\mathbf{P} = \begin{bmatrix} 1 & 6 & 3 \\ 1 & 5 & 3 \\ 1 & 3 & 2 \end{bmatrix}$.

(iii) According to Theorem 4.23, $\mathbf{Q} = \mathbf{P}^{-1} = \begin{bmatrix} 1 & -3 & 3 \\ 1 & -1 & 0 \\ -2 & 3 & -1 \end{bmatrix}$, where we have computed $\mathbf{P}^{-1}$ by row reducing $[\mathbf{P}|\mathbf{I}_3]$ to $[\mathbf{I}_3|\mathbf{P}^{-1}]$.

(iv) We use the Coordinatization Method to find $[\mathbf{v}]_B$. To do this, we form the augmented matrix $[\mathbf{B}|\mathbf{v}]$, where the columns of $\mathbf{B}$ are the vectors in the basis B, and then find the corresponding reduced row echelon form of this matrix:

$$[\mathbf{B}|\mathbf{v}] = \left[\begin{array}{ccc|c} 1 & 6 & 3 & 2 \\ -4 & -24 & -12 & -8 \\ 1 & 5 & 3 & -2 \\ 2 & 8 & 6 & -12 \\ 1 & 3 & 2 & 3 \end{array}\right] \longrightarrow \left[\begin{array}{ccc|c} 1 & 0 & 0 & 17 \\ 0 & 1 & 0 & 4 \\ 0 & 0 & 1 & -13 \\ 0 & 0 & 0 & 0 \\ 0 & 0 & 0 & 0 \end{array}\right]$$

We eliminate the last two rows of the row reduced matrix, since they contain all zeroes. Then, $[\mathbf{v}]_B = [17, 4, -13]$, the remaining entries of the last column.

Next we compute $[\mathbf{v}]_C$. We form the augmented matrix $[\mathbf{C}|\mathbf{v}]$, where the columns of $\mathbf{C}$ are the vectors in the basis C, and then find the corresponding reduced row echelon form of this matrix:

$$[\mathbf{C}|\mathbf{v}] = \left[\begin{array}{ccc|c} 1 & 0 & 0 & 2 \\ -4 & 0 & 0 & -8 \\ 0 & 1 & 0 & -2 \\ -2 & 4 & 0 & -12 \\ 0 & 0 & 1 & 3 \end{array}\right] \longrightarrow \left[\begin{array}{ccc|c} 1 & 0 & 0 & 2 \\ 0 & 1 & 0 & -2 \\ 0 & 0 & 1 & 3 \\ 0 & 0 & 0 & 0 \\ 0 & 0 & 0 & 0 \end{array}\right]$$

We eliminate the last two rows of the row reduced matrix, since they contain all zeroes. Then, $[\mathbf{v}]_C = [2, -2, 3]$, the remaining entries of the last column.

(v) Since $\mathbf{Q}$ is the transition matrix from C to B, Theorem 4.21 implies that $\mathbf{Q}[\mathbf{v}]_C = [\mathbf{v}]_B$. This can be verified for the answers computed above using matrix multiplication..

(c) (i) In accordance with the Simplified Span Method, we row reduce

$$\begin{bmatrix} 3 & -1 & 4 & 6 \\ 6 & 7 & -3 & -2 \\ -4 & -3 & 3 & 4 \\ -2 & 0 & 1 & 2 \end{bmatrix} \text{ to obtain } \mathbf{I}_4.$$

Hence, $C = ([1,0,0,0], [0,1,0,0], [0,0,1,0], [0,0,0,1])$, the standard basis for $\mathbb{R}^4$.

(ii) We use the Transition Matrix Method. To do this, we form the augmented matrix $[\mathbf{C}|\mathbf{B}]$, in which the *columns* of $\mathbf{B}$ are the vectors in the basis B, and the *columns* of $\mathbf{C}$ are the vectors in the basis C. However, since the matrix $\mathbf{C}$ equals $\mathbf{I}_4$, $[\mathbf{C}|\mathbf{B}]$ is already in reduced row echelon form.

Therefore, the transition matrix $\mathbf{P}$ actually equals $\mathbf{B}$. That is, $\mathbf{P} = \begin{bmatrix} 3 & 6 & -4 & -2 \\ -1 & 7 & -3 & 0 \\ 4 & -3 & 3 & 1 \\ 6 & -2 & 4 & 2 \end{bmatrix}$.

(iii) By Theorem 4.23, $\mathbf{Q} = \mathbf{P}^{-1} = \begin{bmatrix} 1 & -4 & -12 & 7 \\ -2 & 9 & 27 & -\frac{31}{2} \\ -5 & 22 & 67 & -\frac{77}{2} \\ 5 & -23 & -71 & 41 \end{bmatrix}$, where we have computed $\mathbf{P}^{-1}$ by row reducing $[\mathbf{P}|\mathbf{I}_3]$ to $[\mathbf{I}_3|\mathbf{P}^{-1}]$.

(iv) We use the Coordinatization Method to find $[\mathbf{v}]_B$. To do this, we form the augmented matrix $[\mathbf{B}|\mathbf{v}]$, where the columns of $\mathbf{B}$ are the vectors in the basis B, and then find the corresponding reduced row echelon form of this matrix:

$$[\mathbf{B}|\mathbf{v}] = \left[\begin{array}{cccc|c} 3 & 6 & -4 & -2 & 10 \\ -1 & 7 & -3 & 0 & 14 \\ 4 & -3 & 3 & 1 & 3 \\ 6 & -2 & 4 & 2 & 12 \end{array}\right] \longrightarrow \left[\begin{array}{c|c} & 2 \\ \mathbf{I}_4 & 1 \\ & -3 \\ & 7 \end{array}\right].$$

Thus, $[\mathbf{v}]_B = [2, 1, -3, 7]$, the last column of the row reduced matrix.

Next we compute $[\mathbf{v}]_C$. But C is the standard basis for $\mathbb{R}^4$, and so $[\mathbf{v}]_C$ actually equals $\mathbf{v}$. That is, $[\mathbf{v}]_C = [10, 14, 3, 12]$.

(v) Since $\mathbf{Q}$ is the transition matrix from C to B, Theorem 4.21 implies that $\mathbf{Q}[\mathbf{v}]_C = [\mathbf{v}]_B$. This can be verified for the answers computed above using matrix multiplication.

(7) (a) To compute each of these transition matrices, we use the Transition Matrix Method. To do this, we form the augmented matrix $[\mathbf{C}_i|\mathbf{B}]$, for $1 \le i \le 5$, in which the columns of $\mathbf{B}$ are the vectors in the basis B, and the columns of $\mathbf{C}_i$ are the vectors in the basis C_i. Then, we convert each of these matrices to reduced row echelon form:

$$[\mathbf{C}_1|\mathbf{B}] = \left[\begin{array}{ccc|ccc} 3 & 2 & -5 & -5 & 3 & 2 \\ -9 & -5 & 9 & 9 & -9 & -5 \\ 2 & 1 & -1 & -1 & 2 & 1 \end{array}\right] \rightarrow \left[\begin{array}{c|ccc} & 0 & 1 & 0 \\ \mathbf{I}_3 & 0 & 0 & 1 \\ & 1 & 0 & 0 \end{array}\right],$$

$$[\mathbf{C}_2|\mathbf{B}] = \left[\begin{array}{ccc|ccc} 2 & -5 & 3 & -5 & 3 & 2 \\ -5 & 9 & -9 & 9 & -9 & -5 \\ 1 & -1 & 2 & -1 & 2 & 1 \end{array}\right] \rightarrow \left[\begin{array}{c|ccc} & 0 & 0 & 1 \\ \mathbf{I}_3 & 1 & 0 & 0 \\ & 0 & 1 & 0 \end{array}\right],$$

$$[\mathbf{C}_3|\mathbf{B}] = \left[\begin{array}{ccc|ccc} -5 & 2 & 3 & -5 & 3 & 2 \\ 9 & -5 & -9 & 9 & -9 & -5 \\ -1 & 1 & 2 & -1 & 2 & 1 \end{array}\right] \rightarrow \left[\begin{array}{c|ccc} & 1 & 0 & 0 \\ \mathbf{I}_3 & 0 & 0 & 1 \\ & 0 & 1 & 0 \end{array}\right],$$

$$[\mathbf{C}_4|\mathbf{B}] = \left[\begin{array}{ccc|ccc} 3 & -5 & 2 & -5 & 3 & 2 \\ -9 & 9 & -5 & 9 & -9 & -5 \\ 2 & -1 & 1 & -1 & 2 & 1 \end{array}\right] \rightarrow \left[\begin{array}{c|ccc} & 0 & 1 & 0 \\ \mathbf{I}_3 & 1 & 0 & 0 \\ & 0 & 0 & 1 \end{array}\right],$$

$$[\mathbf{C}_5|\mathbf{B}] = \left[\begin{array}{ccc|ccc} 2 & 3 & -5 & -5 & 3 & 2 \\ -5 & -9 & 9 & 9 & -9 & -5 \\ 1 & 2 & -1 & -1 & 2 & 1 \end{array}\right] \rightarrow \left[\begin{array}{c|ccc} & 0 & 0 & 1 \\ \mathbf{I}_3 & 0 & 1 & 0 \\ & 1 & 0 & 0 \end{array}\right].$$

Thus, the transition matrices from B to C_1, C_2, C_3, C_4, and C_5, respectively, are

$$\begin{bmatrix} 0 & 1 & 0 \\ 0 & 0 & 1 \\ 1 & 0 & 0 \end{bmatrix}, \begin{bmatrix} 0 & 0 & 1 \\ 1 & 0 & 0 \\ 0 & 1 & 0 \end{bmatrix}, \begin{bmatrix} 1 & 0 & 0 \\ 0 & 0 & 1 \\ 0 & 1 & 0 \end{bmatrix}, \begin{bmatrix} 0 & 1 & 0 \\ 1 & 0 & 0 \\ 0 & 0 & 1 \end{bmatrix}, \begin{bmatrix} 0 & 0 & 1 \\ 0 & 1 & 0 \\ 1 & 0 & 0 \end{bmatrix}.$$

(10) According to Exercise 9, as suggested in the hint, if $\mathbf{P}$ is the matrix whose columns are the vectors in B, and $\mathbf{Q}$ is the matrix whose columns are the vectors in C, then $\mathbf{Q}^{-1}\mathbf{P}$ is the transition matrix from B to C. Now suppose $\mathbf{A}$ is the transition matrix from B to C given in the problem. Then, $\mathbf{A} = \mathbf{Q}^{-1}\mathbf{P}$, implying $\mathbf{QA} = \mathbf{P}$. This yields $\mathbf{Q} = \mathbf{PA}^{-1}$. Hence, we first compute $\mathbf{A}^{-1}$ by row reducing $[\mathbf{A}|\mathbf{I}_3]$ to $[\mathbf{I}_3|\mathbf{A}^{-1}]$. Then we use the result obtained from that computation to find $\mathbf{Q}$, as follows:

$$\mathbf{Q} = \mathbf{PA}^{-1} = \begin{bmatrix} -2 & 1 & -13 \\ 1 & 0 & 5 \\ 3 & 2 & 10 \end{bmatrix} \begin{bmatrix} 49 & 19 & 86 \\ -5 & -2 & -9 \\ 3 & 1 & 5 \end{bmatrix} = \begin{bmatrix} -142 & -53 & -246 \\ 64 & 24 & 111 \\ 167 & 63 & 290 \end{bmatrix}.$$

So, $C = ([-142, 64, 167], [-53, 24, 63], [-246, 111, 290])$, the columns of $\mathbf{Q}$.

(11) (b) First, $\mathbf{Av} = \begin{bmatrix} 14 & -15 & -30 \\ 6 & -7 & -12 \\ 3 & -3 & -7 \end{bmatrix} \begin{bmatrix} 1 \\ 4 \\ -2 \end{bmatrix} = \begin{bmatrix} 14 \\ 2 \\ 5 \end{bmatrix}$.

We use the Coordinatization Method to find $[\mathbf{Av}]_B$. To do this, we form the augmented matrix $[\mathbf{B}|\mathbf{Av}]$, where the columns of $\mathbf{B}$ are the vectors in the basis B, and then find its corresponding reduced row echelon form:

$$[\mathbf{B}|\mathbf{Av}] = \left[\begin{array}{ccc|c} 5 & 1 & 2 & 14 \\ 2 & 1 & 0 & 2 \\ 1 & 0 & 1 & 5 \end{array}\right] \longrightarrow \left[\begin{array}{c|c} & 2 \\ \mathbf{I}_3 & -2 \\ & 3 \end{array}\right].$$

Thus, $[\mathbf{Av}]_B = [2, -2, 3]$, the last column of the row reduced matrix.

Next, we use the Coordinatization Method to find $[\mathbf{v}]_B$. To do this, we form the augmented matrix $[\mathbf{B}|\mathbf{v}]$, and find its corresponding reduced row echelon form:

$$[\mathbf{B}|\mathbf{v}] = \left[\begin{array}{ccc|c} 5 & 1 & 2 & 1 \\ 2 & 1 & 0 & 4 \\ 1 & 0 & 1 & -2 \end{array}\right] \longrightarrow \left[\begin{array}{c|c} & 1 \\ \mathbf{I}_3 & 2 \\ & -3 \end{array}\right].$$

Thus, $[\mathbf{v}]_B = [1, 2, -3]$, the last column of the row reduced matrix.

Finally, we compute $\mathbf{D}[\mathbf{v}]_B = \begin{bmatrix} 2 & 0 & 0 \\ 0 & -1 & 0 \\ 0 & 0 & -1 \end{bmatrix} \begin{bmatrix} 1 \\ 2 \\ -3 \end{bmatrix} = \begin{bmatrix} 2 \\ -2 \\ 3 \end{bmatrix}$.

And so, $\mathbf{D}[\mathbf{v}]_B = [\mathbf{Av}]_B = [2, -2, 3]$.

(12) Let $\mathcal{V}$, B , and the a_i's and $\mathbf{w}$'s be as given in the statement of the theorem.
Proof of part (1): Suppose that $[\mathbf{w}_1]_B = [b_1, \ldots, b_n]$ and $[\mathbf{w}_2]_B = [c_1, \ldots, c_n]$. Then, by definition, $\mathbf{w}_1 = b_1\mathbf{v}_1 + \cdots + b_n\mathbf{v}_n$ and $\mathbf{w}_2 = c_1\mathbf{v}_1 + \cdots + c_n\mathbf{v}_n$. Hence,
$\mathbf{w}_1 + \mathbf{w}_2 = b_1\mathbf{v}_1 + \cdots + b_n\mathbf{v}_n + c_1\mathbf{v}_1 + \cdots + c_n\mathbf{v}_n = (b_1 + c_1)\mathbf{v}_1 + \cdots + (b_n + c_n)\mathbf{v}_n$, implying $[\mathbf{w}_1 + \mathbf{w}_2]_B = [(b_1 + c_1), \ldots, (b_n + c_n)]$, which equals $[\mathbf{w}_1]_B + [\mathbf{w}_2]_B$.
Proof of part (2): Suppose that $[\mathbf{w}_1]_B = [b_1, \ldots, b_n]$. Then, by definition, $\mathbf{w}_1 = b_1\mathbf{v}_1 + \cdots + b_n\mathbf{v}_n$. Hence, $a_1\mathbf{w}_1 = a_1(b_1\mathbf{v}_1 + \cdots + b_n\mathbf{v}_n) = a_1b_1\mathbf{v}_1 + \cdots + a_1b_n\mathbf{v}_n$, implying $[a_1\mathbf{w}_1]_B = [a_1b_1, \ldots, a_1b_n]$, which equals $a_1[b_1, \ldots, b_n]$.
Proof of part (3): Use induction on k.

Base Step ($k = 1$): This is just part (2).
Inductive Step: Assume true for any linear combination of k vectors, and prove true for a linear combination of $k+1$ vectors. Now,
$[a_1\mathbf{w}_1 + \cdots + a_k\mathbf{w}_k + a_{k+1}\mathbf{w}_{k+1}]_B = [a_1\mathbf{w}_1 + \cdots + a_k\mathbf{w}_k]_B + [a_{k+1}\mathbf{w}_{k+1}]_B$ (by part (1))
$= [a_1\mathbf{w}_1 + \cdots + a_k\mathbf{w}_k]_B + a_{k+1}[\mathbf{w}_{k+1}]_B$ (by part (2))
$= a_1[\mathbf{w}_1]_B + \cdots + a_k[\mathbf{w}_k]_B + a_{k+1}[\mathbf{w}_{k+1}]_B$ (by the inductive hypothesis).

(13) Let $\mathbf{v} \in \mathcal{V}$. Then, applying Theorem 4.21 twice yields $(\mathbf{QP})[\mathbf{v}]_B = \mathbf{Q}[\mathbf{v}]_C = [\mathbf{v}]_D$. Therefore, $(\mathbf{QP})[\mathbf{v}]_B = [\mathbf{v}]_D$ for all $\mathbf{v} \in \mathcal{V}$, and so Theorem 4.21 implies that $\mathbf{QP}$ is the (unique) transition matrix from B to D.

(15) (a) False. Typically, changing the order that the vectors appear in an ordered basis also changes the order of the coordinates in the corresponding coordinatization vectors. For example, if $\mathbf{v} = [1,2,3] \in \mathbb{R}^3$, then since $\mathbf{v} = 1\mathbf{i} + 2\mathbf{j} + 3\mathbf{k}$, $[\mathbf{v}]_B = [1,2,3]$. However, since $\mathbf{v} = 2\mathbf{j} + 3\mathbf{k} + 1\mathbf{i}$, $[\mathbf{v}]_C = [2,3,1]$.

(b) True. Now $\mathbf{b}_i = 0\mathbf{b}_1 + \cdots + 0\mathbf{b}_{i-1} + 1\mathbf{b}_i + 0\mathbf{b}_{i+1} + \cdots + 0\mathbf{b}_n$. Thus, $[\mathbf{b}_i]_B$ is the n-vector having the coefficients $0,\ldots,0,1,0,\ldots 0$, respectively, as its entries. Hence, $[\mathbf{b}_i]_B = \mathbf{e}_i$.

(c) True. This is seen by reversing the roles of B and C in the definition of the transition matrix from B to C.

(d) False. This is close to the statement of Theorem 4.21, except that the positions of $[\mathbf{v}]_B$ and $[\mathbf{v}]_C$ are reversed. The correct conclusion is that $\mathbf{P}[\mathbf{v}]_B = [\mathbf{v}]_C$. For a specific counterexample to the statement given in the problem, consider the ordered bases $B = ([12,9],[15,-1])$ and $C = ([3,4],[6,1])$ for $\mathbb{R}^2$ given in Example 8 in Section 4.7 in the textbook. In that example, we found that the transition matrix from B to C is $\mathbf{P} = \begin{bmatrix} 2 & -1 \\ 1 & 3 \end{bmatrix}$, and, for $\mathbf{v} = [3,39]$, $[\mathbf{v}]_B = [4,-3]$ and $[\mathbf{v}]_C = [11,-5]$. However, $\mathbf{P}[\mathbf{v}]_C = \begin{bmatrix} 27 \\ -4 \end{bmatrix} \neq [\mathbf{v}]_B$.

(e) False. This is close to the statement of Theorem 4.22, except that the transition matrix $\mathbf{T}$ from B to D is $\mathbf{QP}$, *not* $\mathbf{PQ}$. For a specific counterexample to the statement given in the problem, consider the ordered bases B, C, and D for $\mathbb{R}^2$ given in Exercise 4(a) of this section, whose solution appears above. Then, as computed in Exercise 4(a), $\mathbf{P} = \begin{bmatrix} 13 & 31 \\ -18 & -43 \end{bmatrix}$, $\mathbf{Q} = \begin{bmatrix} -11 & -8 \\ 29 & 21 \end{bmatrix}$, and $\mathbf{T} = \begin{bmatrix} 1 & 3 \\ -1 & -4 \end{bmatrix}$. But $\mathbf{PQ} = \begin{bmatrix} 756 & 547 \\ -1049 & -759 \end{bmatrix} \neq \mathbf{T}$.

(f) True. This is part of Theorem 4.23.

(g) False. What is true is that the matrix $\mathbf{P}^{-1}$, *not* $\mathbf{D}$, is the transition matrix from standard coordinates to a basis of eigenvectors for $\mathbf{A}$, where $\mathbf{P}$ is the matrix from the Diagonalization Method whose columns are eigenvectors for $\mathbf{A}$. This is explained in Example 13 of Section 4.7 in the textbook. Using $\mathbf{A}$ and $\mathbf{D}$ from that example gives a specific counterexample to the incorrect given statement. For, if $\mathbf{D}$ were the transition matrix from standard coordinates S to a basis B of eigenvectors, then $\mathbf{D}^{-1}$ would be the transition matrix from B to S. Hence, $\mathbf{b}_1 = \mathbf{D}^{-1}[\mathbf{b}_1]_B$ would be an eigenvector for $\mathbf{A}$ in standard coordinates. But $\mathbf{D}^{-1}$ is the diagonal matrix with main diagonal entries $\frac{1}{2}$, -1, -1, and $[\mathbf{b}_1]_B = \mathbf{e}_1$ (see Exercise 15(b), above). So, $\mathbf{b}_1 = \mathbf{D}^{-1}[\mathbf{b}_1]_B = \mathbf{D}^{-1}\mathbf{e}_1 = [\frac{1}{2},0,0]$. But $\mathbf{A}\mathbf{b}_1$ is $[7,3,\frac{3}{2}]$, which is not a scalar multiple of $\mathbf{b}_1$. Hence, $\mathbf{b}_1$ is not an eigenvector for $\mathbf{A}$.

Chapter 5

Section 5.1

(1) (a) The function f is a linear transformation. To prove this, we show that it satisfies the two properties in the definition of a linear transformation.
Property (1): $f([x,y]+[w,z]) = f([x+w,y+z]) = [3(x+w)-4(y+z), -(x+w)+2(y+z)]$
$= [(3x-4y)+(3w-4z), (-x+2y)+(-w+2z)]$
$= [3x-4y, -x+2y] + [3w-4z, -w+2z] = f([x,y]) + f([w,z])$.
Property (2): $f(c[x,y]) = f([cx,cy]) = [3(cx)-4(cy), -(cx)+2(cy)]$
$= [c(3x-4y), c(-x+2y)] = c[3x-4y, -x+2y] = cf([x,y])$.
Also, f is a linear operator because it is a linear transformation whose domain equals its codomain.

(b) The function h is not a linear transformation. If it were, then part (1) of Theorem 5.1 would imply $h(\mathbf{0}) = \mathbf{0}$. However, $h([0,0,0,0]) = [2,-1,0,-3] \neq \mathbf{0}$, so h can not be a linear transformation. Because h is not a linear transformation, it can not be a linear operator.

(d) The function l is a linear transformation. To prove this, we show that it satisfies the two properties in the definition of a linear transformation.

Property (1): $l\left(\begin{bmatrix} a & b \\ c & d \end{bmatrix} + \begin{bmatrix} e & f \\ g & h \end{bmatrix}\right) = l\left(\begin{bmatrix} a+e & b+f \\ c+g & d+h \end{bmatrix}\right)$

$$= \begin{bmatrix} (a+e)-2(c+g)+(d+h) & 3(b+f) \\ -4(a+e) & (b+f)+(c+g)-3(d+h) \end{bmatrix}$$

$$= \begin{bmatrix} (a-2c+d)+(e-2g+h) & 3b+3f \\ -4a-4e & (b+c-3d)+(f+g-3h) \end{bmatrix}$$

$$= \begin{bmatrix} a-2c+d & 3b \\ -4a & b+c-3d \end{bmatrix} + \begin{bmatrix} e-2g+h & 3f \\ -4e & f+g-3h \end{bmatrix}$$

$$= l\left(\begin{bmatrix} a & b \\ c & d \end{bmatrix}\right) + l\left(\begin{bmatrix} e & f \\ g & h \end{bmatrix}\right).$$

Property (2): $l\left(s\begin{bmatrix} a & b \\ c & d \end{bmatrix}\right) = l\left(\begin{bmatrix} sa & sb \\ sc & sd \end{bmatrix}\right)$

$$= \begin{bmatrix} (sa)-2(sc)+(sd) & 3(sb) \\ -4(sa) & (sb)+(sc)-3(sd) \end{bmatrix}$$

$$= \begin{bmatrix} s(a-2c+d) & s(3b) \\ s(-4a) & s(b+c-3d) \end{bmatrix} = s\begin{bmatrix} a-2c+d & 3b \\ -4a & b+c-3d \end{bmatrix} = sl\left(\begin{bmatrix} a & b \\ c & d \end{bmatrix}\right).$$

Also, l is a linear operator because it is a linear transformation whose domain equals its codomain.

(f) The function r is not a linear transformation since Property (2) of the definition fails. For example, $r(8x^3) = 2x^2$, but $8\ r(x^3) = 8(x^2)$. Note that Property (1) also fails, but once we have shown that one of the properties fails it is not necessary to show that the other fails in order to prove that r is not a linear transformation. Also, r is not a linear operator because it is not a linear transformation.

(h) The function t is a linear transformation. To prove this, we show that it satisfies the two properties in the definition of a linear transformation.
Property (1): $t((ax^3+bx^2+cx+d)+(ex^3+fx^2+gx+h)) = t((a+e)x^3+(b+f)x^2+(c+g)x+(d+h))$

$= (a+e)+(b+f)+(c+g)+(d+h) = (a+b+c+d)+(e+f+g+h)$
$= t(ax^3+bx^2+cx+d)+t(ex^3+fx^2+gx+h)$.
Property (2): $t(s(ax^3+bx^2+cx+d)) = t((sa)x^3+(sb)x^2+(sc)x+(sd))$
$= (sa)+(sb)+(sc)+(sd) = s(a+b+c+d) = s\ t(ax^3+bx^2+cx+d)$.
Finally, t is not a linear operator because its domain, $\mathcal{P}_3$, does not equal its codomain, $\mathbb{R}$.

(j) The function v is not a linear transformation because Property (1) of the definition fails to hold. For example, $v(x^2+(x+1)) = v(x^2+x+1) = 1$, but $v(x^2)+v(x+1) = 0+0 = 0$. Note that Property (2) also fails, but once we have shown that one of the properties fails it is not necessary to show that the other fails in order to prove that v is not a linear transformation. Also, v is not a linear operator because it is not a linear transformation.

(l) The function e is not a linear transformation because Property (1) of the definition fails to hold. For example, $e([3,0]+[0,4]) = e([3,4]) = \sqrt{3^2+4^2} = 5$, but $e([3,0])+e([0,4])$ $= \sqrt{3^2+0^2}+\sqrt{0^2+4^2} = 3+4 = 7$. Note that Property (2) also fails (try a negative scalar), but once we have shown that one of the properties fails it is not necessary to show that the other fails in order to prove that e is not a linear transformation. Also, e is not a linear operator because it is not a linear transformation.

(10) (c) We insert the 2×2 matrix from Exercise 9 in the 1st and 3rd rows and columns to perform the rotation in the xz-plane, and keep the 2nd row and 2nd column the same as the identity matrix to indicate that the y-axis stays fixed. Thus we get the matrix $\mathbf{A} = \begin{bmatrix} \cos\theta & 0 & -\sin\theta \\ 0 & 1 & 0 \\ \sin\theta & 0 & \cos\theta \end{bmatrix}$. Let us verify that $\mathbf{A}$ works.

We express vectors in $\mathbb{R}^3$ using cylindrical coordinates, where the cylinders have the y-axis as their axis of rotation, rather than the z-axis, as is usually done. In particular, if r represents the length of a vector $\mathbf{v} = [x,y,z]$, and α represents the angle $[x,0,z]$ makes with the positive y-axis, then $x = r\cos(\alpha)$ and $z = r\sin(\alpha)$. Then
$\mathbf{A}\mathbf{v} = \mathbf{A}[r\cos(\alpha), y, r\sin(\alpha)] = [r\cos(\alpha)\cos(\theta) - r\sin(\alpha)\sin(\theta), y, r\cos(\alpha)\sin(\theta) + r\sin(\alpha)\cos(\theta)]$
$= [r\cos(\alpha+\theta), y, r\sin(\alpha+\theta)]$, where we have used the trigonometric identities
$\cos(\alpha+\theta) = \cos(\alpha)\cos(\theta) - \sin(\alpha)\sin(\theta)$ and $\sin(\alpha+\theta) = \sin(\alpha)\cos(\theta)+\cos(\alpha)\sin(\theta)$. But, from the form $[r\cos(\alpha+\theta), y, r\sin(\alpha+\theta)]$ we see that multiplying $\mathbf{v}$ by $\mathbf{A}$ has added θ to the angle α made in the xz-plane with the positive y-axis. Therefore, multiplying by $\mathbf{A}$ represents a rotation through the angle θ about the y-axis.

(26) We first need to express $\mathbf{i}$ and $\mathbf{j}$ as linear combinations of $\mathbf{i}+\mathbf{j}$ and $-2\mathbf{i}+3\mathbf{j}$. Hence, suppose

$$\begin{cases} a(\mathbf{i}+\mathbf{j}) + b(-2\mathbf{i}+3\mathbf{j}) = \mathbf{i}, & \text{and} \\ c(\mathbf{i}+\mathbf{j}) + d(-2\mathbf{i}+3\mathbf{j}) = \mathbf{j}. \end{cases}$$

Setting the first and second coordinates of each side of these two equations equal to each other produces the linear systems

$$\begin{cases} a - 2b = 1 \\ a + 3b = 0 \end{cases} \text{ and } \begin{cases} c - 2d = 0 \\ c + 3d = 1 \end{cases}.$$

Solving these systems yields $a = \frac{3}{5}$, $b = -\frac{1}{5}$, $c = \frac{2}{5}$, and $d = \frac{1}{5}$. Therefore, $\mathbf{i} = \frac{3}{5}(\mathbf{i}+\mathbf{j}) - \frac{1}{5}(-2\mathbf{i}+3\mathbf{j})$, and so $L(\mathbf{i}) = L\left(\frac{3}{5}(\mathbf{i}+\mathbf{j}) - \frac{1}{5}(-2\mathbf{i}+3\mathbf{j})\right) = L\left(\frac{3}{5}(\mathbf{i}+\mathbf{j})\right) + L\left(-\frac{1}{5}(-2\mathbf{i}+3\mathbf{j})\right) =$
$\frac{3}{5}L(\mathbf{i}+\mathbf{j}) - \frac{1}{5}L(-2\mathbf{i}+3\mathbf{j}) = \frac{3}{5}(\mathbf{i}-3\mathbf{j}) - \frac{1}{5}(-4\mathbf{i}+2\mathbf{j}) = \frac{7}{5}\mathbf{i} - \frac{11}{5}\mathbf{j}$. Similarly, $\mathbf{j} = \frac{2}{5}(\mathbf{i}+\mathbf{j}) + \frac{1}{5}(-2\mathbf{i}+3\mathbf{j})$, and so $L(\mathbf{j}) = L\left(\frac{2}{5}(\mathbf{i}+\mathbf{j}) + \frac{1}{5}(-2\mathbf{i}+3\mathbf{j})\right) = L\left(\frac{2}{5}(\mathbf{i}+\mathbf{j})\right) + L\left(\frac{1}{5}(-2\mathbf{i}+3\mathbf{j})\right) = \frac{2}{5}L(\mathbf{i}+\mathbf{j}) + \frac{1}{5}L(-2\mathbf{i}+3\mathbf{j})$
$= \frac{2}{5}(\mathbf{i}-3\mathbf{j}) + \frac{1}{5}(-4\mathbf{i}+2\mathbf{j}) = -\frac{2}{5}\mathbf{i} - \frac{4}{5}\mathbf{j}$.

(29) Suppose Part (3) of Theorem 5.1 is true for some n. We prove it true for $n+1$. $L(a_1\mathbf{v}_1 + \cdots + a_n\mathbf{v}_n + a_{n+1}\mathbf{v}_{n+1}) = L(a_1\mathbf{v}_1 + \cdots + a_n\mathbf{v}_n) + L(a_{n+1}\mathbf{v}_{n+1})$ (by property (1) of a linear transformation) $= (L(a_1\mathbf{v}_1) + \cdots + L(a_n\mathbf{v}_n)) + L(a_{n+1}\mathbf{v}_{n+1})$ (by the inductive hypothesis) $= L(a_1\mathbf{v}_1) + \cdots + L(a_n\mathbf{v}_n) + L(a_{n+1}\mathbf{v}_{n+1})$, and we are done.

(30) (b) The converse to the statement in part (a) is: If $S = \{\mathbf{v}_1, \ldots, \mathbf{v}_n\}$ is a linearly independent set of vectors in $\mathcal{V}$, then $T = \{L(\mathbf{v}_1), \ldots, L(\mathbf{v}_n)\}$ is a linearly independent set of vectors in $\mathcal{W}$. For a counterexample, consider $S = \{\mathbf{v}_1\}$, where $\mathbf{v}_1$ is any nonzero vector in $\mathcal{V}$. (For instance, we could have $\mathcal{V} = \mathcal{W} = \mathbb{R}^2$, with $\mathbf{v}_1 = \mathbf{e}_1$.) Then S is linearly independent by definition. Let L be the zero linear transformation defined by $L(\mathbf{v}) = \mathbf{0}$, for all $\mathbf{v} \in \mathcal{V}$. Then $T = \{\mathbf{0}\}$, a linearly dependent subset of $\mathcal{W}$.

(31) First, we must show that $L^{-1}(\mathcal{W}')$ is nonempty. Now, $\mathbf{0}_{\mathcal{W}} \in \mathcal{W}'$, so $\mathbf{0}_{\mathcal{V}} \in L^{-1}(\{\mathbf{0}_{\mathcal{W}}\}) \subseteq L^{-1}(\mathcal{W}')$. Hence, $L^{-1}(\mathcal{W}')$ is nonempty. Next, if $\mathbf{x}, \mathbf{y} \in L^{-1}(\mathcal{W}')$ then we must show that $\mathbf{x} + \mathbf{y} \in L^{-1}(\mathcal{W}')$. But $L(\mathbf{x}), L(\mathbf{y}) \in \mathcal{W}'$ by definition of $L^{-1}(\mathcal{W}')$. Hence, $L(\mathbf{x}) + L(\mathbf{y}) \in \mathcal{W}'$, since $\mathcal{W}'$ is closed under addition. Now L is a linear transformation, and so $L(\mathbf{x}) + L(\mathbf{y}) = L(\mathbf{x} + \mathbf{y})$. Therefore, $L(\mathbf{x} + \mathbf{y}) \in \mathcal{W}'$, which implies that $\mathbf{x} + \mathbf{y} \in L^{-1}(\mathcal{W}')$. Finally, if $\mathbf{x} \in L^{-1}(\mathcal{W}')$ then we must show that $a\mathbf{x} \in L^{-1}(\mathcal{W}')$ for all $a \in \mathbb{R}$. Now, $L(\mathbf{x}) \in \mathcal{W}'$. Hence, $aL(\mathbf{x}) \in \mathcal{W}'$ (for any $a \in \mathbb{R}$), since $\mathcal{W}'$ is closed under scalar multiplication. This implies that $L(a\mathbf{x}) \in \mathcal{W}'$. Therefore, $a\mathbf{x} \in L^{-1}(\mathcal{W}')$. Hence $L^{-1}(\mathcal{W}')$ is a subspace by Theorem 4.2.

(36) (a) False. We are only given that property (2) of the definition of a linear transformation holds. Property (1) is also required to hold in order for L to be a linear transformation. For a specific example of a function between vector spaces for which property (2) holds but property (1) fails, consider $f : \mathbb{R}^2 \to \mathbb{R}$ defined by $f([a, b]) = b$ whenever $a \neq 0$ and $f([0, b]) = 0$. To show that property (2) holds for f we consider 3 cases. First, if $a \neq 0$ and $c \neq 0$, $f(c[a, b]) = f([ca, cb]) = cb$ (since $ca \neq 0$) $= cf([a, b])$. Next, if $c = 0$, then $f(c[a, b]) = f([0, 0]) = 0 = 0f([a, b])$. Finally, if $a = 0$, then $f(c[0, b]) = f([0, cb]) = 0 = c(0) = cf([0, b])$. Also, property (1) does not hold because $f([1, 1]) + f([-1, 1]) = 1 + 1 = 2$, but $f([1, 1] + [-1, 1]) = f([0, 2]) = 0$.[1]

(b) True. This is proven in Example 4 in Section 5.1 of the textbook.

(c) False. L is not a linear transformation, and hence, not a linear operator. Since $L(\mathbf{0}) = [1, -2, 3] \neq \mathbf{0}$, L does not satisfy part (1) of Theorem 5.1, and so can not be a linear transformation. (One could also show, instead, that L satisfies neither of the two properties in the definition of a linear transformation.)

(d) False. If $\mathbf{A}$ is a 4×3 matrix, then $L(\mathbf{v}) = \mathbf{A}\mathbf{v}$ is a linear transformation from $\mathbb{R}^3$ to $\mathbb{R}^4$, *not* from $\mathbb{R}^4$ to $\mathbb{R}^3$.

(e) True. This is part (1) of Theorem 5.1.

(f) False. In order for the composition of functions $M_1 \circ M_2$ to be defined, the range of M_2 must be a subset of the domain of M_1. However, in this case, the range of M_2 is a subspace of $\mathcal{X}$, while the domain of M_1 is $\mathcal{V}$. For a specific counterexample, consider $M_1 : \mathcal{M}_{23} \to \mathcal{M}_{32}$ given by $M_1(\mathbf{A}) = \mathbf{A}^T$, and $M_2 : \mathcal{M}_{32} \to \mathbb{R}$ given by $M_2(\mathbf{A}) = a_{12}$. Clearly here, $M_1 \circ M_2$ is not defined. (Note that, in general, the composition $M_2 \circ M_1$ is a well-defined linear transformation, since the codomain of M_1 and the domain of M_2 both equal $\mathcal{W}$.)

(g) True. This is part (1) of Theorem 5.3.

[1] Note that property (1) alone also is insufficient to show that a function between real vector spaces is a linear transformation. However, describing an appropriate counterexample is quite complicated and beyond the level of this course. What makes constructing such a counterexample difficult is that property (1) for a function f does imply that $f(c\mathbf{v}) = cf(\mathbf{v})$ for every *rational* number c, but not for every *real* number.

(h) True. Since $\{\mathbf{0}_{\mathcal{W}}\}$ is a subspace of $\mathcal{W}$, this statement follows directly from part (2) of Theorem 5.3.

Section 5.2

(2) In each part, we will use Theorem 5.5, which states that the ith column of $\mathbf{A}_{BC}$ equals $[L(\mathbf{v}_i)]_C$, where B is an ordered basis for the domain of L, C is an ordered basis for the codomain of L, and $\mathbf{v}_i$ is the ith basis element in B. As directed, we will assume here that B and C are the standard bases in each case.

(a) Substituting each standard basis vector into the given formula for L shows that $L(\mathbf{e}_1) = [-6,-2,3]$, $L(\mathbf{e}_2) = [4,3,-1]$, and $L(\mathbf{e}_3) = [-1,-5,7]$. Using each of these vectors as columns produces the desired matrix for L: $\begin{bmatrix} -6 & 4 & -1 \\ -2 & 3 & -5 \\ 3 & -1 & 7 \end{bmatrix}$.

(c) Substituting each standard basis vector into the given formula for L shows that $L(x^3) = [4,1,-2]$, $L(x^2) = [-1,3,-7]$, $L(x) = [3,-1,5]$ and $L(1) = [3,5,-1]$. Using each of these vectors as columns produces the desired matrix for L: $\begin{bmatrix} 4 & -1 & 3 & 3 \\ 1 & 3 & -1 & 5 \\ -2 & -7 & 5 & -1 \end{bmatrix}$.

(3) In each part, we will use Theorem 5.5, which states that the ith column of $\mathbf{A}_{BC}$ equals $[L(\mathbf{v}_i)]_C$, where B is the given ordered basis for the domain of L, C is the given ordered basis for the codomain of L, and $\mathbf{v}_i$ is the ith basis element in B.

(a) Substituting each vector in B into the given formula for L shows that $L([1,-3,2]) = [4,-7]$, $L([-4,13,-3]) = [-1,25]$, and $L([2,-3,20]) = [56,-24]$. Next, we must express each of these answers in C-coordinates. We use the Coordinatization Method from Section 4.7. Thus, for each vector $\mathbf{v}$ that we need to express in C-coordinates, we could row reduce the augmented matrix whose columns to the left of the bar are the vectors in C, and whose column to the right of the bar is the vector $\mathbf{v}$. But since this process essentially involves solving a linear system, and we have three such systems to solve, we can solve all three simultaneously. Therefore, we row reduce $\left[\begin{array}{cc|ccc} -2 & 5 & 4 & -1 & 56 \\ -1 & 3 & -7 & 25 & -24 \end{array}\right]$ to $\left[\begin{array}{cc|ccc} 1 & 0 & -47 & 128 & -288 \\ 0 & 1 & -18 & 51 & -104 \end{array}\right]$. Hence, $[L([1,-3,2])]_C = [-47,-18]$, $[L([-4,13,-3])]_C = [128,51]$, and $[L([2,-3,20])]_C = [-288,-104]$. Using these coordinatizations as columns produces $\mathbf{A}_{BC} = \begin{bmatrix} -47 & 128 & -288 \\ -18 & 51 & -104 \end{bmatrix}$. (Note that $\mathbf{A}_{BC}$ is just the matrix to the right of the bar in the row reduced matrix, above. This is typical, but in general, any rows of all zeroes must be eliminated first, as in the Coordinatization Method.)

(c) Substituting each vector in B into the given formula for L shows that $L([5,3]) = 10x^2 + 12x + 6$ and $L([3,2]) = 7x^2 + 7x + 4$. Next, we must express each of these answers in C-coordinates. We use the Coordinatization Method from Section 4.7. To do this, we must first convert the above two polynomials, and the three polynomials in C into vectors in $\mathbb{R}^3$ by converting each polynomial of the form $ax^2 + bx + c$ to $[a,b,c]$. Then, for each converted vector $\mathbf{v}$ that we need to express in C-coordinates, we could row reduce the augmented matrix whose columns to the left of the bar are the vectors converted from C, and whose column to the right of the bar is the vector $\mathbf{v}$. But since this process essentially involves solving a linear system, and we have two such systems to solve, we can solve both systems simultaneously. Therefore, we row reduce

$\left[\begin{array}{ccc|cc} 3 & -2 & 1 & 10 & 7 \\ -2 & 2 & -1 & 12 & 7 \\ 0 & -1 & 1 & 6 & 4 \end{array}\right]$ to $\left[\begin{array}{ccc|cc} 1 & 0 & 0 & 22 & 14 \\ 0 & 1 & 0 & 62 & 39 \\ 0 & 0 & 1 & 68 & 43 \end{array}\right]$. Hence, $[L([5,3])]_C = [22, 62, 68]$, and $[L([3,2])]_C = [14, 39, 43]$. Using these coordinatizations as columns produces $\mathbf{A}_{BC} = \begin{bmatrix} 22 & 14 \\ 62 & 39 \\ 68 & 43 \end{bmatrix}$.

(Note that $\mathbf{A}_{BC}$ is just the matrix to the right of the bar in the row reduced matrix, above. This is typical, but in general, any rows of all zeroes must be eliminated first, as in the Coordinatization Method.)

(e) Substituting each vector in B into the given formula for L shows that $L(-5x^2 - x - 1) = \begin{bmatrix} 5 & -3 & -14 \\ -6 & -1 & 10 \end{bmatrix}$, $L(-6x^2+3x+1) = \begin{bmatrix} 6 & 7 & -19 \\ -3 & 1 & 14 \end{bmatrix}$ and $L(2x+1) = \begin{bmatrix} 0 & 5 & -1 \\ 2 & 1 & 1 \end{bmatrix}$. Next, we must express each of these answers in C-coordinates. We use the Coordinatization Method from Section 4.7. To do this, we must first convert the above matrices, and the six matrices in C into vectors in $\mathbb{R}^6$ by converting each matrix $\mathbf{A}$ to $[a_{11}, a_{12}, a_{13}, a_{21}, a_{22}, a_{23}]$. Then, for each converted vector $\mathbf{v}$ that we need to express in C-coordinates, we could row reduce the augmented matrix whose columns to the left of the bar are the vectors converted from C, and whose column to the right of the bar is the vector $\mathbf{v}$. But since this process essentially involves solving a linear system, and we have 3 such systems to solve, we can solve all 3 systems simultaneously. Therefore, we row reduce

$$\left[\begin{array}{cccccc|ccc} 1 & 0 & 0 & 0 & 0 & 0 & 5 & 6 & 0 \\ 0 & -1 & 1 & 0 & 0 & 0 & -3 & 7 & 5 \\ 0 & 0 & 1 & 0 & 0 & 0 & -14 & -19 & -1 \\ 0 & 0 & 0 & -1 & 0 & 0 & -6 & -3 & 2 \\ 0 & 0 & 0 & 0 & 1 & 0 & -1 & 1 & 1 \\ 0 & 0 & 0 & 0 & 1 & 1 & 10 & 14 & 1 \end{array}\right] \text{ to obtain } \left[\begin{array}{c|ccc} & 5 & 6 & 0 \\ & -11 & -26 & -6 \\ \mathbf{I}_6 & -14 & -19 & -1 \\ & 6 & 3 & -2 \\ & -1 & 1 & 1 \\ & 11 & 13 & 0 \end{array}\right].$$

Hence, $[L(-5x^2-x-1)]_C = [5, -11, -14, 6, -1, 11]$, $[L(-6x^2+3x+1)]_C = [6, -26, -19, 3, 1, 13]$, and $[L(2x+1)]_C = [0, -6, -1, -2, 1, 0]$. Using these coordinatizations as columns shows that $\mathbf{A}_{BC}$ is just the 6×3 matrix to the right of the bar in the reduced matrix, above. (This is typical, but in general, any rows of all zeroes must be eliminated first, as in the Coordinatization Method.)

(4) (a) First we find the matrix for L with respect to the standard basis B for $\mathbb{R}^3$. Substituting each vector in B into the given formula for L shows that $L(\mathbf{e}_1) = [-2, 0, 1]$, $L(\mathbf{e}_2) = [1, -1, 0]$, and $L(\mathbf{e}_3) = [0, -1, 3]$. Using each of these vectors as columns produces $\mathbf{A}_{BB} = \begin{bmatrix} -2 & 1 & 0 \\ 0 & -1 & -1 \\ 1 & 0 & 3 \end{bmatrix}$.

Next, we use Theorem 5.6 to compute $\mathbf{A}_{DE}$. To do this, we need the inverse of the transition matrix $\mathbf{P}$ from B to D and the transition matrix $\mathbf{Q}$ from B to E. Now $\mathbf{P}$ is the inverse of the matrix whose columns are the vectors in D, and $\mathbf{Q}$ is the inverse of the matrix whose columns are the vectors in E (see Exercise 8(a) in Section 4.7). Hence, $\mathbf{P}^{-1} = \begin{bmatrix} 15 & 2 & 3 \\ -6 & 0 & -1 \\ 4 & 1 & 1 \end{bmatrix}$ and

$\mathbf{Q}^{-1} = \begin{bmatrix} 1 & 0 & 2 \\ -3 & 3 & -2 \\ 1 & -1 & 1 \end{bmatrix}$. Using row reduction yields $\mathbf{Q} = \left(\mathbf{Q}^{-1}\right)^{-1} = \begin{bmatrix} 1 & -2 & -6 \\ 1 & -1 & -4 \\ 0 & 1 & 3 \end{bmatrix}$.

Finally, using Theorem 5.6 produces $\mathbf{A}_{DE} = \mathbf{Q}\mathbf{A}_{BB}\mathbf{P}^{-1} = \begin{bmatrix} -202 & -32 & -43 \\ -146 & -23 & -31 \\ 83 & 14 & 18 \end{bmatrix}$.

(b) First we find the matrix for L with respect to the standard basis B for $\mathcal{M}_{22}$ and the standard basis C for $\mathbb{R}^2$. Recall that the standard basis B for $\mathcal{M}_{22}$ is $\{\mathbf{\Psi}_{ij} \mid 1 \le i \le 2,\ 1 \le j \le 2\}$, where $\mathbf{\Psi}_{ij}$ is the 2×2 matrix having 1 as its (i, j) entry and zeroes elsewhere. (See Example 3 in Section 4.5.) Substituting each vector in B into the given formula for L shows that $L(\mathbf{\Psi}_{11}) = [6, -2]$, $L(\mathbf{\Psi}_{12}) = [-1, 3]$, $L(\mathbf{\Psi}_{21}) = [3, -1]$ and $L(\mathbf{\Psi}_{22}) = [-2, 4]$. Using each of these vectors as columns produces $\mathbf{A}_{BC} = \begin{bmatrix} 6 & -1 & 3 & -2 \\ -2 & 3 & -1 & 4 \end{bmatrix}$. Next, we use Theorem 5.6 to compute $\mathbf{A}_{DE}$. To do this, we need the inverse of the transition matrix $\mathbf{P}$ from B to D and the transition matrix $\mathbf{Q}$ from C to E. To compute $\mathbf{P}$ (or $\mathbf{P}^{-1}$), we need to convert the matrices in D into vectors in $\mathbb{R}^4$. This is done by converting each 2×2 matrix $\mathbf{A}$ to $[a_{11}, a_{12}, a_{21}, a_{22}]$. After doing this, $\mathbf{P}$ is the inverse of the matrix whose columns are the resulting vectors obtained from D, and $\mathbf{Q}$ is the inverse of the matrix whose columns are the vectors in E (see Exercise 8(a) in Section 4.7). Hence, $\mathbf{P}^{-1} = \begin{bmatrix} 2 & 0 & 1 & 1 \\ 1 & 2 & 1 & 1 \\ 0 & 1 & 2 & 1 \\ 1 & 1 & 1 & 1 \end{bmatrix}$ and $\mathbf{Q}^{-1} = \begin{bmatrix} -2 & -1 \\ 5 & 2 \end{bmatrix}$. Using Theorem 2.13 yields $\mathbf{Q} = \left(\mathbf{Q}^{-1}\right)^{-1} = \begin{bmatrix} 2 & 1 \\ -5 & -2 \end{bmatrix}$. Finally, using Theorem 5.6 produces $\mathbf{A}_{DE} = \mathbf{Q}\mathbf{A}_{BC}\mathbf{P}^{-1} = \begin{bmatrix} 21 & 7 & 21 & 16 \\ -51 & -13 & -51 & -38 \end{bmatrix}$.

(6) (a) First we use the method of Theorem 5.5. Substituting each vector in B into the given formula for L shows that $L([4, -1]) = [9, 7]$ and $L([-7, 2]) = [-16, -13]$. Next, we must express each of these answers in B-coordinates. We use the Coordinatization Method from Section 4.7. Thus, for each vector $\mathbf{v}$ that we need to express in B-coordinates, we could row reduce the augmented matrix whose columns to the left of the bar are the vectors in B, and whose column to the right of the bar is the vector $\mathbf{v}$. But since this process essentially involves solving a linear system, and we have 2 such systems to solve, we can solve both systems simultaneously. Therefore, we row reduce

$$\left[\begin{array}{cc|cc} 4 & -7 & 9 & -16 \\ -1 & 2 & 7 & -13 \end{array}\right] \text{ yielding } \left[\begin{array}{cc|cc} 1 & 0 & 67 & -123 \\ 0 & 1 & 37 & -68 \end{array}\right].$$

Hence, $[L([4, -1])]_B = [67, 37]$ and $[L([-7, 2])]_B = [-123, -68]$. Using these coordinatizations as columns produces $\mathbf{A}_{BB} = \begin{bmatrix} 67 & -123 \\ 37 & -68 \end{bmatrix}$.

Next, we use the method of Theorem 5.6. We begin by finding the matrix for L with respect to the standard basis C for $\mathbb{R}^2$. Substituting each vector in C into the given formula for L shows that $L(\mathbf{i}) = [2, 1]$ and $L(\mathbf{j}) = [-1, -3]$. Using each of these vectors as columns produces $\mathbf{A}_{CC} = \begin{bmatrix} 2 & -1 \\ 1 & -3 \end{bmatrix}$. Next, we use Theorem 5.6 to compute $\mathbf{A}_{BB}$. To do this, we need the transition matrix $\mathbf{P}$ from C to B. Now $\mathbf{P}$ is the inverse of the matrix whose columns are the vectors in B (see Exercise 8(a) in Section 4.7). Hence, $\mathbf{P}^{-1} = \begin{bmatrix} 4 & -7 \\ -1 & 2 \end{bmatrix}$. By Theorem 2.13, $\mathbf{P} = (\mathbf{P}^{-1})^{-1} = \begin{bmatrix} 2 & 7 \\ 1 & 4 \end{bmatrix}$.

Finally, using Theorem 5.6 produces $\mathbf{A}_{BB} = \mathbf{P}\mathbf{A}_{CC}\mathbf{P}^{-1} = \begin{bmatrix} 67 & -123 \\ 37 & -68 \end{bmatrix}$.

(b) First we use the method of Theorem 5.5. Substituting each vector in B into the given formula for L shows that $L(2x^2+2x-1) = 4x^2+3x+1$, $L(x) = x^2-1$, and $L(-3x^2-2x+1) = -4x^2-5x-2$. Next, we must express each of these answers in B-coordinates. We use the Coordinatization Method from Section 4.7. To do this, we must first convert the above polynomials, and the three polynomials in B into vectors in $\mathbb{R}^3$ by converting each polynomial of the form ax^2+bx+c to $[a, b, c]$. Then, for each converted vector $\mathbf{v}$ that we need to express in B-coordinates, we could row reduce the augmented matrix whose columns to the left of the bar are the vectors obtained from B, and whose column to the right of the bar is the vector $\mathbf{v}$. But since this process essentially involves solving a linear system, and we have 3 such systems to solve, we can solve all 3 systems simultaneously. Therefore, we row reduce

$$\left[\begin{array}{ccc|ccc} 2 & 0 & -3 & 4 & 1 & -4 \\ 2 & 1 & -2 & 3 & 0 & -5 \\ -1 & 0 & 1 & 1 & -1 & -2 \end{array}\right] \text{ to } \left[\begin{array}{ccc|ccc} 1 & 0 & 0 & -7 & 2 & 10 \\ 0 & 1 & 0 & 5 & -2 & -9 \\ 0 & 0 & 1 & -6 & 1 & 8 \end{array}\right].$$

Hence, $[L(2x^2+2x-1)]_B = [-7, 5, -6]$, $[L(x)]_B = [2, -2, 1]$, and $[L(-3x^2-2x+1)]_B = [10, -9, 8]$.

Using these coordinatizations as columns produces $\mathbf{A}_{BB} = \begin{bmatrix} -7 & 2 & 10 \\ 5 & -2 & -9 \\ -6 & 1 & 8 \end{bmatrix}$.

Next we use the method of Theorem 5.6. We begin by finding the matrix for L with respect to the standard basis C for $\mathcal{P}_2$. Substituting each vector in C into the given formula for L shows that $L(x^2) = 2x+1$, $L(x) = x^2-1$, and $L(1) = -2x^2+x-1$. We convert these results to vectors in $\mathbb{R}^3$. Then, using each of these vectors as columns produces $\mathbf{A}_{CC} = \begin{bmatrix} 0 & 1 & -2 \\ 2 & 0 & 1 \\ 1 & -1 & -1 \end{bmatrix}$.

Next, we use Theorem 5.6 to compute $\mathbf{A}_{BB}$. To do this, we need the transition matrix $\mathbf{P}$ from C to B. Now $\mathbf{P}$ is the inverse of the matrix whose columns are the polynomials in B, converted into vectors in $\mathbb{R}^3$ (see Exercise 8(a) in Section 4.7). Hence, $\mathbf{P}^{-1} = \begin{bmatrix} 2 & 0 & -3 \\ 2 & 1 & -2 \\ -1 & 0 & 1 \end{bmatrix}$. Using row reduction yields $\mathbf{P} = (\mathbf{P}^{-1})^{-1} = \begin{bmatrix} -1 & 0 & -3 \\ 0 & 1 & 2 \\ -1 & 0 & -2 \end{bmatrix}$. Finally, using Theorem 5.6 produces

$$\mathbf{A}_{BB} = \mathbf{P}\mathbf{A}_{CC}\mathbf{P}^{-1} = \begin{bmatrix} -7 & 2 & 10 \\ 5 & -2 & -9 \\ -6 & 1 & 8 \end{bmatrix}.$$

(7) (a) Computing the derivative of each polynomial in the standard basis B for $\mathcal{P}_3$ shows that $L(x^3) = 3x^2$, $L(x^2) = 2x$, $L(x) = 1$, and $L(1) = 0$. We convert these resulting polynomials in $\mathcal{P}_2$ to vectors in $\mathbb{R}^3$. Then, using each of these vectors as columns produces

$\mathbf{A}_{BC} = \begin{bmatrix} 3 & 0 & 0 & 0 \\ 0 & 2 & 0 & 0 \\ 0 & 0 & 1 & 0 \end{bmatrix}$, where C is the standard basis for $\mathcal{P}_2$. Now, by Theorem 5.5,

$$[L(4x^3 - 5x^2 + 6x - 7)]_C = \mathbf{A}_{BC}[4x^3 - 5x^2 + 6x - 7]_B = \begin{bmatrix} 3 & 0 & 0 & 0 \\ 0 & 2 & 0 & 0 \\ 0 & 0 & 1 & 0 \end{bmatrix} \begin{bmatrix} 4 \\ -5 \\ 6 \\ -7 \end{bmatrix} = \begin{bmatrix} 12 \\ -10 \\ 6 \end{bmatrix}.$$

Converting back from C-coordinates to polynomials yields $(4x^3 - 5x^2 + 6x - 7)' = 12x^2 - 10x + 6$.

· (8) (a) To do this, we need to calculate $L(\mathbf{i})$ and $L(\mathbf{j})$. Rotating the vector $\mathbf{i}$ counterclockwise through an angle of $\frac{\pi}{6}$ produces the vector $[\frac{\sqrt{3}}{2}, \frac{1}{2}]$. Rotating $\mathbf{j}$ counterclockwise through an angle of $\frac{\pi}{6}$ yields $[-\frac{1}{2}, \frac{\sqrt{3}}{2}]$. By Theorem 5.5, the matrix for L with respect to the standard basis is the matrix whose columns are $L(\mathbf{i})$ and $L(\mathbf{j})$; that is, $\begin{bmatrix} \frac{\sqrt{3}}{2} & -\frac{1}{2} \\ \frac{1}{2} & \frac{\sqrt{3}}{2} \end{bmatrix}$.

(9) (b) First we need to compute $L(\mathbf{b}_i)$ for each matrix $\mathbf{b}_i \in B$. Since L is just the transpose transformation, the results of these computations are, respectively

$$\begin{bmatrix} 1 & 0 \\ 0 & 0 \\ 0 & 0 \end{bmatrix}, \begin{bmatrix} 0 & 0 \\ 1 & 0 \\ -1 & 0 \end{bmatrix}, \begin{bmatrix} 0 & 0 \\ 1 & 0 \\ 0 & 0 \end{bmatrix}, \begin{bmatrix} 0 & -1 \\ 0 & 0 \\ 0 & 0 \end{bmatrix}, \begin{bmatrix} 0 & 0 \\ 0 & -1 \\ 0 & -1 \end{bmatrix}, \text{ and } \begin{bmatrix} 0 & 0 \\ 0 & 0 \\ 0 & 1 \end{bmatrix}.$$

Next, we must express each of these answers in C-coordinates. We use the Coordinatization Method from Section 4.7. To do this, we must first convert the above matrices, and the 6 matrices in C into vectors in $\mathbb{R}^6$ by changing each 3×2 matrix $\mathbf{A}$ to the vector $[a_{11}, a_{12}, a_{21}, a_{22}, a_{31}, a_{32}]$. Then, for each resulting vector $\mathbf{v}$ that we need to express in C-coordinates, we could row reduce the augmented matrix whose columns to the left of the bar are the vectors converted from C, and whose column to the right of the bar is the vector $\mathbf{v}$. But since this process essentially involves solving a linear system, and we have 6 such systems to solve, we can solve all 6 systems simultaneously. Therefore, we row reduce

$$\left[\begin{array}{cccccc|cccccc} 1 & 1 & 0 & 0 & 0 & 0 & 1 & 0 & 0 & 0 & 0 & 0 \\ 1 & -1 & 0 & 0 & 0 & 0 & 0 & 0 & 0 & -1 & 0 & 0 \\ 0 & 0 & 1 & 1 & 0 & 0 & 0 & 1 & 1 & 0 & 0 & 0 \\ 0 & 0 & 1 & -1 & 0 & 0 & 0 & 0 & 0 & 0 & -1 & 0 \\ 0 & 0 & 0 & 0 & 1 & 1 & 0 & -1 & 0 & 0 & 0 & 0 \\ 0 & 0 & 0 & 0 & 1 & -1 & 0 & 0 & 0 & 0 & -1 & 1 \end{array}\right]$$

to obtain $\left[\begin{array}{c|cccccc} & \frac{1}{2} & 0 & 0 & -\frac{1}{2} & 0 & 0 \\ & \frac{1}{2} & 0 & 0 & \frac{1}{2} & 0 & 0 \\ \mathbf{I}_6 & 0 & \frac{1}{2} & \frac{1}{2} & 0 & -\frac{1}{2} & 0 \\ & 0 & \frac{1}{2} & \frac{1}{2} & 0 & \frac{1}{2} & 0 \\ & 0 & -\frac{1}{2} & 0 & 0 & -\frac{1}{2} & \frac{1}{2} \\ & 0 & -\frac{1}{2} & 0 & 0 & \frac{1}{2} & -\frac{1}{2} \end{array}\right]$.

The columns of the matrix to the right of the bar gives the C-coordinatizations of the images of the six matrices in B under L. Hence, that 6×6 matrix is $\mathbf{A}_{BC}$.

(10) According to Theorem 5.8, the matrix $\mathbf{A}_{BD}$ for the composition $L_2 \circ L_1$ is the matrix product $\mathbf{A}_{CD}\mathbf{A}_{BC}$.

Hence, $\mathbf{A}_{BD} = \begin{bmatrix} 4 & -1 \\ 2 & 0 \\ -1 & -3 \end{bmatrix} \begin{bmatrix} -2 & 3 & -1 \\ 4 & 0 & -2 \end{bmatrix} = \begin{bmatrix} -12 & 12 & -2 \\ -4 & 6 & -2 \\ -10 & -3 & 7 \end{bmatrix}$.

(13) In each part, we use the fact that, by Theorem 5.5, the ith column of $\mathbf{A}_{BB}$ is the B-coordinatization of $L(\mathbf{v}_i)$.

(a) For each i, $L(\mathbf{v}_i) = \mathbf{v}_i$. But $[\mathbf{v}_i]_B = \mathbf{e}_i$. (See the comment just after Example 4 in Section 4.7.) Hence, the ith column of $\mathbf{A}_{BB}$ is $\mathbf{e}_i$, and so $\mathbf{A}_{BB} = \mathbf{I}_n$.

(c) For each i, $L(\mathbf{v}_i) = c\mathbf{v}_i$. But $[c\mathbf{v}_i]_B = c[\mathbf{v}_i]_B$ (by part (2) of Theorem 4.20) $= c\mathbf{e}_i$ (see the comment just after Example 4 in Section 4.7). Hence, the ith column of $\mathbf{A}_{BB}$ is $c\mathbf{e}_i$, and so $\mathbf{A}_{BB} = c\mathbf{I}_n$.

(e) Now $L(\mathbf{v}_1) = \mathbf{v}_n$, and for each $i > 1$, $L(\mathbf{v}_i) = \mathbf{v}_{i-1}$. But $[\mathbf{v}_j]_B = \mathbf{e}_j$. (See the comment just after Example 4 in Section 4.7.) Hence, the 1st column of $\mathbf{A}_{BB}$ is $\mathbf{e}_n$, and the ith column of $\mathbf{A}_{BB}$ for $i > 1$ is $\mathbf{e}_{i-1}$. Thus, $\mathbf{A}_{BB}$ is the $n \times n$ matrix whose columns are $\mathbf{e}_n, \mathbf{e}_1, \mathbf{e}_2, \ldots, \mathbf{e}_{n-1}$, respectively.

(15) First, note that $[(L_2 \circ L_1)(\mathbf{v})]_D = [L_2(L_1(\mathbf{v}))]_D = \mathbf{A}_{CD}[L_1(\mathbf{v})]_C = \mathbf{A}_{CD}(\mathbf{A}_{BC}[\mathbf{v}]_B) = (\mathbf{A}_{CD}\mathbf{A}_{BC})[\mathbf{v}]_B$. Therefore, by the uniqueness condition in Theorem 5.5, $\mathbf{A}_{CD}\mathbf{A}_{BC}$ is the matrix for $L_2 \circ L_1$ with respect to B and D.

(21) Since $\mathbf{A}$ and $\mathbf{B}$ are similar matrices, there is a nonsingular matrix $\mathbf{P}$ such that $\mathbf{PAP}^{-1} = \mathbf{B}$. Then $p_{\mathbf{B}}(x) = |x\mathbf{I}_n - \mathbf{B}| = |x\mathbf{I}_n - \mathbf{PAP}^{-1}| = |x(\mathbf{PI}_n\mathbf{P}^{-1}) - \mathbf{PAP}^{-1}| = |\mathbf{P}(x\mathbf{I}_n)\mathbf{P}^{-1} - \mathbf{PAP}^{-1}|$
$= |\mathbf{P}(x\mathbf{I}_n - \mathbf{A})\mathbf{P}^{-1}| = |\mathbf{P}||x\mathbf{I}_n - \mathbf{A}||\mathbf{P}^{-1}| = |\mathbf{P}||x\mathbf{I}_n - \mathbf{A}|(\frac{1}{|\mathbf{P}|}) = |x\mathbf{I}_n - \mathbf{A}| = p_{\mathbf{A}}(x)$.

(23) (a) $p_{\mathbf{A}_{BB}}(x) = |x\mathbf{I}_3 - \mathbf{A}_{BB}| = \begin{vmatrix} x - \frac{8}{9} & -\frac{2}{9} & -\frac{2}{9} \\ -\frac{2}{9} & x - \frac{5}{9} & \frac{4}{9} \\ -\frac{2}{9} & \frac{4}{9} & x - \frac{5}{9} \end{vmatrix} = x(x-1)^2$, where we have used basketweaving and algebraic simplification.

(b) For $\lambda_1 = 1$, we row reduce $[\mathbf{I}_3 - \mathbf{A}_{BB}|\mathbf{0}] =$

$$\left[\begin{array}{ccc|c} \frac{1}{9} & -\frac{2}{9} & -\frac{2}{9} & 0 \\ -\frac{2}{9} & \frac{4}{9} & \frac{4}{9} & 0 \\ -\frac{2}{9} & \frac{4}{9} & \frac{4}{9} & 0 \end{array}\right] \text{ to obtain } \left[\begin{array}{ccc|c} 1 & -2 & -2 & 0 \\ 0 & 0 & 0 & 0 \\ 0 & 0 & 0 & 0 \end{array}\right].$$

There are two independent variables, and hence Step 3 of the Diagonalization Method of Section 3.4 produces a 2-element basis for E_1. Following that method, the basis for $E_1 = \{[2,1,0],[2,0,1]\}$.

For $\lambda_2 = 0$, we row reduce $[0\mathbf{I}_3 - \mathbf{A}_{BB}|\mathbf{0}] = [-\mathbf{A}_{BB}|\mathbf{0}] =$

$$\left[\begin{array}{ccc|c} -\frac{8}{9} & -\frac{2}{9} & -\frac{2}{9} & 0 \\ -\frac{2}{9} & -\frac{5}{9} & \frac{4}{9} & 0 \\ -\frac{2}{9} & \frac{4}{9} & -\frac{5}{9} & 0 \end{array}\right] \text{ to obtain } \left[\begin{array}{ccc|c} 1 & 0 & \frac{1}{2} & 0 \\ 0 & 1 & -1 & 0 \\ 0 & 0 & 0 & 0 \end{array}\right].$$

There is one independent variable. Setting it equal to 2 to eliminate fractions produces a 1-element basis for E_0. Following that method, the basis for $E_0 = \{[-1,2,2]\}$.
Combining these bases for E_1 and E_0 yields $C = ([2,1,0],[2,0,1],[-1,2,2])$ (Note: the answer to part (c) will vary if the vectors in C are ordered differently).

(c) The inverse of the transition matrix $\mathbf{P}$ from B to C is the matrix whose columns are the vectors in C (see Exercise 8(a) in Section 4.7). Hence, $\mathbf{P}^{-1} = \begin{bmatrix} 2 & 2 & -1 \\ 1 & 0 & 2 \\ 0 & 1 & 2 \end{bmatrix}$. Using row reduction yields $\mathbf{P} = (\mathbf{P}^{-1})^{-1} = \frac{1}{9}\begin{bmatrix} 2 & 5 & -4 \\ 2 & -4 & 5 \\ -1 & 2 & 2 \end{bmatrix}$.

(30) Let $L : \mathcal{V} \longrightarrow \mathcal{W}$ be a linear transformation such that $L(\mathbf{v}_1) = \mathbf{w}_1$, $L(\mathbf{v}_2) = \mathbf{w}_2, \ldots, L(\mathbf{v}_n) = \mathbf{w}_n$, with $\{\mathbf{v}_1, \mathbf{v}_2, \ldots, \mathbf{v}_n\}$ a basis for $\mathcal{V}$. If $\mathbf{v} \in \mathcal{V}$, then $\mathbf{v} = c_1\mathbf{v}_1 + c_2\mathbf{v}_2 + \cdots + c_n\mathbf{v}_n$, for unique $c_1, c_2, \ldots, c_n \in \mathbb{R}$ (by Theorem 4.10). But then $L(\mathbf{v}) = L(c_1\mathbf{v}_1 + c_2\mathbf{v}_2 + \cdots + c_n\mathbf{v}_n) = c_1L(\mathbf{v}_1) + c_2L(\mathbf{v}_2) + \cdots + c_nL(\mathbf{v}_n)$ $= c_1\mathbf{w}_1 + c_2\mathbf{w}_2 + \cdots + c_n\mathbf{w}_n$. Hence, for each $\mathbf{v} \in \mathcal{V}$, $L(\mathbf{v})$ is completely determined by the given values for $\mathbf{w}_1, \ldots, \mathbf{w}_n$ and the unique values of $c_1, \ldots, c_n$. Therefore, L is uniquely determined.

(31) (a) True. This is shown in Theorem 5.4.

(b) True. This follows from Theorem 5.4.

(c) False. Bases for the domain and codomain must be chosen. Different choices of bases typically produce different matrices for L. For instance, Examples 3 and 4 in Section 5.2 of the textbook show two different matrices for the same linear transformation.

(d) False. This statement has the correct equation scrambled. The correct equation from Theorem 5.5 states that $\mathbf{A}_{BC}[\mathbf{v}]_B = [L(\mathbf{v})]_C$. For a specific counterexample to the scrambled equation in the problem statement, consider $L : \mathbb{R}^2 \to \mathbb{R}$ given by $L([a, b]) = a$ and let $B = (\mathbf{i}, \mathbf{j})$, $C = (1)$ and $\mathbf{v} = [1, 2]$. Then $\mathbf{A}_{BC} = [1, 0]$, a 1×2 matrix, $[\mathbf{v}]_B = [1, 2]$, and $[L(\mathbf{v})]_C = [1]_C = [1]$, a 1-vector. Note that $\mathbf{A}_{BC}[L(\mathbf{v})]_C$ is not even defined, because the matrix and vector are of incompatible sizes. Hence, the scrambled equation can not be correct.

(e) True. This is part of the statement of Theorem 5.5.

(f) False. The given matrix is the matrix (in standard coordinates) for the projection onto the xy-plane. The matrix for the projection L onto the xz-plane (in standard coordinates) is $\begin{bmatrix} 1 & 0 & 0 \\ 0 & 0 & 0 \\ 0 & 0 & 1 \end{bmatrix}$, since $L(\mathbf{i}) = \mathbf{i}$, $L(\mathbf{j}) = \mathbf{0}$, and $L(\mathbf{k}) = \mathbf{k}$.

(g) True. To see this, take the equation $\mathbf{A}_{DE} = \mathbf{Q}\mathbf{A}_{BC}\mathbf{P}^{-1}$ from Theorem 5.6, and multiply both sides on the right by $\mathbf{P}$.

(h) True. This is shown just before the definition of similar matrices in Section 5.2 of the textbook.

(i) True. This is the statement of Theorem 5.7.

(j) False. According to Theorem 5.8, the matrix for $L_2 \circ L_1$ with respect to the standard basis is

$$\begin{bmatrix} 0 & 1 \\ 1 & 0 \end{bmatrix}\begin{bmatrix} 1 & 2 \\ 3 & 4 \end{bmatrix} = \begin{bmatrix} 3 & 4 \\ 1 & 2 \end{bmatrix}, \textit{ not } \begin{bmatrix} 1 & 2 \\ 3 & 4 \end{bmatrix}\begin{bmatrix} 0 & 1 \\ 1 & 0 \end{bmatrix} = \begin{bmatrix} 2 & 1 \\ 4 & 3 \end{bmatrix}.$$

Section 5.3

(1) In each part, let $\mathbf{A}$ represent the given 3×3 matrix such that $L(\mathbf{v}) = \mathbf{A}\mathbf{v}$.

(a) Yes, because $L([1, -2, 3]) = [0, 0, 0]$.

(c) No. If $\mathbf{v} = [2, -1, 4]$ were in range(L), then there would exist a vector $\mathbf{x} = [x_1, x_2, x_3]$ such that $\mathbf{A}\mathbf{x} = \mathbf{v}$. This is equivalent to the system

$$\begin{cases} 5x_1 + x_2 - x_3 = 2 \\ -3x_1 + x_3 = -1 \\ x_1 - x_2 - x_3 = 4 \end{cases}$$, which has no solutions, since

$$\left[\begin{array}{ccc|c} 5 & 1 & -1 & 2 \\ -3 & 0 & 1 & -1 \\ 1 & -1 & -1 & 4 \end{array}\right] \text{ row reduces to } \left[\begin{array}{ccc|c} 1 & 0 & -\frac{1}{3} & \frac{1}{3} \\ 0 & 1 & \frac{2}{3} & \frac{1}{3} \\ 0 & 0 & 0 & 4 \end{array}\right].$$

Note: in the above, we stopped the row reduction process at the augmentation bar. If you use a calculator to perform the row reduction, it may continue row reducing into the column beyond the bar, making that column equal $\mathbf{e}_3$.

(2) (a) No, since $L(x^3 - 5x^2 + 3x - 6) = 6x^3 + 4x - 9 \neq 0$.

(c) Yes. To show this, we need to find $\mathbf{p}(x) = ax^3 + bx^2 + cx + d$ such that $L(\mathbf{p}) = 8x^3 - x - 1$. Equating the coefficients of $L(\mathbf{p})$ to those of $8x^3 - x - 1$ gives the system

$$\begin{cases} 2c = 8 & \leftarrow x^3 \text{ terms} \\ -a - b = -1 & \leftarrow x \text{ terms} \\ -c + d = -1 & \leftarrow \text{constant terms} \end{cases}$$

One solution is $a = 1$, $b = 0$, $c = 4$, $d = 3$. And so, $L(x^3 + 4x + 3) = 8x^3 - x - 1$, showing that $8x^3 - x - 1 \in$ range(L).

(3) In each part, let $\mathbf{A}$ represent the given matrix such that $L(\mathbf{v}) = \mathbf{A}\mathbf{v}$.

(a) The reduced row echelon form of $\mathbf{A}$ is $\mathbf{B} = \begin{bmatrix} 1 & 0 & 2 \\ 0 & 1 & -3 \\ 0 & 0 & 0 \end{bmatrix}$. Then, by the Kernel Method, a basis for ker(L) is found by considering the solution to the system $\mathbf{BX} = \mathbf{0}$ computed by setting the one independent variable in the system equal to 1. This yields the basis $\{[-2, 3, 1]\}$ for ker(L). Since ker(L) has a basis with 1 element, dim(ker(L)) = 1. Next, by the Range Method, a basis for range(L) consists of the first 2 columns of $\mathbf{A}$, since those columns in B contain nonzero pivots. Thus a basis for range(L) = $\{[1, -2, 3], [-1, 3, -3]\}$ and dim(range(L)) = 2. Note that dim(ker(L)) + dim(range(L)) = 1 + 2 = 3 = dim($\mathbb{R}^3$).

(d) The reduced row echelon form of $\mathbf{A}$ is $\mathbf{B} = \begin{bmatrix} 1 & 0 & -1 & 1 \\ 0 & 1 & 3 & -2 \\ 0 & 0 & 0 & 0 \\ 0 & 0 & 0 & 0 \\ 0 & 0 & 0 & 0 \end{bmatrix}$. Then, by the Kernel Method, a basis for ker(L) is found by considering certain particular solutions to the system $\mathbf{BX} = \mathbf{0}$. We note that x_3 and x_4 are independent variables in the system. Letting $x_3 = 1$, $x_4 = 0$ produces the vector $[1, -3, 1, 0]$. Using $x_3 = 0$, $x_4 = 1$ gives us $[-1, 2, 0, 1]$. This yields the basis $\{[1, -3, 1, 0], [-1, 2, 0, 1]\}$ for ker(L). Since ker(L) has a basis with 2 elements, dim(ker(L)) = 2. Next, by the Range Method, a basis for range(L) consists of the first 2 columns of $\mathbf{A}$, since those columns in B contain nonzero pivots. Thus a basis for range(L) = $\{[-14, -4, -6, 3, 4], [-8, -1, 2, -7, 2]\}$ and dim(range(L)) = 2. Note that dim(ker(L)) + dim(range(L)) = 2 + 2 = 4 = dim($\mathbb{R}^4$).

(4) We will solve each part by inspection (mostly).

(a) Now, $\ker(L)$ is the set of vectors sent to $\mathbf{0}$ by L. But, $L([x_1, x_2, x_3]) = [0, x_2] = [0, 0]$ precisely when $x_2 = 0$, and so $\ker(L) = \{[x_1, 0, x_3] \mid x_1, x_3 \in \mathbb{R}\} = \{[a, 0, c] \mid a, c \in \mathbb{R}\}$, which has basis $\{[1, 0, 0], [0, 0, 1]\}$. Thus, $\dim(\ker(L)) = 2$. Also, $\text{range}(L) = \{[0, x_2] \mid x_2 \in \mathbb{R}\}$, the set of images obtained using L, which can also be expressed as $\{[0, b] \mid b \in \mathbb{R}\}$. Hence, a basis for $\text{range}(L) = \{[0, 1]\}$ and $\dim(\text{range}(L)) = 1$. Finally, note that $\dim(\ker(L)) + \dim(\text{range}(L)) = 2 + 1 = 3 = \dim(\mathbb{R}^3)$.

(d) Now, $\ker(L)$ is the set of vectors sent to $\mathbf{0}$ by L. But $L(ax^4+bx^3+cx^2+dx+e) = cx^2+dx+e = 0$ precisely when $c = d = e = 0$, and so $\ker(L) = \{ax^4 + bx^3 \mid a, b \in \mathbb{R}\}$, which has basis $\{x^4, x^3\}$. Thus, $\dim(\ker(L)) = 2$. Also, $\text{range}(L) = \{cx^2 + dx + e \mid c, d, e \in \mathbb{R}\}$, the set of images obtained using L. This is precisely $\mathcal{P}_2$. Hence, a basis for $\text{range}(L) = \{x^2, x, 1\}$ and $\dim(\text{range}(L)) = 3$. Finally, note that $\dim(\ker(L)) + \dim(\text{range}(L)) = 2 + 3 = 5 = \dim(\mathcal{P}_4)$.

(f) Now, $\ker(L)$ is the set of vectors sent to $\mathbf{0}$ by L. But, $L([x_1, x_2, x_3]) = [x_1, 0, x_1 - x_2 + x_3] = [0, 0, 0]$ precisely when $x_1 = 0$ and $x_1 - x_2 + x_3 = 0$. That is when $x_1 = 0$ and $x_2 = x_3$. Hence $\ker(L) = \{[0, x_2, x_2] \mid x_2 \in \mathbb{R}\}$, which can also be expressed as $\{[0, b, b] \mid b \in \mathbb{R}\}$. Thus, $\ker(L)$ has basis $\{[0, 1, 1]\}$, and so $\dim(\ker(L)) = 1$. Also, $\text{range}(L) = \{[x_1, 0, x_1 - x_2 + x_3] \mid x_1, x_2, x_3 \in \mathbb{R}\}$, the set of images obtained using L. Hence,
$\text{range}(L) = \{x_1[1, 0, 1] + x_2[0, 0, -1] + x_3[0, 0, 1] \mid x_1, x_2, x_3 \in \mathbb{R}\}$, and so $\{[1, 0, 1], [0, 0, -1], [0, 0, 1]\}$ spans $\text{range}(L)$. However, this set is not linearly independent, since $[0, 0, 1] = (-1)[0, 0, -1]$. So, if we eliminate the last vector, we get the basis $\{[1, 0, 1], [0, 0, -1]\}$ for $\text{range}(L)$. (A simpler basis for $\text{range}(L)$ can be found using the Simplified Span Method, yielding $\{[1, 0, 0], [0, 0, 1]\}$.) And so, $\dim(\text{range}(L)) = 2$. Finally, note that $\dim(\ker(L)) + \dim(\text{range}(L)) = 1 + 2 = 3 = \dim(\mathbb{R}^3)$.

(g) Now, $\ker(L)$ is the set of vectors sent to $\mathbf{O}_{22}$ by L. But $L(\mathbf{A}) = \mathbf{A}^T = \mathbf{O}_{22}$ precisely when $\mathbf{A} = \mathbf{O}_{22}$. Hence, $\ker(L) = \{\mathbf{O}_{22}\}$, having basis $\{\ \}$ (the empty set), and $\dim(\ker(L)) = 0$. Also, every 2×2 matrix is the image of its transpose (since $(\mathbf{A}^T)^T = \mathbf{A}$), proving $\text{range}(L) = \mathcal{M}_{22}$. Therefore, $\dim(\text{range}(L)) = 4$, and a basis for $\text{range}(L) =$ standard basis for $\mathcal{M}_{22}$. Notice that $\dim(\ker(L)) + \dim(\text{range}(L)) = 0 + 4 = 4 = \dim(\mathcal{M}_{22})$.

(i) Now, $\ker(L)$ is the set of vectors sent to $\mathbf{0}$ by L. Note that $L(\mathbf{p}) = [\mathbf{p}(1), \mathbf{p}'(1)] = [0, 0]$ precisely when $\mathbf{p}(1) = 0$ and $\mathbf{p}'(1) = 0$. But $\mathbf{p}(1) = 0$ implies that $(x - 1)$ is a factor of $\mathbf{p}$. Thus, $\mathbf{p} = (x-1)(ax+b)$. Next, $\mathbf{p}'(1) = 0$ implies $(a(1)+b)(1) + ((1)-1)(a) = 0$ (using the product rule and substituting $x = 1$), which gives $a = -b$. Thus, $\mathbf{p} = (x-1)(ax-a) = a(x^2-2x+1)$. Therefore, a basis for $\ker(L)$ is $\{x^2 - 2x + 1\}$, and $\dim(\ker(L)) = 1$. A spanning set for $\text{range}(L)$ can be found by computing the images of a basis for $\mathcal{P}_2$. Now $L(x^2) = [1, 2]$, $L(x) = [1, 1]$, and $L(1) = [1, 0]$. The first 2 of these form a linearly independent set, and the fact that $\dim(\mathbb{R}^2) = 2$ shows that the 3rd vector is redundant. Hence, a basis for $\text{range}(L)$ is $\{[1, 2], [1, 1]\}$ and $\dim(\text{range}(L)) = 2$. Since $\dim(\text{range}(L)) = \dim(\mathbb{R}^2)$, $\text{range}(L) = \mathbb{R}^2$, and so a simpler basis for $\text{range}(L) =$ standard basis for $\mathbb{R}^2$. Notice that $\dim(\ker(L)) + \dim(\text{range}(L)) = 1 + 2 = 3 = \dim(\mathcal{P}_2)$.

(6) First we show that L is a linear transformation by verifying the two properties.
Property (1): $L(\mathbf{A} + \mathbf{B}) = \text{trace}(\mathbf{A} + \mathbf{B}) = \sum_{i=1}^{3}(a_{ii} + b_{ii})$
$= \sum_{i=1}^{3} a_{ii} + \sum_{i=1}^{3} b_{ii} = \text{trace}(\mathbf{A}) + \text{trace}(\mathbf{B}) = L(\mathbf{A}) + L(\mathbf{B})$.
Property (2): $L(c\mathbf{A}) = \text{trace}(c\mathbf{A}) = \sum_{i=1}^{3} ca_{ii} = c\sum_{i=1}^{3} a_{ii} = c(\text{trace}(\mathbf{A})) = cL(\mathbf{A})$.
Next, $\ker(L)$ is the set of all 3×3 matrices whose trace is 0. Thus,

$\ker(L) = \left\{ \begin{bmatrix} a & b & c \\ d & e & f \\ g & h & -a-e \end{bmatrix} \middle|\ a, b, c, d, e, f, g, h \in \mathbb{R} \right\}$. A detailed computation for a basis for $\ker(L)$

is given in the solution to Exercise 11(b) in Section 4.6, found earlier in this manual. The basis found there has eight elements, and so $\dim(\ker(L)) = 8$. Also, it is possible to obtain any real number a as the trace of some 3×3 matrix since $\text{trace}(\frac{a}{3}\mathbf{I}_3) = a$. Hence, $\text{range}(L) = \mathbb{R}$, and so $\dim(\text{range}(L)) = 1$.

(8) First, $\ker(L)$ is the set of polynomials $\mathbf{p}$ such that $x^2\mathbf{p}(x) = 0$ for all x. But since x^2 is 0 only when $x = 0$, $\mathbf{p}(x)$ must be 0 for all other values of x. However, no nonzero polynomial can have an infinite number of roots. Hence, $\mathbf{p} = \mathbf{0}$. Thus, $\ker(L) = \{\mathbf{0}\}$, and so $\dim(\ker(L)) = 0$. Next, the images under L are precisely those having x^2 as a factor. Thus, $\text{range}(L) = \{ax^4 + bx^3 + cx^2 \mid a, b, c \in \mathbb{R}\}$. Clearly, a basis for $\text{range}(L)$ is $\{x^4, x^3, x^2\}$, and so $\dim(\text{range}(L)) = 3$.

(10) First, let us consider the case $k \le n$. Taking the kth derivative of a polynomial reduces its degree by k, and so the polynomials of degree less than k are precisely those that have kth derivative equal to $\mathbf{0}$. Hence, when $k \le n$, $\ker(L) = \{\text{all polynomials of degree less than } k\} = \mathcal{P}_{k-1}$. Therefore, $\dim(\ker(L)) = k$. Also, computing a kth antiderivative for a polynomial $\mathbf{p}$ of degree $\le (n-k)$ will yield a pre-image under L for $\mathbf{p}$. Hence, $\text{range}(L) = \mathcal{P}_{n-k}$, and $\dim(\text{range}(L)) = n - k + 1$.
When $k > n$, L sends all of $\mathcal{P}_n$ to $\mathbf{0}$, and so $\ker(L) = \mathcal{P}_n$, $\dim(\ker(L)) = n + 1$, $\text{range}(L) = \{\mathbf{0}\}$, and $\dim(\text{range}(L)) = 0$.

(12) First, $\ker(L) = \{\mathbf{0}\}$ because $L(\mathbf{X}) = \mathbf{0} \Rightarrow \mathbf{AX} = \mathbf{0} \Rightarrow \mathbf{A}^{-1}\mathbf{AX} = \mathbf{A}^{-1}\mathbf{0} \Rightarrow \mathbf{X} = \mathbf{0}$. Similarly, every vector $\mathbf{X}$ is in the range of L since $L(\mathbf{A}^{-1}\mathbf{X}) = \mathbf{A}(\mathbf{A}^{-1}\mathbf{X}) = \mathbf{X}$. Thus, $\text{range}(L) = \mathbb{R}^n$.

(16) Consider $L\left(\begin{bmatrix} x \\ y \end{bmatrix}\right) = \begin{bmatrix} 1 & -1 \\ 1 & -1 \end{bmatrix}\begin{bmatrix} x \\ y \end{bmatrix}$. Since $\begin{bmatrix} 1 & -1 \\ 1 & -1 \end{bmatrix}$ row reduces to $\begin{bmatrix} 1 & -1 \\ 0 & 0 \end{bmatrix}$, the Kernel Method and the Range Method show that $\ker(L) = \text{range}(L) = \{[a, a] \mid a \in \mathbb{R}\}$.

(18) (d) The statement to be proven is: Let L: $\mathcal{V} \to \mathcal{W}$ be a linear transformation between finite dimensional vector spaces. Suppose that $\mathbf{A}$ is the matrix for L with respect to some bases for $\mathcal{V}$ and $\mathcal{W}$. Then $\dim(\text{range}(L)) = \text{rank}(\mathbf{A})$, and $\dim(\ker(L)) = n-\text{ rank}(\mathbf{A})$.
To prove this, define T: $\mathbb{R}^n \to \mathbb{R}^m$ to be the linear transformation given by $T(\mathbf{X}) = \mathbf{AX}$. Then $\dim(\text{range}(L)) = \dim(\text{range}(T))$ (by part (b)) $= \text{rank}(\mathbf{A})$ (by part (1) of Theorem 5.12). Finally, $\dim(\ker(L)) = \dim(\ker(T))$ (by part (a)) $= n-\text{ rank}(\mathbf{A})$ (by part (2) of Theorem 5.12).

(20) (a) False. By definition, the set $\{L(\mathbf{v}) \mid \mathbf{v} \in \mathcal{V}\}$ is $\text{range}(L)$, *not* $\ker(L)$. These sets are usually unequal. For specific counterexamples, consider Examples 1, 2, and 3 in Section 5.3 of the textbook.

(b) False. Range(L), in general, is a subspace of the codomain $\mathcal{W}$, *not* of the domain $\mathcal{V}$. A linear transformation such as that given in Example 5 in Section 5.3 of the textbook provides a specific counterexample.

(c) True. This is a direct consequence of the Dimension Theorem.

(d) False. The Dimension Theorem relates the dimensions of $\ker(L)$ and the domain $\mathcal{V}$ of L with $\dim(\text{range}(L))$, *not* with the dimension of the codomain $\mathcal{W}$. For a specific counterexample, consider $L : \mathbb{R}^5 \to \mathbb{R}^3$ given by $L(\mathbf{v}) = \mathbf{0}$ for all $\mathbf{v} \in \mathbb{R}^5$. Then $\ker(L) = \mathbb{R}^5$, and so $\dim(\ker(L)) = 5 \neq 2$.

(e) True. By part (2) of Theorem 5.12, $\dim(\ker(L)) = n - \text{rank}(\mathbf{A})$. Now, $\text{rank}(\mathbf{A})$ is the number of nonzero rows in the reduced row echelon form for $\mathbf{A}$. Hence, it gives the number of pivots obtained from $\mathbf{A}$, which is the number of pivot columns for $\mathbf{A}$. But since $\mathbf{A}$ has n columns in all, it must have $n - \text{rank}(\mathbf{A})$ nonpivot columns.

(f) False. By part (1) of Theorem 5.12, $\dim(\text{range}(L)) = \text{rank}(\mathbf{A})$. See Example 6 in Section 5.3 of the textbook for a specific counterexample. In that example, $\text{rank}(\mathbf{A}) = 3$, and $\dim(\text{range}(L)) = 3 \neq 5 - \text{rank}(\mathbf{A})$.

(g) False. By Corollary 5.13, $\text{rank}(\mathbf{A}) = \text{rank}(\mathbf{A}^T)$. Hence, if $\text{rank}(\mathbf{A}) = 2$, then $\text{rank}(\mathbf{A}^T) = 2$ as well.

(h) False. Even though Corollary 5.13 assures us that dim(row space of $\mathbf{A}$) = dim(column space of $\mathbf{A}$), the subspaces themselves are not necessarily the same. In fact, unless $\mathbf{A}$ is square, they are not even subspaces of the same vector space. See the comments after Example 7 in Section 5.3 of the textbook, which indicate that counterexamples to the statement are provided by Examples 5 through 7 in Section 5.3.

Section 5.4

(1) In each part, we will check whether L is one-to-one using Theorem 5.14, which states that L is one-to-one if and only if $\ker(L) = \{\mathbf{0}\}$.

(a) L is not one-to-one because $L([1,0,0]) = [0,0,0,0]$. Thus, $[1,0,0] \in \ker(L)$, and so $\ker(L) \neq \{\mathbf{0}\}$. Also, L is not onto because $[0,0,0,1]$ is not in range(L). This is because all vectors in range(L) have a zero in their 4th coordinate. Finally, L is not an isomorphism because it is neither one-to-one nor onto.

(c) L is one-to-one because $L([x,y,z]) = [0,0,0]$ implies that $[2x, x+y+z, -y] = [0,0,0]$, which is equivalent to the system

$$\begin{cases} 2x & & & = 0 \\ x + & y + & z & = 0 \\ & -y & & = 0 \end{cases}.$$

A quick inspection shows that $x = y = z = 0$ is the only solution, and hence $\ker(L) = \{\mathbf{0}\}$. Next, L is onto, because every vector $[a,b,c]$ can be expressed as $[2x, x+y+z, -y]$, where $x = \frac{a}{2}$, $y = -c$, and $z = b - \frac{a}{2} + c$. (As an alternative, we could claim that L is onto using Corollary 5.15.) Since L is both one-to-one and onto, L is an isomorphism.

(e) L is one-to-one because $L(ax^2+bx+c) = 0$ implies that $a+b = b+c = a+c = 0$. Thus, $a+b = a+c$, and so $b = c$. Using this and $b+c = 0$ gives $b = c = 0$. Then $a+b = 0$ implies $a = 0$. Hence, $\ker(L) = \{\mathbf{0}\}$. Next, L is onto because every polynomial Ax^2+Bx+C can be expressed as $(a+b)x^2+(b+c)x+(a+c)$, where $a = (A-B+C)/2$, $b = (A+B-C)/2$, and $c = (-A+B+C)/2$. (As an alternative, we could claim that L is onto using Corollary 5.15.) Finally, since L is both one-to-one and onto, L is an isomorphism.

(g) L is not one-to-one because $L\left(\begin{bmatrix} 0 & 1 & 0 \\ 1 & 0 & -1 \end{bmatrix}\right) = \begin{bmatrix} 0 & 0 \\ 0 & 0 \end{bmatrix}$. Hence, $\begin{bmatrix} 0 & 1 & 0 \\ 1 & 0 & -1 \end{bmatrix} \in \ker(L)$, and so $\ker(L) \neq \{\mathbf{O}_{23}\}$. However, L is onto because every 2×2 matrix $\begin{bmatrix} A & B \\ C & D \end{bmatrix}$ can be expressed as $\begin{bmatrix} a & -c \\ 2e & d+f \end{bmatrix}$, where $a = A$, $c = -B$, $e = C/2$, $d = D$, and $f = 0$. Finally, L is not an isomorphism since it is not one-to-one.

(h) L is one-to-one because $L(ax^2+bx+c) = \begin{bmatrix} 0 & 0 \\ 0 & 0 \end{bmatrix}$ implies that $a+c = b-c = -3a = 0$, which gives $a = b = c = 0$. Hence $\ker(L) = \{\mathbf{0}\}$. Next, L is not onto, because $\begin{bmatrix} 0 & 1 \\ 0 & 0 \end{bmatrix}$ is not in range(L), since every matrix in range(L) has a zero for its (1,2) entry. Finally, L is not an isomorphism because it is not onto.

(2) In each part, let $\mathbf{A}$ represent the given matrix such that $L(\mathbf{v}) = \mathbf{A}\mathbf{v}$.

(a) The matrix $\mathbf{A}$ row reduces to $\mathbf{I}_2$. Hence, rank($\mathbf{A}$) = 2. Therefore, Theorem 5.12 implies that $\dim(\ker(L)) = 0$ and $\dim(\text{range}(L)) = 2$. Since $\dim(\ker(L)) = 0$, L is one-to-one. Also, $\dim(\text{range}(L)) = 2 = \dim(\mathbb{R}^2)$ implies that L is onto. Finally, L is an isomorphism, since it is both one-to-one and onto.

(b) The matrix $\mathbf{A}$ row reduces to $\begin{bmatrix} 1 & 0 \\ 0 & 1 \\ 0 & 0 \end{bmatrix}$. Hence, rank($\mathbf{A}$) = 2. Therefore, Theorem 5.12 implies that $\dim(\ker(L)) = 0$ and $\dim(\text{range}(L)) = 2$. Since $\dim(\ker(L)) = 0$, L is one-to-one. Also, $\dim(\text{range}(L)) = 2 < 3 = \dim(\mathbb{R}^3)$ implies that L is not onto. Finally, L is not an isomorphism, since it is not onto.

(c) The matrix $\mathbf{A}$ row reduces to $\begin{bmatrix} 1 & 0 & -\frac{2}{5} \\ 0 & 1 & -\frac{6}{5} \\ 0 & 0 & 0 \end{bmatrix}$. Hence, rank($\mathbf{A}$) = 2. Therefore, Theorem 5.12 implies that $\dim(\ker(L)) = 1$ and $\dim(\text{range}(L)) = 2$. Since $\dim(\ker(L)) \neq 0$, L is not one-to-one. Also, $\dim(\text{range}(L)) = 2 < 3 = \dim(\mathbb{R}^3)$ implies that L is not onto. Finally, L is not an isomorphism, since it is neither one-to-one nor onto.

(3) In each part, let $\mathbf{A}$ represent the given matrix for L.

(a) The matrix $\mathbf{A}$ row reduces to $\mathbf{I}_3$. Hence, rank($\mathbf{A}$) = 3. Therefore, Theorem 5.12 implies that $\dim(\ker(L)) = 0$ and $\dim(\text{range}(L)) = 3$. Since $\dim(\ker(L)) = 0$, L is one-to-one. Also, $\dim(\text{range}(L)) = 3 = \dim(\mathcal{P}_2)$ implies that L is onto. Finally, L is an isomorphism, since it is both one-to-one and onto.

(c) The matrix $\mathbf{A}$ row reduces to $\begin{bmatrix} 1 & 0 & -\frac{10}{11} & \frac{19}{11} \\ 0 & 1 & \frac{3}{11} & -\frac{9}{11} \\ 0 & 0 & 0 & 0 \\ 0 & 0 & 0 & 0 \end{bmatrix}$. Hence, rank($\mathbf{A}$) = 2. Therefore, Theorem 5.12 implies that $\dim(\ker(L)) = 2$ and $\dim(\text{range}(L)) = 2$. Since $\dim(\ker(L)) \neq 0$, L is not one-to-one. Also, $\dim(\text{range}(L)) = 2 < 4 = \dim(\mathcal{P}_3)$ implies that L is not onto. Finally, L is not an isomorphism, since it is neither one-to-one nor onto.

(4) (a) Because $\dim(\mathbb{R}^6) = \dim(\mathcal{P}_5)$, Corollary 5.15 tells us that if L were one-to-one, L would also be onto. But since L is not onto, L is also not one-to-one.

(b) Because $\dim(\mathcal{M}_{22}) = \dim(\mathcal{P}_3)$, Corollary 5.15 tells us that if L were onto, L would also be one-to-one. But since L is not one-to-one, L is also not onto.

(8) In each part, let $\mathbf{A}$ represent the given matrix for L_1 and let $\mathbf{B}$ represent the given matrix for L_2.

(a) (i) By Theorem 5.17, L_1 and L_2 are isomorphisms if and only if $\mathbf{A}$ and $\mathbf{B}$ are nonsingular. But cofactor expansions of each along the 3rd column shows that $|\mathbf{A}| = 1$ and $|\mathbf{B}| = 3$. Since neither of these are zero, both matrices are nonsingular.

(ii) By the remark immediately following Theorem 5.17 in the textbook, $L_1^{-1}(\mathbf{v}) = \mathbf{A}^{-1}\mathbf{v}$ and $L_2^{-1}(\mathbf{v}) = \mathbf{B}^{-1}\mathbf{v}$, where

$\mathbf{A}^{-1} = \begin{bmatrix} 0 & 0 & 1 \\ 0 & -1 & 0 \\ 1 & -2 & 0 \end{bmatrix}$ and $\mathbf{B}^{-1} = \begin{bmatrix} 1 & 0 & 0 \\ 0 & 0 & -\frac{1}{3} \\ 2 & 1 & 0 \end{bmatrix}$ are obtained by row reduction.

(iii) By Theorem 5.8, $(L_2 \circ L_1)(\mathbf{v}) = (\mathbf{BA})\mathbf{v}$. Computing $\mathbf{BA}$ produces

$$(L_2 \circ L_1)\left(\begin{bmatrix} x_1 \\ x_2 \\ x_3 \end{bmatrix}\right) = \begin{bmatrix} 0 & -2 & 1 \\ 1 & 4 & -2 \\ 0 & 3 & 0 \end{bmatrix}\begin{bmatrix} x_1 \\ x_2 \\ x_3 \end{bmatrix}.$$

(iv) By the remark immediately following Theorem 5.17 in the textbook, $(L_2 \circ L_1)^{-1}(\mathbf{v}) = (\mathbf{BA})^{-1}(\mathbf{v})$. Using row reduction to compute the inverse of the matrix we found in part (iii) gives

$$(L_2 \circ L_1)^{-1}\left(\begin{bmatrix} x_1 \\ x_2 \\ x_3 \end{bmatrix}\right) = \begin{bmatrix} 2 & 1 & 0 \\ 0 & 0 & \frac{1}{3} \\ 1 & 0 & \frac{2}{3} \end{bmatrix}\begin{bmatrix} x_1 \\ x_2 \\ x_3 \end{bmatrix}.$$

(v) By Theorem 5.8, $(L_1^{-1} \circ L_2^{-1})(\mathbf{v}) = (\mathbf{A}^{-1}\mathbf{B}^{-1})\mathbf{v}$. Using the matrices $\mathbf{A}^{-1}$ and $\mathbf{B}^{-1}$ we computed in part (ii) yields

$\mathbf{A}^{-1}\mathbf{B}^{-1} = \begin{bmatrix} 2 & 1 & 0 \\ 0 & 0 & \frac{1}{3} \\ 1 & 0 & \frac{2}{3} \end{bmatrix}$. Hence, $(L_1^{-1} \circ L_2^{-1})$ is the same linear operator as $(L_2 \circ L_1)^{-1}$, since parts (iv) and (v) show that they have the same matrix with respect to the standard basis.

(c) (i) By Theorem 5.17, L_1 and L_2 are isomorphisms if and only if $\mathbf{A}$ and $\mathbf{B}$ are nonsingular. But using basketweaving on each shows that $|\mathbf{A}| = -1$ and $|\mathbf{B}| = 1$. Since neither of these are zero, both matrices are nonsingular.

(ii) By the remark immediately following Theorem 5.17 in the textbook, $L_1^{-1}(\mathbf{v}) = \mathbf{A}^{-1}\mathbf{v}$ and $L_2^{-1}(\mathbf{v}) = \mathbf{B}^{-1}\mathbf{v}$, where

$\mathbf{A}^{-1} = \begin{bmatrix} 2 & -4 & -1 \\ 7 & -13 & -3 \\ 5 & -10 & -3 \end{bmatrix}$ and $\mathbf{B}^{-1} = \begin{bmatrix} 1 & 0 & -1 \\ 3 & 1 & -3 \\ -1 & -2 & 2 \end{bmatrix}$ are obtained by row reduction.

(iii) By Theorem 5.8, $(L_2 \circ L_1)(\mathbf{v}) = (\mathbf{BA})\mathbf{v}$. Computing $\mathbf{BA}$ produces

$$(L_2 \circ L_1)\left(\begin{bmatrix} x_1 \\ x_2 \\ x_3 \end{bmatrix}\right) = \begin{bmatrix} 29 & -6 & -4 \\ 21 & -5 & -2 \\ 38 & -8 & -5 \end{bmatrix}\begin{bmatrix} x_1 \\ x_2 \\ x_3 \end{bmatrix}.$$

(iv) By the remark immediately following Theorem 5.17 in the textbook, $(L_2 \circ L_1)^{-1}(\mathbf{v}) = (\mathbf{BA})^{-1}(\mathbf{v})$. Using row reduction to compute the inverse of the matrix we found in part (iii) gives

$$(L_2 \circ L_1)^{-1}\left(\begin{bmatrix} x_1 \\ x_2 \\ x_3 \end{bmatrix}\right) = \begin{bmatrix} -9 & -2 & 8 \\ -29 & -7 & 26 \\ -22 & -4 & 19 \end{bmatrix}\begin{bmatrix} x_1 \\ x_2 \\ x_3 \end{bmatrix}.$$

(v) By Theorem 5.8, $(L_1^{-1} \circ L_2^{-1})(\mathbf{v}) = (\mathbf{A}^{-1}\mathbf{B}^{-1})\mathbf{v}$. Using the matrices $\mathbf{A}^{-1}$ and $\mathbf{B}^{-1}$ we computed in part (ii) yields

$\mathbf{A}^{-1}\mathbf{B}^{-1} = \begin{bmatrix} -9 & -2 & 8 \\ -29 & -7 & 26 \\ -22 & -4 & 19 \end{bmatrix}$. Hence, $(L_1^{-1} \circ L_2^{-1})$ is the same linear operator as $(L_2 \circ L_1)^{-1}$, since parts (iv) and (v) show that they have the same matrix with respect to the standard basis.

(12) (a) Now $R(\mathbf{i}) = \mathbf{j}$ and $R(\mathbf{j}) = \mathbf{i}$. Since the columns of the matrix for R are the images of $\mathbf{i}$ and $\mathbf{j}$, respectively, the matrix is $\begin{bmatrix} 0 & 1 \\ 1 & 0 \end{bmatrix}$.

(15) Suppose L is an isomorphism. Let $\mathbf{M}$ be the matrix for L^{-1} with respect to C and B. Then $(L^{-1} \circ L)(\mathbf{v}) = \mathbf{v}$ for every $\mathbf{v} \in \mathcal{V} \Longrightarrow \mathbf{M}\mathbf{A}_{BC}[\mathbf{v}]_B = [\mathbf{v}]_B$ for every $\mathbf{v} \in \mathcal{V}$ (by Theorem 5.8). Letting $\mathbf{v}$ be the ith basis element in B, $[\mathbf{v}]_B = \mathbf{e}_i$, and so $\mathbf{e}_i = [\mathbf{v}]_B = \mathbf{M}\mathbf{A}_{BC}[\mathbf{v}]_B = \mathbf{M}\mathbf{A}_{BC}\mathbf{e}_i =$ (ith column of $\mathbf{M}\mathbf{A}_{BC}$). Therefore, $\mathbf{M}\mathbf{A}_{BC} = \mathbf{I}_n$, which implies that $\mathbf{M} = \mathbf{A}_{BC}^{-1}$, and so $\mathbf{A}_{BC}$ is nonsingular. Conversely, if $\mathbf{A}_{BC}$ is nonsingular, then $\mathbf{A}_{BC}^{-1}$ exists, and $\mathbf{A}_{BC}^{-1}\mathbf{A}_{BC} = \mathbf{A}_{BC}\mathbf{A}_{BC}^{-1} = \mathbf{I}_n$. Let K be the linear operator from $\mathcal{W}$ to $\mathcal{V}$ whose matrix with respect to C and B is $\mathbf{A}_{BC}^{-1}$. Then, for all $\mathbf{v} \in \mathcal{V}$, $[(K \circ L)(\mathbf{v})]_B = \mathbf{A}_{BC}^{-1}\mathbf{A}_{BC}[\mathbf{v}]_B = \mathbf{I}_n[\mathbf{v}]_B = [\mathbf{v}]_B$. Therefore, $(K \circ L)(\mathbf{v}) = \mathbf{v}$ for all $\mathbf{v} \in \mathcal{V}$, since coordinatizations are unique. Similarly, for all $\mathbf{w} \in \mathcal{W}$, $[(L \circ K)(\mathbf{w})]_C = \mathbf{A}_{BC}\mathbf{A}_{BC}^{-1}[\mathbf{w}]_C = \mathbf{I}_n[\mathbf{w}]_C = [\mathbf{w}]_C$. Hence $(L \circ K)(\mathbf{w}) = \mathbf{w}$ for all $\mathbf{w} \in \mathcal{W}$. Therefore, K acts as an inverse for L. Hence, L is an isomorphism by Theorem 5.16.

(26) (a) To compute $T(\mathbf{v})$, for $\mathbf{v} \in \mathbb{R}^5$, we must first find the polynomial $\mathbf{p}$ in $\mathcal{P}_4$ that produced $\mathbf{v}$. But, $\mathbf{p} = F_1^{-1}(\mathbf{v})$. Next, we calculate $\mathbf{C} = L(\mathbf{p}) \in \mathcal{M}_{22}$. Finally, we find the vector in $\mathbb{R}^4$ corresponding to $\mathbf{C}$. That is, we compute $F_2(\mathbf{C})$. This result is $T(\mathbf{v})$. This composition of transformations yields $T = F_2 \circ L \circ F_1^{-1}$.

(31) (a) False. The second half of the given statement is the converse of what it means for a function to be one-to-one. That is, L is one-to one if and only if $L(\mathbf{v}_1) = L(\mathbf{v}_2)$ implies $\mathbf{v}_1 = \mathbf{v}_2$. Any linear transformation that is not one-to-one, such as Example 2 in Section 5.4 of the textbook, provides a specific counterexample to the given statement. Note that the statement "$\mathbf{v}_1 = \mathbf{v}_2$ implies $L(\mathbf{v}_1) = L(\mathbf{v}_2)$" is actually true for *every* function, not just one-to-one functions.

(b) False. The correct definition of what it means for $L : \mathcal{V} \to \mathcal{W}$ to be onto is that for every $\mathbf{w} \in \mathcal{W}$ there is a $\mathbf{v} \in \mathcal{V}$ such that $L(\mathbf{v}) = \mathbf{w}$. The statement "for all $\mathbf{v} \in \mathcal{V}$, there is some $\mathbf{w} \in \mathcal{W}$ such that $L(\mathbf{v}) = \mathbf{w}$" is true for every function $L : \mathcal{V} \to \mathcal{W}$, not just those that are onto. Thus, any linear transformation that is not onto provides a counterexample. One such function is described just after Example 2 in Section 5.4 of the textbook.

(c) True. This is one direction of the "if and only if" statement in Theorem 5.14.

(d) False. While Corollary 5.15 shows this is true if $\mathcal{V}$ and $\mathcal{W}$ are finite dimensional vector spaces having the same dimension, the statement is not necessarily true otherwise. For example, the linear transformation of Example 2 in Section 5.4 of the textbook is onto, but is not one-to-one. Also, Exercise 2(b) in Section 5.4, whose solution appears in this manual, gives an example of a linear transformation that is one-to-one but not onto.

(e) True. This is part of Theorem 5.16.

(f) True. This is part of Theorem 5.16.

(g) False. In fact, this statement is in direct contradiction with Theorem 5.17, which implies that L is an isomorphism if and only if $\mathbf{A}_{BB}$ is nonsingular.[2] But $\mathbf{A}_{BB}$ is nonsingular if and only if $|\mathbf{A}_{BB}| \neq 0$. Any linear operator on a nontrivial finite dimensional vector space provides a counterexample for the statement given in the problem.

(h) True. Since $\dim(\mathbb{R}^{28}) = \dim(\mathcal{P}_{27}) = \dim(\mathcal{M}_{74}) = 28$, Corollary 5.20 shows that each of these vector spaces is isomorphic to the other two.

[2] We assume here that $\mathcal{V}$ is a nontrivial vector space, or else there is no matrix $\mathbf{A}_{BB}$, since if $\mathcal{V}$ is trivial, B is the empty set.

Section 5.5

(1) In all parts, we just need to perform Step 2 of the Generalized Diagonalization Method. However, this corresponds precisely with the Diagonalization Method from Section 3.4. Also, in each part, let $\mathbf{A}$ represent the given matrix.

(a) First, $p_{\mathbf{A}}(x) = \begin{vmatrix} x-2 & -1 \\ 0 & x-2 \end{vmatrix} = (x-2)^2$. Hence, $\lambda = 2$ is the only eigenvalue for L, and the algebraic multiplicity for λ is 2. To find a basis for E_2, we row reduce

$[2\mathbf{I}_2 - \mathbf{A}|\,\mathbf{0}] = \left[\begin{array}{cc|c} 0 & -1 & 0 \\ 0 & 0 & 0 \end{array}\right]$ to obtain $\left[\begin{array}{cc|c} 0 & 1 & 0 \\ 0 & 0 & 0 \end{array}\right]$.

Setting the independent variable x_1 of the associated system equal to 1 produces the basis $\{[1,0]\}$ for E_2. Since $\dim(E_2) = 1$, the geometric multiplicity of λ is 1.

(c) First, $p_{\mathbf{A}}(x) = \begin{vmatrix} x & -1 & -1 \\ 1 & x-4 & 1 \\ 1 & -5 & x+2 \end{vmatrix} = x^3 - 2x^2 - x + 2$ (using basketweaving) $= (x-1)(x+1)(x-2)$.

Hence, the eigenvalues for L are $\lambda_1 = 1$, $\lambda_2 = -1$, and $\lambda_3 = 2$, each having algebraic multiplicity 1. Next, we solve for a basis for each eigenspace.

For $\lambda_1 = 1$, we row reduce $[1\mathbf{I}_3 - \mathbf{A}|\,\mathbf{0}] = \left[\begin{array}{ccc|c} 1 & -1 & -1 & 0 \\ 1 & -3 & 1 & 0 \\ 1 & -5 & 3 & 0 \end{array}\right]$ to obtain $\left[\begin{array}{ccc|c} 1 & 0 & -2 & 0 \\ 0 & 1 & -1 & 0 \\ 0 & 0 & 0 & 0 \end{array}\right]$.

Setting the independent variable x_3 of the associated system equal to 1 produces the basis $\{[2,1,1]\}$ for E_1.

For $\lambda_2 = -1$, we row reduce $[-1\mathbf{I}_3 - \mathbf{A}|\,\mathbf{0}] = \left[\begin{array}{ccc|c} -1 & -1 & -1 & 0 \\ 1 & -5 & 1 & 0 \\ 1 & -5 & 1 & 0 \end{array}\right]$ to obtain $\left[\begin{array}{ccc|c} 1 & 0 & 1 & 0 \\ 0 & 1 & 0 & 0 \\ 0 & 0 & 0 & 0 \end{array}\right]$.

Setting the independent variable x_3 of the associated system equal to 1 produces the basis $\{[-1,0,1]\}$ for E_{-1}.

For $\lambda_3 = 2$, we row reduce $[2\mathbf{I}_3 - \mathbf{A}|\,\mathbf{0}] = \left[\begin{array}{ccc|c} 2 & -1 & -1 & 0 \\ 1 & -2 & 1 & 0 \\ 1 & -5 & 4 & 0 \end{array}\right]$ to obtain $\left[\begin{array}{ccc|c} 1 & 0 & -1 & 0 \\ 0 & 1 & -1 & 0 \\ 0 & 0 & 0 & 0 \end{array}\right]$.

Setting the independent variable x_3 of the associated system equal to 1 produces the basis $\{[1,1,1]\}$ for E_2.

Since each eigenspace is one-dimensional, each eigenvalue has geometric multiplicity 1.

(d) $p_{\mathbf{A}}(x) = \begin{vmatrix} x-2 & 0 & 0 \\ -4 & x+3 & 6 \\ 4 & -5 & x-8 \end{vmatrix} = (x-2)\Big((x+3)(x-8)+30\Big)$ (using cofactor expansion along the 1st row) $= (x-2)^2(x-3)$. Hence, the eigenvalues for L are $\lambda_1 = 2$ and $\lambda_2 = 3$. The algebraic multiplicities of these eigenvalues are 2 and 1, respectively. Next, we solve for a basis for each eigenspace.

For $\lambda_1 = 2$, we row reduce $[2\mathbf{I}_3 - \mathbf{A}|\,\mathbf{0}] = \left[\begin{array}{ccc|c} 0 & 0 & 0 & 0 \\ -4 & 5 & 6 & 0 \\ 4 & -5 & -6 & 0 \end{array}\right]$ to obtain $\left[\begin{array}{ccc|c} 1 & -\frac{5}{4} & -\frac{3}{2} & 0 \\ 0 & 0 & 0 & 0 \\ 0 & 0 & 0 & 0 \end{array}\right]$.

Setting the independent variables $x_2 = 4$ (to avoid fractions) and $x_3 = 0$ produces the vector

$[5,4,0]$. Setting $x_2 = 0$ and $x_3 = 2$ yields $[3,0,2]$. Hence, $\{[5,4,0],[3,0,2]\}$ is a basis for E_2, and the geometric multiplicity of λ_1 is 2.

For $\lambda_2 = 3$, we row reduce $[3\mathbf{I}_3 - \mathbf{A}|\,\mathbf{0}] = \left[\begin{array}{ccc|c} 1 & 0 & 0 & 0 \\ -4 & 6 & 6 & 0 \\ 4 & -5 & -5 & 0 \end{array}\right]$ to obtain $\left[\begin{array}{ccc|c} 1 & 0 & 0 & 0 \\ 0 & 1 & 1 & 0 \\ 0 & 0 & 0 & 0 \end{array}\right]$.

Setting the independent variable x_3 equal to 1 produces the basis $\{[0,-1,1]\}$ for E_3, and the geometric multiplicity of λ_2 is 1.

(2) (b) Step 1: We let $C = (x^2, x, 1)$, the standard basis for $\mathcal{P}_2$. Then, since $L(x^2) = 2x^2-2x$, $L(x) = x-1$, and $L(1) = 0$, we get the matrix $\mathbf{A} = \begin{bmatrix} 2 & 0 & 0 \\ -2 & 1 & 0 \\ 0 & -1 & 0 \end{bmatrix}$ for L with respect to C.

Step 2: First, $p_{\mathbf{A}}(x) = \begin{vmatrix} x-2 & 0 & 0 \\ 2 & x-1 & 0 \\ 0 & 1 & x \end{vmatrix} = (x-2)(x-1)x$ (since the corresponding matrix is lower triangular). Hence, the eigenvalues for L are $\lambda_1 = 2$, $\lambda_2 = 1$ and $\lambda_3 = 0$. Next, we solve for a basis for each eigenspace.

For $\lambda_1 = 2$, we row reduce $[2\mathbf{I}_3 - \mathbf{A}|\,\mathbf{0}] = \left[\begin{array}{ccc|c} 0 & 0 & 0 & 0 \\ 2 & 1 & 0 & 0 \\ 0 & 1 & 2 & 0 \end{array}\right]$ to obtain $\left[\begin{array}{ccc|c} 1 & 0 & -1 & 0 \\ 0 & 1 & 2 & 0 \\ 0 & 0 & 0 & 0 \end{array}\right]$.

Setting the independent variable x_3 of the associated system equal to 1 produces the basis $\{[1,-2,1]\}$ for E_2 in $\mathbb{R}^3$.

For $\lambda_2 = 1$, we row reduce $[1\mathbf{I}_3 - \mathbf{A}|\,\mathbf{0}] = \left[\begin{array}{ccc|c} -1 & 0 & 0 & 0 \\ 2 & 0 & 0 & 0 \\ 0 & 1 & 1 & 0 \end{array}\right]$ to obtain $\left[\begin{array}{ccc|c} 1 & 0 & 0 & 0 \\ 0 & 1 & 1 & 0 \\ 0 & 0 & 0 & 0 \end{array}\right]$.

Setting the independent variable x_3 of the associated system equal to 1 produces the basis $\{[0,-1,1]\}$ for E_1 in $\mathbb{R}^3$.

For $\lambda_3 = 0$, we row reduce $[0\mathbf{I}_3 - \mathbf{A}|\,\mathbf{0}] = [-\mathbf{A}|\,\mathbf{0}] = \left[\begin{array}{ccc|c} -2 & 0 & 0 & 0 \\ 2 & -1 & 0 & 0 \\ 0 & 1 & 0 & 0 \end{array}\right]$ to obtain $\left[\begin{array}{ccc|c} 1 & 0 & 0 & 0 \\ 0 & 1 & 0 & 0 \\ 0 & 0 & 0 & 0 \end{array}\right]$. Setting the independent variable x_3 of the associated system equal to 1 produces the basis $\{[0,0,1]\}$ for E_0 in $\mathbb{R}^3$.

Since the union of the three bases is $Z = \{[1,-2,1],[0,-1,1],[0,0,1]\}$, which has 3 elements, L is diagonalizable.

Step 3: Converting the vectors in Z into polynomials produces the basis $B = (x^2-2x+1, -x+1, 1)$ for $\mathcal{P}_2$. The matrix for L with respect to B is the diagonal matrix having the eigenvalues for L on its main diagonal; that is, $\mathbf{D} = \begin{bmatrix} 2 & 0 & 0 \\ 0 & 1 & 0 \\ 0 & 0 & 0 \end{bmatrix}$. Let $\mathbf{P} = \begin{bmatrix} 1 & 0 & 0 \\ -2 & -1 & 0 \\ 1 & 1 & 1 \end{bmatrix}$, the matrix whose columns are the vectors in Z. Then row reduction yields $\mathbf{P}^{-1} = \mathbf{P}$. (This does not occur in general.) Matrix multiplication verifies that $\mathbf{D} = \mathbf{P}^{-1}\mathbf{AP}$. (Try it!)

(d) Step 1: We let $C = (x^2, x, 1)$, the standard basis for $\mathcal{P}_2$. Then, since $L(x^2) = -x^2 - 12x + 18$,

$L(x) = -4x$, and $L(1) = -5$, we get the matrix

$\mathbf{A} = \begin{bmatrix} -1 & 0 & 0 \\ -12 & -4 & 0 \\ 18 & 0 & -5 \end{bmatrix}$ for L with respect to C.

Step 2: First, $p_{\mathbf{A}}(x) = \begin{vmatrix} x+1 & 0 & 0 \\ 12 & x+4 & 0 \\ -18 & 0 & x+5 \end{vmatrix} = (x+1)(x+4)(x+5)$ (since the corresponding matrix is lower triangular). Hence, the eigenvalues for L are $\lambda_1 = -1$, $\lambda_2 = -4$ and $\lambda_3 = -5$. Next, we solve for a basis for each eigenspace.

For $\lambda_1 = -1$, we row reduce $[-1\mathbf{I}_3 - \mathbf{A}|\,\mathbf{0}] = \left[\begin{array}{ccc|c} 0 & 0 & 0 & 0 \\ 12 & 3 & 0 & 0 \\ -18 & 0 & 4 & 0 \end{array}\right]$ to obtain $\left[\begin{array}{ccc|c} 1 & 0 & -\frac{2}{9} & 0 \\ 0 & 1 & \frac{8}{9} & 0 \\ 0 & 0 & 0 & 0 \end{array}\right]$.

Setting the independent variable x_3 of the associated system equal to 9 (to eliminate fractions) produces the basis $\{[2,-8,9]\}$ for E_{-1} in $\mathbb{R}^3$.

For $\lambda_2 = -4$, we row reduce $[-4\mathbf{I}_3 - \mathbf{A}|\,\mathbf{0}] = \left[\begin{array}{ccc|c} -3 & 0 & 0 & 0 \\ 12 & 0 & 0 & 0 \\ -18 & 0 & 1 & 0 \end{array}\right]$ to obtain $\left[\begin{array}{ccc|c} 1 & 0 & 0 & 0 \\ 0 & 0 & 1 & 0 \\ 0 & 0 & 0 & 0 \end{array}\right]$.

Setting the independent variable x_2 of the associated system equal to 1 produces the basis $\{[0,1,0]\}$ for E_{-4} in $\mathbb{R}^3$.

For $\lambda_3 = -5$, we row reduce $[-5\mathbf{I}_3 - \mathbf{A}|\,\mathbf{0}] = \left[\begin{array}{ccc|c} -4 & 0 & 0 & 0 \\ 12 & -1 & 0 & 0 \\ -18 & 0 & 0 & 0 \end{array}\right]$ to obtain $\left[\begin{array}{ccc|c} 1 & 0 & 0 & 0 \\ 0 & 1 & 0 & 0 \\ 0 & 0 & 0 & 0 \end{array}\right]$.

Setting the independent variable x_3 of the associated system equal to 1 produces the basis $\{[0,0,1]\}$ for E_{-5} in $\mathbb{R}^3$.

Since the union of the three bases is $Z = \{[2,-8,9],[0,1,0],[0,0,1]\}$, which has 3 elements, L is diagonalizable.

Step 3: Converting the vectors in Z into polynomials produces the basis $B = (2x^2 - 8x + 9, x, 1)$ for $\mathcal{P}_2$. The matrix for L with respect to B is the diagonal matrix having the eigenvalues for L on its main diagonal; that is, $\mathbf{D} = \begin{bmatrix} -1 & 0 & 0 \\ 0 & -4 & 0 \\ 0 & 0 & -5 \end{bmatrix}$. Let $\mathbf{P} = \begin{bmatrix} 2 & 0 & 0 \\ -8 & 1 & 0 \\ 9 & 0 & 1 \end{bmatrix}$, the matrix whose columns are the vectors in Z. Then row reduction yields $\mathbf{P}^{-1} = \begin{bmatrix} \frac{1}{2} & 0 & 0 \\ 4 & 1 & 0 \\ -\frac{9}{2} & 0 & 1 \end{bmatrix}$. Matrix multiplication verifies that $\mathbf{D} = \mathbf{P}^{-1}\mathbf{A}\mathbf{P}$. (Try it!)

(e) Step 1: We let $C = (\mathbf{i},\mathbf{j})$, the standard basis for $\mathbb{R}^2$. Then, since $L(\mathbf{i}) = [\frac{1}{2}, \frac{\sqrt{3}}{2}]$, and $L(\mathbf{j}) = [-\frac{\sqrt{3}}{2}, \frac{1}{2}]$, we get the matrix $\mathbf{A} = \begin{bmatrix} \frac{1}{2} & -\frac{\sqrt{3}}{2} \\ \frac{\sqrt{3}}{2} & \frac{1}{2} \end{bmatrix}$ for L with respect to C.

Step 2: First, $p_{\mathbf{A}}(x) = \begin{vmatrix} x-\frac{1}{2} & \frac{\sqrt{3}}{2} \\ -\frac{\sqrt{3}}{2} & x-\frac{1}{2} \end{vmatrix} = x^2 - x + 1$, which has no real roots. Hence, L has no eigenvalues, and so is not diagonalizable.

(h) Step 1: Let C be the standard basis for $\mathcal{M}_{22}$; that is, $C = \{\mathbf{\Psi}_{ij} \mid 1 \le i \le 2,\ 1 \le j \le 2\}$, where $\mathbf{\Psi}_{ij}$ is the 2×2 matrix having 1 in its (i,j) entry and zeroes elsewhere. (See Example 3 in Section 4.5.) Now $L(\mathbf{\Psi}_{11}) = \begin{bmatrix} -4 & 0 \\ -10 & 0 \end{bmatrix}$, $L(\mathbf{\Psi}_{12}) = \begin{bmatrix} 0 & -4 \\ 0 & -10 \end{bmatrix}$, $L(\mathbf{\Psi}_{21}) = \begin{bmatrix} 3 & 0 \\ 7 & 0 \end{bmatrix}$, and

$L(\mathbf{\Psi}_{22}) = \begin{bmatrix} 0 & 3 \\ 0 & 7 \end{bmatrix}$, and so we get the matrix $\mathbf{A} = \begin{bmatrix} -4 & 0 & 3 & 0 \\ 0 & -4 & 0 & 3 \\ -10 & 0 & 7 & 0 \\ 0 & -10 & 0 & 7 \end{bmatrix}$

for L with respect to C.

Step 2: First, $p_{\mathbf{A}}(x) = \begin{vmatrix} x+4 & 0 & -3 & 0 \\ 0 & x+4 & 0 & -3 \\ 10 & 0 & x-7 & 0 \\ 0 & 10 & 0 & x-7 \end{vmatrix}$. Using cofactor expansion along the 4th row, and cofactor expansion on the 2nd row for each of the resulting 3×3 determinants yields $p_{\mathbf{A}}(x) = 10\left(3\left((x+4)(x-7)+30\right)\right) + (x-7)(x+4)\left((x+4)(x-7)+30\right)$

$= \left((x+4)(x-7)+30\right)^2 = (x-1)^2(x-2)^2$. Hence, the eigenvalues for L are $\lambda_1 = 1$, and $\lambda_2 = 2$. Next, we solve for a basis for each eigenspace.

For $\lambda_1 = 1$, we row reduce $[1\mathbf{I}_3 - \mathbf{A}|\,\mathbf{0}] =$

$$\left[\begin{array}{cccc|c} 5 & 0 & -3 & 0 & 0 \\ 0 & 5 & 0 & -3 & 0 \\ 10 & 0 & -6 & 0 & 0 \\ 0 & 10 & 0 & -6 & 0 \end{array}\right] \text{ to obtain } \left[\begin{array}{cccc|c} 1 & 0 & -\frac{3}{5} & 0 & 0 \\ 0 & 1 & 0 & -\frac{3}{5} & 0 \\ 0 & 0 & 0 & 0 & 0 \\ 0 & 0 & 0 & 0 & 0 \end{array}\right].$$

Setting the independent variables $x_3 = 5$ (to avoid fractions) and $x_4 = 0$ produces the vector $[3,0,5,0]$. Setting $x_3 = 0$ and $x_4 = 5$ yields $[0,3,0,5]$. Hence, $\{[3,0,5,0],[0,3,0,5]\}$ is a basis for E_1 in $\mathbb{R}^4$.

For $\lambda_2 = 2$, we row reduce $[2\mathbf{I}_3 - \mathbf{A}|\,\mathbf{0}] =$

$$\left[\begin{array}{cccc|c} 6 & 0 & -3 & 0 & 0 \\ 0 & 6 & 0 & -3 & 0 \\ 10 & 0 & -5 & 0 & 0 \\ 0 & 10 & 0 & -5 & 0 \end{array}\right] \text{ to obtain } \left[\begin{array}{cccc|c} 1 & 0 & -\frac{1}{2} & 0 & 0 \\ 0 & 1 & 0 & -\frac{1}{2} & 0 \\ 0 & 0 & 0 & 0 & 0 \\ 0 & 0 & 0 & 0 & 0 \end{array}\right].$$

Setting the independent variables $x_3 = 2$ (to avoid fractions) and $x_4 = 0$ produces the vector $[1,0,2,0]$. Setting $x_3 = 0$ and $x_4 = 2$ yields $[0,1,0,2]$. Hence, $\{[1,0,2,0],[0,1,0,2]\}$ is a basis for E_2 in $\mathbb{R}^4$.

The union of the three bases is

$Z = \{[3,0,5,0],[0,3,0,5],[1,0,2,0],[0,1,0,2]\}$. Since $|Z| = 4$, L is diagonalizable.

Step 3: Converting the vectors in Z into matrices produces the basis

$B = \left(\begin{bmatrix} 3 & 0 \\ 5 & 0 \end{bmatrix}, \begin{bmatrix} 0 & 3 \\ 0 & 5 \end{bmatrix}, \begin{bmatrix} 1 & 0 \\ 2 & 0 \end{bmatrix}, \begin{bmatrix} 0 & 1 \\ 0 & 2 \end{bmatrix}\right)$ for $\mathcal{M}_{22}$. The matrix for L with respect to B is the diagonal matrix having the eigenvalues for L on its main diagonal; that is,

$\mathbf{D} = \begin{bmatrix} 1 & 0 & 0 & 0 \\ 0 & 1 & 0 & 0 \\ 0 & 0 & 2 & 0 \\ 0 & 0 & 0 & 2 \end{bmatrix}$. Let $\mathbf{P} = \begin{bmatrix} 3 & 0 & 1 & 0 \\ 0 & 3 & 0 & 1 \\ 5 & 0 & 2 & 0 \\ 0 & 5 & 0 & 2 \end{bmatrix}$, the matrix whose columns are the vectors in Z. Then row reduction yields $\mathbf{P}^{-1} = \begin{bmatrix} 2 & 0 & -1 & 0 \\ 0 & 2 & 0 & -1 \\ -5 & 0 & 3 & 0 \\ 0 & -5 & 0 & 3 \end{bmatrix}$. Matrix multiplication verifies that $\mathbf{D} = \mathbf{P}^{-1}\mathbf{A}\mathbf{P}$. (Try it!)

(4) (a) Consider $C = (x^2, x, 1)$, the standard basis for $\mathcal{P}_2$. Since $a = 1$ in this part, $L(dx^2 + ex + f) = d(x+1)^2 + e(x+1) + f$. Therefore, $L(x^2) = (x+1)^2 = x^2 + 2x + 1$, $L(x) = x + 1$, and $L(1) = 1$. Hence, we get the matrix $\mathbf{A} = \begin{bmatrix} 1 & 0 & 0 \\ 2 & 1 & 0 \\ 1 & 1 & 1 \end{bmatrix}$ for L with respect to C.

Now, $p_{\mathbf{A}}(x) = \begin{vmatrix} x-1 & 0 & 0 \\ -2 & x-1 & 0 \\ -1 & -1 & x-1 \end{vmatrix} = (x-1)^3$ (since the corresponding matrix is lower triangular). Hence, the only eigenvalue for L is $\lambda = 1$. Next, we solve for a basis for E_1.

We row reduce $[1\mathbf{I}_3 - \mathbf{A} \,|\, \mathbf{0}] = \left[\begin{array}{ccc|c} 0 & 0 & 0 & 0 \\ -2 & 0 & 0 & 0 \\ -1 & -1 & 0 & 0 \end{array}\right]$ to obtain $\left[\begin{array}{ccc|c} 1 & 0 & 0 & 0 \\ 0 & 1 & 0 & 0 \\ 0 & 0 & 0 & 0 \end{array}\right]$.

Setting the independent variable x_3 of the associated system equal to 1 produces the basis $\{[0,0,1]\}$ for E_1 in $\mathbb{R}^3$. Converting the vector in this basis back to polynomial form yields the basis $E_1 = \{1\}$ in $\mathcal{P}_2$. (Note that our process has produced only one eigenvector for L, and so L is not diagonalizable.)

(7) (a) Let $\mathbf{A} = \begin{bmatrix} 1 & 1 & -1 \\ 0 & 1 & 0 \\ 0 & 0 & 1 \end{bmatrix}$. Then, $p_{\mathbf{A}}(x) = \begin{vmatrix} x-1 & -1 & 1 \\ 0 & x-1 & 0 \\ 0 & 0 & x-1 \end{vmatrix} = (x-1)^3$ (since the corresponding matrix is upper triangular). Hence, the only eigenvalue for $\mathbf{A}$ is $\lambda = 1$, which has algebraic multiplicity 3. Next, we solve for a basis for E_1. We row reduce

$[1\mathbf{I}_3 - \mathbf{A} \,|\, \mathbf{0}] = \left[\begin{array}{ccc|c} 0 & -1 & 1 & 0 \\ 0 & 0 & 0 & 0 \\ 0 & 0 & 0 & 0 \end{array}\right]$ to obtain $\left[\begin{array}{ccc|c} 0 & 1 & -1 & 0 \\ 0 & 0 & 0 & 0 \\ 0 & 0 & 0 & 0 \end{array}\right]$.

Setting the independent variables $x_1 = 1$ and $x_3 = 0$ produces the vector $[1,0,0]$. Setting $x_1 = 0$ and $x_3 = 1$ yields $[0,1,1]$. Hence, $\{[1,0,0],[0,1,1]\}$ is a basis for E_1. Since this basis has 2 elements, $\lambda = 1$ has geometric multiplicity 2.

(b) Let $\mathbf{A} = \begin{bmatrix} 1 & 0 & 0 \\ 0 & 1 & 0 \\ 0 & 0 & 0 \end{bmatrix}$. Then, $p_{\mathbf{A}}(x) = \begin{vmatrix} x-1 & 0 & 0 \\ 0 & x-1 & 0 \\ 0 & 0 & x \end{vmatrix} = (x-1)^2 x$ (since the corresponding matrix is upper triangular). Hence, $\lambda = 1$ is an eigenvalue for $\mathbf{A}$ which has algebraic multiplicity 2. (Zero is also an eigenvalue for $\mathbf{A}$.) Next, we solve for a basis for E_1. Performing the row operation $\langle 1\rangle \leftrightarrow \langle 3\rangle$ puts $[1\mathbf{I}_3 - \mathbf{A}|\,\mathbf{0}] = \left[\begin{array}{ccc|c} 0 & 0 & 0 & 0 \\ 0 & 0 & 0 & 0 \\ 0 & 0 & 1 & 0 \end{array}\right]$ into reduced row echelon form, yielding $\left[\begin{array}{ccc|c} 0 & 0 & 1 & 0 \\ 0 & 0 & 0 & 0 \\ 0 & 0 & 0 & 0 \end{array}\right]$. Setting the independent variables $x_1 = 1$ and $x_2 = 0$ produces the vector $[1,0,0]$. Setting $x_1 = 0$ and $x_2 = 1$ yields $[0,1,0]$. Hence, $\{[1,0,0],[0,1,0]\}$ is a basis for E_1. Since this basis has 2 elements, $\lambda = 1$ has geometric multiplicity 2.

(15) Suppose $\mathbf{v} \in B_1 \cap B_2$. Then $\mathbf{v} \in B_1$ and $\mathbf{v} \in B_2$. Now $\mathbf{v} \in B_1 \Longrightarrow \mathbf{v} \in E_{\lambda_1} \Longrightarrow L(\mathbf{v}) = \lambda_1\mathbf{v}$. Similarly, $\mathbf{v} \in B_2 \Longrightarrow \mathbf{v} \in E_{\lambda_2} \Longrightarrow L(\mathbf{v}) = \lambda_2\mathbf{v}$. Hence, $\lambda_1\mathbf{v} = L(\mathbf{v}) = \lambda_2\mathbf{v}$, and so $(\lambda_1-\lambda_2)\mathbf{v} = \mathbf{0}$. Since $\lambda_1 \neq \lambda_2$, $\mathbf{v} = \mathbf{0}$. But $\mathbf{0}$ can not be an element of any basis. This contradiction shows that $B_1 \cap B_2$ must be empty.

(16) (a) E_{λ_i} is a subspace, hence it is closed under linear combinations. Therefore, since $\mathbf{u}_i$ is a linear combination of basis vectors for E_{λ_i}, it must also be in E_{λ_i}.

(b) First, substitute $\mathbf{u}_i$ for $\sum_{j=1}^{k_i} a_{ij}\mathbf{v}_{ij}$ in the given double sum equation. This proves $\sum_{i=1}^{n} \mathbf{u}_i = \mathbf{0}$. Now, the set of all nonzero $\mathbf{u}_i$'s is linearly independent by Theorem 5.22, since they are eigenvectors corresponding to distinct eigenvalues. But then the nonzero terms in $\sum_{i=1}^{n} \mathbf{u}_i$ would give a nontrivial linear combination from a linearly independent set equal to the zero vector. This contradiction shows that all of the $\mathbf{u}_i$'s must equal $\mathbf{0}$.

(c) Using part (b) and the definition of $\mathbf{u}_i$, $\mathbf{0} = \sum_{j=1}^{k_i} a_{ij}\mathbf{v}_{ij}$, for each i. But $\{\mathbf{v}_{i1}, \ldots, \mathbf{v}_{ik_i}\}$ is linearly independent, since it is a basis for E_{λ_i}. Hence, for each i, $a_{i1} = \cdots = a_{ik_i} = 0$.

(d) Apply Theorem 4.7 to B.

(17) (a) False. The correct definition for the eigenspace E_λ is $E_\lambda = \{\mathbf{v} \in \mathcal{V} \mid L(\mathbf{v}) = \lambda\mathbf{v}\}$. Note that the set $S = \{\lambda L(\mathbf{v}) \mid \mathbf{v} \in \mathcal{V}\}$ is the set of images of L multiplied by λ. But since $\lambda L(\mathbf{v}) = L(\lambda\mathbf{v})$, it is not hard to show that S equals range(L). For a specific example in which $E_\lambda \neq S$, consider Example 7 in Section 5.5 of the textbook, using $\lambda = 2$. In that example, L is an isomorphism, so $S = \text{range}(L) = \mathbb{R}^3$. However, the example shows that E_λ has dimension 1, since the algebraic multiplicity of $\lambda = 2$ is only 1.

(b) True. This is the definition of the characteristic polynomial of a linear operator. (The characteristic polynomial for L is independent of the particular ordered basis chosen for $\mathcal{V}$ (because of Theorems 5.5 and 5.6). Therefore, the definition of a characteristic polynomial of a linear operator makes sense as given in the text.)

(c) True. This follows directly from Corollary 5.23.

(d) False. While one direction of this "if and only if" statement is just Theorem 5.22, the converse of that statement (the other half of the "if and only if") is false. For a specific counterexample, consider the zero linear operator $L : \mathbb{R}^2 \to \mathbb{R}^2$. Then $\lambda = 0$ is the only eigenvalue, and $E_0 = \mathbb{R}^2$. Thus, $\{\mathbf{i},\mathbf{j}\}$ is a set of linearly independent eigenvectors for the same eigenvalue $\lambda = 0$.

(e) True. This is Theorem 5.24.

(f) True. This follows from Theorem 5.27 (which indicates that a union of the bases for the distinct eigenspaces contains 6 vectors) and Theorem 5.24 (which indicates that the 6 vectors in the union of these bases are linearly independent, and hence form a basis for $\mathbb{R}^6$).

(g) True. This is the purpose of the Method.

(h) False. For a specific example in which the algebraic multiplicity of an eigenvalue is strictly larger than its geometric multiplicity, see Example 12 in Section 5.5 of the textbook. (Note that by Theorem 5.25, the algebraic multiplicity of an eigenvalue λ is always greater than or equal to its geometric multiplicity.)

(i) False. According to Theorem 5.27, each geometric multiplicity must also equal its corresponding algebraic multiplicity in order for L to be diagonalizable. For a specific counterexample, consider $L : \mathbb{R}^7 \to \mathbb{R}^7$ given by $L(\mathbf{e}_i) = \mathbf{e}_{i+1}$, for $1 \leq i \leq 6$, and $L(\mathbf{e}_7) = \mathbf{0}$. Then it can be shown that $p_L(x) = x^7$, and so $\lambda = 0$ is the only eigenvalue for L, having algebraic multiplicity 7. However, it can be shown that E_0 has basis $\{\mathbf{e}_7\}$, and so the geometric multiplicity of $\lambda = 0$ is 1. Hence, by Theorem 5.27, L is not diagonalizable. You should verify all of the claims made in this example.

(j) True. Since $p_{\mathbf{A}}(x) = (x-1)(x-4) = (1-x)(4-x)$, this follows directly from the Cayley-Hamilton Theorem.

Chapter 6

Section 6.1

(1) (a) The vectors are orthogonal because $[3,-2]\cdot[4,6]=0$. The vectors are not orthonormal because $\|[3,-2]\|=\sqrt{13}\neq 1$.

(c) The vectors are not orthogonal because $\left[\frac{3}{\sqrt{13}},-\frac{2}{\sqrt{13}}\right]\cdot\left[\frac{1}{\sqrt{10}},-\frac{3}{\sqrt{10}}\right]=\frac{9}{\sqrt{130}}\neq 0$. Since the vectors are not orthogonal, they are also not orthonormal.

(f) The set of vectors is orthogonal because $[2,-3,1,2]\cdot[-1,2,8,0]=0$, $[2,-3,1,2]\cdot[6,-1,1,-8]=0$, and $[-1,2,8,0]\cdot[6,-1,1,-8]=0$. The vectors are not orthonormal because $\|[2,-3,1,2]\|=\sqrt{18}\neq 0$.

(2) In each part, let $\mathbf{A}$ represent the given matrix.

(a) Straightforward computation shows that $\mathbf{A}\mathbf{A}^T=\mathbf{I}_2$. Therefore, $\mathbf{A}^{-1}=\mathbf{A}^T$, and so $\mathbf{A}$ is an orthogonal matrix.

(c) $\mathbf{A}\mathbf{A}^T=\begin{bmatrix}109 & 27 & -81\\ 27 & 19 & -27\\ -81 & -27 & 91\end{bmatrix}\neq\mathbf{I}_3$. Therefore, $\mathbf{A}$ is not an orthogonal matrix.

(e) Straightforward computation shows that $\mathbf{A}\mathbf{A}^T=\mathbf{I}_4$. Therefore, $\mathbf{A}^{-1}=\mathbf{A}^T$, and so $\mathbf{A}$ is an orthogonal matrix.

(3) (a) First, B is an orthonormal basis for $\mathbb{R}^2$ because it has 2 elements and $\left[-\frac{\sqrt{3}}{2},\frac{1}{2}\right]\cdot\left[\frac{1}{2},\frac{\sqrt{3}}{2}\right]=0$, $\left\|\left[-\frac{\sqrt{3}}{2},\frac{1}{2}\right]\right\|=1$, and $\left\|\left[\frac{1}{2},\frac{\sqrt{3}}{2}\right]\right\|=1$. So, by Theorem 6.3,
$[\mathbf{v}]_B=\left[\left([-2,3]\cdot\left[-\frac{\sqrt{3}}{2},\frac{1}{2}\right]\right),\left([-2,3]\cdot\left[\frac{1}{2},\frac{\sqrt{3}}{2}\right]\right)\right]=\left[\frac{2\sqrt{3}+3}{2},\frac{3\sqrt{3}-2}{2}\right]$.

(c) First, B is an orthonormal basis for $\mathbb{R}^4$ because it has 4 elements and
$\left[\frac{1}{2},-\frac{1}{2},\frac{1}{2},\frac{1}{2}\right]\cdot\left[\frac{3}{2\sqrt{3}},\frac{1}{2\sqrt{3}},-\frac{1}{2\sqrt{3}},-\frac{1}{2\sqrt{3}}\right]=0$,
$\left[\frac{1}{2},-\frac{1}{2},\frac{1}{2},\frac{1}{2}\right]\cdot\left[0,\frac{2}{\sqrt{6}},\frac{1}{\sqrt{6}},\frac{1}{\sqrt{6}}\right]=0$, $\left[\frac{1}{2},-\frac{1}{2},\frac{1}{2},\frac{1}{2}\right]\cdot\left[0,0,-\frac{1}{\sqrt{2}},\frac{1}{\sqrt{2}}\right]=0$,
$\left[\frac{3}{2\sqrt{3}},\frac{1}{2\sqrt{3}},-\frac{1}{2\sqrt{3}},-\frac{1}{2\sqrt{3}}\right]\cdot\left[0,\frac{2}{\sqrt{6}},\frac{1}{\sqrt{6}},\frac{1}{\sqrt{6}}\right]=0$, $\left[\frac{3}{2\sqrt{3}},\frac{1}{2\sqrt{3}},-\frac{1}{2\sqrt{3}},-\frac{1}{2\sqrt{3}}\right]\cdot\left[0,0,-\frac{1}{\sqrt{2}},\frac{1}{\sqrt{2}}\right]=0$,
$\left[0,\frac{2}{\sqrt{6}},\frac{1}{\sqrt{6}},\frac{1}{\sqrt{6}}\right]\cdot\left[0,0,-\frac{1}{\sqrt{2}},\frac{1}{\sqrt{2}}\right]=0$, $\left\|\left[\frac{1}{2},-\frac{1}{2},\frac{1}{2},\frac{1}{2}\right]\right\|=1$,
$\left\|\left[\frac{3}{2\sqrt{3}},\frac{1}{2\sqrt{3}},-\frac{1}{2\sqrt{3}},-\frac{1}{2\sqrt{3}}\right]\right\|=1$, $\left\|\left[0,\frac{2}{\sqrt{6}},\frac{1}{\sqrt{6}},\frac{1}{\sqrt{6}}\right]\right\|=1$, and $\left\|\left[0,0,-\frac{1}{\sqrt{2}},\frac{1}{\sqrt{2}}\right]\right\|=1$. So, by Theorem 6.3, $[\mathbf{v}]_B=\left[\left([8,4,-3,5]\cdot\left[\frac{1}{2},-\frac{1}{2},\frac{1}{2},\frac{1}{2}\right]\right),\left([8,4,-3,5]\cdot\left[\frac{3}{2\sqrt{3}},\frac{1}{2\sqrt{3}},-\frac{1}{2\sqrt{3}},-\frac{1}{2\sqrt{3}}\right]\right),\right.$
$\left.\left([8,4,-3,5]\cdot\left[0,\frac{2}{\sqrt{6}},\frac{1}{\sqrt{6}},\frac{1}{\sqrt{6}}\right]\right),\left([8,4,-3,5]\cdot\left[0,0,-\frac{1}{\sqrt{2}},\frac{1}{\sqrt{2}}\right]\right)\right]=\left[3,\frac{13\sqrt{3}}{3},\frac{5\sqrt{6}}{3},4\sqrt{2}\right]$.

(4) (a) First, let $\mathbf{v}_1=[5,-1,2]$.
Next, $\mathbf{v}_2=[2,-1,-4]-\left(\frac{[2,-1,-4]\cdot[5,-1,2]}{[5,-1,2]\cdot[5,-1,2]}\right)[5,-1,2]=[2,-1,-4]-\frac{3}{30}[5,-1,2]=\left[\frac{3}{2},-\frac{9}{10},-\frac{21}{5}\right]$.
We multiply $\mathbf{v}_2$ by $\frac{10}{3}$ to simplify its form, yielding $\mathbf{v}_2=[5,-3,-14]$. Hence, an orthogonal basis for the subspace is $\{[5,-1,2],[5,-3,-14]\}$.

(c) First, let $\mathbf{v}_1 = [2, 1, 0, -1]$.
Next, $\mathbf{v}_2 = [1, 1, 1, -1] - \left(\frac{[1,1,1,-1]\cdot[2,1,0,-1]}{[2,1,0,-1]\cdot[2,1,0,-1]}\right)[2, 1, 0, -1] = [1, 1, 1, -1] - \frac{4}{6}[2, 1, 0, -1]$
$= \left[-\frac{1}{3}, \frac{1}{3}, 1, -\frac{1}{3}\right]$. Multiplying by 3 to eliminate fractions yields $\mathbf{v}_2 = [-1, 1, 3, -1]$.
Finally, $\mathbf{v}_3 = [1, -2, 1, 1] - \left(\frac{[1,-2,1,1]\cdot[2,1,0,-1]}{[2,1,0,-1]\cdot[2,1,0,-1]}\right)[2, 1, 0, -1] - \left(\frac{[1,-2,1,1]\cdot[-1,1,3,-1]}{[-1,1,3,-1]\cdot[-1,1,3,-1]}\right)[-1, 1, 3, -1]$
$= [1, -2, 1, 1] - \frac{(-1)}{6}[2, 1, 0, -1] - \frac{(-1)}{12}[-1, 1, 3, -1] = \left[\frac{5}{4}, -\frac{7}{4}, \frac{5}{4}, \frac{3}{4}\right]$. Multiplying by 4 to eliminate fractions produces $\mathbf{v}_3 = [5, -7, 5, 3]$. Hence, an orthogonal basis for the subspace is $\{[2, 1, 0, -1], [-1, 1, 3, -1], [5, -7, 5, 3]\}$.

(5) (a) First, we must find an ordinary basis for $\mathbb{R}^3$ containing $[2, 2, -3]$. To do this, we use the Enlarging Method from Section 4.6. We add the standard basis for $\mathbb{R}^3$ to the vector we have and use the Independence Test Method. Hence, we row reduce

$\begin{bmatrix} 2 & 1 & 0 & 0 \\ 2 & 0 & 1 & 0 \\ -3 & 0 & 0 & 1 \end{bmatrix}$ to $\begin{bmatrix} 1 & 0 & 0 & -\frac{1}{3} \\ 0 & 1 & 0 & \frac{2}{3} \\ 0 & 0 & 1 & \frac{2}{3} \end{bmatrix}$, giving the basis $\{[2, 2, -3], [1, 0, 0], [0, 1, 0]\}$. We perform the Gram-Schmidt Process on this set.
Let $\mathbf{v}_1 = [2, 2, -3]$. Then $\mathbf{v}_2 = [1, 0, 0] - \left(\frac{[1,0,0]\cdot[2,2,-3]}{[2,2,-3]\cdot[2,2,-3]}\right)[2, 2, -3] = [1, 0, 0] - \frac{2}{17}[2, 2, -3] = \left[\frac{13}{17}, -\frac{4}{17}, \frac{6}{17}\right]$. Multiplying by 17 to eliminate fractions yields $\mathbf{v}_2 = [13, -4, 6]$. Finally,
$\mathbf{v}_3 = [0, 1, 0] - \left(\frac{[0,1,0]\cdot[2,2,-3]}{[2,2,-3]\cdot[2,2,-3]}\right)[2, 2, -3] - \left(\frac{[0,1,0]\cdot[13,-4,6]}{[13,-4,6]\cdot[13,-4,6]}\right)[13, -4, 6]$
$= [0, 1, 0] - \frac{2}{17}[2, 2, -3] - \frac{(-4)}{221}[13, -4, 6] = \left[0, \frac{9}{13}, \frac{6}{13}\right]$. Multiplying by $\frac{13}{3}$ to simplify the form of the vector gives us $\mathbf{v}_3 = [0, 3, 2]$. Hence, an orthogonal basis for $\mathbb{R}^3$ containing $[2, 2, -3]$ is $\{[2, 2, -3], [13, -4, 6], [0, 3, 2]\}$.

(c) First, we must find an ordinary basis for $\mathbb{R}^3$ containing $[1, -3, 1]$ and $[2, 5, 13]$. To do this, we use the Enlarging Method from Section 4.6. We add the standard basis for $\mathbb{R}^3$ to the given vector, and use the Independence Test Method. Hence, we row reduce

$\begin{bmatrix} 1 & 2 & 1 & 0 & 0 \\ -3 & 5 & 0 & 1 & 0 \\ 1 & 13 & 0 & 0 & 1 \end{bmatrix}$ to $\begin{bmatrix} 1 & 0 & 0 & -\frac{13}{44} & \frac{5}{44} \\ 0 & 1 & 0 & \frac{1}{44} & \frac{3}{44} \\ 0 & 0 & 1 & \frac{1}{4} & -\frac{1}{4} \end{bmatrix}$, giving the basis $\{[1, -3, 1], [2, 5, 13], [1, 0, 0]\}$.

We perform the Gram-Schmidt Process on this set.
Let $\mathbf{v}_1 = [1, -3, 1]$. Since we know that the given set of vectors is orthogonal, we may simply use $\mathbf{v}_2 = [2, 5, 13]$. (If the two given vectors were not already orthogonal, we would need to apply Gram-Schmidt to obtain $\mathbf{v}_2$.) Finally,
$\mathbf{v}_3 = [1, 0, 0] - \left(\frac{[1,0,0]\cdot[1,-3,1]}{[1,-3,1]\cdot[1,-3,1]}\right)[1, -3, 1] - \left(\frac{[1,0,0]\cdot[2,5,13]}{[2,5,13]\cdot[2,5,13]}\right)[2, 5, 13]$
$= [1, 0, 0] - \frac{1}{11}[1, -3, 1] - \frac{2}{198}[2, 5, 13] = \left[\frac{8}{9}, \frac{2}{9}, -\frac{2}{9}\right]$. Multiplying by $\frac{9}{2}$ to simplify the form of the vector gives $\mathbf{v}_3 = [4, 1, -1]$. Hence, an orthogonal basis for $\mathbb{R}^3$ containing the two given vectors is $\{[1, -3, 1], [2, 5, 13], [4, 1, -1]\}$.

(e) First, we must find an ordinary basis for $\mathbb{R}^4$ containing $[2, 1, -2, 1]$. To do this, we use the Enlarging Method from Section 4.6. We add the standard basis for $\mathbb{R}^4$ to the given vector and use the Independence Test Method. Hence, we row reduce

$\begin{bmatrix} 2 & 1 & 0 & 0 & 0 \\ 1 & 0 & 1 & 0 & 0 \\ -2 & 0 & 0 & 1 & 0 \\ 1 & 0 & 0 & 0 & 1 \end{bmatrix}$ to $\begin{bmatrix} 1 & 0 & 0 & 0 & 1 \\ 0 & 1 & 0 & 0 & -2 \\ 0 & 0 & 1 & 0 & -1 \\ 0 & 0 & 0 & 1 & 2 \end{bmatrix}$, giving the basis $\{[2,1,-2,1],[1,0,0,0],[0,1,0,0],[0,0,1,0]\}$. We perform the Gram-Schmidt Process on this set. Let $\mathbf{v}_1 = [2,1,-2,1]$. Then, $\mathbf{v}_2 = [1,0,0,0] - \left(\frac{[1,0,0,0]\cdot[2,1,-2,1]}{[2,1,-2,1]\cdot[2,1,-2,1]}\right)[2,1,-2,1] = [1,0,0,0] - \frac{2}{10}[2,1,-2,1] = \left[\frac{3}{5},-\frac{1}{5},\frac{2}{5},-\frac{1}{5}\right]$. Multiplying by 5 to eliminate fractions yields $\mathbf{v}_2 = [3,-1,2,-1]$. Next, $\mathbf{v}_3 = [0,1,0,0] - \left(\frac{[0,1,0,0]\cdot[2,1,-2,1]}{[2,1,-2,1]\cdot[2,1,-2,1]}\right)[2,1,-2,1] - \left(\frac{[0,1,0,0]\cdot[3,-1,2,-1]}{[3,-1,2,-1]\cdot[3,-1,2,-1]}\right)[3,-1,2,-1] = [0,1,0,0] - \frac{1}{10}[2,1,-2,1] - \frac{(-1)}{15}[3,-1,2,-1] = \left[0,\frac{5}{6},\frac{1}{3},-\frac{1}{6}\right]$. Multiplying by 6 to eliminate fractions gives us $\mathbf{v}_3 = [0,5,2,-1]$. Finally, $\mathbf{v}_4 = [0,0,1,0] - \left(\frac{[0,0,1,0]\cdot[2,1,-2,1]}{[2,1,-2,1]\cdot[2,1,-2,1]}\right)[2,1,-2,1] - \left(\frac{[0,0,1,0]\cdot[3,-1,2,-1]}{[3,-1,2,-1]\cdot[3,-1,2,-1]}\right)[3,-1,2,-1] - \left(\frac{[0,0,1,0]\cdot[0,5,2,-1]}{[0,5,2,-1]\cdot[0,5,2,-1]}\right)[0,5,2,-1]$
$= [0,0,1,0] - \frac{(-2)}{10}[2,1,-2,1] - \frac{2}{15}[3,-1,2,-1] - \frac{2}{30}[0,5,2,-1] = \left[0,0,\frac{1}{5},\frac{2}{5}\right]$. Multiplying by 5 to eliminate fractions produces $\mathbf{v}_4 = [0,0,1,2]$. Hence, an orthogonal basis for $\mathbb{R}^4$ containing the given vector is $\{[2,1,-2,1],[3,-1,2,-1],[0,5,2,-1],[0,0,1,2]\}$.

(7) In each part, let $\mathbf{A}$ represent the given matrix. Then, according to the directions for the problem, a vector in the direction of the axis of rotation is found by solving for an eigenvector for $\mathbf{A}$ corresponding to the eigenvalue $\lambda = 1$.

(a) First, we must verify that $\mathbf{A}$ is an orthogonal matrix. Calculating $\mathbf{A}^T$ and performing matrix multiplication easily shows that $\mathbf{A}\mathbf{A}^T = \mathbf{I}_3$. Therefore, $\mathbf{A}^{-1} = \mathbf{A}^T$, and so $\mathbf{A}$ is orthogonal. Basketweaving quickly verifies that $|\mathbf{A}| = 1$. Finally, we need to find an eigenvector for $\mathbf{A}$ corresponding to $\lambda = 1$. To do this, we row reduce

$$[1\mathbf{I}_3 - \mathbf{A}|\mathbf{0}] = \left[\begin{array}{ccc|c} \frac{9}{11} & -\frac{6}{11} & \frac{9}{11} & 0 \\ \frac{9}{11} & \frac{5}{11} & -\frac{2}{11} & 0 \\ -\frac{6}{11} & -\frac{7}{11} & \frac{5}{11} & 0 \end{array}\right] \text{ to } \left[\begin{array}{ccc|c} 1 & 0 & \frac{1}{3} & 0 \\ 0 & 1 & -1 & 0 \\ 0 & 0 & 0 & 0 \end{array}\right].$$

Setting the independent variable x_3 equal to 3 to eliminate fractions produces the eigenvector $[-1,3,3]$. This is a vector along the desired axis of rotation.

(c) First, we must verify that $\mathbf{A}$ is an orthogonal matrix. Calculating $\mathbf{A}^T$ and performing matrix multiplication easily shows that $\mathbf{A}\mathbf{A}^T = \mathbf{I}_3$. Therefore, $\mathbf{A}^{-1} = \mathbf{A}^T$, and so $\mathbf{A}$ is orthogonal. Basketweaving quickly verifies that $|\mathbf{A}| = 1$. Finally, we need to find an eigenvector for $\mathbf{A}$ corresponding to $\lambda = 1$. To do this, we row reduce

$$[1\mathbf{I}_3 - \mathbf{A}|\mathbf{0}] = \left[\begin{array}{ccc|c} \frac{1}{7} & -\frac{2}{7} & -\frac{3}{7} & 0 \\ -\frac{3}{7} & \frac{13}{7} & \frac{2}{7} & 0 \\ -\frac{2}{7} & -\frac{3}{7} & \frac{13}{7} & 0 \end{array}\right] \text{ to } \left[\begin{array}{ccc|c} 1 & 0 & -5 & 0 \\ 0 & 1 & -1 & 0 \\ 0 & 0 & 0 & 0 \end{array}\right].$$

Setting the independent variable x_3 equal to 1 produces the eigenvector $[5,1,1]$. This is a vector along the desired axis of rotation.

(8) (b) No. The vectors might not be unit vectors after scalar multiplication. For example, if we multiply every vector in the orthonormal set $\{[1,0],[0,1]\}$ by 2, we get $\{[2,0],[0,2]\}$. This new set is orthogonal, but it is not orthonormal because each of its elements has length 2.

(16) (b) By part (1) of Theorem 6.6, a 3×3 matrix is orthogonal if and only if its rows form an orthonormal basis for $\mathbb{R}^3$. Thus, we must find an orthonormal basis containing $\frac{1}{\sqrt{6}}[1,2,1]$. To do this, we will first find an orthogonal basis containing $[1,2,1]$ and then normalize the vectors afterwards.
First, we must find an ordinary basis for $\mathbb{R}^3$ containing $[1,2,1]$. To do this, we use the Enlarging Method from Section 4.6. We add the standard basis for $\mathbb{R}^3$ to the vector we have and then use the Independence Test Method. Hence, we row reduce
$\begin{bmatrix} 1 & 1 & 0 & 0 \\ 2 & 0 & 1 & 0 \\ 1 & 0 & 0 & 1 \end{bmatrix}$ to $\begin{bmatrix} 1 & 0 & 0 & 1 \\ 0 & 1 & 0 & -1 \\ 0 & 0 & 1 & -2 \end{bmatrix}$, giving the basis $\{[1,2,1],[1,0,0],[0,1,0]\}$. Next, we perform the Gram-Schmidt Process on this set.
Let $\mathbf{v}_1 = [1,2,1]$. Then $\mathbf{v}_2 = [1,0,0] - \left(\frac{[1,0,0]\cdot[1,2,1]}{[1,2,1]\cdot[1,2,1]}\right)[1,2,1] = [1,0,0] - \frac{1}{6}[1,2,1] = \left[\frac{5}{6}, -\frac{1}{3}, -\frac{1}{6}\right]$.
Multiplying by 6 to eliminate fractions yields $\mathbf{v}_2 = [5,-2,-1]$. Next,
$\mathbf{v}_3 = [0,1,0] - \left(\frac{[0,1,0]\cdot[1,2,1]}{[1,2,1]\cdot[1,2,1]}\right)[1,2,1] - \left(\frac{[0,1,0]\cdot[5,-2,-1]}{[5,-2,-1]\cdot[5,-2,-1]}\right)[5,-2,-1] = [0,1,0] - \frac{2}{6}[1,2,1] - \frac{(-2)}{30}[5,-2,-1] = \left[0, \frac{1}{5}, -\frac{2}{5}\right]$. Multiplying by 5 to eliminate fractions gives us $\mathbf{v}_3 = [0,1,-2]$.
Normalizing $\mathbf{v}_1$, $\mathbf{v}_2$, and $\mathbf{v}_3$ produces the orthonormal basis
$\left\{\left[\frac{1}{\sqrt{6}}, \frac{2}{\sqrt{6}}, \frac{1}{\sqrt{6}}\right], \left[\frac{5}{\sqrt{30}}, -\frac{2}{\sqrt{30}}, -\frac{1}{\sqrt{30}}\right], \left[0, \frac{1}{\sqrt{5}}, -\frac{2}{\sqrt{5}}\right]\right\}$ for $\mathbb{R}^3$. Using these vectors as rows for a matrix, rationalizing denominators and simplifying gives us the orthogonal matrix

$$\begin{bmatrix} \frac{\sqrt{6}}{6} & \frac{\sqrt{6}}{3} & \frac{\sqrt{6}}{6} \\ \frac{\sqrt{30}}{6} & -\frac{\sqrt{30}}{15} & -\frac{\sqrt{30}}{30} \\ 0 & \frac{\sqrt{5}}{5} & -\frac{2\sqrt{5}}{5} \end{bmatrix},$$ whose first row is the given vector $\frac{1}{\sqrt{6}}[1,2,1]$.

(17) Proof of the other half of part (1) of Theorem 6.6: Suppose that the rows of $\mathbf{A}$ form an orthonormal basis for $\mathbb{R}^n$. Then the (i,j) entry of $\mathbf{A}\mathbf{A}^T$ = (ith row of $\mathbf{A}$)$\cdot$(jth column of $\mathbf{A}^T$)
= (ith row of $\mathbf{A}$)$\cdot$(jth row of $\mathbf{A}$) $= \begin{cases} 1 & i=j \\ 0 & i \neq j \end{cases}$ (since the rows of $\mathbf{A}$ form an orthonormal set). Hence $\mathbf{A}\mathbf{A}^T = \mathbf{I}_n$, and $\mathbf{A}$ is orthogonal.

Proof of part (2): $\mathbf{A}$ is orthogonal if and only if $\mathbf{A}^T$ is orthogonal (by Exercise 11(a)) if and only if the rows of $\mathbf{A}^T$ form an orthonormal basis for $\mathbb{R}^n$ (by part (1) of Theorem 6.6) if and only if the columns of $\mathbf{A}$ form an orthonormal basis for $\mathbb{R}^n$.

(21) (a) False. The remaining vectors in the set must also be perpendicular to each other. For example, the set $\{[0,0],[1,1],[2,3]\}$ contains the zero vector but is not orthogonal because $[1,1]\cdot[2,3] = 5 \neq 0$.

(b) True. This is because $\mathbf{e}_i \cdot \mathbf{e}_j = 0$ for $i \neq j$, and $\|\mathbf{e}_i\| = 1$ for all i.

(c) True. This is part of Theorem 6.3.

(d) False. If the set of vectors $\{\mathbf{w}_1, \ldots, \mathbf{w}_k\}$ is not orthogonal at the outset, adding more vectors will not make the original vectors perpendicular to each other. For example, the linearly independent set $\{[1,1,0],[1,0,1]\}$ can not be enlarged to an orthogonal basis for $\mathbb{R}^3$, because any such enlargement $\{[1,1,0],[1,0,1],\mathbf{v}_3\}$ will still have $[1,1,0]\cdot[1,0,1] = 1 \neq 0$.

(e) True. By Theorem 4.17, $\mathcal{W}$ has a finite basis B. Also, $|B| > 0$, since $\mathcal{W}$ is nontrivial. Then, by Theorem 6.4, the Gram-Schmidt Process can be used on the basis B to find an orthogonal basis for $\mathcal{W}$.

(f) True. Since $\mathbf{A}^T\mathbf{A} = \mathbf{I}_n$, Theorem 2.9 tells us that $\mathbf{A}^{-1} = \mathbf{A}^T$. Hence, $\mathbf{A}$ is orthogonal by definition.

(g) True. Just after the definition of an orthogonal matrix in Section 6.1 in the textbook, it is stated that the product of orthogonal matrices is orthogonal. (Also see Exercise 11(b).) Hence, $\mathbf{BA}$ is orthogonal. It is also proved in the textbook (just after the definition of an orthogonal matrix) that the determinant of an orthogonal matrix is always ± 1. Therefore, since $\mathbf{BA}$ is orthogonal, $|\mathbf{BA}| = \pm 1$.

(h) False. According to Theorem 6.6, the rows or columns of a matrix must be *orthonormal*, not just orthogonal, in order to guarantee that the matrix is orthogonal. So, for example, the rows of $\mathbf{A} = \begin{bmatrix} 1 & 2 \\ 2 & -1 \end{bmatrix}$ form an orthogonal set, but $\mathbf{A}$ is not orthogonal, since $\mathbf{A}^T\mathbf{A} = 5\mathbf{I}_2 \neq \mathbf{I}_2$.

(i) True. By part (1) of Theorem 6.6, the rows of $\mathbf{A}$ form an orthonormal basis for $\mathbb{R}^n$. However, the set of rows of $R(\mathbf{A})$ is the same as the set of rows of $\mathbf{A}$, only listed in a different order. However, these vectors form an orthonormal basis for $\mathbb{R}^n$ regardless of the order in which they are listed. Hence, part (1) of Theorem 6.6 asserts that $R(\mathbf{A})$ is also an orthogonal matrix.

(j) True. This is the statement of Theorem 6.7.

Section 6.2

(1) If you need assistance combining the Gram-Schmidt Process with the Enlarging Method of Section 4.6, please review Example 5 in Section 6.1 of the textbook and Exercise 5 in Section 6.1. Solutions for the starred parts for Exercise 5 appear above in this manual.

(a) The vector $[2,3]$, found by inspection, is clearly orthogonal to $[3,-2]$. Hence $[2,3] \in \mathcal{W}^{\perp}$. Now, $\mathcal{W}^{\perp}$ can not be 2-dimensional, or it would be all of $\mathbb{R}^2$, and yet $[3,-2] \notin \mathcal{W}^{\perp}$. Therefore, $\mathcal{W}^{\perp}$ must be 1-dimensional, and so $\mathcal{W}^{\perp} = \text{span}(\{[2,3]\})$. Note that $\dim(\mathcal{W}) + \dim(\mathcal{W}^{\perp}) = 1 + 1 = 2 = \dim(\mathbb{R}^2)$.

(c) First, we find an orthogonal basis for $\mathcal{W}$ by performing the Gram-Schmidt Process on its basis $\{[1,4,-2],[2,1,-1]\}$. This results in the orthogonal basis $B = \{[1,4,-2],[34,-11,-5]\}$ for $\mathcal{W}$. Next, we enlarge this set to a basis for $\mathbb{R}^3$ using the Enlarging Method of Section 4.6. This results in the basis $C = \{[1,4,-2],[34,-11,-5],[1,0,0]\}$ for $\mathbb{R}^3$. Now we perform the Gram-Schmidt Process on the basis C to find an orthogonal basis D for $\mathbb{R}^3$. Note that since the first two vectors in C are already perpendicular to each other, they will not be changed by the Gram-Schmidt Process. Completing the Gram-Schmidt Process produces the basis $D = \{[1,4,-2],[34,-11,-5],[2,3,7]\}$. Since B is an orthogonal basis for $\mathcal{W}$, Theorem 6.11 shows that $\{[2,3,7]\}$ is an orthogonal basis for $\mathcal{W}^{\perp}$. Hence, $\mathcal{W}^{\perp} = \text{span}(\{[2,3,7]\})$. Note that $\dim(\mathcal{W}) + \dim(\mathcal{W}^{\perp}) = 2 + 1 = 3 = \dim(\mathbb{R}^3)$.

(e) Looking at the given equation for the plane $\mathcal{W}$, we see that $\mathcal{W}$ is defined to be $\{[x,y,z] \mid [x,y,z]\cdot[-2,5,-1] = 0\}$. That is, $\mathcal{W}$ is the set of vectors orthogonal to $[-2,5,-1]$, and hence orthogonal to the subspace $\mathcal{Y} = \text{span}(\{[-2,5,-1]\})$ of $\mathbb{R}^3$. So, by definition of orthogonal complement, $\mathcal{W} = \mathcal{Y}^{\perp}$. Hence, by Corollary 6.13, $\mathcal{W}^{\perp} = \left(\mathcal{Y}^{\perp}\right)^{\perp} = \mathcal{Y} = \text{span}(\{[-2,5,-1]\})$. Note that $\dim(\mathcal{W}) + \dim(\mathcal{W}^{\perp}) = 2 + 1 = 3 = \dim(\mathbb{R}^3)$.

(f) First, we find an orthogonal basis for $\mathcal{W}$ by performing the Gram-Schmidt Process on its basis $\{[1,-1,0,2],[0,1,2,-1]\}$. This results in the orthogonal basis $B = \{[1,-1,0,2],[1,1,4,0]\}$ for $\mathcal{W}$. Next, we enlarge this set to a basis for $\mathbb{R}^4$ using the Enlarging Method of Section 4.6. This results in the basis $C = \{[1,-1,0,2],[1,1,4,0],[1,0,0,0],[0,1,0,0]\}$ for $\mathbb{R}^4$. Now we perform the Gram-Schmidt Process on the basis C to find an orthogonal basis D for $\mathbb{R}^4$. Note that since the first two vectors in C are already perpendicular to each other, they will not be changed by the Gram-Schmidt Process. Completing the Gram-Schmidt Process produces the basis $D = \{[1,-1,0,2],[1,1,4,0],[7,1,-2,-3],[0,4,-1,2]\}$. Since B is an orthogonal basis for $\mathcal{W}$, Theorem 6.11 shows that $\{[7,1,-2,-3],[0,4,-1,2]\}$ is an orthogonal basis for $\mathcal{W}^\perp$. Hence, $\mathcal{W}^\perp = \text{span}(\{[7,1,-2,-3],[0,4,-1,2]\})$. Note that $\dim(\mathcal{W}) + \dim(\mathcal{W}^\perp) = 2+2 = 4 = \dim(\mathbb{R}^4)$.

(2) (a) First we use the Gram-Schmidt Process on the basis $\{[1,-2,-1],[3,-1,0]\}$ for $\mathcal{W}$ to find the following orthogonal basis for $\mathcal{W}$: $\{[1,-2,-1],[13,4,5]\}$. Normalizing this orthogonal basis for $\mathcal{W}$ produces the orthonormal basis $\left\{\frac{1}{\sqrt{6}}[1,-2,-1], \frac{1}{\sqrt{210}}[13,4,5]\right\}$. Hence, by definition,
$\mathbf{w}_1 = \mathbf{proj}_{\mathcal{W}}\mathbf{v} = \left([-1,3,2]\cdot\left(\frac{1}{\sqrt{6}}[1,-2,-1]\right)\right)\left(\frac{1}{\sqrt{6}}[1,-2,-1]\right)$
$+\left([-1,3,2]\cdot\left(\frac{1}{\sqrt{210}}[13,4,5]\right)\right)\left(\frac{1}{\sqrt{210}}[13,4,5]\right)$
$= \frac{1}{6}(-9)[1,-2,-1] + \frac{1}{210}(9)[13,4,5] = \left[-\frac{33}{35}, \frac{111}{35}, \frac{12}{7}\right]$.
Finally, $\mathbf{w}_2 = \mathbf{proj}_{\mathcal{W}^\perp}\mathbf{v} = \mathbf{v} - \mathbf{proj}_{\mathcal{W}}\mathbf{v} = [-1,3,2] - \left[-\frac{33}{35}, \frac{111}{35}, \frac{12}{7}\right] = \left[-\frac{2}{35}, -\frac{6}{35}, \frac{2}{7}\right]$.

(b) Looking at the given equation for the plane $\mathcal{W}$, we see that $\mathcal{W}$ is defined to be $\{[x,y,z] \mid [x,y,z]\cdot[2,-2,1] = 0\}$. That is, $\mathcal{W}$ is the set of vectors orthogonal to $[2,-2,1]$, and hence orthogonal to the subspace $\mathcal{Y} = \text{span}(\{[2,-2,1]\})$ of $\mathbb{R}^3$. So, by definition of orthogonal complement, $\mathcal{W} = \mathcal{Y}^\perp$. Now, by Corollary 6.13, $\mathcal{W}^\perp = \left(\mathcal{Y}^\perp\right)^\perp = \mathcal{Y} = \text{span}\left(\{\frac{1}{3}[2,-2,1]\}\right)$, where we have normalized $[2,-2,1]$ to create an orthonormal basis for $\mathcal{Y}$. Hence, by definition of $\mathbf{proj}_{\mathcal{Y}}\mathbf{v}$, $\mathbf{w}_2 = \mathbf{proj}_{\mathcal{W}^\perp}\mathbf{v} = \mathbf{proj}_{\mathcal{Y}}\mathbf{v} = \left([1,-4,3]\cdot\left(\frac{1}{3}[2,-2,1]\right)\right)\left(\frac{1}{3}[2,-2,1]\right)$
$= \frac{1}{9}(13)[2,-2,1] = \left[\frac{26}{9}, -\frac{26}{9}, \frac{13}{9}\right]$.
Finally, $\mathbf{w}_1 = \mathbf{proj}_{\mathcal{W}}\mathbf{v} = \mathbf{v} - \mathbf{proj}_{\mathcal{W}^\perp}\mathbf{v} = [1,-4,3] - \left[\frac{26}{9}, -\frac{26}{9}, \frac{13}{9}\right] = \left[-\frac{17}{9}, -\frac{10}{9}, \frac{14}{9}\right]$. Note that in this problem we found it easier to compute $\mathbf{proj}_{\mathcal{W}^\perp}\mathbf{v}$ first and use it to calculate $\mathbf{proj}_{\mathcal{W}}\mathbf{v}$ because it was easier to determine an orthonormal basis for $\mathcal{W}^\perp$ than for $\mathcal{W}$.

(4) (a) Let $\mathbf{v} = [-2,3,1]$. By Theorem 6.16, the minimum distance from P to $\mathcal{W}$ is $\|\mathbf{v} - \mathbf{proj}_{\mathcal{W}}\mathbf{v}\|$. Thus, we need to compute $\mathbf{proj}_{\mathcal{W}}\mathbf{v}$. To do this, we need an orthonormal basis for $\mathcal{W}$. We begin with the given basis $\{[-1,4,4],[2,-1,0]\}$ for $\mathcal{W}$ and apply the Gram-Schmidt Process to obtain $\{[-1,4,4],[20,-3,8]\}$. Normalizing the vectors in this basis produces the orthonormal basis $\left\{\frac{1}{\sqrt{33}}[-1,4,4], \frac{1}{\sqrt{473}}[20,-3,8]\right\}$ for $\mathcal{W}$. Thus, by definition,
$\mathbf{proj}_{\mathcal{W}}\mathbf{v} = \left([-2,3,1]\cdot\left(\frac{1}{\sqrt{33}}[-1,4,4]\right)\right)\left(\frac{1}{\sqrt{33}}[-1,4,4]\right)$
$+\left([-2,3,1]\cdot\left(\frac{1}{\sqrt{473}}[20,-3,8]\right)\right)\left(\frac{1}{\sqrt{473}}[20,-3,8]\right)$
$= \frac{1}{33}(18)[-1,4,4] + \frac{1}{473}(-41)[20,-3,8] = \left[-\frac{98}{43}, \frac{105}{43}, \frac{64}{43}\right]$.
Hence, the minimum distance from P to $\mathcal{W}$ is
$\|\mathbf{v} - \mathbf{proj}_{\mathcal{W}}\mathbf{v}\| = \left\|[-2,3,1] - \left[-\frac{98}{43}, \frac{105}{43}, \frac{64}{43}\right]\right\| = \left\|\left[\frac{12}{43}, \frac{24}{43}, -\frac{21}{43}\right]\right\| = \frac{3\sqrt{129}}{43}$.

(d) Let $\mathbf{v} = [-1,4,-2,2]$. By Theorem 6.16, the minimum distance from P to $\mathcal{W}$ is $\|\mathbf{v} - \mathbf{proj}_{\mathcal{W}}\mathbf{v}\| = \|\mathbf{proj}_{\mathcal{W}^\perp}\mathbf{v}\|$. To compute this, we will find an orthonormal basis for $\mathcal{W}^\perp$. Now $\mathcal{W}$ is defined to be $\{[x,y,z,w] \mid [x,y,z,w]\cdot[2,0,-3,2] = 0\}$. That is, $\mathcal{W}$ is the set of vectors orthogonal to $[2,0,-3,2]$, and hence orthogonal to the subspace $\mathcal{Y} = \text{span}(\{[2,0,-3,2]\})$. So, by definition of orthogonal complement, $\mathcal{W} = \mathcal{Y}^\perp$. Thus, by Corollary 6.13, $\mathcal{W}^\perp = \left(\mathcal{Y}^\perp\right)^\perp = \mathcal{Y}$. Therefore, a basis for $\mathcal{W}^\perp$ is $\{[2,0,-3,2]\}$. Normalizing, we get the orthonormal basis $\left\{\frac{1}{\sqrt{17}}[2,0,-3,2]\right\}$ for $\mathcal{W}^\perp$. So, by definition, $\mathbf{proj}_{\mathcal{W}^\perp}\mathbf{v} = \left([-1,4,-2,2]\cdot\left(\frac{1}{\sqrt{17}}[2,0,-3,2]\right)\right)\left(\frac{1}{\sqrt{17}}[2,0,-3,2]\right) = \frac{1}{17}(8)[2,0,-3,2]$.
Hence, the minimum distance from P to $\mathcal{W}$ is $\|\mathbf{proj}_{\mathcal{W}^\perp}\mathbf{v}\| = \left\|\frac{8}{17}[2,0,-3,2]\right\| = \frac{8\sqrt{17}}{17}$.

(5) In each part, let $\mathbf{A}$ represent the given matrix. This problem utilizes the methods illustrated in Examples 8 and 9 and the surrounding discussion in Section 6.2 of the textbook.

(a) The first step is to find the eigenvalues of $\mathbf{A}$. So,

$p_{\mathbf{A}}(x) = \begin{vmatrix} x - \frac{2}{11} & \frac{3}{11} & \frac{3}{11} \\ \frac{3}{11} & x - \frac{10}{11} & \frac{1}{11} \\ \frac{3}{11} & \frac{1}{11} & x - \frac{10}{11} \end{vmatrix} = x^3 - 2x^2 + x$ (by basketweaving) $= x(x-1)^2$. Hence, $\mathbf{A}$ has eigenvalues $\lambda_1 = 0$ and $\lambda_2 = 1$, as required for L to be an orthogonal projection. (L can not be an orthogonal reflection since neither eigenvalue is -1.) Next, by row reducing $[0\mathbf{I}_3 - \mathbf{A} \mid \mathbf{0}]$ and $[1\mathbf{I}_3 - \mathbf{A} \mid \mathbf{0}]$, we find the following bases for the associated eigenspaces: basis for $E_0 = \{[3,1,1]\}$, basis for $E_1 = \{[-1,3,0],[-1,0,3]\}$. Since the basis vector $[3,1,1]$ for E_0 is perpendicular to both basis vectors for E_1, the operator L whose matrix is $\mathbf{A}$ with respect to the standard basis is an orthogonal projection onto the plane determined by E_1. But since $E_1 = E_0^{\perp}$, E_1 is the plane consisting of the vectors orthogonal to $[3,1,1]$. Hence, L is the orthogonal projection onto the plane $3x + y + z = 0$.

(d) The first step is to find the eigenvalues of $\mathbf{A}$. So,

$p_{\mathbf{A}}(x) = \begin{vmatrix} x - \frac{7}{15} & \frac{2}{15} & \frac{14}{15} \\ \frac{4}{15} & x - \frac{14}{15} & \frac{7}{15} \\ \frac{4}{5} & \frac{1}{5} & x + \frac{2}{5} \end{vmatrix} = x^3 - x^2 - x + 1$ (by basketweaving) $= (x-1)^2(x+1)$.

Hence, $\mathbf{A}$ has eigenvalues $\lambda_1 = 1$ and $\lambda_2 = -1$, as required for L to be an orthogonal reflection. (L can not be an orthogonal projection since neither eigenvalue is 0.) Next, by row reducing $[1\mathbf{I}_3 - \mathbf{A} \mid \mathbf{0}]$ and $[-1\mathbf{I}_3 - \mathbf{A} \mid \mathbf{0}]$, we find the following bases for the associated eigenspaces: basis for $E_1 = \{[-1,4,0],[-7,0,4]\}$, basis for $E_{-1} = \{[2,1,3]\}$. However, $[2,1,3] \cdot [-1,4,0] = 2 \neq 0$. Therefore, the eigenspaces are not orthogonal complements. Hence, the operator L is neither an orthogonal projection nor an orthogonal reflection.

(6) In this problem, we utilize the ideas presented in Example 8 of Section 6.2 in the textbook. Let $\mathcal{W} = \{[x,y,z] \mid 2x - y + 2z = 0\} = \{[x,y,z] \mid [x,y,z] \cdot [2,-1,2] = 0\}$. Then, if $\mathcal{Y} = \text{span}(\{[2,-1,2]\})$, $\mathcal{W} = \mathcal{Y}^{\perp}$, and, by Corollary 6.13, $\mathcal{W}^{\perp} = \mathcal{Y}$. So, $\{[2,-1,2]\}$ is a basis for $\mathcal{W}^{\perp}$, and by the Inspection Method of Section 4.6, $\{[1,0,-1],[1,2,0]\}$ is a basis for $\mathcal{W}$. Now $\mathcal{W} = E_1$, the eigenspace corresponding to the eigenvalue $\lambda_1 = 1$, for L, since the plane $\mathcal{W}$ is left fixed by L. Also, $\mathcal{W}^{\perp} = E_0$, the eigenspace corresponding to the eigenvalue $\lambda_2 = 0$ for L, since all vectors perpendicular to $\mathcal{W}$ are sent to $\mathbf{0}$ by L. So, we know by the Generalized Diagonalization Method of Section 5.5 that if $\mathbf{A}$ is the matrix for L with respect to the standard basis for $\mathbb{R}^3$, then $\mathbf{D} = \mathbf{P}^{-1}\mathbf{A}\mathbf{P}$, where $\mathbf{P}$ is a matrix whose columns are a basis for $\mathbb{R}^3$ consisting of eigenvectors for L, and $\mathbf{D}$ is a diagonal matrix with the eigenvalues for L on its main diagonal. Note that $\mathbf{D} = \mathbf{P}^{-1}\mathbf{A}\mathbf{P}$ implies that $\mathbf{A} = \mathbf{P}\mathbf{D}\mathbf{P}^{-1}$. Setting $\mathbf{D} = \begin{bmatrix} 0 & 0 & 0 \\ 0 & 1 & 0 \\ 0 & 0 & 1 \end{bmatrix}$ and, using the bases for $\mathcal{W}^{\perp} = E_0$ and $\mathcal{W} = E_1$ from above, $\mathbf{P} = \begin{bmatrix} 2 & 1 & 1 \\ -1 & 0 & 2 \\ 2 & -1 & 0 \end{bmatrix}$. Row reduction yields $\mathbf{P}^{-1} = \frac{1}{9}\begin{bmatrix} 2 & -1 & 2 \\ 4 & -2 & -5 \\ 1 & 4 & 1 \end{bmatrix}$. Hence, $\mathbf{A} = \mathbf{P}\mathbf{D}\mathbf{P}^{-1} = \frac{1}{9}\begin{bmatrix} 5 & 2 & -4 \\ 2 & 8 & 2 \\ -4 & 2 & 5 \end{bmatrix}$.

(9) (a) The following argument works for any orthogonal projection onto a plane through the origin in $\mathbb{R}^3$.
An orthogonal projection onto a plane in $\mathbb{R}^3$ has two eigenvalues, $\lambda_1 = 1$ and $\lambda_2 = 0$. The eigenspace E_1 is the given plane, and thus is 2-dimensional. Hence, the geometric multiplicity of λ_1 is 2. The eigenspace E_0 is the orthogonal complement of the plane, since the vectors perpendicular to the plane are sent to $\mathbf{0}$ by L. By Corollary 6.12, $\dim(E_0) = 3 - \dim(E_1) = 1$. Hence, the geometric multiplicity of λ_2 is 1. Since the geometric multiplicities of the eigenvalues add up to 3, L is diagonalizable, and each geometric multiplicity must equal its corresponding algebraic multiplicity. Therefore, the factor $(x-1)$ appears in $p_L(x)$ raised to the 2nd power, and the factor of x appears in $p_L(x)$ raised to the 1st power. Hence, $p_L(x) = (x-1)^2x = x^3 - 2x^2 + x$.

(c) The following argument works for any orthogonal reflection through a plane through the origin in $\mathbb{R}^3$.
An orthogonal reflection through a plane in $\mathbb{R}^3$ has two eigenvalues, $\lambda_1 = 1$ and $\lambda_2 = -1$. The eigenspace E_1 is the given plane, and thus is 2-dimensional. Hence, the geometric multiplicity of λ_1 is 2. The eigenspace E_{-1} is the orthogonal complement of the plane, since the vectors perpendicular to the plane are reflected to their opposites by L. By Corollary 6.12, $\dim(E_0) = 3 - \dim(E_1) = 1$. Hence, the geometric multiplicity of λ_2 is 1. Since the geometric multiplicities of the eigenvalues add up to 3, L is diagonalizable, and each geometric multiplicity must equal its corresponding algebraic multiplicity. Therefore, the factor $(x-1)$ appears in $p_L(x)$ raised to the 2nd power, and the factor of $(x+1)$ appears in $p_L(x)$ raised to the 1st power. Hence, $p_L(x) = (x-1)^2(x+1) = x^3 - x^2 - x + 1$.

(10) (a) We will consider three different approaches to this problem. The third method is the easiest, but it involves using the cross product of vectors in $\mathbb{R}^3$, and so can only be used in $\mathbb{R}^3$ and when $\dim(\mathcal{W}) = 2$.
Method #1: First we find an orthonormal basis for $\mathcal{W}$. To do this, we need an orthogonal basis, which we get by applying the Gram-Schmidt Process to the given basis $\{[2,-1,1],[1,0,-3]\}$. So, let $\mathbf{v}_1 = [2,-1,1]$. Then $\mathbf{v}_2 = [1,0,-3] - \left(\frac{[1,0,-3]\cdot[2,-1,1]}{[2,-1,1]\cdot[2,-1,1]}\right)[2,-1,1] = [1,0,-3] - \frac{(-1)}{6}[2,-1,1] = \left[\frac{4}{3}, -\frac{1}{6}, -\frac{17}{6}\right]$. Multiplying by 6 to eliminate fractions yields $\mathbf{v}_2 = [8,-1,-17]$. We normalize these vectors to find the following orthonormal basis for $\mathcal{W}$: $\left\{\frac{1}{\sqrt{6}}[2,-1,1], \frac{1}{\sqrt{354}}[8,-1,-17]\right\}$.

Now the matrix for L with respect to the standard basis is the 3×3 matrix whose columns are $L(\mathbf{i})$, $L(\mathbf{j})$, and $L(\mathbf{k})$. Using the definition of the projection operator we get
$L(\mathbf{i}) = \mathbf{proj}_{\mathcal{W}}\mathbf{i} = \left(\mathbf{i}\cdot\left(\frac{1}{\sqrt{6}}[2,-1,1]\right)\right)\left(\frac{1}{\sqrt{6}}[2,-1,1]\right)$
$+\left(\mathbf{i}\cdot\left(\frac{1}{\sqrt{354}}[8,-1,-17]\right)\right)\left(\frac{1}{\sqrt{354}}[8,-1,-17]\right) = \frac{1}{3}[2,-1,1] + \frac{4}{177}[8,-1,-17] = \frac{1}{59}[50,-21,-3]$,
$L(\mathbf{j}) = \mathbf{proj}_{\mathcal{W}}\mathbf{j} = \left(\mathbf{j}\cdot\left(\frac{1}{\sqrt{6}}[2,-1,1]\right)\right)\left(\frac{1}{\sqrt{6}}[2,-1,1]\right)$
$+\left(\mathbf{j}\cdot\left(\frac{1}{\sqrt{354}}[8,-1,-17]\right)\right)\left(\frac{1}{\sqrt{354}}[8,-1,-17]\right) = \frac{(-1)}{6}[2,-1,1] + \frac{(-1)}{354}[8,-1,-17]$
$= \frac{1}{59}[-21,10,-7]$, and $L(\mathbf{k}) = \mathbf{proj}_{\mathcal{W}}\mathbf{k} = \left(\mathbf{k}\cdot\left(\frac{1}{\sqrt{6}}[2,-1,1]\right)\right)\left(\frac{1}{\sqrt{6}}[2,-1,1]\right)$
$+\left(\mathbf{k}\cdot\left(\frac{1}{\sqrt{354}}[8,-1,-17]\right)\right)\left(\frac{1}{\sqrt{354}}[8,-1,-17]\right) = \frac{1}{6}[2,-1,1] + \frac{(-17)}{354}[8,-1,-17] = \frac{1}{59}[-3,-7,58]$.

Therefore, the matrix for L with respect to the standard basis is $\mathbf{A} = \frac{1}{59}\begin{bmatrix} 50 & -21 & -3 \\ -21 & 10 & -7 \\ -3 & -7 & 58 \end{bmatrix}$.

Method #2: The subspace $\mathcal{W}$ is a plane in $\mathbb{R}^3$, and L is the orthogonal projection onto that plane. The plane $\mathcal{W}$ is also the eigenspace E_1 for L, corresponding to the eigenvalue $\lambda_1 = 1$. The eigenspace E_0, corresponding to the eigenvalue $\lambda_2 = 0$ is $\mathcal{W}^\perp$. We are given the basis $B = \{[2,-1,1],[1,0,-3]\}$ for $\mathcal{W}$, but we also need a basis for $\mathcal{W}^\perp$. To find this basis for $\mathcal{W}^\perp$, we first expand the basis B for $\mathcal{W}$ to a basis C for $\mathbb{R}^3$ using the Enlarging Method, producing $C = \{[2,-1,1],[1,0,-3],[1,0,0]\}$. Next, apply the Gram-Schmidt Process to C, obtaining the orthogonal basis $D = \{[2,-1,1],[8,-1,-17],[3,7,1]\}$ for $\mathbb{R}^3$. Now $\{[2,-1,1],[8,-1,-17]\}$ is a basis for $\mathcal{W} = E_1$, and so Theorem 6.11 shows that $\{[3,7,1]\}$ is a basis for $\mathcal{W}^\perp = E_0$. So each vector in D is an eigenvector for L. Thus, the matrix $\mathbf{A}$ for L with respect to the standard basis is given by $\mathbf{A} = \mathbf{PDP}^{-1}$, where $\mathbf{D} = \begin{bmatrix} 1 & 0 & 0 \\ 0 & 1 & 0 \\ 0 & 0 & 0 \end{bmatrix}$ and $\mathbf{P} = \begin{bmatrix} 2 & 8 & 3 \\ -1 & -1 & 7 \\ 1 & -17 & 1 \end{bmatrix}$. Using row reduction to compute $\mathbf{P}^{-1}$ produces $\mathbf{P}^{-1} = \frac{1}{354}\begin{bmatrix} 118 & -59 & 59 \\ 8 & -1 & -17 \\ 18 & 42 & 6 \end{bmatrix}$. Performing the matrix product $\mathbf{PDP}^{-1}$ gives the matrix $\mathbf{A}$ computed above using Method #1.

Method #3: The subspace $\mathcal{W}$ is a plane is $\mathbb{R}^3$, and L is the orthogonal projection onto that plane. The plane $\mathcal{W}$ is also the eigenspace E_1 for L, corresponding to the eigenvalue $\lambda_1 = 1$. The eigenspace E_0 corresponding to the eigenvalue $\lambda_2 = 0$ is $\mathcal{W}^\perp$. We are given the basis $B = \{[2,-1,1],[1,0,-3]\}$ for $\mathcal{W}$, but we also need a basis for $\mathcal{W}^\perp$. In Exercise 8 of Section 3.1 we defined the cross product of two vectors in $\mathbb{R}^3$ ($\mathbf{v} \times \mathbf{w}$). In that exercise it is proven that $\mathbf{v} \times \mathbf{w}$ is orthogonal to both $\mathbf{v}$ and $\mathbf{w}$. Thus, if we take the cross product of the two vectors in the basis B for $\mathcal{W}$, we will get a vector orthogonal to both of these vectors, which places it in $\mathcal{W}^\perp$. Hence, $[2,-1,1] \times [1,0,-3] = [3,7,1] \in \mathcal{W}^\perp$. But since $\dim(\mathcal{W}^\perp) = 1$ (by Corollary 6.12), $\{[3,7,1]\}$ must be a basis for $\mathcal{W}^\perp$. So now we have a basis $C = \{[3,7,1],[2,-1,1],[1,0,-3]\}$ for $\mathbb{R}^3$ consisting of eigenvectors. Thus, the matrix $\mathbf{A}$ for L with respect to the standard basis is given by $\mathbf{A} = \mathbf{PDP}^{-1}$, where $\mathbf{D} = \begin{bmatrix} 0 & 0 & 0 \\ 0 & 1 & 0 \\ 0 & 0 & 1 \end{bmatrix}$ and $\mathbf{P} = \begin{bmatrix} 3 & 2 & 1 \\ 7 & -1 & 0 \\ 1 & 1 & -3 \end{bmatrix}$. Using row reduction to compute $\mathbf{P}^{-1}$ produces $\mathbf{P}^{-1} = \frac{1}{59}\begin{bmatrix} 3 & 7 & 1 \\ 21 & -10 & 7 \\ 8 & -1 & -17 \end{bmatrix}$. Performing the matrix product $\mathbf{PDP}^{-1}$ gives the matrix $\mathbf{A}$ computed above using Method #1.

(c) (Note: Three solutions by three different methods were presented for part (a), above. However, the third method used there will not work in part (c) because the cross product is not defined in $\mathbb{R}^4$. Hence, we will only solve this problem two ways, corresponding to the first two methods presented in part (a).)

Method #1: First we find an orthonormal basis for $\mathcal{W}$. To do this, we need an orthogonal basis, which we get by applying the Gram-Schmidt Process to the given basis $\{[1,2,1,0],[-1,0,-2,1]\}$. So, let $\mathbf{v}_1 = [1,2,1,0]$. Then $\mathbf{v}_2 = [-1,0,-2,1] - \left(\frac{[-1,0,-2,1]\cdot[1,2,1,0]}{[1,2,1,0]\cdot[1,2,1,0]}\right)[1,2,1,0] = [-1,0,-2,1] - \frac{(-3)}{6}[1,2,1,0] = \left[-\frac{1}{2},1,-\frac{3}{2},1\right]$. Multiplying by 2 to eliminate fractions yields $\mathbf{v}_2 = [-1,2,-3,2]$. We normalize these vectors to find the following orthonormal basis for $\mathcal{W}$:
$\left\{\frac{1}{\sqrt{6}}[1,2,1,0], \frac{1}{3\sqrt{2}}[-1,2,-3,2]\right\}$ Now the matrix for L with respect to the standard basis is the 4×4 matrix whose columns are $L(\mathbf{e}_1)$, $L(\mathbf{e}_2)$, $L(\mathbf{e}_3)$ and $L(\mathbf{e}_4)$. Using the definition of the pro-

jection operator we get

$L(\mathbf{e}_1) = \mathbf{proj}_{\mathcal{W}}\mathbf{e}_1 = \left(\mathbf{e}_1 \cdot \left(\frac{1}{\sqrt{6}}[1,2,1,0]\right)\right)\left(\frac{1}{\sqrt{6}}[1,2,1,0]\right)$
$+\left(\mathbf{e}_1 \cdot \left(\frac{1}{3\sqrt{2}}[-1,2,-3,2]\right)\right)\left(\frac{1}{3\sqrt{2}}[-1,2,-3,2]\right) = \frac{1}{6}[1,2,1,0] + \frac{(-1)}{18}[-1,2,-3,2] = \frac{1}{9}[2,2,3,-1]$,
$L(\mathbf{e}_2) = \mathbf{proj}_{\mathcal{W}}\mathbf{e}_2 = \left(\mathbf{e}_2 \cdot \left(\frac{1}{\sqrt{6}}[1,2,1,0]\right)\right)\left(\frac{1}{\sqrt{6}}[1,2,1,0]\right)$
$+\left(\mathbf{e}_2 \cdot \left(\frac{1}{3\sqrt{2}}[-1,2,-3,2]\right)\right)\left(\frac{1}{3\sqrt{2}}[-1,2,-3,2]\right) = \frac{1}{3}[1,2,1,0] + \frac{1}{9}[-1,2,-3,2] = \frac{1}{9}[2,8,0,2]$,
$L(\mathbf{e}_3) = \mathbf{proj}_{\mathcal{W}}\mathbf{e}_3 = \left(\mathbf{e}_3 \cdot \left(\frac{1}{\sqrt{6}}[1,2,1,0]\right)\right)\left(\frac{1}{\sqrt{6}}[1,2,1,0]\right)$
$+\left(\mathbf{e}_3 \cdot \left(\frac{1}{3\sqrt{2}}[-1,2,-3,2]\right)\right)\left(\frac{1}{3\sqrt{2}}[-1,2,-3,2]\right) = \frac{1}{6}[1,2,1,0] + \frac{(-1)}{6}[-1,2,-3,2] = \frac{1}{3}[1,0,2,-1]$,
and $L(\mathbf{e}_4) = \mathbf{proj}_{\mathcal{W}}\mathbf{e}_4 = \left(\mathbf{e}_4 \cdot \left(\frac{1}{\sqrt{6}}[1,2,1,0]\right)\right)\left(\frac{1}{\sqrt{6}}[1,2,1,0]\right)$
$+\left(\mathbf{e}_4 \cdot \left(\frac{1}{3\sqrt{2}}[-1,2,-3,2]\right)\right)\left(\frac{1}{3\sqrt{2}}[-1,2,-3,2]\right) = (0)[1,2,1,0] + \frac{1}{9}[-1,2,-3,2] = \frac{1}{9}[-1,2,-3,2]$.

Therefore, the matrix for L with respect to the standard basis is $\mathbf{A} = \frac{1}{9}\begin{bmatrix} 2 & 2 & 3 & -1 \\ 2 & 8 & 0 & 2 \\ 3 & 0 & 6 & -3 \\ -1 & 2 & -3 & 2 \end{bmatrix}$.

Method #2: The subspace $\mathcal{W}$ is a two-dimensional subspace of $\mathbb{R}^4$, and L is the orthogonal projection onto that subspace. The subspace $\mathcal{W}$ is also the eigenspace E_1 for L, corresponding to the eigenvalue $\lambda_1 = 1$. The eigenspace E_0 corresponding to the eigenvalue $\lambda_2 = 0$ is $\mathcal{W}^\perp$. We are given the basis $B = \{[1,2,1,0], [-1,0,-2,1]\}$ for $\mathcal{W}$, but we also need a basis for $\mathcal{W}^\perp$. To find this basis for $\mathcal{W}^\perp$, we first expand the basis B for $\mathcal{W}$ to a basis C for $\mathbb{R}^4$ using the Enlarging Method, producing $C = \{[1,2,1,0], [-1,0,-2,1], [1,0,0,0], [0,1,0,0]\}$. Next, we apply the Gram-Schmidt Process to C, obtaining the orthogonal basis $D = \{[1,2,1,0], [-1,2,-3,2], [7,-2,-3,1], [0,1,-2,-4]\}$ for $\mathbb{R}^4$. Now $\{[1,2,1,0], [-1,2,-3,2]\}$ is a basis for $\mathcal{W} = E_1$, and so Theorem 6.11 shows that $\{[7,-2,-3,1], [0,1,-2,-4]\}$ is a basis for $\mathcal{W}^\perp = E_0$. So each vector in D is an eigenvector for L. Thus, the matrix $\mathbf{A}$ for L with respect to the standard basis is given by $\mathbf{A} = \mathbf{PDP}^{-1}$, where $\mathbf{D} = \begin{bmatrix} 1 & 0 & 0 & 0 \\ 0 & 1 & 0 & 0 \\ 0 & 0 & 0 & 0 \\ 0 & 0 & 0 & 0 \end{bmatrix}$ and $\mathbf{P} = \begin{bmatrix} 1 & -1 & 7 & 0 \\ 2 & 2 & -2 & 1 \\ 1 & -3 & -3 & -2 \\ 0 & 2 & 1 & -4 \end{bmatrix}$. Using row reduction to compute $\mathbf{P}^{-1}$ produces $\mathbf{P}^{-1} = \frac{1}{126}\begin{bmatrix} 21 & 42 & 21 & 0 \\ -7 & 14 & -21 & 14 \\ 14 & -4 & -6 & 2 \\ 0 & 6 & -12 & -24 \end{bmatrix}$. Performing the matrix product $\mathbf{PDP}^{-1}$ gives the matrix $\mathbf{A}$ computed above using Method #1.

(18) Let S be a spanning set for $\mathcal{W}$, and let $\mathbf{v} \in \mathcal{W}^\perp$. If $\mathbf{u} \in S$, then $\mathbf{u} \in \mathcal{W}$, and so by definition of $\mathcal{W}^\perp$, $\mathbf{v} \cdot \mathbf{u} = 0$. Hence $\mathbf{v}$ is orthogonal to every vector in S.
Conversely, suppose $\mathbf{v} \cdot \mathbf{u} = 0$ for all $\mathbf{u} \in S$. Let $\mathbf{w} \in \mathcal{W}$. Then $\mathbf{w} = a_1\mathbf{u}_1 + \cdots + a_n\mathbf{u}_n$ for some $\mathbf{u}_1, \ldots, \mathbf{u}_n \in S$. Hence $\mathbf{v} \cdot \mathbf{w} = a_1(\mathbf{v} \cdot \mathbf{u}_1) + \cdots + a_n(\mathbf{v} \cdot \mathbf{u}_n) = 0 + \cdots + 0 = 0$. Thus $\mathbf{v} \in \mathcal{W}^\perp$.

(20) We first prove that L is a linear operator. Suppose $\{\mathbf{u}_1, \ldots, \mathbf{u}_k\}$ is an orthonormal basis for $\mathcal{W}$. Then
$L(\mathbf{v}_1 + \mathbf{v}_2) = \mathbf{proj}_{\mathcal{W}}(\mathbf{v}_1 + \mathbf{v}_2) = ((\mathbf{v}_1 + \mathbf{v}_2) \cdot \mathbf{u}_1)\mathbf{u}_1 + \cdots + ((\mathbf{v}_1 + \mathbf{v}_2) \cdot \mathbf{u}_k)\mathbf{u}_k$
$= ((\mathbf{v}_1 \cdot \mathbf{u}_1) + (\mathbf{v}_2 \cdot \mathbf{u}_1))\mathbf{u}_1 + \cdots + ((\mathbf{v}_1 \cdot \mathbf{u}_k) + (\mathbf{v}_2 \cdot \mathbf{u}_k))\mathbf{u}_k$

$= (\mathbf{v}_1 \cdot \mathbf{u}_1)\mathbf{u}_1 + \cdots + (\mathbf{v}_1 \cdot \mathbf{u}_k)\mathbf{u}_k + (\mathbf{v}_2 \cdot \mathbf{u}_1)\mathbf{u}_1 + \cdots + (\mathbf{v}_2 \cdot \mathbf{u}_k)\mathbf{u}_k$
$= \mathbf{proj}_{\mathcal{W}}(\mathbf{v}_1) + \mathbf{proj}_{\mathcal{W}}(\mathbf{v}_2) = L(\mathbf{v}_1) + L(\mathbf{v}_2)$.
Similarly, $L(c\mathbf{v}) = \mathbf{proj}_{\mathcal{W}}(c\mathbf{v}) = ((c\mathbf{v}) \cdot \mathbf{u}_1)\mathbf{u}_1 + \cdots + ((c\mathbf{v}) \cdot \mathbf{u}_k)\mathbf{u}_k$
$= c(\mathbf{v} \cdot \mathbf{u}_1)\mathbf{u}_1 + \cdots + c(\mathbf{v} \cdot \mathbf{u}_k)\mathbf{u}_k = c((\mathbf{v} \cdot \mathbf{u}_1)\mathbf{u}_1 + \cdots + (\mathbf{v} \cdot \mathbf{u}_k)\mathbf{u}_k)$
$= c\mathbf{proj}_{\mathcal{W}}(\mathbf{v}) = cL(\mathbf{v})$. Hence, L is a linear operator.
To show $\mathcal{W}^\perp = \ker(L)$, we first prove that $\mathcal{W}^\perp \subseteq \ker(L)$. If $\mathbf{v} \in \mathcal{W}^\perp$, then since $\mathbf{v} \cdot \mathbf{w} = 0$ for every $\mathbf{w} \in \mathcal{W}$, $L(\mathbf{v}) = \mathbf{proj}_{\mathcal{W}}(c\mathbf{v}) = (\mathbf{v} \cdot \mathbf{u}_1)\mathbf{u}_1 + \cdots + (\mathbf{v} \cdot \mathbf{u}_k)\mathbf{u}_k = 0\mathbf{u}_1 + \cdots + 0\mathbf{u}_k = \mathbf{0}$. Hence, $\mathbf{v} \in \ker(L)$.
To finish the proof, we use the Dimension Theorem to help show that $\dim(\mathcal{W}^\perp) = \dim(\ker(L))$. Notice that $\text{range}(L) = \mathcal{W}$. To see this, first note that if $\mathbf{w} \in \mathcal{W}$, then $\mathbf{w} = \mathbf{w} + \mathbf{0}$, where $\mathbf{w} \in \mathcal{W}$ and $\mathbf{0} \in \mathcal{W}^\perp$. So, by the statement in the text just after Example 6 in Section 6.2, $L(\mathbf{w}) = \mathbf{proj}_{\mathcal{W}}\mathbf{w} = \mathbf{w}$. Therefore, every $\mathbf{w}$ in $\mathcal{W}$ is in the range of L. Similarly, the same statement after Example 6 implies $\mathbf{proj}_{\mathcal{W}}\mathbf{v} \in \mathcal{W}$ for every $\mathbf{v} \in \mathbb{R}^n$. Hence, $\text{range}(L) = \mathcal{W}$.
Now $\dim(\ker(L)) = n - \dim(\text{range}(L)) = n - \dim(\mathcal{W}) = \dim(\mathcal{W}^\perp)$. Since $\mathcal{W}^\perp \subseteq \ker(L)$, Theorem 4.17 implies that $\mathcal{W}^\perp = \ker(L)$.

(23) Suppose that T is any point in $\mathcal{W}$ and $\mathbf{w}$ is the vector from the origin to T. We need to show that $\|\mathbf{v} - \mathbf{w}\| \geq \|\mathbf{v} - \mathbf{proj}_{\mathcal{W}}\mathbf{v}\|$; that is, the distance from P to T is at least as large as the distance from P to the terminal point of $\mathbf{proj}_{\mathcal{W}}\mathbf{v}$. Let $\mathbf{a} = \mathbf{v} - \mathbf{proj}_{\mathcal{W}}\mathbf{v}$ and $\mathbf{b} = (\mathbf{proj}_{\mathcal{W}}\mathbf{v}) - \mathbf{w}$.
First, we show that $\mathbf{a} \in \mathcal{W}^\perp$. This follows from the statement just after Example 6 in Section 6.2 of the text.
Next we show that $\mathbf{b} \in \mathcal{W}$. Now, by the statement after Example 6, $\mathbf{proj}_{\mathcal{W}}\mathbf{v} \in \mathcal{W}$. Also, $\mathbf{w} \in \mathcal{W}$ (since T is in $\mathcal{W}$), and so $\mathbf{b} = (\mathbf{proj}_{\mathcal{W}}\mathbf{v}) - \mathbf{w} \in \mathcal{W}$.
Finally, $\|\mathbf{v} - \mathbf{w}\|^2 = \|\mathbf{v} - \mathbf{proj}_{\mathcal{W}}\mathbf{v} + (\mathbf{proj}_{\mathcal{W}}\mathbf{v}) - \mathbf{w}\|^2 = \|\mathbf{a} + \mathbf{b}\|^2 = (\mathbf{a} + \mathbf{b}) \cdot (\mathbf{a} + \mathbf{b})$
$= \mathbf{a} \cdot \mathbf{a} + \mathbf{a} \cdot \mathbf{b} + \mathbf{b} \cdot \mathbf{a} + \mathbf{b} \cdot \mathbf{b} = \mathbf{a} \cdot \mathbf{a} + \mathbf{0} + \mathbf{0} + \mathbf{b} \cdot \mathbf{b} = \|\mathbf{a}\|^2 + \|\mathbf{b}\|^2 \geq \|\mathbf{a}\|^2 = \|\mathbf{v} - \mathbf{proj}_{\mathcal{W}}\mathbf{v}\|^2$,
which completes the proof.

(25) (a) True. This is the definition of $\mathcal{W}^\perp$.

(b) True. This follows from Theorem 6.9.

(c) False. By Theorem 6.10, $\mathcal{W} \cap \mathcal{W}^\perp = \{\mathbf{0}\}$, for every subspace $\mathcal{W}$ of $\mathbb{R}^n$.

(d) False. If the original basis $\{\mathbf{b}_1, \ldots, \mathbf{b}_7\}$ is an orthogonal basis then this is true, but it is false in general. So, for example, if $\mathbf{b}_i = \mathbf{e}_i$ for $1 \leq i \leq 6$, and $\mathbf{b}_7 = \mathbf{e}_1 + \mathbf{e}_7$, then $\{\mathbf{b}_1, \ldots, \mathbf{b}_7\}$ is a basis for $\mathbb{R}^7$. If $\mathcal{W} = \text{span}(\{\mathbf{b}_1, \ldots, \mathbf{b}_4\})$, then $\mathbf{b}_7 \notin \mathcal{W}^\perp$ because $\mathbf{b}_7 \cdot \mathbf{b}_1 = 1 \neq 0$. In this case, $\{\mathbf{b}_5, \mathbf{b}_6, \mathbf{b}_7\}$ is not a basis for $\mathcal{W}^\perp$.

(e) True. This follows from Corollary 6.12.

(f) False. The orthogonal complement of a subspace is not a setwise complement of the subspace. For example, if $\mathcal{W}$ is the subspace $\{[a, 0] \mid a \in \mathbb{R}\}$ of $\mathbb{R}^2$, then $\mathcal{W}^\perp = \{[0, b] \mid b \in \mathbb{R}\}$. However, the vector $[1, 1]$ of $\mathbb{R}^2$ is in neither of these subspaces.

(g) True. This is Corollary 6.13.

(h) True. This was discussed in Example 5 and illustrated in Figure 6.2 in Section 6.2 of the textbook.

(i) True. This is part of Theorem 6.15.

(j) False. The eigenvalues for an orthogonal projection are 0 and 1, not -1 and 1. Vectors perpendicular to the plane are sent to $\mathbf{0}$ by the projection. No vector is sent to its opposite. The matrix for the projection diagonalizes to a matrix with 0, 1, and 1 on the main diagonal. See Example 8 in Section 6.2 of the textbook. (Keep in mind that orthogonal *reflections* have matrices that diagonalize to a diagonal matrix with -1, 1, 1 on the main diagonal.)

(k) True. This follows from Theorem 6.16 and the "boxed" comment just after Example 6 in Section 6.2 of the textbook.

(l) True. This follows from the "boxed" comment just after Example 6 in Section 6.2 of the textbook.

Section 6.3

(1) (a) The matrix for L with respect to the standard basis is $\mathbf{A} = \begin{bmatrix} 3 & 2 \\ 2 & 5 \end{bmatrix}$. Since $\mathbf{A}$ is symmetric, L is a symmetric operator by Theorem 6.17.

(d) Let $\mathbf{v}_1 = \frac{[a,b,c]}{\|[a,b,c]\|}$ and let $\{\mathbf{v}_2, \mathbf{v}_3\}$ be an orthonormal basis for for the plane $ax + by + cz = 0$. Then $B = \{\mathbf{v}_1, \mathbf{v}_2, \mathbf{v}_3\}$ is an orthonormal basis for $\mathbb{R}^3$, and the matrix for L with respect to B is the diagonal matrix with 0, 1, 1 on its main diagonal. Hence, L is an orthogonally diagonalizable linear operator. Therefore, L is symmetric by Theorem 6.19.

(e) Rotations in $\mathbb{R}^3$ about an axis through an angle of $\frac{\pi}{3}$ have $\lambda = 1$ as their only eigenvalue. Also, the eigenspace E_1 is 1-dimensional, containing vectors parallel to the axis of rotation. Hence, the sum of the geometric multiplicities of the eigenvalues of L is 1. Since this is less than 3, L is not diagonalizable. Hence L is not orthogonally diagonalizable. Therefore, Theorem 6.19 implies that L is not symmetric.

(g) Direct computation shows that $L(\mathbf{e}_1) = [4,0,3,0]$, $L(\mathbf{e}_2) = [0,4,0,3]$, $L(\mathbf{e}_3) = [3,0,9,0]$, and $L(\mathbf{e}_4) = [0,3,0,9]$. The matrix for L with respect to the standard basis is the 4×4 matrix whose columns are these four vectors in the order listed. Since this matrix is symmetric, L is a symmetric operator by Theorem 6.17.

(2) (a) Let $\mathbf{P} = \begin{bmatrix} \frac{3}{5} & \frac{4}{5} \\ \frac{4}{5} & -\frac{3}{5} \end{bmatrix}$, the matrix whose columns are the basis vectors for the eigenspaces E_1 and E_{-1}. Then the desired matrix $\mathbf{A}$ equals $\mathbf{PDP}^{-1}$, where $\mathbf{D}$ is the diagonal matrix with the eigenvalues 1 and -1 on its main diagonal. Theorem 2.13 shows that $\mathbf{P}^{-1} = \mathbf{P}$, and so

$$\mathbf{A} = \begin{bmatrix} \frac{3}{5} & \frac{4}{5} \\ \frac{4}{5} & -\frac{3}{5} \end{bmatrix} \begin{bmatrix} 1 & 0 \\ 0 & -1 \end{bmatrix} \begin{bmatrix} \frac{3}{5} & \frac{4}{5} \\ \frac{4}{5} & -\frac{3}{5} \end{bmatrix} = \frac{1}{25} \begin{bmatrix} -7 & 24 \\ 24 & 7 \end{bmatrix}, \text{ which is symmetric.}$$

(d) Let $\mathbf{P} = \begin{bmatrix} 12 & 12 & -3 & -2 \\ 3 & -1 & 12 & 24 \\ 4 & 7 & 0 & -12 \\ 0 & 12 & 4 & 11 \end{bmatrix}$, the matrix whose columns are the basis vectors for the eigenspaces E_{-1} and E_1. Then the desired matrix $\mathbf{A}$ equals $\mathbf{PDP}^{-1}$, where $\mathbf{D}$ is the diagonal matrix with the eigenvalues -1, -1, 1 and 1 on its main diagonal. Using row reduction to find $\mathbf{P}^{-1}$ shows that $\mathbf{P}^{-1} = \frac{1}{169} \begin{bmatrix} 12 & 7 & 1 & -12 \\ 0 & -4 & 3 & 12 \\ -11 & 12 & 24 & -2 \\ 4 & 0 & -12 & 3 \end{bmatrix}$.

Hence, $\mathbf{A} = \mathbf{PDP}^{-1} = \frac{1}{169} \begin{bmatrix} -119 & -72 & -96 & 0 \\ -72 & 119 & 0 & 96 \\ -96 & 0 & 119 & -72 \\ 0 & 96 & -72 & -119 \end{bmatrix}$.

(3) In each part, we follow Steps 2 and 3 of the Orthogonal Diagonalization Method of Section 6.3.

(a) Step 2(a): $p_{\mathbf{A}}(x) = \begin{vmatrix} x-144 & 60 \\ 60 & x-25 \end{vmatrix} = x^2 - 169x = x(x-169)$, giving us eigenvalues $\lambda_1 = 0$ and $\lambda_2 = 169$. Using row reduction to find particular solutions to the systems $[0\mathbf{I}_2 - \mathbf{A}|\mathbf{0}]$ and $[169\mathbf{I}_2 - \mathbf{A}|\mathbf{0}]$ as in the Diagonalization Method of Section 3.4 produces the basis $\{[5,12]\}$ for E_0 and $\{[-12,5]\}$ for E_{169}.

Step 2(b): Since E_0 and E_{169} are 1-dimensional, the given bases are already orthogonal. To create orthonormal bases, we need to normalize the vectors, yielding $\{\frac{1}{13}[5,12]\}$ for E_0 and $\{\frac{1}{13}[-12,5]\}$ for E_{169}.

Step 2(c): $Z = \left(\frac{1}{13}[5,12], \frac{1}{13}[-12,5]\right)$.

Step 3: Since C is the standard basis for $\mathbb{R}^2$, $B = Z = \left(\frac{1}{13}[5,12], \frac{1}{13}[-12,5]\right)$. Now $\mathbf{P}$ is the matrix whose columns are the vectors in Z, and $\mathbf{D}$ is the diagonal matrix whose main diagonal entries are the eigenvalues 0 and 169. Hence, $\mathbf{P} = \frac{1}{13}\begin{bmatrix} 5 & -12 \\ 12 & 5 \end{bmatrix}$, and $\mathbf{D} = \begin{bmatrix} 0 & 0 \\ 0 & 169 \end{bmatrix}$.

Now $\mathbf{P}^{-1} = \mathbf{P}^T$, and so $\mathbf{D} = \mathbf{P}^T\mathbf{A}\mathbf{P}$, which is easily verified by performing the matrix product.

(c) Step 2(a): $p_{\mathbf{A}}(x) = \begin{vmatrix} x-\frac{17}{9} & -\frac{8}{9} & \frac{4}{9} \\ -\frac{8}{9} & x-\frac{17}{9} & \frac{4}{9} \\ \frac{4}{9} & \frac{4}{9} & x-\frac{11}{9} \end{vmatrix} = x^3 - 5x^2 + 7x - 3 = (x-1)^2(x-3)$, giving us eigenvalues $\lambda_1 = 1$ and $\lambda_2 = 3$. Using row reduction to find particular solutions to the systems $[1\mathbf{I}_3 - \mathbf{A}|\mathbf{0}]$ and $[3\mathbf{I}_3 - \mathbf{A}|\mathbf{0}]$ as in the Diagonalization Method of Section 3.4 produces the basis $\{[-1,1,0],[1,0,2]\}$ for E_1 and $\{[-2,-2,1]\}$ for E_3.

Step 2(b): Since E_3 is 1-dimensional, the given basis is already orthogonal. To make it an orthonormal basis, we need to normalize the vector, yielding $\left\{\frac{1}{3}[-2,-2,1]\right\}$ for E_3. Now E_1 is 2-dimensional, and so we must use the Gram-Schmidt Process to turn the basis we have into an orthogonal basis. We let $\mathbf{v}_1 = [-1,1,0]$. Then $\mathbf{v}_2 = [1,0,2] - \left(\frac{[1,0,2]\cdot[-1,1,0]}{[-1,1,0]\cdot[-1,1,0]}\right)[-1,1,0] = \left[\frac{1}{2},\frac{1}{2},2\right]$. Multiplying by 2 to simplify produces $\mathbf{v}_2 = [1,1,4]$. Thus we have the orthogonal basis $\{\mathbf{v}_1,\mathbf{v}_2\}$ for E_1. Normalizing these vectors gives us the orthonormal basis $\left\{\frac{1}{\sqrt{2}}[-1,1,0], \frac{1}{3\sqrt{2}}[1,1,4]\right\}$.

Step 2(c): $Z = \left(\frac{1}{\sqrt{2}}[-1,1,0], \frac{1}{3\sqrt{2}}[1,1,4], \frac{1}{3}[-2,-2,1]\right)$.

Step 3: Since C is the standard basis for $\mathbb{R}^3$, $B = Z = \left(\frac{1}{\sqrt{2}}[-1,1,0], \frac{1}{3\sqrt{2}}[1,1,4], \frac{1}{3}[-2,-2,1]\right)$. Now $\mathbf{P}$ is the matrix whose columns are the vectors in Z, and $\mathbf{D}$ is the diagonal matrix whose main diagonal entries are the eigenvalues 1, 1 and 3. Hence,

$$\mathbf{P} = \begin{bmatrix} -\frac{1}{\sqrt{2}} & \frac{1}{3\sqrt{2}} & -\frac{2}{3} \\ \frac{1}{\sqrt{2}} & \frac{1}{3\sqrt{2}} & -\frac{2}{3} \\ 0 & \frac{4}{3\sqrt{2}} & \frac{1}{3} \end{bmatrix} \text{ and } \mathbf{D} = \begin{bmatrix} 1 & 0 & 0 \\ 0 & 1 & 0 \\ 0 & 0 & 3 \end{bmatrix}.$$

Now $\mathbf{P}^{-1} = \mathbf{P}^T$, and so $\mathbf{D} = \mathbf{P}^T\mathbf{A}\mathbf{P}$, which is verified by performing the matrix product.

(e) Step 2(a): The hint in the textbook indicates that $p_{\mathbf{A}}(x) = (x-2)^2(x+3)(x-5)$, giving eigenvalues $\lambda_1 = 2$, $\lambda_2 = -3$, and $\lambda_3 = 5$. Using row reduction to find particular solutions to the systems $[2\mathbf{I}_4 - \mathbf{A}|\mathbf{0}]$, $[-3\mathbf{I}_4 - \mathbf{A}|\mathbf{0}]$ and $[5\mathbf{I}_4 - \mathbf{A}|\mathbf{0}]$ as in the Diagonalization Method of Section 3.4 produces the basis $\{[3,2,1,0],[-2,3,0,1]\}$ for E_2, $\{[1,0,-3,2]\}$ for E_{-3} and $\{[0,-1,2,3]\}$ for E_5.

Step 2(b): Since E_{-3} and E_5 are 1-dimensional, the given bases are already orthogonal. To make these bases orthonormal, we need to normalize, yielding $\left\{\frac{1}{\sqrt{14}}[1,0,-3,2]\right\}$ for E_3 and

$\left\{\frac{1}{\sqrt{14}}[0,-1,2,3]\right\}$ for E_5. Now E_2 is 2-dimensional, and so typically, we would use the Gram-Schmidt Process to turn the basis for E_2 into an orthogonal basis. However, quick inspection shows us that the basis we have is already orthogonal! (This is not the typical situation.) Thus, we only need to normalize to get the orthonormal basis $\left\{\frac{1}{\sqrt{14}}[3,2,1,0], \frac{1}{\sqrt{14}}[-2,3,0,1]\right\}$ for E_2.

Step 2(c): $Z = \left(\frac{1}{\sqrt{14}}[3,2,1,0], \frac{1}{\sqrt{14}}[-2,3,0,1], \frac{1}{\sqrt{14}}[1,0,-3,2], \ \frac{1}{\sqrt{14}}[0,-1,2,3]\right)$.

Step 3: Since C is the standard basis for $\mathbb{R}^3$,

$B = Z = \left(\frac{1}{\sqrt{14}}[3,2,1,0], \frac{1}{\sqrt{14}}[-2,3,0,1], \frac{1}{\sqrt{14}}[1,0,-3,2], \ \frac{1}{\sqrt{14}}[0,-1,2,3]\right)$. Now $\mathbf{P}$ is the matrix whose columns are the vectors in Z, and $\mathbf{D}$ is the diagonal matrix whose main diagonal entries are the eigenvalues 2, 2, -3, and 5. Hence,

$$\mathbf{P} = \frac{1}{\sqrt{14}}\begin{bmatrix} 3 & -2 & 1 & 0 \\ 2 & 3 & 0 & -1 \\ 1 & 0 & -3 & 2 \\ 0 & 1 & 2 & 3 \end{bmatrix} \text{ and } \mathbf{D} = \begin{bmatrix} 2 & 0 & 0 & 0 \\ 0 & 2 & 0 & 0 \\ 0 & 0 & -3 & 0 \\ 0 & 0 & 0 & 5 \end{bmatrix}.$$

Now $\mathbf{P}^{-1} = \mathbf{P}^T$, and so $\mathbf{D} = \mathbf{P}^T\mathbf{A}\mathbf{P}$, which is easily verified by performing the matrix product.

(g) Step 2(a): $p_{\mathbf{A}}(x) = \begin{vmatrix} x-11 & -2 & 10 \\ -2 & x-14 & -5 \\ 10 & -5 & x+10 \end{vmatrix} = x^3 - 15x^2 - 225x + 3375 = (x-15)^2(x+15)$,

giving us eigenvalues $\lambda_1 = 15$ and
$\lambda_2 = -15$. Using row reduction to find particular solutions to the systems $[15\mathbf{I}_3 - \mathbf{A}|\mathbf{0}]$ and $[-15\mathbf{I}_3 - \mathbf{A}|\mathbf{0}]$ as in the Diagonalization Method of Section 3.4 produces the basis $\{[1,2,0],[-5,0,2]\}$ for E_{15} and $\{[2,-1,5]\}$ for E_{-15}.

Step 2(b): Since E_{-15} is 1-dimensional, the given basis is already orthogonal. To make the basis orthonormal, we need to normalize, yielding $\left\{\frac{1}{\sqrt{30}}[2,-1,5]\right\}$ for E_3. Now E_{15} is 2-dimensional, and so we must use the Gram-Schmidt Process to turn the basis for E_{15} into an orthogonal basis. We let $\mathbf{v}_1 = [1,2,0]$. Then $\mathbf{v}_2 = [-5,0,2] - \left(\frac{[-5,0,2]\cdot[1,2,0]}{[1,2,0]\cdot[1,2,0]}\right)[1,2,0] = [-4,2,2]$. Thus we have the orthogonal basis $\{\mathbf{v}_1, \mathbf{v}_2\}$ for E_1. Normalizing these vectors gives the orthonormal basis $\left\{\frac{1}{\sqrt{5}}[1,2,0], \frac{1}{\sqrt{6}}[-2,1,1]\right\}$.

Step 2(c): $Z = \left(\frac{1}{\sqrt{5}}[1,2,0], \frac{1}{\sqrt{6}}[-2,1,1], \frac{1}{\sqrt{30}}[2,-1,5]\right)$.

Step 3: Since C is the standard basis for $\mathbb{R}^3$,

$B = Z = \left(\frac{1}{\sqrt{5}}[1,2,0], \frac{1}{\sqrt{6}}[-2,1,1], \frac{1}{\sqrt{30}}[2,-1,5]\right)$. Now $\mathbf{P}$ is the matrix whose columns are the vectors in Z, and $\mathbf{D}$ is the diagonal matrix whose main diagonal entries are the eigenvalues 15,

15, and -15. Hence, $\mathbf{P} = \begin{bmatrix} \frac{1}{\sqrt{5}} & -\frac{2}{\sqrt{6}} & \frac{2}{\sqrt{30}} \\ \frac{2}{\sqrt{5}} & \frac{1}{\sqrt{6}} & -\frac{1}{\sqrt{30}} \\ 0 & \frac{1}{\sqrt{6}} & \frac{5}{\sqrt{30}} \end{bmatrix}$ and $\mathbf{D} = \begin{bmatrix} 15 & 0 & 0 \\ 0 & 15 & 0 \\ 0 & 0 & -15 \end{bmatrix}$.

Now $\mathbf{P}^{-1} = \mathbf{P}^T$, and so $\mathbf{D} = \mathbf{P}^T\mathbf{A}\mathbf{P}$, which is easily verified by performing the matrix product.

(4) (a) We use the Orthogonal Diagonalization Method from Section 6.3:
Step 1: The plane $\mathcal{V}$ is 2-dimensional, and we are given the images under L of two vectors in $\mathcal{V}$. Since the set $\{[-10,15,6],[15,6,10]\}$ containing these two vectors is linearly independent, it forms a basis for $\mathcal{V}$. A quick inspection shows this basis is already orthogonal. Thus, to find an

orthonormal basis for $\mathcal{V}$, we only need to normalize. Doing so produces the orthonormal basis $C = \left(\frac{1}{19}[-10,15,6], \frac{1}{19}[15,6,10]\right)$ for $\mathcal{V}$.
Next, we need to compute the matrix $\mathbf{A}$ for L with respect to C. Since $\mathcal{V}$ is 2-dimensional, $\mathbf{A}$ will be a 2×2 matrix. Applying L to each vector in C yields $L\left(\frac{1}{19}[-10,15,6]\right) = \frac{1}{19}L([-10,15,6]) = \frac{1}{19}[50,-18,8]$ and $L\left(\frac{1}{19}[15,6,10]\right) = \frac{1}{19}L([15,6,10]) = \frac{1}{19}[-5,36,22]$. We need to express each of these results in C-coordinates. Hence, we row reduce

$$\left[\begin{array}{cc|cc} -\frac{10}{19} & \frac{15}{19} & \frac{50}{19} & -\frac{5}{19} \\ \frac{15}{19} & \frac{6}{19} & -\frac{18}{19} & \frac{36}{19} \\ \frac{6}{19} & \frac{10}{19} & \frac{8}{19} & \frac{22}{19} \end{array}\right] \text{ to obtain } \left[\begin{array}{cc|cc} 1 & 0 & -2 & 2 \\ 0 & 1 & 2 & 1 \\ 0 & 0 & 0 & 0 \end{array}\right].$$

(Note: In performing the above row reduction by hand, or if entering the matrix into a computer or calculator, it is easier to begin by first performing three type (I) operations to mentally multiply each row by 19.) Crossing out the last row of all zeroes in the above result and using the matrix to the right of the bar gives the desired matrix $\mathbf{A} = \begin{bmatrix} -2 & 2 \\ 2 & 1 \end{bmatrix}$.

Step 2(a): Now $p_{\mathbf{A}}(x) = \begin{vmatrix} x+2 & -2 \\ -2 & x-1 \end{vmatrix} = x^2 + x - 6 = (x-2)(x+3)$, giving eigenvalues $\lambda_1 = 2$ and $\lambda_2 = -3$. Using row reduction to find particular solutions to the systems $[2\mathbf{I}_2 - \mathbf{A}|\mathbf{0}]$ and $[(-3)\mathbf{I}_2 - \mathbf{A}|\mathbf{0}]$ as in the Diagonalization Method of Section 3.4 produces the basis $\{[1,2]\}$ for E_2 and $\{[-2,1]\}$ for E_{-3}.
Step 2(b): Since E_2 and E_{-3} are 1-dimensional, the given bases are already orthogonal. To make them orthonormal bases, we need to normalize, yielding $\left\{\frac{1}{\sqrt{5}}[1,2]\right\}$ for E_2 and $\left\{\frac{1}{\sqrt{5}}[-2,1]\right\}$ for E_{-3}.
Step 2(c): $Z = \left(\frac{1}{\sqrt{5}}[1,2], \frac{1}{\sqrt{5}}[-2,1]\right)$.
Step 3: Since $C = \left(\frac{1}{19}[-10,15,6], \frac{1}{19}[15,6,10]\right)$, we need to reverse the coordinatization isomorphism to calculate the vectors in a basis B for $\mathcal{V}$. First,
$\frac{1}{\sqrt{5}}[1,2] \mapsto \frac{1}{\sqrt{5}}\left(1\left(\frac{1}{19}[-10,15,6]\right) + 2\left(\frac{1}{19}[15,6,10]\right)\right) = \frac{1}{19\sqrt{5}}[20,27,26]$. Similarly,
$\frac{1}{\sqrt{5}}[-2,1] \mapsto \frac{1}{\sqrt{5}}\left((-2)\left(\frac{1}{19}[-10,15,6]\right) + 1\left(\frac{1}{19}[15,6,10]\right)\right) = \frac{1}{19\sqrt{5}}[35,-24,-2]$. Hence, an orthonormal basis for $\mathcal{V}$ consisting of eigenvectors for L is $B = \left(\frac{1}{19\sqrt{5}}[20,27,26], \frac{1}{19\sqrt{5}}[35,-24,-2]\right)$.
Now $\mathbf{P}$ is the matrix whose columns are the vectors in Z, and $\mathbf{D}$, the matrix for L with respect to B, is the diagonal matrix whose main diagonal entries are the eigenvalues 2 and -3. Hence,
$\mathbf{P} = \frac{1}{\sqrt{5}}\begin{bmatrix} 1 & -2 \\ 2 & 1 \end{bmatrix}$, and $\mathbf{D} = \begin{bmatrix} 2 & 0 \\ 0 & -3 \end{bmatrix}$.
Now $\mathbf{P}^{-1} = \mathbf{P}^T$, and so $\mathbf{D} = \mathbf{P}^T\mathbf{A}\mathbf{P}$, which is easily verified by performing the matrix product.

(5) In each part, it is important to orthogonally diagonalize the matrix involved, not just diagonalize it. Otherwise, the matrix $\mathbf{A}$ obtained might not be orthogonally diagonalizable, and so will not correspond to a symmetric operator (by Theorem 6.19), implying that it will not be a symmetric matrix (by Theorem 6.17), as desired.

(a) Let $\mathbf{B} = \frac{1}{25}\begin{bmatrix} 119 & -108 \\ -108 & 56 \end{bmatrix}$ and let C be the standard basis for $\mathbb{R}^2$. First, we must orthogonally diagonalize $\mathbf{B}$. We use Steps 2 and 3 of the Orthogonal Diagonalization Method of Section 6.3.

Step 2(a): Now $p_{\mathbf{B}}(x) = \begin{vmatrix} x - \frac{119}{25} & \frac{108}{25} \\ \frac{108}{25} & x - \frac{56}{25} \end{vmatrix} = x^2 - 7x - 8 = (x-8)(x+1)$, giving eigenvalues $\lambda_1 = 8$ and $\lambda_2 = -1$. Using row reduction to find particular solutions to the systems $[8\mathbf{I}_2 - \mathbf{B}|\mathbf{0}]$ and $[(-1)\mathbf{I}_2 - \mathbf{B}|\mathbf{0}]$ as in the Diagonalization Method of Section 3.4 produces the basis $\{[-4,3]\}$ for E_8 and $\{[3,4]\}$ for E_{-1}.

Step 2(b): Since E_8 and E_{-1} are 1-dimensional, the given bases are already orthogonal. To make them orthonormal bases, we need to normalize, yielding $\left\{\frac{1}{5}[-4,3]\right\}$ for E_8 and $\left\{\frac{1}{5}[3,4]\right\}$ for E_{-1}.

Step 2(c): $Z = \left(\frac{1}{5}[-4,3], \frac{1}{5}[3,4]\right)$.

Step 3: Since C is the standard basis for $\mathbb{R}^2$, $B = Z = \left(\frac{1}{5}[-4,3], \frac{1}{5}[3,4]\right)$. Now $\mathbf{P}$ is the matrix whose columns are the vectors in Z, and $\mathbf{D}$ is the diagonal matrix whose main diagonal entries are the eigenvalues 8 and -1. Hence, $\mathbf{P} = \frac{1}{5}\begin{bmatrix} -4 & 3 \\ 3 & 4 \end{bmatrix}$, and $\mathbf{D} = \begin{bmatrix} 8 & 0 \\ 0 & -1 \end{bmatrix}$.

Now $\mathbf{P}^{-1} = \mathbf{P}^T$, and so $\mathbf{B} = \mathbf{PDP}^T$. Letting $\mathbf{D}_2 = \begin{bmatrix} 2 & 0 \\ 0 & -1 \end{bmatrix}$, whose diagonal entries are the cube roots of the eigenvalues 8 and -1 gives $\mathbf{B} = \mathbf{PD}_2^3\mathbf{P}^T = \mathbf{PD}_2(\mathbf{P}^T\mathbf{P})\mathbf{D}_2(\mathbf{P}^T\mathbf{P})\mathbf{D}_2\mathbf{P}^T = (\mathbf{PD}_2\mathbf{P}^T)^3$. Therefore, if we let $\mathbf{A} = \mathbf{PD}_2\mathbf{P}^T = \frac{1}{25}\begin{bmatrix} 23 & -36 \\ -36 & 2 \end{bmatrix}$, then $\mathbf{A}^3 = \mathbf{B}$, as desired.

(c) Let $\mathbf{B} = \begin{bmatrix} 17 & 16 & -16 \\ 16 & 41 & -32 \\ -16 & -32 & 41 \end{bmatrix}$ and let C be the standard basis for $\mathbb{R}^3$. First, we must orthogonally diagonalize $\mathbf{B}$. We use Steps 2 and 3 of the Orthogonal Diagonalization Method of Section 6.3.

Step 2(a): Now $p_{\mathbf{B}}(x) = \begin{vmatrix} x-17 & -16 & 16 \\ -16 & x-41 & 32 \\ 16 & 32 & x-41 \end{vmatrix} = x^3 - 99x^2 + 1539x - 6561 = (x-81)(x-9)^2$, giving eigenvalues $\lambda_1 = 81$ and $\lambda_2 = 9$. Using row reduction to find particular solutions to the systems $[81\mathbf{I}_3 - \mathbf{B}|\mathbf{0}]$ and $[9\mathbf{I}_3 - \mathbf{B}|\mathbf{0}]$ as in the Diagonalization Method of Section 3.4 produces the basis $\{[-1,-2,2]\}$ for E_{81} and $\{[-2,1,0],[2,0,1]\}$ for E_9.

Step 2(b): Since E_{81} is 1-dimensional, the given basis is already orthogonal. To make the basis orthonormal, we need to normalize, yielding $\left\{\frac{1}{3}[-1,-2,2]\right\}$ for E_{81}. Now E_9 is 2-dimensional, and so we must use the Gram-Schmidt Process to turn the basis for E_9 into an orthogonal basis. We let $\mathbf{v}_1 = [-2,1,0]$. Then $\mathbf{v}_2 = [2,0,1] - \left(\frac{[2,0,1]\cdot[-2,1,0]}{[-2,1,0]\cdot[-2,1,0]}\right)[-2,1,0] = \left[\frac{2}{5}, \frac{4}{5}, 1\right]$. Multiplying by 5 to eliminate fractions produces $\mathbf{v}_2 = [2,4,5]$. Thus we have an orthogonal basis $\{\mathbf{v}_1, \mathbf{v}_2\}$ for E_9. Normalizing gives the orthonormal basis $\left\{\frac{1}{\sqrt{5}}[-2,1,0], \frac{1}{3\sqrt{5}}[2,4,5]\right\}$ for E_9.

Step 2(c): $Z = \left(\frac{1}{3}[-1,-2,2], \frac{1}{\sqrt{5}}[-2,1,0], \frac{1}{3\sqrt{5}}[2,4,5]\right)$.

Step 3: Since C is the standard basis for $\mathbb{R}^3$, $B = Z = \left(\frac{1}{3}[-1,-2,2], \frac{1}{\sqrt{5}}[-2,1,0], \frac{1}{3\sqrt{5}}[2,4,5]\right)$. Now $\mathbf{P}$ is the matrix whose columns are the vectors in Z, and $\mathbf{D}$ is the diagonal matrix whose main diagonal entries are the eigenvalues 81, 9 and 9. Hence,

$$\mathbf{P} = \begin{bmatrix} -\frac{1}{3} & -\frac{2}{\sqrt{5}} & \frac{2}{3\sqrt{5}} \\ -\frac{2}{3} & \frac{1}{\sqrt{5}} & \frac{4}{3\sqrt{5}} \\ \frac{2}{3} & 0 & \frac{5}{3\sqrt{5}} \end{bmatrix}, \text{ and } \mathbf{D} = \begin{bmatrix} 81 & 0 & 0 \\ 0 & 9 & 0 \\ 0 & 0 & 9 \end{bmatrix}.$$

Now $\mathbf{P}^{-1} = \mathbf{P}^T$, and so $\mathbf{B} = \mathbf{PDP}^T$. Letting $\mathbf{D}_2 = \begin{bmatrix} 9 & 0 & 0 \\ 0 & 3 & 0 \\ 0 & 0 & 3 \end{bmatrix}$, whose diagonal entries are the square roots of the eigenvalues 81, 9 and 9 gives $\mathbf{B} = \mathbf{PD}_2^2\mathbf{P}^T = \mathbf{PD}_2(\mathbf{P}^T\mathbf{P})\mathbf{D}_2\mathbf{P}^T = \left(\mathbf{PD}_2\mathbf{P}^T\right)^2$. Therefore, if we let $\mathbf{A} = \mathbf{PD}_2\mathbf{P}^T = \frac{1}{3}\begin{bmatrix} 11 & 4 & -4 \\ 4 & 17 & -8 \\ -4 & -8 & 17 \end{bmatrix}$, then $\mathbf{A}^2 = \mathbf{B}$, as desired.

(6) For example, the matrix $\mathbf{A}$ in Example 7 of Section 5.5 is diagonalizable but not symmetric and hence not orthogonally diagonalizable.

(7) If $\mathbf{A}=\begin{bmatrix} a & b \\ b & c \end{bmatrix}$, then $p_{\mathbf{A}}(x) = \begin{vmatrix} x-a & -b \\ -b & x-c \end{vmatrix} = (x-a)(x-c) - b^2 = x^2 - (a+c)x + (ac-b^2)$. Using the quadratic formula and simplifying yields the two eigenvalues $\lambda_1 = \frac{1}{2}\left(a+c+\sqrt{(a-c)^2+4b^2}\right)$ and $\lambda_2 = \frac{1}{2}\left(a+c-\sqrt{(a-c)^2+4b^2}\right)$. Now, $\mathbf{D}$ is the diagonal matrix with these eigenvalues on its main diagonal. Hence, $\mathbf{D} = \frac{1}{2}\begin{bmatrix} a+c+\sqrt{(a-c)^2+4b^2} & 0 \\ 0 & a+c-\sqrt{(a-c)^2+4b^2} \end{bmatrix}$.

(8) (b) Since L is symmetric, it is orthogonally diagonalizable by Theorem 6.19. But if 0 is the only eigenvalue of L, the geometric multiplicity of $\lambda = 0$ must be $\dim(\mathcal{V})$. Therefore, the eigenspace for 0 must be all of $\mathcal{V}$, and so $L(\mathbf{v}) = 0\mathbf{v} = \mathbf{0}$ for all $\mathbf{v} \in \mathcal{V}$. Therefore L must be the zero linear operator.

(13) (a) True. The given condition on L is the definition of a symmetric operator. The statement then follows from Lemma 6.18.

(b) False. A symmetric operator is only guaranteed to have a symmetric matrix with respect to an ordered orthonormal basis, not every ordered basis. For example, the symmetric operator on $\mathbb{R}^2$ given by $L\left(\begin{bmatrix} x \\ y \end{bmatrix}\right) = \begin{bmatrix} 1 & 0 \\ 0 & 2 \end{bmatrix}\begin{bmatrix} x \\ y \end{bmatrix}$ has matrix $\mathbf{A} = \begin{bmatrix} 1 & -1 \\ 0 & 2 \end{bmatrix}$ with respect to the basis $\{[1,0],[1,1]\}$. But $\mathbf{A}$ is not symmetric.

(c) True. This is part of Theorem 6.17.

(d) True. This follows from the definition of orthogonally diagonalizable and Theorem 6.19.

(e) True. The Orthogonal Diagonalization Method states that $\mathbf{D} = \mathbf{P}^{-1}\mathbf{AP}$, and that the matrix $\mathbf{P}$ is orthogonal. Thus, $\mathbf{P}^{-1} = \mathbf{P}^T$. Hence, $\mathbf{D} = \mathbf{P}^{-1}\mathbf{AP} \Rightarrow \mathbf{PDP}^{-1} = \mathbf{P}(\mathbf{P}^{-1}\mathbf{AP})\mathbf{P}^{-1} \Rightarrow \mathbf{PDP}^{-1} = \mathbf{I}_n\mathbf{AI}_n \Rightarrow \mathbf{PDP}^T = \mathbf{A}$.

(f) True. In the Orthogonal Diagonalization Method, $\mathbf{P}$ is the transition matrix from an ordered orthonormal basis of eigenvectors for $\mathbf{A}$ to standard coordinates. (We can assume we are using standard coordinates here because the problem refers only to a matrix rather than a linear transformation.)

Chapter 7

Section 7.1

(1) For extra help with adding and multiplying complex numbers, see Appendix C.

(a) $[2+i, 3, -i] + [-1+3i, -2+i, 6] = [2+i-1+3i, 3-2+i, -i+6] = [1+4i, 1+i, 6-i]$

(b) $(-8+3i)[4i, 2-3i, -7+i] = [(-8+3i)(4i), (-8+3i)(2-3i), (-8+3i)(-7+i)]$
$= [-12-32i, -7+30i, 53-29i]$

(d) $\overline{(-4)[6-3i, 7-2i, -8i]} = \overline{[-24+12i, -28+8i, 32i]} = [\overline{-24+12i}, \overline{-28+8i}, \overline{32i}]$
$= [-24-12i, -28-8i, -32i]$

(e) $[-2+i, 5-2i, 3+4i] \cdot [1+i, 4-3i, -6i] = (-2+i)\left(\overline{1+i}\right) + (5-2i)\left(\overline{4-3i}\right) + (3+4i)\left(\overline{-6i}\right)$
$= (-2+i)(1-i) + (5-2i)(4+3i) + (3+4i)(6i) = (-1+3i) + (26+7i) + (-24+18i) = 1+28i$

(3) For extra help with adding and multiplying complex numbers, see Appendix C.

(a) $$\begin{bmatrix} 2+5i & -4+i \\ -3-6i & 8-3i \end{bmatrix} + \begin{bmatrix} 9-i & -3i \\ 5+2i & 4+3i \end{bmatrix} = \begin{bmatrix} (2+5i)+(9-i) & (-4+i)+(-3i) \\ (-3-6i)+(5+2i) & (8-3i)+(4+3i) \end{bmatrix}$$
$$= \begin{bmatrix} 11+4i & -4-2i \\ 2-4i & 12 \end{bmatrix}$$

(c) $$\mathbf{C}^* = \left(\overline{\mathbf{C}}\right)^T = \begin{bmatrix} \overline{1+i} & \overline{-2i} & \overline{6+4i} \\ \overline{0} & \overline{3+i} & \overline{5} \\ \overline{-10i} & \overline{0} & \overline{7-3i} \end{bmatrix}^T = \begin{bmatrix} 1-i & 2i & 6-4i \\ 0 & 3-i & 5 \\ 10i & 0 & 7+3i \end{bmatrix}^T$$
$$= \begin{bmatrix} 1-i & 0 & 10i \\ 2i & 3-i & 0 \\ 6-4i & 5 & 7+3i \end{bmatrix}$$

(d) $$(-3i)\mathbf{D} = \begin{bmatrix} (-3i)(5-i) & (-3i)(-i) & (-3i)(-3) \\ (-3i)(2+3i) & (-3i)(0) & (-3i)(-4+i) \end{bmatrix} = \begin{bmatrix} -3-15i & -3 & 9i \\ 9-6i & 0 & 3+12i \end{bmatrix}$$

(f) $$\begin{bmatrix} 2+5i & -4+i \\ -3-6i & 8-3i \end{bmatrix} \begin{bmatrix} 9-i & -3i \\ 5+2i & 4+3i \end{bmatrix} =$$
$$\begin{bmatrix} (2+5i)(9-i)+(-4+i)(5+2i) & (2+5i)(-3i)+(-4+i)(4+3i) \\ (-3-6i)(9-i)+(8-3i)(5+2i) & (-3-6i)(-3i)+(8-3i)(4+3i) \end{bmatrix}$$
$$= \begin{bmatrix} (23+43i)+(-22-3i) & (15-6i)+(-19-8i) \\ (-33-51i)+(46+i) & (-18+9i)+(41+12i) \end{bmatrix} = \begin{bmatrix} 1+40i & -4-14i \\ 13-50i & 23+21i \end{bmatrix}$$

(i) $$\mathbf{C}^T\mathbf{D}^* = \mathbf{C}^T\left(\overline{\mathbf{D}}\right)^T = \mathbf{C}^T \begin{bmatrix} \overline{5-i} & \overline{-i} & \overline{-3} \\ \overline{2+3i} & \overline{0} & \overline{-4+i} \end{bmatrix}^T = \mathbf{C}^T \begin{bmatrix} \overline{5-i} & \overline{2+3i} \\ \overline{-i} & \overline{0} \\ \overline{-3} & \overline{-4+i} \end{bmatrix}$$
$$= \begin{bmatrix} 1+i & 0 & -10i \\ -2i & 3+i & 0 \\ 6+4i & 5 & 7-3i \end{bmatrix} \begin{bmatrix} 5+i & 2-3i \\ i & 0 \\ -3 & -4-i \end{bmatrix} =$$
$$\begin{bmatrix} (1+i)(5+i)+0(i)+(-10i)(-3) & (1+i)(2-3i)+0(0)+(-10i)(-4-i) \\ (-2i)(5+i)+(3+i)(i)+0(-3) & (-2i)(2-3i)+(3+i)(0)+0(-4-i) \\ (6+4i)(5+i)+5(i)+(7-3i)(-3) & (6+4i)(2-3i)+5(0)+(7-3i)(-4-i) \end{bmatrix} =$$

$$\begin{bmatrix} (4+6i)+0+30i & (5-i)+0+(-10+40i) \\ (2-10i)+(-1+3i)+0 & (-6-4i)+0+0 \\ (26+26i)+(5i)+(-21+9i) & (24-10i)+0+(-31+5i) \end{bmatrix} = \begin{bmatrix} 4+36i & -5+39i \\ 1-7i & -6-4i \\ 5+40i & -7-5i \end{bmatrix}$$

(4) (b) $(\mathbf{ZW})^* = \overline{(\mathbf{ZW})^T} = \overline{\mathbf{W}^T\mathbf{Z}^T}$ (by part (4) of Theorem 7.2) $= (\overline{\mathbf{W}^T})(\overline{\mathbf{Z}^T}) = \mathbf{W}^*\mathbf{Z}^*$

(5) In each part, let $\mathbf{A}$ represent the given matrix.

(a) The matrix $\mathbf{A}$ is skew-Hermitian because all of its main diagonal entries are pure imaginary (their real parts all equal zero) and, for $i \neq j$, $a_{ij} = -\overline{a_{ji}}$.

(b) The matrix $\mathbf{A}$ is not Hermitian because a_{11} is not a real number. Also, $\mathbf{A}$ is not skew-Hermitian because a_{11} is not a pure imaginary number.

(c) The matrix $\mathbf{A}$ is Hermitian because all of its main diagonal entries are real, and, since all other entries are zero (and hence, for $i \neq j$, $a_{ij} = \overline{a_{ji}}$).

(d) The matrix $\mathbf{A}$ is skew-Hermitian because all of its main diagonal entries are pure imaginary, and, since all other entries are zero (and hence, for $i \neq j$, $a_{ij} = -\overline{a_{ji}}$).

(e) The matrix $\mathbf{A}$ is Hermitian because all of its main diagonal entries are real, and, for $i \neq j$, $a_{ij} = -\overline{a_{ji}}$.

(9) Let $\mathbf{Z}$ be Hermitian. Then, since $\mathbf{Z}^* = \mathbf{Z}$, $\mathbf{ZZ}^* = \mathbf{ZZ} = \mathbf{Z}^*\mathbf{Z}$. Similarly, if $\mathbf{Z}$ is skew-Hermitian, $\mathbf{Z}^* = -\mathbf{Z}$. Therefore, $\mathbf{ZZ}^* = \mathbf{Z}(-\mathbf{Z}) = -(\mathbf{ZZ}) = (-\mathbf{Z})\mathbf{Z} = \mathbf{Z}^*\mathbf{Z}$.

(11) (a) False. Consider that $[i,-1]\cdot[1,i] = i\left(\overline{1}\right) + (-1)\left(\overline{i}\right) = i(1) + (-1)(-i) = 2i$.

(b) False. When computing the complex dot product, we find the complex conjugates of the entries in the second vector before multiplying corresponding coordinates. However, this is not done when computing each entry of the matrix product of complex matrices. For a specific example, consider the matrices $\mathbf{Z}$ and $\mathbf{W}$ in Example 2 in Section 7.1 of the textbook. In that example, the (1,1) entry of $\mathbf{ZW}$ is $-5-8i$. However, the complex dot product of the 1st row of $\mathbf{Z}$ with the 1st column of $\mathbf{W}$ is $(1-i)\,\overline{(-2i)} + (2i)\,\overline{(-1+3i)} + (-2+i)\,\overline{(-2+i)} =$ $(1-i)\,(2i) + (2i)\,(-1-3i) + (-2+i)\,(-2-i) = (2+2i) + (6-2i) + 5 = 13$.

(c) True. Both are equal to the conjugate transpose of $\mathbf{Z}$. This is stated in Section 7.1 of the textbook just before Example 3.

(d) False. By part (5) of Theorem 7.1, $\mathbf{v}_1 \cdot (k\mathbf{v}_2) = \overline{k}(\mathbf{v}_1 \cdot \mathbf{v}_2)$, which does not necessarily equal $k(\mathbf{v}_1 \cdot \mathbf{v}_2)$ when k is complex. For example, $[1,1]\cdot(i[1,1]) = [1,1]\cdot[i,i] = 1(\overline{i}) + 1(\overline{i}) = -2i$, but $i([1,1]\cdot[1,1]) = i(1(\overline{1}) + 1(\overline{1})) = i(2) = 2i$.

(e) False. The matrix $\mathbf{H}$ in Example 4 in Section 7.1 of the textbook is Hermitian but not symmetric. In fact, a Hermitian matrix is symmetric if and only if all of its entries are real.

(f) True. In fact, the transpose of a skew-Hermitian matrix $\mathbf{K}$ is skew-Hermitian, since $\left(\mathbf{K}^T\right)^* = \left(\overline{\left(\mathbf{K}^T\right)}\right)^T = \left(\mathbf{K}^*\right)^T = \left(-\mathbf{K}\right)^T = -\left(\mathbf{K}^T\right)$. Hence, $\mathbf{K}^T$ is normal by Theorem 7.4.

Section 7.2

(1) Remember: if z is a complex number, $\frac{1}{z} = \frac{\overline{z}}{|z|^2}$. For further help with complex arithmetic, review Appendix C.

(a) Start with the augmented matrix $\left[\begin{array}{cc|c} 3+i & 5+5i & 29+33i \\ 1+i & 6-2i & 30-12i \end{array}\right]$.

Pivoting in the first column, we perform the following operations:
$\langle 1\rangle \leftarrow \frac{1}{3+i}\langle 1\rangle$, which is the same as $\langle 1\rangle \leftarrow \left(\frac{3}{10}-\frac{1}{10}i\right)\langle 1\rangle$
$\langle 2\rangle \leftarrow -(1+i)\langle 1\rangle + \langle 2\rangle$. This results in the matrix
$\left[\begin{array}{cc|c} 1 & 2+i & 12+7i \\ 0 & 5-5i & 25-31i \end{array}\right]$. Next, we pivot in column 2:
$\langle 2\rangle \leftarrow \frac{1}{5-5i}\langle 2\rangle$, which is the same as $\langle 2\rangle \leftarrow \left(\frac{1}{10}+\frac{1}{10}i\right)\langle 2\rangle$
This gives $\left[\begin{array}{cc|c} 1 & 2+i & 12+7i \\ 0 & 1 & \frac{28}{5}-\frac{3}{5}i \end{array}\right]$. Hence, $z = \frac{28}{5}-\frac{3}{5}i$.
Back substitution yields $w = (12+7i)-(2+i)\left(\frac{28}{5}-\frac{3}{5}i\right) = \frac{1}{5}+\frac{13}{5}i$.

(c) Begin with the augmented matrix $\left[\begin{array}{ccc|c} 3i & -6+3i & 12+18i & -51+9i \\ 3+2i & 1+7i & 25-2i & -13+56i \\ 1+i & 2i & 9+i & -7+17i \end{array}\right]$,

we row reduce to $\left[\begin{array}{cccc} 1 & 0 & 4-3i & 2+5i \\ 0 & 1 & -i & 5+2i \\ 0 & 0 & 0 & 0 \end{array}\right]$.

Thus, the solution set is $\{(\,(2+5i)-(4-3i)c,\ (5+2i)+ic,\ c) \mid c \in \mathbb{C}\}$.

(e) After applying the row reduction method to the first column, the matrix changes from
$\left[\begin{array}{cc|c} 3-2i & 12+5i & 3+11i \\ 5+4i & -2+23i & -14+15i \end{array}\right]$ to $\left[\begin{array}{ccc} 1 & 2+3i & -1+3i \\ 0 & 0 & 3+4i \end{array}\right]$.
The second row shows that the system has no solutions.

(2) (b) Using a cofactor expansion on the first row yields
$|\mathbf{A}| = i\Big((1-i)(2-i)-(i)(-2)\Big) - (2)\Big((1+i)(2-i)-(i)(4)\Big) + (5i)\Big((1+i)(-2)-(1-i)(4)\Big)$
$= i(1-i)-(2)(3-3i)+(5i)(-6+2i) = -15-23i$. Hence $\overline{|\mathbf{A}|} = -15+23i$.

Next, $|\mathbf{A}^*| = \left|\begin{array}{ccc} -i & 1-i & 4 \\ 2 & 1+i & -2 \\ -5i & -i & 2+i \end{array}\right|$.

Again, using a cofactor expansion on the first row, this determinant equals
$(-i)\Big((1+i)(2+i)-(-2)(-i)\Big) - (1-i)\Big((2)(2+i)-(-2)(-5i)\Big) + 4\Big((2)(-i)-(1+i)(-5i)\Big)$
$= (-i)(1+i)-(1-i)(4-8i)+4(-5+3i) = -15+23i$.

(3) In each part, let $\mathbf{A}$ represent the given matrix. Your computations may produce complex scalar multiples of the vectors given here, although they might not be immediately recognized as such.

(a) $p_{\mathbf{A}}(x) = \left|\begin{array}{cc} x-(4+3i) & 1+3i \\ -8+2i & x+(5+2i) \end{array}\right| = (x-(4+3i))(x+(5+2i)) - (-8+2i)(1+3i) =$ $x^2+(1-i)x-i = (x-i)(x+1)$. Hence, the eigenvalues are $\lambda_1 = i$ and $\lambda_2 = -1$. Now we solve for corresponding eigenvectors.

For $\lambda_1 = i$, we row reduce $[i\mathbf{I}_2 - \mathbf{A}|\mathbf{0}] = \left[\begin{array}{cc|c} -4-2i & 1+3i & 0 \\ -8+2i & 5+3i & 0 \end{array}\right]$ to $\left[\begin{array}{cc|c} 1 & -\frac{1}{2}-\frac{1}{2}i & 0 \\ 0 & 0 & 0 \end{array}\right]$.

Setting the independent variable z_2 in the associated system equal to 2 to eliminate fractions yields the eigenvector $[1+i, 2]$. Hence, $E_i = \{c[1+i, 2] \mid c \in \mathbb{C}\}$.

For $\lambda_2 = -1$, we row reduce $[(-1)\mathbf{I}_2 - \mathbf{A}|\mathbf{0}] = \left[\begin{array}{cc|c} -5-3i & 1+3i & 0 \\ -8+2i & 4+2i & 0 \end{array}\right]$ to $\left[\begin{array}{cc|c} 1 & -\frac{7}{17}-\frac{6}{17}i & 0 \\ 0 & 0 & 0 \end{array}\right]$.

Setting the independent variable z_2 in the associated system equal to 17 to eliminate fractions yields the eigenvector $[7+6i, 17]$. Hence, $E_{-1} = \{c[7+6i, 17] \mid c \in \mathbb{C}\}$.

(c) Using cofactor expansion along the first row yields

$$p_{\mathbf{A}}(x) = \begin{vmatrix} x-(4+3i) & 4+2i & -4-7i \\ -2+4i & x-(-2+5i) & -7+4i \\ 4+2i & -4-2i & x+(4+6i) \end{vmatrix}$$

$$= (x-(4+3i))\Big((x-(-2+5i))(x+(4+6i)) - (-7+4i)(-4-2i)\Big)$$
$$-(4+2i)\Big((-2+4i)(x+(4+6i)) - (-7+4i)(4+2i)\Big)$$
$$+(-4-7i)\Big((-2+4i)(-4-2i) - (x-(-2+5i))(4+2i)\Big)$$
$$= (x-(4+3i))(x^2+(6+i)x+(2-6i)) - (4+2i)((-2+4i)x+(4+2i))$$
$$+(-4-7i)((-4-2i)x+(-2+4i)) = \left(x^3+(2-2i)x^2+(-19-28i)x+(-26+18i)\right)$$
$$-((-16+12i)x+(12+16i)) + ((2+36i)x+(36-2i))$$
$$= x^3+(2-2i)x^2+(-1-4i)x-2 = (x-i)^2(x+2).$$ (A computer algebra system would help here.)

Hence, the eigenvalues are $\lambda_1 = i$ and $\lambda_2 = -2$. Now we solve for corresponding eigenvectors.

For $\lambda_1 = i$, we row reduce $[i\mathbf{I}_3 - \mathbf{A}|\mathbf{0}] =$

$$\left[\begin{array}{ccc|c} -4-2i & 4+2i & -4-7i & 0 \\ -2+4i & 2-4i & -7+4i & 0 \\ 4+2i & -4-2i & 4+7i & 0 \end{array}\right] \text{ to } \left[\begin{array}{ccc|c} 1 & -1 & \frac{3}{2}+i & 0 \\ 0 & 0 & 0 & 0 \\ 0 & 0 & 0 & 0 \end{array}\right].$$

Setting the independent variables z_2 and z_3 in the associated system equal to 1 and 0, respectively, yields the eigenvector $[1,1,0]$. Then setting the independent variables z_2 and z_3 in the associated system equal to 0 and 2 (to eliminate fractions), respectively, gives the eigenvector $[-3-2i, 0, 2]$.

For $\lambda_2 = -2$, we row reduce $[(-2)\mathbf{I}_3 - \mathbf{A}|\mathbf{0}] =$

$$\left[\begin{array}{ccc|c} -6-3i & 4+2i & -4-7i & 0 \\ -2+4i & -5i & -7+4i & 0 \\ 4+2i & -4-2i & 2+6i & 0 \end{array}\right] \text{ to } \left[\begin{array}{ccc|c} 1 & 0 & 1 & 0 \\ 0 & 1 & -i & 0 \\ 0 & 0 & 0 & 0 \end{array}\right].$$

Setting the independent variable z_3 in the associated system equal to 1 yields the eigenvector $[-1, i, 1]$.

Hence $E_i = \{c[1,1,0] + d[(-3-2i), 0, 2] \mid c, d \in \mathbb{C}\}$ and $E_{-2} = \{c[-1, i, 1] \mid c \in \mathbb{C}\}$.

(4) (a) We use the solution to Exercise 3(a), above. Let $\mathbf{A}$ be the matrix given in Exercise 3(a). $\mathbf{A}$ is diagonalizable since two eigenvectors were found while following the Diagonalization Method in Section 3.4. The matrix $\mathbf{P}$ is the matrix whose columns are the eigenvectors we found. The matrix $\mathbf{D}$ is the diagonal matrix whose main diagonal entries are the eigenvalues i and -1. Hence,

$\mathbf{P} = \begin{bmatrix} 1+i & 7+6i \\ 2 & 17 \end{bmatrix}$ and $\mathbf{D} = \begin{bmatrix} i & 0 \\ 0 & -1 \end{bmatrix}$. Now $\mathbf{P}^{-1} = \frac{1}{3+5i}\begin{bmatrix} 17 & -7-6i \\ -2 & 1+i \end{bmatrix}$, by using Theorem 2.13. Using $\frac{1}{3+5i} = \frac{1}{34}(3-5i)$, we can verify that $\mathbf{D} = \mathbf{P}^{-1}\mathbf{A}\mathbf{P}$.

(6) (a) True. Performing the row operation $\langle 2\rangle \leftarrow -i\,\langle 1\rangle + \langle 2\rangle$ on $\mathbf{A}$ produces a matrix $\mathbf{B}$ whose second row is zero. Hence, $|\mathbf{B}| = 0$ (perform a cofactor expansion along its second row). But since $\mathbf{B}$ was obtained from $\mathbf{A}$ by performing a single type (II) row operation, $|\mathbf{A}| = |\mathbf{B}|$. Therefore, $|\mathbf{A}| = 0$.

(b) False. Example 2 in Section 7.2 of the textbook illustrates a 3×3 matrix $\mathbf{A}$ having eigenvalue 3 having algebraic multiplicity 1. What is true, however, is that the *sum* of all the algebraic multiplicities of all the eigenvalues of a complex $n \times n$ matrix must equal n.

(c) False. If the number of eigenvectors produced by the Diagonalization Method for some eigenvalue does not equal its algebraic multiplicity, the matrix is still not diagonalizable when thought of as a complex matrix. For example, the real matrix given in Example 6 of Section 3.4 has -2 as an eigenvalue with algebraic multiplicity 2 but only one eigenvector is produced for this eigenvalue by the Diagonalization Method. Hence, this matrix is not diagonalizable, even when thought of as a complex matrix.

(d) False. The Fundamental Theorem of Algebra only guarantees that an nth degree complex polynomial factors into n linear factors. However, if these factors are not different, the polynomial will not have n distinct roots. For example, $x^2 - 2ix - 1 = (x - i)^2$ has only one root, namely i.

Section 7.3

(2) In each part, let S represent the given set of vectors.

(b) The set S is not linearly independent because $[-3+6i, 3, 9i] = 3i[2+i, -i, 3]$. Thus, $\{[2+i, -i, 3]\}$ is a basis for span(S), and so dim(span(S)) $= 1$.

(d) The matrix formed using the vectors in S as columns row reduces to $\begin{bmatrix} 1 & 0 & i \\ 0 & 1 & -2i \\ 0 & 0 & 0 \end{bmatrix}$. Therefore, by the Independence Test Method, $\{[3 - i, 1 + 2i, -i], [1 + i, -2, 4 + i]\}$ is a basis for span(S). Hence, dim(span(S)) $= 2$.

(3) In each part, let S represent the given set of vectors.

(b) Since $[-3+6i, 3, 9i]$ is not a *real* scalar multiple of $[2+i, -i, 3]$, S is linearly independent. (Consider the 3rd coordinate: $9i \neq 3c$ for any $c \in \mathbb{R}$.) Hence, S is a basis for span(S), and dim(span(S)) $= 2$.

(d) Separating the real and imaginary parts of each vector in S, we can think of them as vectors in $\mathbb{R}^6$. That is, rewrite each complex vector $[a + bi, c + di, e + fi]$ as $[a, b, c, d, e, f]$. We then create a matrix using each 6-vector as a column. This gives

$$\begin{bmatrix} 3 & 1 & 3 \\ -1 & 1 & 1 \\ 1 & -2 & -2 \\ 2 & 0 & 5 \\ 0 & 4 & 3 \\ -1 & 1 & -8 \end{bmatrix} \text{ which row reduces to } \begin{bmatrix} 1 & 0 & 0 \\ 0 & 1 & 0 \\ 0 & 0 & 1 \\ 0 & 0 & 0 \\ 0 & 0 & 0 \\ 0 & 0 & 0 \end{bmatrix}.$$

The Independence Test Method then shows that S is linearly independent. Therefore, S is a basis for span(S), and dim(span(S)) $= 3$.

(4) (b) The matrix $\left[\begin{array}{ccc|c} 2i & 3+i & -3+5i & 3-i \\ -1+3i & -2 & 2i & -5-5i \\ 4 & 1-i & -5+3i & 7+i \end{array}\right]$ row reduces to $\left[\begin{array}{ccc|c} 1 & 0 & 0 & i \\ 0 & 1 & 0 & 1+i \\ 0 & 0 & 1 & -1 \end{array}\right]$.

Hence, $[\mathbf{z}]_B = [i, 1+i, -1]$.

(5) With $\mathbb{C}^2$ considered as a real vector space, we use the ordered basis $B = ([1,0],[i,0],[0,1],[0,i])$. Now $[L([1,0])]_B = [0,1]_B = [0,0,1,0]$, $[L([i,0])]_B = [0,-i]_B = [0,0,0,-1]$, $[L([0,1])]_B = [1,0]_B = [1,0,0,0]$, and $[L([0,i])]_B = [-i,0]_B = [0,-1,0,0]$. Therefore, the matrix for L with respect to B is $\begin{bmatrix} 0 & 0 & 1 & 0 \\ 0 & 0 & 0 & -1 \\ 1 & 0 & 0 & 0 \\ 0 & -1 & 0 & 0 \end{bmatrix}$.

(8) We present two methods for solving this problem:
Method 1: Let B be the standard basis for $\mathbb{C}^2$, C be the basis $\{[1+i,-1+3i],[1-i,1+2i]\}$ for $\mathbb{C}^2$, and D be the standard basis for $\mathbb{C}^3$. Then, by the given values for L on C, $\mathbf{A}_{CD} = \begin{bmatrix} 3-i & 2+i \\ 5 & 1-3i \\ -i & 3 \end{bmatrix}$.

Next, we compute the transition matrix $\mathbf{P}$ from B to C. Since B is the standard basis, we can do this by computing the inverse of $\begin{bmatrix} 1+i & 1-i \\ -1+3i & 1+2i \end{bmatrix}$. Theorem 2.13 yields $\mathbf{P} = \frac{-3+i}{10}\begin{bmatrix} 1+2i & -1+i \\ 1-3i & 1+i \end{bmatrix}$.

Finally, by Theorem 5.6, $\mathbf{A}_{BD} = \mathbf{A}_{CD}\mathbf{P} = \begin{bmatrix} -3+i & -\frac{2}{5}-\frac{11}{5}i \\ \frac{1}{2}-\frac{3}{2}i & -i \\ -\frac{1}{2}+\frac{7}{2}i & -\frac{8}{5}-\frac{4}{5}i \end{bmatrix}$.

Method 2: We begin by expressing $[1,0]$ and $[0,1]$ as linear combinations of $[1+i,-1+3i]$ and $[1-i,1+2i]$ by row reducing $\left[\begin{array}{cc|cc} 1+i & 1-i & 1 & 0 \\ -1+3i & 1+2i & 0 & 1 \end{array}\right]$ to obtain $\left[\begin{array}{cc|cc} 1 & 0 & -\frac{1}{2}-\frac{1}{2}i & \frac{1}{5}-\frac{2}{5}i \\ 0 & 1 & i & -\frac{2}{5}-\frac{1}{5}i \end{array}\right]$.

Therefore, $[1,0] = \left(-\frac{1}{2}-\frac{1}{2}i\right)[1+i,-1+3i] + i[1-i,1+2i]$ and
$[0,1] = \left(\frac{1}{5}-\frac{2}{5}i\right)[1+i,-1+3i] + \left(-\frac{2}{5}-\frac{1}{5}i\right)[1-i,1+2i]$.
Hence, $L([1,0]) = \left(-\frac{1}{2}-\frac{1}{2}i\right)[3-i,5,-i] + i\,[2+i,1-3i,3] = \left[-3+i, \frac{1}{2}-\frac{3}{2}i, -\frac{1}{2}+\frac{7}{2}i\right]$,
and $L([0,1]) = \left(\frac{1}{5}-\frac{2}{5}i\right)[3-i,5,-i] + \left(-\frac{2}{5}-\frac{1}{5}i\right)[2+i,1-3i,3] = \left[-\frac{2}{5}-\frac{11}{5}i, -i, -\frac{8}{5}-\frac{4}{5}i\right]$. Using $L([1,0])$ and $L([0,1])$ as columns produces the 3×2 matrix $\mathbf{A}_{BD}$ obtained using Method 1.

(9) (a) True. This is the complex analog of Theorem 4.19.

(b) False. If $\mathbf{v} = i$, then $L(i\mathbf{v}) = L(i(i)) = L(-1) = \overline{(-1)} = -1$.
However, $iL(\mathbf{v}) = iL(i) = i\left(\overline{i}\right) = i(-i) = 1$.

(c) True. This follows from the complex analog of Theorem 4.20.

(d) False. Complex vector spaces (and their complex subspaces) must have even dimension as *real* vector spaces, not necessarily as complex vector spaces. For example, the subspace $\mathcal{W} = \text{span}(\{[1,0,0]\})$ of $\mathbb{C}^3$ has complex dimension 1.

Section 7.4

(1) Remember to use the conjugates of the coordinates of the second vector when computing a complex dot product.

(a) The given set is not orthogonal because $[1+2i, -3-i] \cdot [4-2i, 3+i] = -10+10i \neq 0$.

(c) The given set is orthogonal because $[2i, -1, i] \cdot [1, -i, -1] = 0$, $[2i, -1, i] \cdot [0, 1, i] = 0$, and $[1, -i, -1] \cdot [0, 1, i] = 0$.

(3) (a) First, we must find an ordinary basis for $\mathbb{C}^3$ containing $[1+i, i, 1]$. To do this, we use the Enlarging Method from Section 4.6. We add the standard basis for $\mathbb{C}^3$ to the vector we have and use the Independence Test Method. Hence, we row reduce

$$\begin{bmatrix} 1+i & 1 & 0 & 0 \\ i & 0 & 1 & 0 \\ 1 & 0 & 0 & 1 \end{bmatrix} \text{ to } \begin{bmatrix} 1 & 0 & 0 & 1 \\ 0 & 1 & 0 & -1-i \\ 0 & 0 & 1 & -i \end{bmatrix}, \text{ yielding the basis } \{[1+i, i, 1], [1,0,0], [0,1,0]\}.$$

We perform the Gram-Schmidt Process on this set.
First, let $\mathbf{v}_1 = [1+i, i, 1]$.
Next, $\mathbf{v}_2 = [1,0,0] - \left(\frac{[1,0,0]\cdot[1+i,i,1]}{[1+i,i,1]\cdot[1+i,i,1]}\right)[1+i,i,1] = [1,0,0] - \frac{1-i}{4}[1+i,i,1] = \left[\frac{1}{2}, -\frac{1}{4}-\frac{1}{4}i, -\frac{1}{4}+\frac{1}{4}i\right]$.
Multiplying by 4 to eliminate fractions yields $\mathbf{v}_2 = [2, -1-i, -1+i]$.
Finally, $\mathbf{v}_3 = [0,1,0] - \left(\frac{[0,1,0]\cdot[1+i,i,1]}{[1+i,i,1]\cdot[1+i,i,1]}\right)[1+i,i,1] - \left(\frac{[0,1,0]\cdot[2,-1-i,-1+i]}{[2,-1-i,-1+i]\cdot[2,-1-i,-1+i]}\right)[2,-1-i,-1+i]$
$= [0,1,0] - \frac{(-i)}{4}[1+i,i,1] - \frac{(-1+i)}{8}[2,-1-i,-1+i] = \left[0, \frac{1}{2}, \frac{1}{2}i\right]$. Multiplying by 2 to eliminate fractions produces $\mathbf{v}_3 = [0,1,i]$. Hence, the desired orthogonal set is $\{[1+i,i,1], [2,-1-i,-1+i], [0,1,i]\}$.

(b) According to part (1) of Theorem 7.6, a 3×3 matrix is unitary if and only if its rows form an orthonormal basis for $\mathbb{C}^3$. Normalizing the orthogonal basis we computed in part (a) yields an orthonormal basis whose first vector is a scalar multiple of $[1+i, i, 1]$. Using the vectors from this orthonormal basis as the rows of a matrix produces the unitary matrix $\begin{bmatrix} \frac{1+i}{2} & \frac{i}{2} & \frac{1}{2} \\ \frac{2}{\sqrt{8}} & \frac{-1-i}{\sqrt{8}} & \frac{-1+i}{\sqrt{8}} \\ 0 & \frac{1}{\sqrt{2}} & \frac{i}{\sqrt{2}} \end{bmatrix}$.

(7) (a) $\mathbf{A}$ is unitary if and only if $\mathbf{A}^* = \mathbf{A}^{-1}$ if and only if $\overline{\mathbf{A}^T} = \mathbf{A}^{-1}$ if and only if $(\overline{\mathbf{A}^T})^T = (\mathbf{A}^{-1})^T$ if and only if $(\mathbf{A}^T)^* = (\mathbf{A}^{-1})^T$ if and only if $(\mathbf{A}^T)^* = (\mathbf{A}^T)^{-1}$ if and only if $\mathbf{A}^T$ is unitary.

(10) (b) $p_{\mathbf{A}}(x) = \begin{vmatrix} x-(1-6i) & 10+2i \\ -2+10i & x-5 \end{vmatrix} = (x-(1-6i))(x-5) - (10+2i)(-2+10i)$
$= x^2 + (-6+6i)x + (45-126i)$. The quadratic formula produces the eigenvalues $\lambda_1 = 9+6i$ and $\lambda_2 = -3-12i$. Now we solve for corresponding eigenvectors.
For $\lambda_1 = 9+6i$, we row reduce $[(9+6i)\mathbf{I}_2 - \mathbf{A}|\mathbf{0}] =$

$$\left[\begin{array}{cc|c} 8+12i & 10+2i & 0 \\ -2+10i & 4+6i & 0 \end{array}\right] \text{ to } \left[\begin{array}{cc|c} 1 & \frac{1}{2}-\frac{1}{2}i & 0 \\ 0 & 0 & 0 \end{array}\right].$$

Setting the independent variable z_2 in the associated system equal to 2 to eliminate fractions yields the eigenvector $[-1+i, 2]$.

For $\lambda_2 = -3 - 12i$, we row reduce $[(-3-12i)\mathbf{I}_2 - \mathbf{A}|\mathbf{0}] =$

$$\left[\begin{array}{cc|c} -4-6i & 10+2i & 0 \\ -2+10i & -8-12i & 0 \end{array}\right] \text{ to } \left[\begin{array}{cc|c} 1 & -1+i & 0 \\ 0 & 0 & 0 \end{array}\right].$$

Setting the independent variable z_2 in the associated system equal to 1 yields the eigenvector $[1-i, 1]$.

Note that the two eigenvectors we have computed form an orthogonal basis for $\mathbb{C}^2$. Normalizing each of these vectors and using them as columns for a matrix produces the unitary matrix

$\mathbf{P} = \begin{bmatrix} \frac{-1+i}{\sqrt{6}} & \frac{1-i}{\sqrt{3}} \\ \frac{2}{\sqrt{6}} & \frac{1}{\sqrt{3}} \end{bmatrix}$. Direct computation shows that $\mathbf{P}^{-1}\mathbf{AP} = \mathbf{P}^*\mathbf{AP} = \begin{bmatrix} 9+6i & 0 \\ 0 & -3-12i \end{bmatrix}$, a diagonal matrix having the eigenvalues of $\mathbf{A}$ on its main diagonal.

(13) Let $\mathbf{A}$ represent the given matrix. Then, $p_{\mathbf{A}}(x) = \begin{vmatrix} x-1 & -2-i & -1+2i \\ -2+i & x+3 & i \\ -1-2i & -i & x-2 \end{vmatrix}$. By basketweaving, this determinant equals $(x-1)(x+3)(x-2) + (-2-i)(i)(-1-2i) + (-1+2i)(-2+i)(-i) - (x-1)(i)(-i) - (-2-i)(-2+i)(x-2) - (-1+2i)(x+3)(-1-2i) = (x^3 - 7x + 6) + (-5) + (-5) - (x-1) - (5x-10) - (5x+15) = x^3 - 18x - 8$. Testing the divisors of (-8), the constant term, shows that -4 is a root of $p_{\mathbf{A}}(x)$. Then, dividing $p_{\mathbf{A}}(x)$ by $(x+4)$ yields $p_{\mathbf{A}}(x) = (x+4)(x^2 - 4x - 2)$. Finally, using the quadratic formula on $x^2 - 4x - 2$ gives the roots $2 + \sqrt{6}$ and $2 - \sqrt{6}$. Hence, the eigenvalues of $\mathbf{A}$ are -4, $2+\sqrt{6}$, and $2-\sqrt{6}$, which are all real. (Using a computer algebra system makes these computations much easier.)

(15) (a) False. Instead, every Hermitian matrix is unitarily diagonalizable, by Corollary 7.9. But, for example, the Hermitian matrix given in Exercise 13 of Section 7.4 is not unitary, since its rows do not form an orthonormal basis for $\mathbb{C}^3$.

(b) True. Since the dot product of two vectors in $\mathbb{R}^n$ is the same as the complex dot product of those vectors (because a real number is its own complex conjugate), then any orthonormal set of n vectors in $\mathbb{R}^n$ is also an orthonormal set of n vectors in $\mathbb{C}^n$.

(c) True. The definition of unitarily diagonalizable is that $\mathbf{A}$ is unitarily diagonalizable if and only if there is a unitary matrix $\mathbf{P}$ such that $\mathbf{P}^{-1}\mathbf{AP}$ is diagonal. But, a matrix $\mathbf{U}$ is unitary if and only if $\mathbf{U}^{-1} = \mathbf{U}^*$. Hence, $\mathbf{A}$ is unitarily diagonalizable if and only if there is a unitary matrix $\mathbf{P}$ such that $\mathbf{P}^*\mathbf{AP}$ is diagonal. Now notice that $\mathbf{P}$ is unitary if and only if $\overline{\mathbf{P}}$ is unitary (by Exercise 6(a)) if and only if $\left(\overline{\mathbf{P}}\right)^T$ is unitary (by Exercise 7(a)) if and only if $\mathbf{P}^*$ is unitary. Therefore, letting $\mathbf{Q} = \mathbf{P}^*$ then shows that $\mathbf{A}$ is unitarily diagonalizable if and only if there is a unitary matrix $\mathbf{Q}$ such that $\mathbf{QAQ}^*$ is diagonal, which is what we needed to show.

(d) True. If the columns of $\mathbf{A}$ form an orthonormal basis for $\mathbb{C}^n$, then part (2) of Theorem 7.6 implies that $\mathbf{A}$ is unitary. Hence, part (1) of Theorem 7.6 shows that the rows of $\mathbf{A}$ form an orthonormal basis for $\mathbb{C}^n$.

(e) False. In general, $\mathbf{A}^*$, not $\mathbf{A}^T$, is the matrix for the adjoint of L with respect to the standard basis. This is explained in Section 7.4 of the textbook, just after Corollary 7.9.

Section 7.5

(1) (b) $\langle \mathbf{x}, \mathbf{y} \rangle = (\mathbf{A}\mathbf{x}) \cdot (\mathbf{A}\mathbf{y}) =$

$$\left(\begin{bmatrix} 5 & 4 & 2 \\ -2 & 3 & 1 \\ 1 & -1 & 0 \end{bmatrix} \begin{bmatrix} 3 \\ -2 \\ 4 \end{bmatrix} \right) \cdot \left(\begin{bmatrix} 5 & 4 & 2 \\ -2 & 3 & 1 \\ 1 & -1 & 0 \end{bmatrix} \begin{bmatrix} -2 \\ 1 \\ -1 \end{bmatrix} \right) = [15, -8, 5] \cdot [-8, 6, -3] = -183.$$

Also, $\|\mathbf{x}\| = \sqrt{\langle \mathbf{x}, \mathbf{x} \rangle}$. But $\langle \mathbf{x}, \mathbf{x} \rangle = (\mathbf{A}\mathbf{x}) \cdot (\mathbf{A}\mathbf{x}) =$

$$\left(\begin{bmatrix} 5 & 4 & 2 \\ -2 & 3 & 1 \\ 1 & -1 & 0 \end{bmatrix} \begin{bmatrix} 3 \\ -2 \\ 4 \end{bmatrix} \right) \cdot \left(\begin{bmatrix} 5 & 4 & 2 \\ -2 & 3 & 1 \\ 1 & -1 & 0 \end{bmatrix} \begin{bmatrix} 3 \\ -2 \\ 4 \end{bmatrix} \right) = [15, -8, 5] \cdot [15, -8, 5] = 314.$$

Hence, $\|\mathbf{x}\| = \sqrt{314}$.

(3) (b) $\langle \mathbf{f}, \mathbf{g} \rangle = \int_0^\pi e^t \sin t \, dt = (-e^t \cos t)\Big|_0^\pi - \int_0^\pi (-e^t \cos t) \, dt$ (using integration by parts)

$= (-e^t \cos t)\Big|_0^\pi + (e^t \sin t)\Big|_0^\pi - \int_0^\pi e^t \sin t \, dt$ (using integration by parts again)

$= (-e^t \cos t)\Big|_0^\pi + (e^t \sin t)\Big|_0^\pi - \langle \mathbf{f}, \mathbf{g} \rangle$. Adding $\langle \mathbf{f}, \mathbf{g} \rangle$ to both ends of the equation produces

$2 \langle \mathbf{f}, \mathbf{g} \rangle = (-e^t \cos t)\Big|_0^\pi + (e^t \sin t)\Big|_0^\pi = \Big((-e^\pi \cos \pi) - (-e^0 \cos 0)\Big) + \Big((e^\pi \sin \pi) - (e^0 \sin 0)\Big)$

$= e^\pi + 1$. Dividing by 2 yields $\langle \mathbf{f}, \mathbf{g} \rangle = \frac{1}{2}(e^\pi + 1)$.

Next, $\|\mathbf{f}\| = \sqrt{\langle \mathbf{f}, \mathbf{f} \rangle}$. But, $\langle \mathbf{f}, \mathbf{f} \rangle = \int_0^\pi e^t e^t \, dt = \int_0^\pi e^{2t} \, dt = \frac{1}{2} e^{2t}\Big|_0^\pi = \frac{1}{2}\left(e^{2\pi} - 1\right)$. Hence, $\|\mathbf{f}\| = \sqrt{\frac{1}{2}(e^{2\pi} - 1)}$.

(6) $\|k\mathbf{x}\| = \sqrt{< k\mathbf{x}, k\mathbf{x} >} = \sqrt{k < \mathbf{x}, k\mathbf{x} >}$ (by property (5) of an inner product space) $= \sqrt{k\bar{k} < \mathbf{x}, \mathbf{x} >}$ (by part (3) of Theorem 7.11) $= \sqrt{|k|^2 < \mathbf{x}, \mathbf{x} >} = |k|\sqrt{< \mathbf{x}, \mathbf{x} >} = |k| \, \|\mathbf{x}\|$.

(9) (a) The distance between $\mathbf{f}$ and $\mathbf{g}$ is defined to be $\|\mathbf{f} - \mathbf{g}\| = \sqrt{\langle (\mathbf{f} - \mathbf{g}), (\mathbf{f} - \mathbf{g}) \rangle}$. Now,

$\langle (\mathbf{f} - \mathbf{g}), (\mathbf{f} - \mathbf{g}) \rangle = \int_0^\pi (t - \sin t)(t - \sin t) \, dt = \int_0^\pi \left(t^2 - 2t \sin t + \sin^2 t\right) \, dt$

$= \int_0^\pi t^2 \, dt - 2 \int_0^\pi t \sin t \, dt + \int_0^\pi \sin^2 t \, dt$. We compute each integral individually.

First, $\int_0^\pi t^2 \, dt = \frac{1}{3} t^3\Big|_0^\pi = \frac{\pi^3}{3}$. Next, using integration by parts,

$\int_0^\pi t \sin t \, dt = (-t \cos t)\Big|_0^\pi - \int_0^\pi (-\cos t) \, dt = (-t \cos t)\Big|_0^\pi + \sin t\Big|_0^\pi = (\pi - 0) - (0 - 0) = \pi$. Finally,

using the identity $\sin^2 t = \frac{1}{2}(1 - \cos 2t)$, $\int_0^\pi \sin^2 t \, dt = \int_0^\pi \frac{1}{2}(1 - \cos 2t) \, dt = \frac{1}{2}(t - \frac{1}{2} \sin 2t)\Big|_0^\pi$

$= \left(\frac{1}{2}\pi - 0\right) - (0 - 0) = \frac{1}{2}\pi$. Combining these produces $\langle (\mathbf{f} - \mathbf{g}), (\mathbf{f} - \mathbf{g}) \rangle = \frac{\pi^3}{3} - 2\pi + \frac{1}{2}\pi = \frac{\pi^3}{3} - \frac{3\pi}{2}$.

Hence, $\|\mathbf{f} - \mathbf{g}\| = \sqrt{\frac{\pi^3}{3} - \frac{3\pi}{2}}$.

(10) (b) The angle θ between $\mathbf{x}$ and $\mathbf{y}$ is defined to be $\arccos\left(\frac{\langle \mathbf{x} \cdot \mathbf{y} \rangle}{\|\mathbf{x}\| \, \|\mathbf{y}\|}\right)$. First we compute $\langle \mathbf{x}, \mathbf{y} \rangle$. Now

$$\langle \mathbf{x}, \mathbf{y} \rangle = (\mathbf{A}\mathbf{x}) \cdot (\mathbf{A}\mathbf{y}) = \left(\begin{bmatrix} -2 & 0 & 1 \\ 1 & -1 & 2 \\ 3 & -1 & -1 \end{bmatrix} \begin{bmatrix} 2 \\ -1 \\ 3 \end{bmatrix} \right) \cdot \left(\begin{bmatrix} -2 & 0 & 1 \\ 1 & -1 & 2 \\ 3 & -1 & -1 \end{bmatrix} \begin{bmatrix} 5 \\ -2 \\ 2 \end{bmatrix} \right)$$

$= [-1,9,4] \cdot [-8,11,15] = 167$. Next, $\|\mathbf{x}\|^2 = \langle \mathbf{x},\mathbf{x}\rangle = (\mathbf{Ax})\cdot(\mathbf{Ax}) =$

$$\left(\begin{bmatrix} -2 & 0 & 1 \\ 1 & -1 & 2 \\ 3 & -1 & -1 \end{bmatrix}\begin{bmatrix} 2 \\ -1 \\ 3 \end{bmatrix}\right) \cdot \left(\begin{bmatrix} -2 & 0 & 1 \\ 1 & -1 & 2 \\ 3 & -1 & -1 \end{bmatrix}\begin{bmatrix} 2 \\ -1 \\ 3 \end{bmatrix}\right)$$

$= [-1,9,4] \cdot [-1,9,4] = 98$. Similarly, $\|\mathbf{y}\|^2 = \langle \mathbf{y},\mathbf{y}\rangle = (\mathbf{Ay})\cdot(\mathbf{Ay}) =$

$$\left(\begin{bmatrix} -2 & 0 & 1 \\ 1 & -1 & 2 \\ 3 & -1 & -1 \end{bmatrix}\begin{bmatrix} 5 \\ -2 \\ 2 \end{bmatrix}\right) \cdot \left(\begin{bmatrix} -2 & 0 & 1 \\ 1 & -1 & 2 \\ 3 & -1 & -1 \end{bmatrix}\begin{bmatrix} 5 \\ -2 \\ 2 \end{bmatrix}\right)$$

$= [-8,11,15] \cdot [-8,11,15] = 410$. Therefore, $\|\mathbf{x}\| = \sqrt{98}$ and $\|\mathbf{y}\| = \sqrt{410}$, and so $\theta = \arccos\left(\frac{167}{\sqrt{98}\sqrt{410}}\right) \approx \arccos(0.83313) \approx 0.586$ radians, or, $33.6°$.

(14) To prove that a set of vectors $\{\mathbf{v}_1,\ldots,\mathbf{v}_n\}$ is orthogonal, we must show that $\langle \mathbf{v}_i,\mathbf{v}_j\rangle = 0$ for $i \neq j$.

(a) Now $\langle t^2, t+1\rangle = (1)(0) + (0)(1) + (0)(1) = 0$, $\langle t^2, t-1\rangle = (1)(0) + (0)(1) + (0)(-1) = 0$, and $\langle t+1, t-1\rangle = (0)(0) + (1)(1) + (1)(-1) = 0$. Hence, the given set is orthogonal.

(c) Now $\langle [5,-2],[3,4]\rangle = (5)(3) - (5)(4) - (-2)(3) + (2)(-2)(4) = -15 \neq 0$. Therefore, the given set is not orthogonal.

(19) We begin with the basis $\{\mathbf{w}_1,\mathbf{w}_2,\mathbf{w}_3\}$ for $\mathcal{P}_2$, with $\mathbf{w}_1 = t^2 - t + 1$, $\mathbf{w}_2 = 1$, and $\mathbf{w}_3 = t$, and apply the Gram-Schmidt Process. (Note: The Enlarging Method is not needed here since it can be seen by inspection that $\{\mathbf{w}_1,\mathbf{w}_2,\mathbf{w}_3\} = \{t^2-t+1, 1, t\}$ is linearly independent, and since $\dim(\mathcal{P}_2) = 3$, $\{\mathbf{w}_1,\mathbf{w}_2,\mathbf{w}_3\}$ is a basis for $\mathcal{P}_2$.) First, let $\mathbf{v}_1 = \mathbf{w}_1 = t^2 - t + 1$. Next, $\mathbf{v}_2 = \mathbf{w}_2 - \frac{\langle \mathbf{w}_2,\mathbf{v}_1\rangle}{\langle \mathbf{v}_1,\mathbf{v}_1\rangle}\mathbf{v}_1$. But $\langle \mathbf{w}_2,\mathbf{v}_1\rangle = \int_{-1}^{1}(1)(t^2-t+1)\,dt = \left(\frac{1}{3}t^3 - \frac{1}{2}t^2 + t\right)\Big|_{-1}^{1} = \frac{8}{3}$, and $\langle \mathbf{v}_1,\mathbf{v}_1\rangle = \int_{-1}^{1}(t^2-t+1)(t^2-t+1)\,dt = \int_{-1}^{1}(t^4-2t^3+3t^2-2t+1)\,dt = \left(\frac{1}{5}t^5 - \frac{1}{2}t^4 + t^3 - t^2 + t\right)\Big|_{-1}^{1} = \frac{22}{5}$. Hence, $\mathbf{v}_2 = (1) - \frac{(8/3)}{(22/5)}(t^2-t+1)$.
Multiplying by 33 to eliminate fractions yields $\mathbf{v}_2 = -20t^2 + 20t + 13$. Finally,
$\mathbf{v}_3 = \mathbf{w}_3 - \frac{\langle \mathbf{w}_3,\mathbf{v}_1\rangle}{\langle \mathbf{v}_1,\mathbf{v}_1\rangle}\mathbf{v}_1 - \frac{\langle \mathbf{w}_3,\mathbf{v}_2\rangle}{\langle \mathbf{v}_2,\mathbf{v}_2\rangle}\mathbf{v}_2$. However, $\langle \mathbf{w}_3,\mathbf{v}_1\rangle = \int_{-1}^{1}(t)(t^2-t+1)\,dt = \int_{-1}^{1}(t^3-t^2+t)\,dt =$

$\left(\frac{1}{4}t^4 - \frac{1}{3}t^3 + \frac{1}{2}t^2\right)\Big|_{-1}^{1} = -\frac{2}{3}$, $\langle \mathbf{w}_3,\mathbf{v}_2\rangle = \int_{-1}^{1}(t)(-20t^2+20t+13)\,dt = \int_{-1}^{1}(-20t^3+20t^2+13t)\,dt =$

$\left(-5t^4 + \frac{20}{3}t^3 + \frac{13}{2}t^2\right)\Big|_{-1}^{1} = \frac{40}{3}$, and $\langle \mathbf{v}_2,\mathbf{v}_2\rangle = \int_{-1}^{1}(-20t^2+20t+13)(-20t^2+20t+13)\,dt$

$= \int_{-1}^{1}(400t^4 - 800t^3 - 120t^2 + 520t + 169)\,dt = \left(80t^5 - 200t^4 - 40t^3 + 260t^2 + 169t\right)\Big|_{-1}^{1} = 418$. Hence, $\mathbf{v}_3 = (t) - \frac{(-2/3)}{(22/5)}(t^2-t+1) - \frac{(40/3)}{(418)}(-20t^2+20t+13)$. Multiplying by 19 to eliminate fractions gives $\mathbf{v}_3 = 15t^2 + 4t - 5$. Thus, we get the orthogonal basis $\{\mathbf{v}_1,\mathbf{v}_2,\mathbf{v}_3\}$, with $\mathbf{v}_1 = t^2 - t + 1$, $\mathbf{v}_2 = -20t^2 + 20t + 13$, and $\mathbf{v}_3 = 15t^2 + 4t - 5$.

(22) (b) First, we prove prove part (4) of Theorem 7.18 as follows, using a proof similar to the proof of Theorem 6.11:

Let $\{\mathbf{v}_1,\ldots,\mathbf{v}_n\}$ be an orthogonal basis for $\mathcal{V}$, with $\mathcal{W} = \text{span}(\{\mathbf{v}_1,\ldots,\mathbf{v}_k\})$. Let $\mathcal{X} = \text{span}(\{\mathbf{v}_{k+1},\ldots,\mathbf{v}_n\})$. Since $\{\mathbf{v}_{k+1},\ldots,\mathbf{v}_n\}$ is linearly independent, it is a basis for $\mathcal{W}^\perp$ if $\mathcal{X} = \mathcal{W}^\perp$. We will show that $\mathcal{X} \subseteq \mathcal{W}^\perp$ and $\mathcal{W}^\perp \subseteq \mathcal{X}$.

To show $\mathcal{X} \subseteq \mathcal{W}^\perp$, we must prove that any vector $\mathbf{x}$ of the form $d_{k+1}\mathbf{v}_{k+1} + \cdots + d_n\mathbf{v}_n$ (for some scalars $d_{k+1}, \ldots, d_n$) is orthogonal to every vector $\mathbf{w} \in \mathcal{W}$. Now if $\mathbf{w} \in \mathcal{W}$, then $\mathbf{w} = c_1\mathbf{v}_1 + \cdots + c_k\mathbf{v}_k$, for some scalars $c_1, \ldots, c_k$. Hence,

$$\mathbf{x} \cdot \mathbf{w} = (d_{k+1}\mathbf{v}_{k+1} + \cdots + d_n\mathbf{v}_n) \cdot (c_1\mathbf{v}_1 + \cdots + c_k\mathbf{v}_k),$$

which equals zero when expanded because each vector in $\{\mathbf{v}_{k+1}, \ldots, \mathbf{v}_n\}$ is orthogonal to every vector in $\{\mathbf{v}_1, \ldots, \mathbf{v}_k\}$. Hence, $\mathbf{x} \in \mathcal{W}^\perp$, and so $\mathcal{X} \subseteq \mathcal{W}^\perp$.
To show $\mathcal{W}^\perp \subseteq \mathcal{X}$, we must show that any vector $\mathbf{x}$ in $\mathcal{W}^\perp$ is also in span($\{\mathbf{v}_{k+1}, \ldots, \mathbf{v}_n\}$). Let $\mathbf{x} \in \mathcal{W}^\perp$. Since $\{\mathbf{v}_1, \ldots, \mathbf{v}_n\}$ is an orthogonal basis for $\mathcal{V}$, Theorem 7.15 tells us that

$$\mathbf{x} = \frac{(\mathbf{x}\cdot\mathbf{v}_1)}{(\mathbf{v}_1\cdot\mathbf{v}_1)}\mathbf{v}_1 + \cdots + \frac{(\mathbf{x}\cdot\mathbf{v}_k)}{(\mathbf{v}_k\cdot\mathbf{v}_k)}\mathbf{v}_k + \frac{(\mathbf{x}\cdot\mathbf{v}_{k+1})}{(\mathbf{v}_{k+1}\cdot\mathbf{v}_{k+1})}\mathbf{v}_{k+1} + \cdots + \frac{(\mathbf{x}\cdot\mathbf{v}_n)}{(\mathbf{v}_n\cdot\mathbf{v}_n)}\mathbf{v}_n.$$

However, since each of $\mathbf{v}_1, \ldots, \mathbf{v}_k$ is in $\mathcal{W}$, we know that $\mathbf{x}\cdot\mathbf{v}_1 = \cdots = \mathbf{x}\cdot\mathbf{v}_k = 0$. Hence,

$$\mathbf{x} = \frac{(\mathbf{x}\cdot\mathbf{v}_{k+1})}{(\mathbf{v}_{k+1}\cdot\mathbf{v}_{k+1})}\mathbf{v}_{k+1} + \cdots + \frac{(\mathbf{x}\cdot\mathbf{v}_n)}{(\mathbf{v}_n\cdot\mathbf{v}_n)}\mathbf{v}_n,$$

and so $\mathbf{x} \in$ span($\{\mathbf{v}_{k+1}, \ldots, \mathbf{v}_n\}$). Thus, $\mathcal{W}^\perp \subseteq \mathcal{X}$.
Next, we prove part (5) of Theorem 7.18 as follows:
Let $\mathcal{W}$ be a subspace of $\mathcal{V}$ of dimension k. By Theorem 7.16, $\mathcal{W}$ has an orthogonal basis $\{\mathbf{v}_1, \ldots, \mathbf{v}_k\}$. Expand this basis to an orthogonal basis for all of $\mathcal{V}$. (That is, first expand to any basis for $\mathcal{V}$ by Theorem 4.19, then use the Gram-Schmidt Process. Since the first k vectors are already orthogonal, this expands $\{\mathbf{v}_1, \ldots, \mathbf{v}_k\}$ to an orthogonal basis $\{\mathbf{v}_1, \ldots, \mathbf{v}_n\}$ for $\mathcal{V}$.) Then, by part (4) of Theorem 7.18, $\{\mathbf{v}_{k+1}, \ldots, \mathbf{v}_n\}$ is a basis for $\mathcal{W}^\perp$, and so $\dim(\mathcal{W}^\perp) = n - k$. Hence $\dim(\mathcal{W}) + \dim(\mathcal{W}^\perp) = n$.

(d) By part (3) of Theorem 7.18, $\mathcal{W} \subseteq (\mathcal{W}^\perp)^\perp$. Next, by part (5) of Theorem 7.18, $\dim(\mathcal{W}) = n - \dim(\mathcal{W}^\perp) = n - (n - \dim((\mathcal{W}^\perp)^\perp)) = \dim((\mathcal{W}^\perp)^\perp)$. Thus, by Theorem 4.17, or its complex analog, $\mathcal{W} = (\mathcal{W}^\perp)^\perp$.

(23) We begin with the basis $B = \{t^3 + t^2, t - 1, 1, t^2\}$ for $\mathcal{P}_3$ containing the polynomials $t^3 + t^2$ and $t - 1$. (Notice that the Enlarging Method is not needed here since it can be seen by inspection that $B = \{t^3 + t^2, t - 1, 1, t^2\}$ is linearly independent (since each element is a polynomial of a different degree), and since $\dim(\mathcal{P}_3) = 4$, B is a basis for $\mathcal{P}_3$.) We use the Gram-Schmidt Process on B. Let $\mathbf{v}_1 = t^3 + t^2$. Next, $\mathbf{v}_2 = (t-1) - \frac{\langle t-1, t^3+t^2\rangle}{\langle t^3+t^2, t^3+t^2\rangle}(t^3+t^2) = (t-1) - \frac{0}{2}(t^3+t^2) = t - 1$. Also, $\mathbf{v}_3 = (1) - \frac{\langle 1, t^3+t^2\rangle}{\langle t^3+t^2, t^3+t^2\rangle}(t^3+t^2) - \frac{\langle 1, t-1\rangle}{\langle t-1, t-1\rangle}(t-1) = (1) - \frac{0}{2}(t^3+t^2) - \frac{(-1)}{2}(t-1) = \frac{1}{2}t + \frac{1}{2}$. Multiplying by 2 to eliminate fractions yields $\mathbf{v}_3 = t + 1$. Finally,
$\mathbf{v}_4 = (t^2) - \frac{\langle t^2, t^3+t^2\rangle}{\langle t^3+t^2, t^3+t^2\rangle}(t^3+t^2) - \frac{\langle t^2, t-1\rangle}{\langle t-1, t-1\rangle}(t-1) - \frac{\langle t^2, t+1\rangle}{\langle t+1, t+1\rangle}(t+1)$
$= (t^2) - \frac{1}{2}(t^3+t^2) - \frac{0}{2}(t-1) - \frac{0}{2}(t+1) = -\frac{1}{2}t^3 + \frac{1}{2}t^2$. Multiplying by -2 to simplify the form of the result produces $\mathbf{v}_4 = t^3 - t^2$. Hence, by part (4) of Theorem 7.18, $\mathcal{W}^\perp = \text{span}(\{t+1, t^3 - t^2\})$.

(25) As in the hint, let $\{\mathbf{v}_1, \ldots, \mathbf{v}_k\}$ be an orthonormal basis for $\mathcal{W}$. Now if $\mathbf{v} \in \mathcal{V}$, let $\mathbf{w}_1 = \langle \mathbf{v}, \mathbf{v}_1\rangle \mathbf{v}_1 + \cdots + \langle \mathbf{v}, \mathbf{v}_k\rangle \mathbf{v}_k$ and $\mathbf{w}_2 = \mathbf{v} - \mathbf{w}_1$. Then, $\mathbf{w}_1 \in \mathcal{W}$ because $\mathbf{w}_1$ is a linear combination of basis vectors for $\mathcal{W}$. We claim that $\mathbf{w}_2 \in \mathcal{W}^\perp$. To see this, let $\mathbf{u} \in \mathcal{W}$. Then $\mathbf{u} = a_1\mathbf{v}_1 + \cdots + a_k\mathbf{v}_k$ for some $a_1, \ldots, a_k$. Then
$\langle \mathbf{u}, \mathbf{w}_2\rangle = \langle \mathbf{u}, \mathbf{v} - \mathbf{w}_1\rangle = \langle a_1\mathbf{v}_1 + \cdots + a_k\mathbf{v}_k,\ \mathbf{v} - (\langle \mathbf{v}, \mathbf{v}_1\rangle \mathbf{v}_1 + \cdots + \langle \mathbf{v}, \mathbf{v}_k\rangle \mathbf{v}_k)\rangle =$

$\langle a_1\mathbf{v}_1 + \cdots + a_k\mathbf{v}_k,\ \mathbf{v}\rangle - \langle a_1\mathbf{v}_1 + \cdots + a_k\mathbf{v}_k,\ \langle\mathbf{v},\mathbf{v}_1\rangle\mathbf{v}_1 + \cdots + \langle\mathbf{v},\mathbf{v}_k\rangle\mathbf{v}_k\rangle$
$= \sum_{i=1}^k a_i\langle\mathbf{v}_i,\mathbf{v}\rangle - \sum_{i=1}^k\sum_{j=1}^k a_i\overline{\langle\mathbf{v},\mathbf{v}_j\rangle}\langle\mathbf{v}_i,\mathbf{v}_j\rangle$. But $\langle\mathbf{v}_i,\mathbf{v}_j\rangle = 0$ when $i \neq j$ and $\langle\mathbf{v}_i,\mathbf{v}_j\rangle = 1$ when $i = j$, since $\{\mathbf{v}_1,\ldots,\mathbf{v}_k\}$ is an orthonormal set. Hence, $\langle\mathbf{u},\mathbf{w}_2\rangle = \sum_{i=1}^k a_i\langle\mathbf{v}_i,\mathbf{v}\rangle - \sum_{i=1}^k a_i\overline{\langle\mathbf{v},\mathbf{v}_i\rangle} = \sum_{i=1}^k a_i\langle\mathbf{v}_i,\mathbf{v}\rangle - \sum_{i=1}^k a_i\langle\mathbf{v}_i,\mathbf{v}\rangle = 0$. Since this is true for every $\mathbf{u}\in\mathcal{W}$, we conclude that $\mathbf{w}_2\in\mathcal{W}^\perp$. Finally, we want to show uniqueness of decomposition. Suppose that $\mathbf{v} = \mathbf{w}_1 + \mathbf{w}_2$ and $\mathbf{v} = \mathbf{w}_1' + \mathbf{w}_2'$, where $\mathbf{w}_1,\mathbf{w}_1' \in \mathcal{W}$ and $\mathbf{w}_2,\ \mathbf{w}_2' \in \mathcal{W}^\perp$. We want to show that $\mathbf{w}_1 = \mathbf{w}_1'$ and $\mathbf{w}_2 = \mathbf{w}_2'$. Now, $\mathbf{w}_1 - \mathbf{w}_1' = \mathbf{w}_2' - \mathbf{w}_2$. Also, $\mathbf{w}_1 - \mathbf{w}_1' \in \mathcal{W}$, but $\mathbf{w}_2' - \mathbf{w}_2 \in \mathcal{W}^\perp$. Thus, $\mathbf{w}_1 - \mathbf{w}_1' = \mathbf{w}_2' - \mathbf{w}_2 \in \mathcal{W}\cap\mathcal{W}^\perp$. By part (2) of Theorem 7.18, $\mathbf{w}_1 - \mathbf{w}_1' = \mathbf{w}_2' - \mathbf{w}_2 = \mathbf{0}$. Hence, $\mathbf{w}_1 = \mathbf{w}_1'$ and $\mathbf{w}_2 = \mathbf{w}_2'$.

(26) From Example 8 in Section 7.5 of the textbook we know that $\left\{\frac{1}{\sqrt{\pi}}\cos t, \frac{1}{\sqrt{\pi}}\sin t\right\}$ is an orthonormal basis for $\mathcal{W} = \text{span}(\{\cos t, \sin t\})$. (We quickly verify this: $\langle\cos t,\sin t\rangle = \int_{-\pi}^{\pi}(\cos t\sin t)\,dt = \left(\frac{1}{2}\sin^2 t\right)\Big|_{-\pi}^{\pi} = (0-0) = 0$. Hence, $\{\cos t,\sin t\}$ is an orthogonal basis for $\mathcal{W}$. We normalize these vectors to get an orthonormal basis. Now

$\langle\cos t,\cos t\rangle = \int_{-\pi}^{\pi}\cos^2 t\,dt = \int_{-\pi}^{\pi}\frac{1}{2}(1+\cos 2t)\,dt = \left(\frac{1}{2}t + \sin 2t\right)\Big|_{-\pi}^{\pi} = \left(\frac{\pi}{2}+0\right) - \left(-\frac{\pi}{2}+0\right) = \pi$, and

$\langle\sin t,\sin t\rangle = \int_{-\pi}^{\pi}\sin^2 t\,dt = \int_{-\pi}^{\pi}\frac{1}{2}(1-\cos 2t)\,dt = \left(\frac{1}{2}t - \sin 2t\right)\Big|_{-\pi}^{\pi} = \left(\frac{\pi}{2}-0\right) - \left(-\frac{\pi}{2}-0\right) = \pi$.

Therefore, $\|\cos t\| = \sqrt{\pi}$ and $\|\sin t\| = \sqrt{\pi}$, giving us the orthonormal basis $\left\{\frac{1}{\sqrt{\pi}}\cos t, \frac{1}{\sqrt{\pi}}\sin t\right\}$ for $\mathcal{W}$.)

Using $\mathbf{f} = \frac{1}{k}e^t$, $\mathbf{w}_1 = \mathbf{proj}_{\mathcal{W}}\mathbf{f} = \left\langle\mathbf{f}, \frac{1}{\sqrt{\pi}}\cos t\right\rangle\left(\frac{1}{\sqrt{\pi}}\cos t\right) + \left\langle\mathbf{f}, \frac{1}{\sqrt{\pi}}\sin t\right\rangle\left(\frac{1}{\sqrt{\pi}}\sin t\right)$. So we have two more inner products to compute. First, $\left\langle\mathbf{f}, \frac{1}{\sqrt{\pi}}\cos t\right\rangle = \frac{1}{k\sqrt{\pi}}\int_{-\pi}^{\pi}e^t\cos t\,dt =$

$\left(\frac{1}{k\sqrt{\pi}}e^t\sin t\right)\Big|_{-\pi}^{\pi} - \frac{1}{k\sqrt{\pi}}\int_{-\pi}^{\pi}(e^t\sin t)\,dt$, using integration by parts,

$= \left(\frac{1}{k\sqrt{\pi}}e^t\sin t\right)\Big|_{-\pi}^{\pi} - \left(\frac{-1}{k\sqrt{\pi}}e^t\cos t\right)\Big|_{-\pi}^{\pi} + \frac{1}{k\sqrt{\pi}}\int_{-\pi}^{\pi}(-e^t\cos t)\,dt$, using integration by parts again.

Adding $\left\langle\mathbf{f}, \frac{1}{\sqrt{\pi}}\cos t\right\rangle = \frac{1}{k\sqrt{\pi}}\int_{-\pi}^{\pi}e^t\cos t\,dt$ to both ends of this equation produces $2\left\langle\mathbf{f}, \frac{1}{\sqrt{\pi}}\cos t\right\rangle = \left(\frac{1}{k\sqrt{\pi}}e^t\sin t\right)\Big|_{-\pi}^{\pi} - \left(\frac{-1}{k\sqrt{\pi}}e^t\cos t\right)\Big|_{-\pi}^{\pi} = (0-0) - \left(\frac{e^\pi}{k\sqrt{\pi}} - \frac{e^{-\pi}}{k\sqrt{\pi}}\right) = -\frac{1}{\sqrt{\pi}}$, since $k = e^\pi - e^{-\pi}$. Dividing by 2 yields $\left\langle\mathbf{f}, \frac{1}{\sqrt{\pi}}\cos t\right\rangle = -\frac{1}{2\sqrt{\pi}}$. Similarly, $\left\langle\mathbf{f}, \frac{1}{\sqrt{\pi}}\sin t\right\rangle = \frac{1}{k\sqrt{\pi}}\int_{-\pi}^{\pi}e^t\sin t\,dt$

$= \left(\frac{-1}{k\sqrt{\pi}}e^t\cos t\right)\Big|_{-\pi}^{\pi} - \frac{1}{k\sqrt{\pi}}\int_{-\pi}^{\pi}(-e^t\cos t)\,dt$, using integration by parts,

$= \left(\frac{-1}{k\sqrt{\pi}}e^t\cos t\right)\Big|_{-\pi}^{\pi} + \left(\frac{1}{k\sqrt{\pi}}e^t\sin t\right)\Big|_{-\pi}^{\pi} - \frac{1}{k\sqrt{\pi}}\int_{-\pi}^{\pi}(e^t\sin t)\,dt$, using integration by parts again.

Adding $\left\langle\mathbf{f}, \frac{1}{\sqrt{\pi}}\sin t\right\rangle = \frac{1}{k\sqrt{\pi}}\int_{-\pi}^{\pi}e^t\sin t\,dt$ to both ends of this equation produces

$2\left\langle\mathbf{f}, \frac{1}{\sqrt{\pi}}\sin t\right\rangle = \left(\frac{-1}{k\sqrt{\pi}}e^t\cos t\right)\Big|_{-\pi}^{\pi} + \left(\frac{1}{k\sqrt{\pi}}e^t\sin t\right)\Big|_{-\pi}^{\pi} = \left(\frac{e^\pi}{k\sqrt{\pi}} - \frac{e^{-\pi}}{k\sqrt{\pi}}\right) + (0-0) = \frac{1}{\sqrt{\pi}}$.

Dividing by 2 yields $\left\langle \mathbf{f}, \frac{1}{\sqrt{\pi}}\sin t\right\rangle = \frac{1}{2\sqrt{\pi}}$. Therefore, $\mathbf{w}_1 = \mathbf{proj}_{\mathcal{W}}\mathbf{f}$
$= \left(-\frac{1}{2\sqrt{\pi}}\right)\left(\frac{1}{\sqrt{\pi}}\cos t\right) + \left(\frac{1}{2\sqrt{\pi}}\right)\left(\frac{1}{\sqrt{\pi}}\sin t\right) = \frac{1}{2\pi}(\sin t - \cos t)$.
Also, $\mathbf{w}_2 = \mathbf{f} - \mathbf{w}_1 = \frac{1}{k}e^t - \frac{1}{2\pi}\sin t + \frac{1}{2\pi}\cos t$. Finally, we must verify that $\langle \mathbf{w}_1, \mathbf{w}_2\rangle = 0$.
Now $\langle \mathbf{w}_1, \mathbf{w}_2\rangle = \int_{-\pi}^{\pi}\left(\frac{1}{2\pi}(\sin t - \cos t)\right)\left(\frac{1}{k}e^t - \frac{1}{2\pi}\sin t + \frac{1}{2\pi}\cos t\right)\,dt$
$= \frac{1}{2k\pi}\int_{-\pi}^{\pi}(e^t\sin t)\,dt - \frac{1}{4\pi^2}\int_{-\pi}^{\pi}\sin^2 t\,dt - \frac{1}{2k\pi}\int_{-\pi}^{\pi}(e^t\cos t)\,dt - \frac{1}{4\pi^2}\int_{-\pi}^{\pi}\cos^2 t\,dt + \frac{1}{2\pi^2}\int_{-\pi}^{\pi}\sin t\cos t\,dt$.
We consider each integral separately. First, $\frac{1}{2k\pi}\int_{-\pi}^{\pi}(e^t\sin t)\,dt = \frac{1}{2\sqrt{\pi}}\left\langle \mathbf{f}, \frac{1}{\sqrt{\pi}}\sin t\right\rangle = \frac{1}{2\sqrt{\pi}}\left(\frac{1}{2\sqrt{\pi}}\right)$ (from our earlier computations) $= \frac{1}{4\pi}$. Next, $\frac{1}{4\pi^2}\int_{-\pi}^{\pi}\sin^2 t\,dt = \frac{1}{4\pi^2}\langle\sin t, \sin t\rangle = \frac{1}{4\pi^2}(\pi)$ (from our earlier computations) $= \frac{1}{4\pi}$. Similarly $\frac{1}{2k\pi}\int_{-\pi}^{\pi}(e^t\cos t)\,dt = \frac{1}{2\sqrt{\pi}}\left\langle \mathbf{f}, \frac{1}{\sqrt{\pi}}\cos t\right\rangle = \frac{1}{2\sqrt{\pi}}\left(\frac{-1}{2\sqrt{\pi}}\right) = -\frac{1}{4\pi}$ and $\frac{1}{4\pi^2}\int_{-\pi}^{\pi}\cos^2 t\,dt = \frac{1}{4\pi^2}(\pi) = \frac{1}{4\pi}$. Finally, $\frac{1}{2\pi^2}\int_{-\pi}^{\pi}\sin t\cos t\,dt = \frac{1}{2\pi^2}\langle\sin t, \cos t\rangle = \frac{1}{2\pi^2}\langle\cos t, \sin t\rangle = 0$ (from our earlier computations). Hence, $\langle \mathbf{w}_1, \mathbf{w}_2\rangle = \frac{1}{4\pi} - \frac{1}{4\pi} - (-\frac{1}{4\pi}) - \frac{1}{4\pi} = 0$, completing the problem.

(29) (b) First, we establish that range$(L) = \mathcal{W}$. We start by showing that $\mathcal{W} \subseteq$ range(L). Let $\mathbf{v} \in \mathcal{W}$. Then we claim that $L(\mathbf{v}) = \mathbf{v}$. Note that $\mathbf{v} \in \mathcal{W}$ implies that the unique decomposition of $\mathbf{v}$ into $\mathbf{v} = \mathbf{w}_1 + \mathbf{w}_2$ with $\mathbf{w}_1 \in \mathcal{W}$ and $\mathbf{w}_2 \in \mathcal{W}^\perp$ is $\mathbf{w}_1 = \mathbf{v}$ and $\mathbf{w}_2 = \mathbf{0}$. (This is because for these vectors $\mathbf{w}_1$ and $\mathbf{w}_2$, we have $\mathbf{w}_1 + \mathbf{w}_2 = \mathbf{v}$, and because by Theorem 7.19 this decomposition is unique.) Hence, since $L(\mathbf{v}) = \mathbf{proj}_{\mathcal{W}}\mathbf{v} = \mathbf{w}_1$, we get $L(\mathbf{v}) = \mathbf{v}$. From this we can conclude that $\mathcal{W} \subseteq$ range(L), since every vector in $\mathcal{W}$ is the image of itself. Also, since the codomain of L is $\mathcal{W}$, range$(L) \subseteq \mathcal{W}$. Hence, range$(L) = \mathcal{W}$.
Next, we establish that $\ker(L) = \mathcal{W}^\perp$. First we show that $\mathcal{W}^\perp \subseteq \ker(L)$. Let $\mathbf{v} \in \mathcal{W}^\perp$. We claim that $L(\mathbf{v}) = \mathbf{0}$. Note that $\mathbf{v} \in \mathcal{W}^\perp$ implies that the unique decomposition of $\mathbf{v}$ into $\mathbf{v} = \mathbf{w}_1 + \mathbf{w}_2$ with $\mathbf{w}_1 \in \mathcal{W}$ and $\mathbf{w}_2 \in \mathcal{W}^\perp$ is $\mathbf{w}_1 = \mathbf{0}$ and $\mathbf{w}_2 = \mathbf{v}$. (This is because for these vectors $\mathbf{w}_1$ and $\mathbf{w}_2$, we have $\mathbf{w}_1 + \mathbf{w}_2 = \mathbf{v}$, and because by Theorem 7.19 this decomposition is unique.) Hence, since $L(\mathbf{v}) = \mathbf{proj}_{\mathcal{W}}\mathbf{v} = \mathbf{w}_1$, we get $L(\mathbf{v}) = \mathbf{0}$. From this we can conclude that $\mathcal{W}^\perp \subseteq \ker(L)$. We still need to prove that $\ker(L) \subseteq \mathcal{W}^\perp$. So, let $\mathbf{x} \in \ker(L)$. Then $\mathbf{x} = \mathbf{u}_1 + \mathbf{u}_2$ with $\mathbf{u}_1 \in \mathcal{W}$ and $\mathbf{u}_2 \in \mathcal{W}^\perp$. Also, $L(\mathbf{x}) = \mathbf{u}_1$. But, $\mathbf{x} \in \ker(L)$ implies $\mathbf{u}_1 = L(\mathbf{x}) = \mathbf{0}$. Therefore, $\mathbf{x} = \mathbf{u}_1 + \mathbf{u}_2 = \mathbf{0} + \mathbf{u}_2 = \mathbf{u}_2 \in \mathcal{W}^\perp$. This completes the proof.

(30) (a) False. The correct equation is $\|k\mathbf{x}\| = |k|\,\|\mathbf{x}\|$. For example, in $\mathbb{C}^2$ with the usual complex dot product, consider $\mathbf{x} = [1,0]$ and $k = i$. Then $\|k\mathbf{x}\| = \|i[1,0]\| = \|[i,0]\| = \sqrt{[i,0]\cdot[i,0]} = \sqrt{(i)(\overline{i}) + (0)(0)} = \sqrt{1} = 1$. Also, $\|\mathbf{x}\| = \sqrt{[1,0]\cdot[1,0]} = \sqrt{1} = 1$. Hence, $|k|\,\|\mathbf{x}\| = |i|(1) = (1)(1) = 1 = \|k\mathbf{x}\|$, but $\overline{k}\,\|\mathbf{x}\| = \overline{i}(1) = (-i)(1) = -i \neq \|k\mathbf{x}\|$.

(b) False. The distance between two vectors $\mathbf{x}$ and $\mathbf{y}$ is defined to be $\|\mathbf{x}-\mathbf{y}\| = \sqrt{\langle \mathbf{x}-\mathbf{y}, \mathbf{x}-\mathbf{y}\rangle}$, which is always real and positive (since $\mathbf{x}$ and $\mathbf{y}$ are distinct) by parts (1) and (2) of the definition of an inner product. Hence, the distance can not be pure imaginary.

(c) False. For a set S of vectors to be orthonormal, S must be an orthogonal set. This implies that S is linearly independent, but the converse is not true. There are many sets of linearly independent unit vectors that are not orthogonal sets. For example, $S = \left\{[1,0], \left[\frac{\sqrt{2}}{2}, \frac{\sqrt{2}}{2}\right]\right\}$ in $\mathbb{R}^2$ using the usual dot product is such a set.

(d) True. For example, $\mathbb{R}^2$ has the usual dot product as one inner product, and the inner product of Example 2 in Section 7.5 in the textbook as a different inner product.

(e) False. The uniqueness assertion is that, using a single, fixed inner product, and a single, fixed subspace $\mathcal{W}$, the decomposition of a particular vector described in the Projection Theorem is unique. However, different inner products could result in different decompositions. For example, consider the subspace $\mathcal{W} = \text{span}(\{[1,0]\})$ in $\mathbb{R}^2$. Let $\mathbf{v} = [0,1]$. Using the dot product as the inner product on $\mathbb{R}^2$, the decomposition of $\mathbf{v}$ is $\mathbf{v} = [0,0] + [0,1]$, with $[0,0] \in \mathcal{W}$ and $[0,1] \in \mathcal{W}^{\perp}$ (since $[0,1] \cdot [1,0] = 0$). However, consider the inner product from Example 2 in Section 7.5 in the textbook. It is easily verified that $\{[1,0]\}$ is an orthonormal basis for $\mathcal{W}$ with this inner product, and so, $\mathbf{proj}_{\mathcal{W}}\mathbf{v} = \langle [0,1],[1,0] \rangle [1,0] = (-1)[1,0] = [-1,0]$. Therefore, the decomposition of $\mathbf{v}$ using this inner product is $\mathbf{v} = [-1,0] + [1,1]$.

Chapter 8

Section 8.1

(1) In parts (a) through (d), use the fact the adjacency matrix for a graph is defined to be the matrix whose (i, j) entry is 1 if there is an edge between the vertices P_i and P_j and is 0 otherwise. In parts (e) through (h), use the fact the adjacency matrix for a digraph is defined to be the matrix whose (i, j) entry is 1 if there is an edge directed from the vertex P_i to the vertex P_j and is 0 otherwise. Inspection shows that the matrices in parts (a), (b), (c), (d), and (g) are symmetric, while those in parts (e), (f) and (h) are not.

(a) G_1: $\begin{bmatrix} 0 & 1 & 1 & 1 \\ 1 & 0 & 1 & 1 \\ 1 & 1 & 0 & 1 \\ 1 & 1 & 1 & 0 \end{bmatrix}$ (b) G_2 : $\begin{bmatrix} 1 & 1 & 0 & 0 & 1 \\ 1 & 0 & 0 & 1 & 0 \\ 0 & 0 & 1 & 0 & 1 \\ 0 & 1 & 0 & 0 & 1 \\ 1 & 0 & 1 & 1 & 1 \end{bmatrix}$ (c) G_3: $\begin{bmatrix} 0 & 0 & 0 \\ 0 & 0 & 0 \\ 0 & 0 & 0 \end{bmatrix}$

(d) G_4: $\begin{bmatrix} 0 & 1 & 0 & 1 & 0 & 0 \\ 1 & 0 & 1 & 1 & 0 & 0 \\ 0 & 1 & 0 & 0 & 1 & 1 \\ 1 & 1 & 0 & 0 & 1 & 0 \\ 0 & 0 & 1 & 1 & 0 & 1 \\ 0 & 0 & 1 & 0 & 1 & 0 \end{bmatrix}$ (e) D_1: $\begin{bmatrix} 0 & 1 & 0 & 0 \\ 0 & 0 & 1 & 0 \\ 1 & 0 & 0 & 1 \\ 1 & 1 & 0 & 0 \end{bmatrix}$ (f) D_2: $\begin{bmatrix} 0 & 1 & 1 & 0 \\ 0 & 1 & 1 & 1 \\ 0 & 1 & 0 & 1 \\ 0 & 0 & 0 & 1 \end{bmatrix}$

(g) D_3: $\begin{bmatrix} 0 & 1 & 1 & 0 & 0 \\ 1 & 0 & 0 & 0 & 1 \\ 1 & 0 & 0 & 1 & 0 \\ 0 & 0 & 1 & 0 & 1 \\ 0 & 1 & 0 & 1 & 0 \end{bmatrix}$ (h) D_4: $\begin{bmatrix} 0 & 1 & 0 & 0 & 0 & 0 & 0 & 0 \\ 0 & 0 & 1 & 0 & 0 & 0 & 0 & 0 \\ 0 & 0 & 0 & 1 & 0 & 0 & 0 & 0 \\ 0 & 0 & 0 & 0 & 1 & 0 & 0 & 0 \\ 0 & 0 & 0 & 0 & 0 & 1 & 0 & 0 \\ 0 & 0 & 0 & 0 & 0 & 0 & 1 & 0 \\ 0 & 0 & 0 & 0 & 0 & 0 & 0 & 1 \\ 1 & 0 & 0 & 0 & 0 & 0 & 0 & 0 \end{bmatrix}$

(2) For a matrix to be the adjacency matrix for a simple graph or digraph, it must be a square matrix, all of whose entries are either 0 or 1. In addition, the adjacency matrix for a *graph* must be symmetric.
A can not be the adjacency matrix for either a graph or digraph because it is not square.
B can not be the adjacency matrix for either a graph or digraph because its (1,1) entry equals 2, which is not 0 or 1.
C can not be the adjacency matrix for either a graph or digraph because its (1,1) entry equals 6, which is not 0 or 1.
D can not be the adjacency matrix for either a graph or digraph because it is not square.
E can not be the adjacency matrix for either a graph or digraph because its (1,4) entry equals 6, which is not 0 or 1.
F can be the adjacency matrix for either a graph or digraph (see Figure 5 on next page).
G can be the adjacency matrix for a digraph only, since **G** is not symmetric (see Figure 6 on next page).
H can be the adjacency matrix for a digraph only, since **H** is not symmetric (see Figure 7 on next page).
I can be the adjacency matrix for a graph or digraph (see Figure 8 on next page).

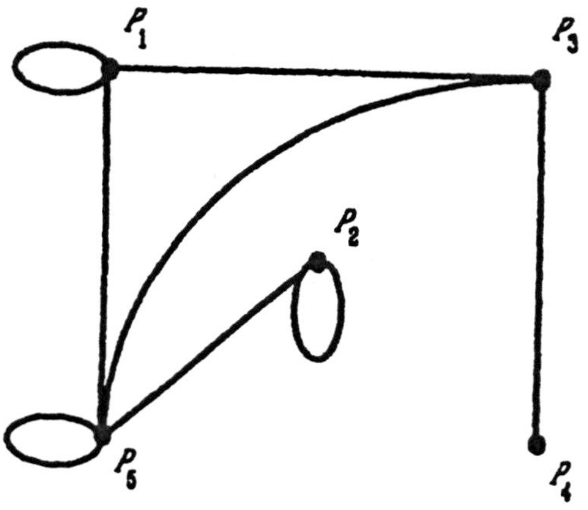

Figure 5: Graph for F

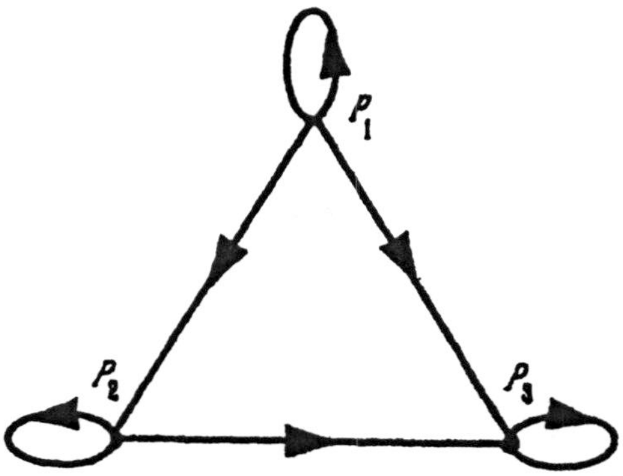

Figure 6: Digraph for G

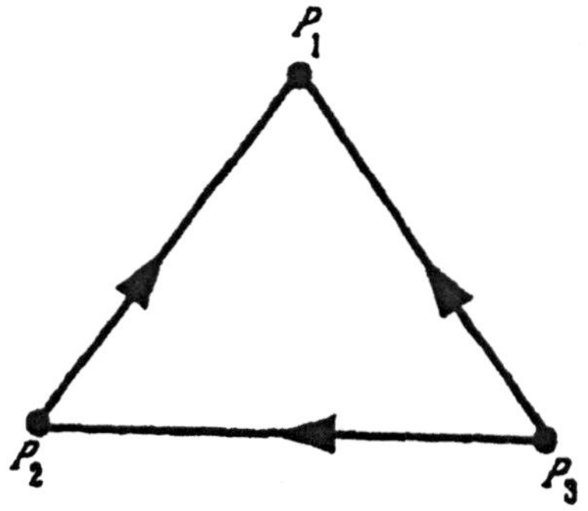

Figure 7: Digraph for H

Figure 8: Graph for I

Figure 9: Graph for K

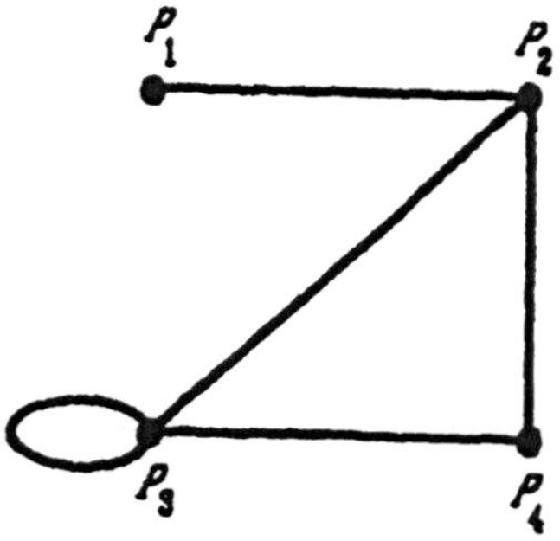

Figure 10: Graph for L

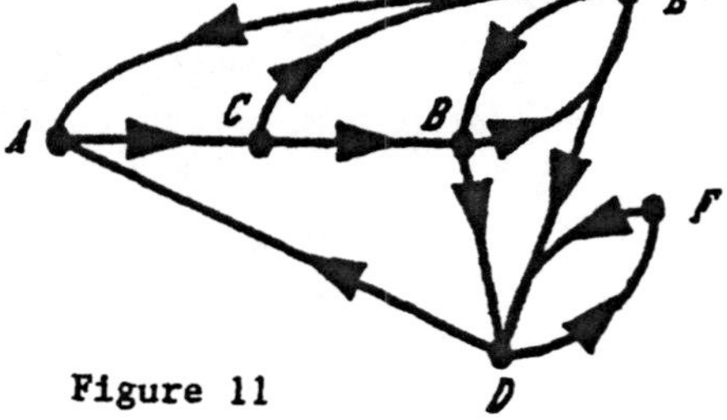

Figure 11

J can not be the adjacency matrix for either a graph or digraph because its (1,2) entry equals 2, which is not 0 or 1.
K can be the adjacency matrix for a graph or digraph (see Figure 9 on previous page).
L can be the adjacency matrix for a graph or digraph (see Figure 10 on previous page).
M can not be the adjacency matrix for either a graph or digraph because its (1,1) entry equals -2, which is not 0 or 1.

(3) The digraph is shown in Figure 11 on the previous page, and the adjacency matrix is

$$\begin{array}{c} \\ A \\ B \\ C \\ D \\ E \\ F \end{array} \begin{array}{c} \begin{array}{cccccc} A & B & C & D & E & F \end{array} \\ \left[\begin{array}{cccccc} 0 & 0 & 1 & 0 & 0 & 0 \\ 0 & 0 & 0 & 1 & 1 & 0 \\ 0 & 1 & 0 & 0 & 1 & 0 \\ 1 & 0 & 0 & 0 & 0 & 1 \\ 1 & 1 & 0 & 1 & 0 & 0 \\ 0 & 0 & 0 & 1 & 0 & 0 \end{array}\right] \end{array},$$

where we have placed a 1 in the (i,j) entry of the matrix as well as a corresponding directed edge in the digraph, if author i influences author j. The transpose gives no new information. But it does suggest a different interpretation of the results: namely, the (i,j) entry of the transpose equals 1 if author j influences author i (see Figure 11 on previous page).

(4) The adjacency matrix for the digraph in Figure 8.5 in Section 8.1 of the textbook is

$$\mathbf{A} = \begin{bmatrix} 0 & 1 & 1 & 0 & 1 \\ 1 & 1 & 1 & 0 & 0 \\ 0 & 1 & 0 & 1 & 0 \\ 0 & 0 & 1 & 0 & 0 \\ 1 & 1 & 0 & 1 & 0 \end{bmatrix},$$

whose (i,j) entry is 1 if there is an edge directed from the vertex P_i to the vertex P_j and is 0 otherwise.

Direct computation shows that $\mathbf{A}^2 = \begin{bmatrix} 2 & 3 & 1 & 2 & 0 \\ 1 & 3 & 2 & 1 & 1 \\ 1 & 1 & 2 & 0 & 0 \\ 0 & 1 & 0 & 1 & 0 \\ 1 & 2 & 3 & 0 & 1 \end{bmatrix}$ and $\mathbf{A}^3 = \begin{bmatrix} 3 & 6 & 7 & 1 & 2 \\ 4 & 7 & 5 & 3 & 1 \\ 1 & 4 & 2 & 2 & 1 \\ 1 & 1 & 2 & 0 & 0 \\ 3 & 7 & 3 & 4 & 1 \end{bmatrix}$.

(a) The number of paths of length 3 from P_2 to P_4 is the (2,4) entry of $\mathbf{A}^3$. Hence, there are 3 such paths.

(c) The number of paths of length 1, 2, and 3 from P_3 to P_2 is the (3,2) entry of $\mathbf{A}$, $\mathbf{A}^2$, and $\mathbf{A}^3$, respectively. Adding these three matrix entries yields $1+1+4=6$. Hence, the number of paths of length ≤ 3 from P_3 to P_2 is 6.

(e) The number of paths of length k from P_4 to P_5 is the (4,5) entry of $\mathbf{A}^k$. Thus, to find the shortest path from P_4 to P_5, we look for the smallest positive integer power k for which the (4,5) entry of $\mathbf{A}^k$ is nonzero, indicating that a path of length k exists. Inspecting the matrices $\mathbf{A}$, $\mathbf{A}^2$,and $\mathbf{A}^3$, above, we see that there is no path of length up to 3 from P_4 to P_5, since each of these matrices has a 0 in its (4,5) entry. However, the (4,5) entry of $\mathbf{A}^4$ is the dot product of the 4th row of $\mathbf{A}$ with the 5th column of $\mathbf{A}^3$, which is 1. Therefore, there is one path of length 4 from P_4 to P_5, and the shortest path from P_4 to P_5 has length 4.

(5) The adjacency matrix for the digraph in Figure 8.6 in Section 8.1 of the textbook is

$$\mathbf{A} = \begin{bmatrix} 0 & 1 & 1 & 0 & 0 \\ 0 & 0 & 1 & 1 & 0 \\ 1 & 1 & 0 & 1 & 0 \\ 0 & 0 & 1 & 1 & 0 \\ 1 & 1 & 0 & 0 & 0 \end{bmatrix},$$

whose (i, j) entry is 1 if there is an edge directed from the vertex P_i to the vertex P_j and is 0 otherwise.

Direct computation shows that $\mathbf{A}^2 = \begin{bmatrix} 1 & 1 & 1 & 2 & 0 \\ 1 & 1 & 1 & 2 & 0 \\ 0 & 1 & 3 & 2 & 0 \\ 1 & 1 & 1 & 2 & 0 \\ 0 & 1 & 2 & 1 & 0 \end{bmatrix}$ and $\mathbf{A}^3 = \begin{bmatrix} 1 & 2 & 4 & 4 & 0 \\ 1 & 2 & 4 & 4 & 0 \\ 3 & 3 & 3 & 6 & 0 \\ 1 & 2 & 4 & 4 & 0 \\ 2 & 2 & 2 & 4 & 0 \end{bmatrix}$.

(a) The number of paths of length 3 from P_2 to P_4 is the (2,4) entry of $\mathbf{A}^3$. Hence, there are 4 such paths.

(c) The number of paths of length 1, 2, and 3 from P_3 to P_2 is the (3,2) entry of $\mathbf{A}$, $\mathbf{A}^2$, and $\mathbf{A}^3$, respectively. Adding these three matrix entries yields $1 + 1 + 3 = 5$. Hence, the number of paths of length ≤ 3 from P_3 to P_2 is 5.

(e) An inspection of the digraph shows that no edges go to P_5. Hence, there are no paths from any vertex of any length going to P_5. This can also be seen in the adjacency matrix. Since the 5th column of $\mathbf{A}$ has all zeroes, every power $\mathbf{A}^k$ of $\mathbf{A}$ (for integer $k > 0$) will also have all zeroes in its 5th column (because the 5th column of $\mathbf{A}^k$ equals $\mathbf{A}^{k-1}$ multiplied by the 5th column of $\mathbf{A}$).

(6) In each part, we use the adjacency matrices for the digraphs and their powers as computed above for Exercises 4 and 5.

(a) The number of cycles of length 3 connecting P_2 to itself is the (2,2) entry of $\mathbf{A}^3$. Using the matrix $\mathbf{A}^3$ from Exercise 4 yields 7 cycles of length 3 connecting P_2 to itself in Figure 8.5. Using the matrix $\mathbf{A}^3$ from Exercise 5 yields 2 cycles of length 3 connecting P_2 to itself in Figure 8.6.

(c) The number of cycles of length ≤ 4 connecting P_4 to itself is the sum of the (4,4) entries of $\mathbf{A}$, $\mathbf{A}^2$, $\mathbf{A}^3$, and $\mathbf{A}^4$. For Figure 8.5, since we did not compute $\mathbf{A}^4$ in Exercise 4, we still need the (4,4) entry of $\mathbf{A}^4$. This is the dot product of the 4th row of $\mathbf{A}$ with the 4th column of $\mathbf{A}^3$, resulting in 2. Using this and the (4,4) entries from the matrices from Exercise 4 yields $0+1+0+2 = 3$ cycles of length ≤ 4 connecting P_4 to itself in Figure 8.5. For Figure 8.6, since we did not compute $\mathbf{A}^4$ in Exercise 5, we still need the (4,4) entry of $\mathbf{A}^4$. This is the dot product of the 4th row of $\mathbf{A}$ with the 4th column of $\mathbf{A}^3$, resulting in 10. Using this and the (4,4) entries from the matrices from Exercise 5 yields $1 + 2 + 4 + 10 = 17$ cycles of length ≤ 4 connecting P_4 to itself in Figure 8.6.

(7) (a) If the vertex is the ith vertex, then the ith row and ith column entries of the adjacency matrix all equal zero, except possibly for the (i, i) entry, which would equal 1 if there is a loop at the ith vertex.

(b) If the vertex is the ith vertex, then the ith row entries of the adjacency matrix all equal zero, except possibly for the (i, i) entry. (Note: The ith column entries may be nonzero since there may be edges connecting other vertices *to* the ith vertex.)

(8) (a) Since a 1 in the (i, i) entry indicates the existence of a loop at the ith vertex, the trace, which is the sum of the main diagonal entries, equals the total number of loops in the graph or digraph.

(9) (a) The digraph in Figure 8.5 is strongly connected because there is a path of length ≤ 4 from any vertex to any other. In the solution to Exercise 4, above, we computed the adjacency matrix $\mathbf{A}$ for this digraph, and its powers $\mathbf{A}^2$ and $\mathbf{A}^3$. Note that only the (4,5) entry is zero in all of the matrices $\mathbf{A}$, $\mathbf{A}^2$, and $\mathbf{A}^3$ simulataneously, showing that there is a path of length ≤ 3 from each vertex to every other vertex, except from P_4 to P_5. However, in the solution to Exercise 4(e), above, we computed the (4,5) entry of $\mathbf{A}^4$ to be 1. Hence, there is one path of length 4 from P_4 to P_5, which completes the verification that the digraph is strongly connected. The digraph in Figure 8.6 is not strongly connected since there is no path directed to P_5 from any other vertex (see answer to Exercise 5(e), above).

(10) (b) Yes, it is a dominance digraph, because no tie games are possible and because each team plays every other team. Thus if P_i and P_j are two given teams, either P_i defeats P_j or vice versa.

(11) We prove Theorem 8.1 by induction on k. For the base step, $k = 1$. This case is true by the definition of the adjacency matrix. For the inductive step, we assume the theorem is true for k, and prove that it is true for $k+1$. Let $\mathbf{B} = \mathbf{A}^k$. Then $\mathbf{A}^{k+1} = \mathbf{BA}$. Note that the (i,j) entry of $\mathbf{A}^{k+1}$ = (ith row of $\mathbf{B}$)·(jth column of $\mathbf{A}$) $= \sum_{q=1}^{n} b_{iq}a_{qj} = \sum_{q=1}^{n}$ (number of paths of length k from P_i to P_q)·(number of paths of length 1 from P_q to P_j) = (number of paths of length $k+1$ from P_i to P_j). The last equation is true because any path of length $k+1$ from P_i to P_j must be composed of a path of length k from P_i to P_q for some q followed by a path of length 1 from P_q to P_j.

(12) (a) True. In a simple graph, no distinction is made regarding the direction of any edge. Thus, there is an edge connecting P_i to P_j if and only if there is an edge connecting P_j to P_i. The same edge is used in both cases. Hence, the (i,j) entry of the adjacency matrix must have the same value as the (j,i) entry of the matrix.

(b) False. By definition, the adjacency matrix of a simple digraph can contain only the numbers 0 and 1 as entries.

(c) True. The (1,2) entry of $\mathbf{A}^n$ gives the number of paths of length n from P_1 to P_2. If this entry is zero for all $n \geq 1$, then there are no paths of any length from P_1 to P_2.

(d) False. In a simple graph, a single edge connecting P_i and P_j for $i \neq j$ places a 1 in both the (i,j) and the (j,i) entries of the adjacency matrix. Hence, summing 1's in the adjacency matrix would count all such edges twice. For example, the graph G_1 in Example 1 of Section 8.1 of the textbook has 4 edges, but its adjacency matrix contains eight 1's. Note that G_2 in that same example has 6 edges, and its adjacency matrix has eleven 1's. (The loop connecting P_2 to itself generates only one 1 in the adjacency matrix.)

(e) True. In a simple digraph, since the edges are directed, each edge corresponds to exactly one nonzero entry in the associated adjacency matrix. That is, an edge from P_i to P_j places a 1 in the (i,j) entry of the matrix, and in no other entry.

(f) True. To get from P_1 to P_3 in $k+j$ steps, first follow the path of length k from P_1 to P_2, and then follow the path of length j from P_2 to P_3.

(g) True. The (k,i) entry of the adjacency matrix gives the number of edges connecting P_k to P_i (either 0 or 1). Adding these numbers for $1 \leq k \leq n$ gives the total the number of edges connecting all vertices to P_i.

Section 8.2

(1) (a) The given circuit has two junctions and two loops. Both junctions produce the same equation (by Kirchhoff's First Law): $I_1 = I_2 + I_3$. The two loops starting and ending at the voltage source

$36V$ are:

(1) $I_1 \to I_2 \to I_1$
(2) $I_1 \to I_3 \to I_1$.

Kirchhoff's Second Law gives an Ohm's Law equation for each of these two loops:

$36V - I_2(4\Omega) - I_1(2\Omega) = 0$ (loop 1)
$36V + 7V - I_3(9\Omega) - I_1(2\Omega) = 0$ (loop 2)

These relationships lead to the following system of linear equations:

$$\begin{cases} I_1 - I_2 - I_3 = 0 \\ 2I_1 + 4I_2 = 36 \\ 2I_1 + 9I_3 = 43 \end{cases}.$$

Thus, we row reduce

$$\left[\begin{array}{ccc|c} 1 & -1 & -1 & 0 \\ 2 & 4 & 0 & 36 \\ 2 & 0 & 9 & 43 \end{array}\right] \text{ to obtain } \left[\begin{array}{ccc|c} 1 & 0 & 0 & 8 \\ 0 & 1 & 0 & 5 \\ 0 & 0 & 1 & 3 \end{array}\right].$$

This gives the solution $I_1 = 8$, $I_2 = 5$, and $I_3 = 3$.

(c) The given circuit has four junctions and four loops. By Kirchhoff's First Law, the junctions produce the following equations:

$$\begin{aligned} I_1 &= I_2 + I_6 \\ I_6 &= I_3 + I_4 + I_5 \\ I_6 &= I_3 + I_4 + I_5 \\ I_1 &= I_2 + I_6 \end{aligned}.$$

The last two of these equations are redundant. Next, the four loops starting and ending at the voltage source $42V$ are:

(1) $I_1 \to I_6 \to I_3 \to I_6 \to I_1$
(2) $I_1 \to I_6 \to I_4 \to I_6 \to I_1$
(3) $I_1 \to I_6 \to I_5 \to I_6 \to I_1$
(4) $I_1 \to I_2 \to I_1$

Kirchhoff's Second Law gives an Ohm's Law equation for each of these four loops:

$42V - I_3(6\Omega) - I_1(2\Omega) = 0$ (loop 1)
$42V - I_4(9\Omega) - I_1(2\Omega) = 0$ (loop 2)
$42V - I_5(9\Omega) - I_1(2\Omega) = 0$ (loop 3)
$42V + 7V - I_2(5\Omega) - I_1(2\Omega) = 0$ (loop 4)

These relationships lead to the following system of linear equations:

$$\begin{cases} I_1 - I_2 - I_6 = 0 \\ I_3 + I_4 + I_5 - I_6 = 0 \\ 2I_1 + 6I_3 = 42 \\ 2I_1 + 9I_4 = 42 \\ 2I_1 + 9I_5 = 42 \\ 2I_1 + 5I_2 = 49 \end{cases}$$

Thus, we row reduce

$$\left[\begin{array}{cccccc|c} 1 & -1 & 0 & 0 & 0 & -1 & 0 \\ 0 & 0 & 1 & 1 & 1 & -1 & 0 \\ 2 & 0 & 6 & 0 & 0 & 0 & 42 \\ 2 & 0 & 0 & 9 & 0 & 0 & 42 \\ 2 & 0 & 0 & 0 & 9 & 0 & 42 \\ 2 & 5 & 0 & 0 & 0 & 0 & 49 \end{array}\right] \text{ to obtain } \left[\begin{array}{cccccc|c} 1 & 0 & 0 & 0 & 0 & 0 & 12 \\ 0 & 1 & 0 & 0 & 0 & 0 & 5 \\ 0 & 0 & 1 & 0 & 0 & 0 & 3 \\ 0 & 0 & 0 & 1 & 0 & 0 & 2 \\ 0 & 0 & 0 & 0 & 1 & 0 & 2 \\ 0 & 0 & 0 & 0 & 0 & 1 & 7 \end{array}\right].$$

This gives the solution $I_1 = 12$, $I_2 = 5$, $I_3 = 3$, $I_4 = 2$, $I_5 = 2$, and $I_6 = 7$.

(2) (a) True. However, some of the equations produced by the junctions may be redundant, as in the solution to Exercise 1(c), above.

(b) True. This is Ohm's law: $V = IR$.

Section 8.3

(1) In addition to the notation in Section 8.3 of the textbook, we use $\mathbf{B}$ to represent the column matrix containing the y-coordinates of the given points.

(a) Using the given data, the matrices $\mathbf{A}$ and $\mathbf{B}$ are

$$\mathbf{A} = \begin{bmatrix} 1 & 3 \\ 1 & 1 \\ 1 & 0 \\ 1 & 2 \end{bmatrix} \text{ and } \mathbf{B} = \begin{bmatrix} -8 \\ -5 \\ -4 \\ -1 \end{bmatrix}.$$

Direct computation yields $\mathbf{A}^T\mathbf{A} = \begin{bmatrix} 4 & 6 \\ 6 & 14 \end{bmatrix}$ and $\mathbf{A}^T\mathbf{B} = \begin{bmatrix} -18 \\ -31 \end{bmatrix}$.

Row reducing the augmented matrix

$$\left[\begin{array}{cc|c} 4 & 6 & -18 \\ 6 & 14 & -31 \end{array}\right] \text{ gives } \left[\begin{array}{cc|c} 1 & 0 & -3.3 \\ 0 & 1 & -0.8 \end{array}\right],$$

and so the line of best fit is $y = -0.8x - 3.3$. Plugging in $x = 5$ produces $y \approx -7.3$.

(c) Using the given data, the matrices $\mathbf{A}$ and $\mathbf{B}$ are

$$\mathbf{A} = \begin{bmatrix} 1 & -4 \\ 1 & -3 \\ 1 & -2 \\ 1 & -1 \\ 1 & 0 \end{bmatrix} \text{ and } \mathbf{B} = \begin{bmatrix} 10 \\ 8 \\ 7 \\ 5 \\ 4 \end{bmatrix}.$$

Direct computation yields $\mathbf{A}^T\mathbf{A} = \begin{bmatrix} 5 & -10 \\ -10 & 30 \end{bmatrix}$ and $\mathbf{A}^T\mathbf{B} = \begin{bmatrix} 34 \\ -83 \end{bmatrix}$.

Row reducing the augmented matrix

$$\left[\begin{array}{cc|c} 5 & -10 & 34 \\ -10 & 30 & -83 \end{array}\right] \text{ gives } \left[\begin{array}{cc|c} 1 & 0 & 3.8 \\ 0 & 1 & -1.5 \end{array}\right],$$

and so the line of best fit is $y = -1.5x + 3.8$. Plugging in $x = 5$ produces $y \approx -3.7$.

(2) In addition to the notation in Section 8.3 of the textbook, we use **B** to represent the column matrix containing the y-coordinates of the given points.

(a) Using the given data, the matrices **A** and **B** are

$$\mathbf{A} = \begin{bmatrix} 1 & -4 & 16 \\ 1 & -2 & 4 \\ 1 & 0 & 0 \\ 1 & 2 & 4 \end{bmatrix} \text{ and } \mathbf{B} = \begin{bmatrix} 8 \\ 5 \\ 3 \\ 6 \end{bmatrix}.$$

Direct computation yields $\mathbf{A}^T\mathbf{A} = \begin{bmatrix} 4 & -4 & 24 \\ -4 & 24 & -64 \\ 24 & -64 & 288 \end{bmatrix}$ and $\mathbf{A}^T\mathbf{B} = \begin{bmatrix} 22 \\ -30 \\ 172 \end{bmatrix}$.

Row reducing the augmented matrix

$$\left[\begin{array}{ccc|c} 4 & -4 & 24 & 22 \\ -4 & 24 & -64 & -30 \\ 24 & -64 & 288 & 172 \end{array}\right] \text{ gives } \left[\begin{array}{ccc|c} 1 & 0 & 0 & 3.600 \\ 0 & 1 & 0 & 0.350 \\ 0 & 0 & 1 & 0.375 \end{array}\right],$$

and so the least squares quadratic polynomial is $y = 0.375x^2 + 0.35x + 3.6$.

(c) Using the given data, the matrices **A** and **B** are

$$\mathbf{A} = \begin{bmatrix} 1 & -4 & 16 \\ 1 & -3 & 9 \\ 1 & -2 & 4 \\ 1 & 0 & 0 \\ 1 & 1 & 1 \end{bmatrix} \text{ and } \mathbf{B} = \begin{bmatrix} -3 \\ -2 \\ -1 \\ 0 \\ 1 \end{bmatrix}.$$

Direct computation yields $\mathbf{A}^T\mathbf{A} = \begin{bmatrix} 5 & -8 & 30 \\ -8 & 30 & -98 \\ 30 & -98 & 354 \end{bmatrix}$ and $\mathbf{A}^T\mathbf{B} = \begin{bmatrix} -5 \\ 21 \\ -69 \end{bmatrix}$.

Row reducing the augmented matrix

$$\left[\begin{array}{ccc|c} 5 & -8 & 30 & -5 \\ -8 & 30 & -98 & 21 \\ 30 & -98 & 354 & -69 \end{array}\right] \text{ gives } \left[\begin{array}{ccc|c} 1 & 0 & 0 & 0.266 \\ 0 & 1 & 0 & 0.633 \\ 0 & 0 & 1 & -0.042 \end{array}\right],$$

and so the least-squares quadratic polynomial is $y = -0.042x^2 + 0.633x + 0.266$.

(3) In addition to the notation in Section 8.3 of the textbook, we use **B** to represent the column matrix containing the y-coordinates of the given points.

(a) Using the given data, the matrices **A** and **B** are

$$\mathbf{A} = \begin{bmatrix} 1 & -3 & 9 & -27 \\ 1 & -2 & 4 & -8 \\ 1 & -1 & 1 & -1 \\ 1 & 0 & 0 & 0 \\ 1 & 1 & 1 & 1 \end{bmatrix} \text{ and } \mathbf{B} = \begin{bmatrix} -3 \\ -1 \\ 0 \\ 1 \\ 4 \end{bmatrix}.$$

Direct computation yields $\mathbf{A}^T\mathbf{A} = \begin{bmatrix} 5 & -5 & 15 & -35 \\ -5 & 15 & -35 & 99 \\ 15 & -35 & 99 & -275 \\ -35 & 99 & -275 & 795 \end{bmatrix}$ and $\mathbf{A}^T\mathbf{B} = \begin{bmatrix} 1 \\ 15 \\ -27 \\ 93 \end{bmatrix}$.

Row reducing the augmented matrix

$$\left[\begin{array}{cccc|c} 5 & -5 & 15 & -35 & 1 \\ -5 & 15 & -35 & 99 & 15 \\ 15 & -35 & 99 & -275 & -27 \\ -35 & 99 & -275 & 795 & 93 \end{array}\right] \text{ gives } \left[\begin{array}{cccc|c} 1 & 0 & 0 & 0 & \frac{37}{35} \\ 0 & 1 & 0 & 0 & \frac{25}{14} \\ 0 & 0 & 1 & 0 & \frac{25}{28} \\ 0 & 0 & 0 & 1 & \frac{1}{4} \end{array}\right],$$

and so the least-squares cubic polynomial is $y = \frac{1}{4}x^3 + \frac{25}{28}x^2 + \frac{25}{14}x + \frac{37}{35}$.

(4) In addition to the notation in Section 8.3 of the textbook, we use $\mathbf{B}$ to represent the column matrix containing the y-coordinates of the given points.

(a) First, we use the function $y = x^4$ and the given x-values to generate the following data points: $(-2, 16)$, $(-1, 1)$, $(0, 0)$, $(1, 1)$, $(2, 16)$. Using this data, the matrices $\mathbf{A}$ and $\mathbf{B}$ are

$$\mathbf{A} = \begin{bmatrix} 1 & -2 & 4 \\ 1 & -1 & 1 \\ 1 & 0 & 0 \\ 1 & 1 & 1 \\ 1 & 2 & 4 \end{bmatrix} \text{ and } \mathbf{B} = \begin{bmatrix} 16 \\ 1 \\ 0 \\ 1 \\ 16 \end{bmatrix}.$$

Direct computation yields $\mathbf{A}^T\mathbf{A} = \begin{bmatrix} 5 & 0 & 10 \\ 0 & 10 & 0 \\ 10 & 0 & 34 \end{bmatrix}$ and $\mathbf{A}^T\mathbf{B} = \begin{bmatrix} 34 \\ 0 \\ 130 \end{bmatrix}$.

Row reducing the augmented matrix

$$\left[\begin{array}{ccc|c} 5 & 0 & 10 & 34 \\ 0 & 10 & 0 & 0 \\ 10 & 0 & 34 & 130 \end{array}\right] \text{ gives } \left[\begin{array}{ccc|c} 1 & 0 & 0 & -2.0571 \\ 0 & 1 & 0 & 0.0000 \\ 0 & 0 & 1 & 4.4286 \end{array}\right],$$

and so the least-squares quadratic polynomial is $y = 4.4286x^2 - 2.0571$.

(c) First, we use the function $y = \ln x$ and the given x-values to generate the following data points: $(1, 0)$, $(2, 0.6931)$, $(3, 1.0986)$, $(4, 1.3863)$. Note that, while we only list 4 digits after the decimal point here, all computations were actually performed on a calculator displaying 12 significant digits of accuracy. Using this data, the matrices $\mathbf{A}$ and $\mathbf{B}$ are

$$\mathbf{A} = \begin{bmatrix} 1 & 1 & 1 \\ 1 & 2 & 4 \\ 1 & 3 & 9 \\ 1 & 4 & 16 \end{bmatrix} \text{ and } \mathbf{B} = \begin{bmatrix} 0.0000 \\ 0.6931 \\ 1.0986 \\ 1.3863 \end{bmatrix}.$$

Direct computation yields $\mathbf{A}^T\mathbf{A} = \begin{bmatrix} 4 & 10 & 30 \\ 10 & 30 & 100 \\ 30 & 100 & 354 \end{bmatrix}$ and $\mathbf{A}^T\mathbf{B} = \begin{bmatrix} 3.1781 \\ 10.2273 \\ 34.8408 \end{bmatrix}$.

Row reducing the augmented matrix

$$\left[\begin{array}{ccc|c} 4 & 10 & 30 & 3.1781 \\ 10 & 30 & 100 & 10.2273 \\ 30 & 100 & 354 & 34.8408 \end{array}\right] \text{ gives } \left[\begin{array}{ccc|c} 1 & 0 & 0 & -0.8534 \\ 0 & 1 & 0 & 0.9633 \\ 0 & 0 & 1 & -0.1014 \end{array}\right],$$

and so the least-squares quadratic polynomial is $y = -0.1014x^2 + 0.9633x - 0.8534$.

(e) First, we use the function $y = \cos x$ and the given x-values to generate the following data points:

$(-\frac{\pi}{2}, 0)$, $(-\frac{\pi}{4}, \frac{\sqrt{2}}{2})$, $(0,1)$, $(\frac{\pi}{4}, \frac{\sqrt{2}}{2})$, $(\frac{\pi}{2}, 0)$. Making decimal approximations yields $(-1.5708, 0)$, $(-0.7854, 0.7071)$, $(0,1)$, $(0.7854, 0.7071)$, $(1.5708, 0)$. Note that, while we only list 4 digits after the decimal point here, all computations were actually performed on a calculator displaying 12 significant digits of accuracy. Using this data, the matrices $\mathbf{A}$ and $\mathbf{B}$ are

$$\mathbf{A} = \begin{bmatrix} 1 & -1.5708 & 2.4674 & -3.8758 \\ 1 & -0.7854 & 0.6169 & -0.4845 \\ 1 & 0 & 0 & 0 \\ 1 & 0.7854 & 0.6169 & 0.4845 \\ 1 & 1.5708 & 2.4674 & 3.8758 \end{bmatrix} \text{ and } \mathbf{B} = \begin{bmatrix} 0 \\ 0.7071 \\ 1 \\ 0.7071 \\ 0 \end{bmatrix}.$$ Direct computation yields

$$\mathbf{A}^T\mathbf{A} = \begin{bmatrix} 5 & 0 & 6.1685 & 0 \\ 0 & 6.1685 & 0 & 12.9371 \\ 6.1685 & 0 & 12.9371 & 0 \\ 0 & 12.9371 & 0 & 30.5128 \end{bmatrix} \text{ and } \mathbf{A}^T\mathbf{B} = \begin{bmatrix} 2.4142 \\ 0 \\ 0.8724 \\ 0 \end{bmatrix}.$$

Row reducing the augmented matrix

$$\left[\begin{array}{cccc|c} 5 & 0 & 6.1685 & 0 & 2.4142 \\ 0 & 6.1685 & 0 & 12.9371 & 0 \\ 6.1685 & 0 & 12.9371 & 0 & 0.8724 \\ 0 & 12.9371 & 0 & 30.5128 & 0 \end{array}\right] \text{ gives } \left[\begin{array}{c|c} \mathbf{I}_4 & \begin{array}{c} 0.9706 \\ 0 \\ -0.3954 \\ 0 \end{array} \end{array}\right],$$

and so the least-squares cubic polynomial is $y = 0x^3 - 0.3954x^2 + 0.9706$.

(5) In addition to the notation in Section 8.3 of the textbook, we use $\mathbf{B}$ to represent the column matrix containing the y-coordinates of the given points. This problem uses the following set of data points: $(1,3)$, $(2,3.3)$, $(3,3.7)$, $(4,4.1)$, $(5,4.6)$.

(a) Using the given data, the matrices $\mathbf{A}$ and $\mathbf{B}$ are $\mathbf{A} = \begin{bmatrix} 1 & 1 \\ 1 & 2 \\ 1 & 3 \\ 1 & 4 \\ 1 & 5 \end{bmatrix}$ and $\mathbf{B} = \begin{bmatrix} 3.0 \\ 3.3 \\ 3.7 \\ 4.1 \\ 4.6 \end{bmatrix}$.

Direct computation yields $\mathbf{A}^T\mathbf{A} = \begin{bmatrix} 5 & 15 \\ 15 & 55 \end{bmatrix}$ and $\mathbf{A}^T\mathbf{B} = \begin{bmatrix} 18.7 \\ 60.1 \end{bmatrix}$.

Row reducing the augmented matrix

$$\left[\begin{array}{cc|c} 5 & 15 & 18.7 \\ 15 & 55 & 60.1 \end{array}\right] \text{ gives } \left[\begin{array}{cc|c} 1 & 0 & 2.54 \\ 0 & 1 & 0.40 \end{array}\right],$$

and so the line of best fit is $y = 0.4x + 2.54$. Next, to find out when the angle reaches 20°, set $y = 20$ in this equation and solve for x, producing $x = \frac{17.46}{0.4} = 43.65$. Hence, the angle reaches 20° in the 44th month.

(b) Using the given data, the matrices $\mathbf{A}$ and $\mathbf{B}$ are

$$\mathbf{A} = \begin{bmatrix} 1 & 1 & 1 \\ 1 & 2 & 4 \\ 1 & 3 & 9 \\ 1 & 4 & 16 \\ 1 & 5 & 25 \end{bmatrix} \text{ and } \mathbf{B} = \begin{bmatrix} 3.0 \\ 3.3 \\ 3.7 \\ 4.1 \\ 4.6 \end{bmatrix}.$$

Direct computation yields $\mathbf{A}^T\mathbf{A} = \begin{bmatrix} 5 & 15 & 55 \\ 15 & 55 & 225 \\ 55 & 225 & 979 \end{bmatrix}$ and $\mathbf{A}^T\mathbf{B} = \begin{bmatrix} 18.7 \\ 60.1 \\ 230.1 \end{bmatrix}$.

Row reducing the augmented matrix

$$\left[\begin{array}{ccc|c} 5 & 15 & 55 & 18.7 \\ 15 & 55 & 225 & 60.1 \\ 55 & 225 & 979 & 230.1 \end{array}\right] \text{ gives } \left[\begin{array}{ccc|c} 1 & 0 & 0 & 2.74 \\ 0 & 1 & 0 & 0.2286 \\ 0 & 0 & 1 & 0.02857 \end{array}\right],$$

and so the least-squares quadratic polynomial is $y = 0.02857x^2 + 0.2286x + 2.74$. Next, to find out when the angle reaches 20°, set $y = 20$ in this equation and solve for x. The Quadratic Formula produces the positive solution $x \approx 20.9$. Hence, the angle reaches 20° in the 21st month.

(7) In addition to the notation in Section 8.3 of the textbook, we use **B** to represent the column matrix containing the y-coordinates of the given points. Using the given data, the matrices **A** and **B** are

$$\mathbf{A} = \begin{bmatrix} 1 & -2 & 4 \\ 1 & 0 & 0 \\ 1 & 3 & 9 \end{bmatrix} \text{ and } \mathbf{B} = \begin{bmatrix} 6 \\ 2 \\ 8 \end{bmatrix}.$$

Direct computation yields $\mathbf{A}^T\mathbf{A} = \begin{bmatrix} 3 & 1 & 13 \\ 1 & 13 & 19 \\ 13 & 19 & 97 \end{bmatrix}$ and $\mathbf{A}^T\mathbf{B} = \begin{bmatrix} 16 \\ 12 \\ 96 \end{bmatrix}$.

Row reducing the augmented matrix

$\left[\begin{array}{ccc|c} 3 & 1 & 13 & 16 \\ 1 & 13 & 19 & 12 \\ 13 & 19 & 97 & 96 \end{array}\right]$ gives $\left[\begin{array}{ccc|c} 1 & 0 & 0 & 2 \\ 0 & 1 & 0 & -\frac{2}{5} \\ 0 & 0 & 1 & \frac{4}{5} \end{array}\right]$, and so the least-squares quadratic polynomial is $y = \frac{4}{5}x^2 - \frac{2}{5}x + 2$. Since $y = 6$ when $x = -2$, $y = 2$ when $x = 0$, and $y = 8$ when $x = 3$, this polynomial is the *exact* quadratic through the given three points.

(9) (a) The matrices **A** and **B** are $\mathbf{A} = \begin{bmatrix} 4 & -3 \\ 2 & 5 \\ 3 & 1 \end{bmatrix}$ and $\mathbf{B} = \begin{bmatrix} 12 \\ 32 \\ 21 \end{bmatrix}$.

Direct computation yields $\mathbf{A}^T\mathbf{A} = \begin{bmatrix} 29 & 1 \\ 1 & 35 \end{bmatrix}$ and $\mathbf{A}^T\mathbf{B} = \begin{bmatrix} 175 \\ 145 \end{bmatrix}$.

Row reducing the augmented matrix

$$\left[\begin{array}{cc|c} 29 & 1 & 175 \\ 1 & 35 & 145 \end{array}\right] \text{ gives } \left[\begin{array}{cc|c} 1 & 0 & \frac{230}{39} \\ 0 & 1 & \frac{155}{39} \end{array}\right],$$

giving the least-squares solutions $x_1 = \frac{230}{39}$, $x_2 = \frac{155}{39}$. Note that

$$\begin{cases} 4x_1 - 3x_2 = 11\frac{2}{3}, & \text{which is almost 12} \\ 2x_1 + 5x_2 = 31\frac{2}{3}, & \text{which is almost 32} \\ 3x_1 + x_2 = 21\frac{2}{3}, & \text{which is close to 21} \end{cases}.$$

Hence, the least-squares solution comes close to satisfying the original inconsistent system.

(10) (a) True. Since the distance of each data point to the line is 0, the sum of the squares of these distances is also 0. That is the minimum possible value for the sum S_f, since all such sums are nonnegative.

(b) False. The computed coefficient of x^3 in the degree three least-squares polynomial might be zero. For example, if all of the data points lie on a straight line, that line will be the degree t least-squares polynomial for every $t \geq 1$.

(c) False. For example, it is easy to check that the least-squares line computed in Example 1 of Section 8.3 of the textbook does not pass through any of the data points used to compute the line.

(d) False. Theorem 8.2 states that $\mathbf{A}$ is an $n \times (t+1)$ matrix. Hence, $\mathbf{A}^T$ is a $(t+1) \times n$ matrix. Therefore, $\mathbf{A}^T\mathbf{A}$ is a $(t+1) \times (t+1)$ matrix.

Section 8.4

(1) $\mathbf{A}$ is not stochastic, since $\mathbf{A}$ is not square; $\mathbf{A}$ is not regular, since $\mathbf{A}$ is not stochastic.
$\mathbf{B}$ is not stochastic, since the entries of column 2 do not sum to 1; $\mathbf{B}$ is not regular, since $\mathbf{B}$ is not stochastic.
$\mathbf{C}$ is stochastic; $\mathbf{C}$ is regular, since $\mathbf{C}$ is stochastic and has all nonzero entries.
$\mathbf{D}$ is stochastic; $\mathbf{D}$ is not regular, since every positive power of $\mathbf{D}$ is a matrix whose rows are the rows of $\mathbf{D}$ permuted in some order, and hence every such power contains zero entries.
$\mathbf{E}$ is not stochastic, since the entries of column 1 do not sum to 1; $\mathbf{E}$ is not regular, since $\mathbf{E}$ is not stochastic.
$\mathbf{F}$ is stochastic; $\mathbf{F}$ is not regular, since every positive power of $\mathbf{F}$ has all second row entries zero.
$\mathbf{G}$ is not stochastic, since $\mathbf{G}$ is not square; $\mathbf{G}$ is not regular, since $\mathbf{G}$ is not stochastic.

$\mathbf{H}$ is stochastic; $\mathbf{H}$ is regular, since $\mathbf{H}^2 = \begin{bmatrix} \frac{1}{2} & \frac{1}{4} & \frac{1}{4} \\ \frac{1}{4} & \frac{1}{2} & \frac{1}{4} \\ \frac{1}{4} & \frac{1}{4} & \frac{1}{2} \end{bmatrix}$, which has no zero entries.

(2) (a) First, $\mathbf{p}_1 = \mathbf{M}\mathbf{p} = \begin{bmatrix} \frac{1}{4} & \frac{1}{3} \\ \frac{3}{4} & \frac{2}{3} \end{bmatrix} \begin{bmatrix} \frac{2}{3} \\ \frac{1}{3} \end{bmatrix} = \begin{bmatrix} \frac{5}{18} \\ \frac{13}{18} \end{bmatrix}$. Then $\mathbf{p}_2 = \mathbf{M}\mathbf{p}_1 = \begin{bmatrix} \frac{1}{4} & \frac{1}{3} \\ \frac{3}{4} & \frac{2}{3} \end{bmatrix} \begin{bmatrix} \frac{5}{18} \\ \frac{13}{18} \end{bmatrix} = \begin{bmatrix} \frac{67}{216} \\ \frac{149}{216} \end{bmatrix}$ $\approx \begin{bmatrix} 0.3102 \\ 0.6898 \end{bmatrix}$.

(c) First, $\mathbf{p}_1 = \mathbf{M}\mathbf{p} = \begin{bmatrix} \frac{1}{4} & \frac{1}{3} & \frac{1}{2} \\ \frac{1}{2} & \frac{1}{3} & \frac{1}{6} \\ \frac{1}{4} & \frac{1}{3} & \frac{1}{3} \end{bmatrix} \begin{bmatrix} \frac{1}{4} \\ \frac{1}{2} \\ \frac{1}{4} \end{bmatrix} = \begin{bmatrix} \frac{17}{48} \\ \frac{1}{3} \\ \frac{5}{16} \end{bmatrix}$.

Then $\mathbf{p}_2 = \mathbf{M}\mathbf{p}_1 = \begin{bmatrix} \frac{1}{4} & \frac{1}{3} & \frac{1}{2} \\ \frac{1}{2} & \frac{1}{3} & \frac{1}{6} \\ \frac{1}{4} & \frac{1}{3} & \frac{1}{3} \end{bmatrix} \begin{bmatrix} \frac{17}{48} \\ \frac{1}{3} \\ \frac{5}{16} \end{bmatrix} = \begin{bmatrix} \frac{205}{576} \\ \frac{49}{144} \\ \frac{175}{576} \end{bmatrix} \approx \begin{bmatrix} 0.3559 \\ 0.3403 \\ 0.3038 \end{bmatrix}$.

(3) (a) We must solve the system $(\mathbf{M} - \mathbf{I}_2)\mathbf{x} = \mathbf{0}$, under the additional condition that $x_1 + x_2 = 1$. Thus, we row reduce $\left[\begin{array}{cc|c} -\frac{1}{2} & \frac{1}{3} & 0 \\ \frac{1}{2} & -\frac{1}{3} & 0 \\ 1 & 1 & 1 \end{array}\right]$ to $\left[\begin{array}{cc|c} 1 & 0 & \frac{2}{5} \\ 0 & 1 & \frac{3}{5} \\ 0 & 0 & 0 \end{array}\right]$, giving the steady state vector $\mathbf{p}_f = \left[\frac{2}{5}, \frac{3}{5}\right]$.

(4) (a) We will round numbers to four places after the decimal point. Then,

$$\mathbf{M}^2 = \begin{bmatrix} 0.4167 & 0.3889 \\ 0.5833 & 0.6111 \end{bmatrix}, \mathbf{M}^3 = \begin{bmatrix} 0.4028 & 0.3981 \\ 0.5972 & 0.6019 \end{bmatrix}, \mathbf{M}^4 = \begin{bmatrix} 0.4005 & 0.3997 \\ 0.5995 & 0.6003 \end{bmatrix},$$

$$\mathbf{M}^5 = \begin{bmatrix} 0.4001 & 0.3999 \\ 0.5999 & 0.6001 \end{bmatrix}, \mathbf{M}^6 = \begin{bmatrix} 0.4000 & 0.4000 \\ 0.6000 & 0.6000 \end{bmatrix}, \mathbf{M}^7 = \begin{bmatrix} 0.4000 & 0.4000 \\ 0.6000 & 0.6000 \end{bmatrix}.$$

Since all entries of $\mathbf{M}^6$ and $\mathbf{M}^7$ agree to 4 places after the decimal point, we stop here. Hence, the steady state vector for this Markov chain is given by any of the columns of $\mathbf{M}^7$, namely $[0.4000, 0.6000]$, rounded to 4 significant digits.

(5) (a) The initial probability vector $\mathbf{p}$ is $[0.3, 0.15, 0.45, 0.1]$. (Note that 10% of the citizens did not vote in the last election.) Then

$$\mathbf{p}_1 = \mathbf{M}\mathbf{p} = \begin{bmatrix} 0.7 & 0.2 & 0.2 & 0.1 \\ 0.1 & 0.6 & 0.1 & 0.1 \\ 0.1 & 0.2 & 0.6 & 0.1 \\ 0.1 & 0.0 & 0.1 & 0.7 \end{bmatrix} \begin{bmatrix} 0.30 \\ 0.15 \\ 0.45 \\ 0.10 \end{bmatrix} = \begin{bmatrix} 0.340 \\ 0.175 \\ 0.340 \\ 0.145 \end{bmatrix}.$$

Thus, in the next election, 34% will vote for Party A, 17.5% will vote for Party B, 34% will vote for Party C, and 14.5% will not vote at all.
For the election after that,

$$\mathbf{p}_2 = \mathbf{M}\mathbf{p}_1 = \begin{bmatrix} 0.7 & 0.2 & 0.2 & 0.1 \\ 0.1 & 0.6 & 0.1 & 0.1 \\ 0.1 & 0.2 & 0.6 & 0.1 \\ 0.1 & 0.0 & 0.1 & 0.7 \end{bmatrix} \begin{bmatrix} 0.340 \\ 0.175 \\ 0.340 \\ 0.145 \end{bmatrix} = \begin{bmatrix} 0.3555 \\ 0.1875 \\ 0.2875 \\ 0.1695 \end{bmatrix}.$$

Hence, in that election, 35.55% will vote for Party A, 18.75% will vote for Party B, 28.75% will vote for Party C, and 16.95% will not vote at all.

(b) A lengthy computation shows that $\mathbf{M}^{100}\mathbf{p}_1 \approx [0.36, 0.20, 0.24, 0.20]$. Hence, in a century, the votes would be 36% for Party A and 24% for Party C.
For an alternate approach, we assume that by the end of a century the system will have achieved its steady state. Hence, we can solve for the steady-state vector. Since the transition matrix $\mathbf{M}$ is regular (you can easily verify that $\mathbf{M}^2$ has no zero entries), part (5) of Theorem 8.5 shows that we can solve for $\mathbf{p}_f$ by solving the system $(\mathbf{M} - \mathbf{I}_4)\mathbf{x} = \mathbf{0}$, under the additional condition that $x_1 + x_2 + x_3 + x_4 = 1$. Hence, we row reduce

$$\left[\begin{array}{cccc|c} -0.3 & 0.2 & 0.2 & 0.1 & 0 \\ 0.1 & -0.4 & 0.1 & 0.1 & 0 \\ 0.1 & 0.2 & -0.4 & 0.1 & 0 \\ 0.1 & 0.0 & 0.1 & -0.3 & 0 \\ 1.0 & 1.0 & 1.0 & 1.0 & 1 \end{array}\right] \text{ to } \left[\begin{array}{cccc|c} 1 & 0 & 0 & 0 & 0.36 \\ 0 & 1 & 0 & 0 & 0.20 \\ 0 & 0 & 1 & 0 & 0.24 \\ 0 & 0 & 0 & 1 & 0.20 \\ 0 & 0 & 0 & 0 & 0.00 \end{array}\right].$$

Therefore, the steady-state vector is $[0.36, 0.20, 0.24, 0.20]$. Again, we get that, in a century, the votes would be 36% for Party A and 24% for Party C.

(6) (a) To compute the transition matrix $\mathbf{M}$, we use the fact that the (i, j) entry m_{ij} of $\mathbf{M}$ is the probability that the rat moves from room j to room i, where we use the labels A, B, C, D, and E and 1, 2, 3, 4, and 5 for the rooms interchangeably. So, for example, $m_{ii} = \frac{1}{2}$ for each i because the rat has a 50% chance of staying in the same room. To compute m_{21}, note that each of the 4 doorways leaving room A has the same probablility of being used when a doorway is used, namely, $\frac{1}{4}$. But, since the rat only uses a doorway only $\frac{1}{2}$ of the time, $m_{21} = \frac{1}{2} \times \frac{1}{4} = \frac{1}{8}$. Similarly, to

compute m_{41}, again note that each door out of room A has probability $\frac{1}{4}$ of being used any time a door is used. Now combining the facts that a doorway is used only half the time, and that there are two doors from room A to room D implies that $m_{41} = \left(\frac{1}{4} \times 2\right) \times \frac{1}{2} = \frac{1}{4}$. Similar computations produce all of the other entries of $\mathbf{M}$ giving

$$\mathbf{M} = \begin{bmatrix} \frac{1}{2} & \frac{1}{6} & \frac{1}{6} & \frac{1}{5} & 0 \\ \frac{1}{8} & \frac{1}{2} & 0 & 0 & \frac{1}{5} \\ \frac{1}{8} & 0 & \frac{1}{2} & \frac{1}{10} & \frac{1}{10} \\ \frac{1}{4} & 0 & \frac{1}{6} & \frac{1}{2} & \frac{1}{5} \\ 0 & \frac{1}{3} & \frac{1}{6} & \frac{1}{5} & \frac{1}{2} \end{bmatrix}.$$

(b) If $\mathbf{M}$ is the stochastic matrix in part (a), then

$$\mathbf{M}^2 = \begin{bmatrix} \frac{41}{120} & \frac{1}{6} & \frac{1}{5} & \frac{13}{60} & \frac{9}{100} \\ \frac{1}{8} & \frac{27}{80} & \frac{13}{240} & \frac{13}{200} & \frac{1}{5} \\ \frac{3}{20} & \frac{13}{240} & \frac{73}{240} & \frac{29}{200} & \frac{3}{25} \\ \frac{13}{48} & \frac{13}{120} & \frac{29}{120} & \frac{107}{300} & \frac{13}{60} \\ \frac{9}{80} & \frac{1}{3} & \frac{1}{5} & \frac{13}{60} & \frac{28}{75} \end{bmatrix},$$

which has all entries nonzero. Thus, $\mathbf{M}$ is regular.

(c) If the rat starts in room C, its initial state vector is $\mathbf{p} = [0, 0, 1, 0, 0]$. By Theorem 8.4, the state vector $\mathbf{p}_2$ is given by $\mathbf{M}^2\mathbf{p}$. Using the matrix $\mathbf{M}^2$ computed in part (b) and the fact that $\mathbf{p}$ is the vector $\mathbf{e}_3$, we see that $\mathbf{M}^2\mathbf{p}$ is just the third column of $\mathbf{M}^2$, namely $\left[\frac{1}{5}, \frac{13}{240}, \frac{73}{240}, \frac{29}{120}, \frac{1}{5}\right]$. Hence, the probability of the rat being in room D after two time intervals is $\frac{29}{120}$.

(d) Since the transition matrix $\mathbf{M}$ is regular, part (5) of Theorem 8.5 shows that we can solve for $\mathbf{p}_f$ by solving the system $(\mathbf{M}-\mathbf{I}_5)\mathbf{x} = \mathbf{0}$, under the additional condition that $x_1+x_2+x_3+x_4+x_5 = 1$. Hence, we row reduce

$$\left[\begin{array}{ccccc|c} -\frac{1}{2} & \frac{1}{6} & \frac{1}{6} & \frac{1}{5} & 0 & 0 \\ \frac{1}{8} & -\frac{1}{2} & 0 & 0 & \frac{1}{5} & 0 \\ \frac{1}{8} & 0 & -\frac{1}{2} & \frac{1}{10} & \frac{1}{10} & 0 \\ \frac{1}{4} & 0 & \frac{1}{6} & -\frac{1}{2} & \frac{1}{5} & 0 \\ 0 & \frac{1}{3} & \frac{1}{6} & \frac{1}{5} & -\frac{1}{2} & 0 \\ 1 & 1 & 1 & 1 & 1 & 1 \end{array}\right] \text{ to } \left[\begin{array}{ccccc|c} 1 & 0 & 0 & 0 & 0 & \frac{1}{5} \\ 0 & 1 & 0 & 0 & 0 & \frac{3}{20} \\ 0 & 0 & 1 & 0 & 0 & \frac{3}{20} \\ 0 & 0 & 0 & 1 & 0 & \frac{1}{4} \\ 0 & 0 & 0 & 0 & 1 & \frac{1}{4} \\ 0 & 0 & 0 & 0 & 0 & 0 \end{array}\right].$$

Therefore, the steady-state vector is $\left[\frac{1}{5}, \frac{3}{20}, \frac{3}{20}, \frac{1}{4}, \frac{1}{4}\right]$. Over time, the rat frequents rooms B and C the least and rooms D and E the most.

(9) We need to show that the product of any k stochastic matrices is stochastic. We prove this by induction on k. The Base Step is $k = 2$. Let $\mathbf{A}$ and $\mathbf{B}$ be $n \times n$ stochastic matrices. Then the entries of $\mathbf{AB}$ are nonnegative, since the entries of both $\mathbf{A}$ and $\mathbf{B}$ are nonnegative. Furthermore, the sum of the entries in the jth column of $\mathbf{AB} = \sum_{i=1}^{n}(\mathbf{AB})_{ij} = \sum_{i=1}^{n}$ (ith row of $\mathbf{A}$)$\cdot$(jth column of $\mathbf{B}$) $= \sum_{i=1}^{n} a_{i1}b_{1j} + a_{i2}b_{2j} + \cdots + a_{in}b_{nj} = \left(\sum_{i=1}^{n} a_{i1}\right) b_{1j} + \left(\sum_{i=1}^{n} a_{i2}\right) b_{2j} + \cdots + \left(\sum_{i=1}^{n} a_{in}\right) b_{nj} = (1)b_{1j} + (1)b_{2j} + \cdots + (1)b_{nj}$ (since $\mathbf{A}$ is stochastic, and each of the summations is a sum of the entries of a column of $\mathbf{A}$) $= 1$ (since $\mathbf{B}$ is stochastic). Hence $\mathbf{AB}$ is stochastic.

For the Inductive Step, we assume that the product of any k stochastic $n \times n$ matrices is stochastic, and we must show that the product of any $k+1$ stochastic $n \times n$ matrices is stochastic. So, let $\mathbf{A}_1, \ldots, \mathbf{A}_{k+1}$ be stochastic $n \times n$ matrices. Then $\mathbf{A}_1\mathbf{A}_2 \cdots \mathbf{A}_k\mathbf{A}_{k+1} = (\mathbf{A}_1\mathbf{A}_2 \cdots \mathbf{A}_k)\mathbf{A}_{k+1}$. Now $\mathbf{A}_1\mathbf{A}_2 \cdots \mathbf{A}_k$ is stochastic by the inductive hypothesis. Hence, $(\mathbf{A}_1\mathbf{A}_2 \cdots \mathbf{A}_k)\mathbf{A}_{k+1}$ is stochastic by the Base Step.

(10) We prove Theorem 8.4 by induction on n. Suppose $\mathbf{M}$ is a $k \times k$ matrix.
Base Step ($n = 1$): ith entry in $\mathbf{Mp}$ = (ith row of $\mathbf{M}$)$\cdot\mathbf{p} = \sum_{j=1}^{k} m_{ij}p_j = \sum_{j=1}^{k}$ (probability of moving from state S_j to S_i)(probability of being in state S_j) = probability of being in state S_i after 1 step of the process = ith entry of $\mathbf{p}_1$.
Inductive Step: Assume $\mathbf{p}_k = \mathbf{M}^k\mathbf{p}$ is the probability vector after k steps of the process. Then, after an additional step of the process, the computation in the Base Step shows that probability vector $\mathbf{p}_{k+1} = \mathbf{M}\mathbf{p}_k = \mathbf{M}(\mathbf{M}^k\mathbf{p}) = \mathbf{M}^{k+1}\mathbf{p}$.

(11) (a) False. While the sum of the entries in each column of a stochastic matrix must equal 1, the sum of the entries in each row are not required to equal 1. For example, this is true of the stochastic matrix $\mathbf{M}$ in Example 1 of Section 8.4 of the textbook. Now the columns of the transpose of a matrix equal the rows of the original matrix. Thus, the transpose of a stochastic matrix might not have the sum of its entries along each column equal to 1, and so might not be stochastic.

(b) True. If $\mathbf{A}$ is upper triangular, then $\mathbf{A}^k$ is also upper triangular for all $k \geq 1$. Hence, $\mathbf{A}^k$ will have zeroes below the main diagonal.

(c) True. By part (5) of Theorem 8.5, the given statement is true for the vector $\mathbf{p} = \mathbf{p}_f$.

(d) True. By part (3) of Theorem 8.5, if the Markov chain has a regular transition matrix, then there is a limit vector for every initial vector (and all of these limit vectors are equal). However, starting with the initial vector $\mathbf{p}$, the given condition implies that $\mathbf{M}^k\mathbf{p} = \mathbf{q}$ when k is odd, and $\mathbf{M}^k\mathbf{p} = \mathbf{p}$ when k is even. Hence, $\lim_{k\to\infty} \mathbf{M}^k\mathbf{p}$ can not exist, since $\mathbf{M}^k\mathbf{p}$ oscillates between $\mathbf{p}$ and $\mathbf{q}$. Therefore, $\mathbf{M}$ can not be regular.

(e) False. Each entry of the transition matrix gives the probability of changing from one specific state to another.

Section 8.5

(1) (a) First, we must convert the letters in the message "PROOF BY INDUCTION" into numbers by listing each letter's position in the alphabet:

P	R	O	O	F	B	Y	I	N	D	U	C	T	I	O	N
16	18	15	15	6	2	25	9	14	4	21	3	20	9	15	14

Next, we take these numbers in pairs to form eight 2-vectors, and multiply the matrix
$\mathbf{A} = \begin{bmatrix} 3 & -4 \\ 5 & -7 \end{bmatrix}$ times each of these 2-vectors in turn. This yields:

$$\mathbf{A}\begin{bmatrix} 16 \\ 18 \end{bmatrix} = \begin{bmatrix} -24 \\ -46 \end{bmatrix}, \quad \mathbf{A}\begin{bmatrix} 15 \\ 15 \end{bmatrix} = \begin{bmatrix} -15 \\ -30 \end{bmatrix}, \quad \mathbf{A}\begin{bmatrix} 6 \\ 2 \end{bmatrix} = \begin{bmatrix} 10 \\ 16 \end{bmatrix}, \quad \mathbf{A}\begin{bmatrix} 25 \\ 9 \end{bmatrix} = \begin{bmatrix} 39 \\ 62 \end{bmatrix},$$

$$\mathbf{A}\begin{bmatrix} 14 \\ 4 \end{bmatrix} = \begin{bmatrix} 26 \\ 42 \end{bmatrix}, \quad \mathbf{A}\begin{bmatrix} 21 \\ 3 \end{bmatrix} = \begin{bmatrix} 51 \\ 84 \end{bmatrix}, \quad \mathbf{A}\begin{bmatrix} 20 \\ 9 \end{bmatrix} = \begin{bmatrix} 24 \\ 37 \end{bmatrix}, \quad \mathbf{A}\begin{bmatrix} 15 \\ 14 \end{bmatrix} = \begin{bmatrix} -11 \\ -23 \end{bmatrix}.$$

Therefore, the encoded message is

$-24 \quad -46 \quad -15 \quad -30 \quad 10 \quad 16 \quad 39 \quad 62 \quad 26 \quad 42 \quad 51 \quad 84 \quad 24 \quad 37 \quad -11 \quad -23$.

(2) (a) If $\mathbf{A}$ is the given key matrix, by row reducing $[\mathbf{A}|\mathbf{I}_3]$ we find $\mathbf{A}^{-1}$ to be $\mathbf{A}^{-1} = \begin{bmatrix} -2 & -1 & 0 \\ -6 & -4 & 1 \\ 9 & 4 & 1 \end{bmatrix}$.

Next, we take the numbers representing the message to be decoded in triples to form eight 3-vectors, and multiply the matrix $\mathbf{A}^{-1}$ times each of these 3-vectors in turn. This yields:

$$\mathbf{A}^{-1}\begin{bmatrix} -62 \\ 116 \\ 107 \end{bmatrix} = \begin{bmatrix} 8 \\ 15 \\ 13 \end{bmatrix}, \quad \mathbf{A}^{-1}\begin{bmatrix} -32 \\ 59 \\ 67 \end{bmatrix} = \begin{bmatrix} 5 \\ 23 \\ 15 \end{bmatrix}, \quad \mathbf{A}^{-1}\begin{bmatrix} -142 \\ 266 \\ 223 \end{bmatrix} = \begin{bmatrix} 18 \\ 11 \\ 9 \end{bmatrix},$$

$$\mathbf{A}^{-1}\begin{bmatrix} -160 \\ 301 \\ 251 \end{bmatrix} = \begin{bmatrix} 19 \\ 7 \\ 15 \end{bmatrix}, \quad \mathbf{A}^{-1}\begin{bmatrix} -122 \\ 229 \\ 188 \end{bmatrix} = \begin{bmatrix} 15 \\ 4 \\ 6 \end{bmatrix}, \quad \mathbf{A}^{-1}\begin{bmatrix} -122 \\ 229 \\ 202 \end{bmatrix} = \begin{bmatrix} 15 \\ 18 \\ 20 \end{bmatrix},$$

$$\mathbf{A}^{-1}\begin{bmatrix} -78 \\ 148 \\ 129 \end{bmatrix} = \begin{bmatrix} 8 \\ 5 \\ 19 \end{bmatrix}, \quad \mathbf{A}^{-1}\begin{bmatrix} -111 \\ 207 \\ 183 \end{bmatrix} = \begin{bmatrix} 15 \\ 21 \\ 12 \end{bmatrix}.$$

We take the results and associate with each number its corresponding letter in the alphabet:

8	15	13	5	23	15	18	11	9	19	7	15
H	O	M	E	W	O	R	K	I	S	G	O

15	4	6	15	18	20	8	5	19	15	21	12
O	D	F	O	R	T	H	E	S	O	U	L

Therefore, the message is "HOMEWORK IS GOOD FOR THE SOUL."

(3) (a) True. This is because the code used for a given letter is dependent on the letters that surround it in the message, not just upon the letter itself. Thus, the same letter may be replaced by several different numbers within the same message.

(b) True. If the encoding matrix is singular, then there is no inverse matrix that can be used to decode the message. Decoding would involve solving systems having an infinite number of solutions.

(c) False. The message is typically broken up into shorter strings of length k, where k is less than the length of the entire message (see Example 1 in Section 8.5 of the textbook). Of course, the higher the value of k, the more difficult it generally is to break the code. However, large values for k also make it more time consuming to invert the key matrix and to perform the necessary matrix products.

Section 8.6

(1) (a) The Jacobian matrix $\mathbf{J} = \begin{bmatrix} \frac{\partial x}{\partial u} & \frac{\partial x}{\partial v} \\ \frac{\partial y}{\partial u} & \frac{\partial y}{\partial v} \end{bmatrix} = \begin{bmatrix} 1 & 1 \\ 1 & -1 \end{bmatrix}$. Therefore, $dx\,dy = \left||\mathbf{J}|\right|\,du\,dv = 2\,du\,dv$.

(c) The Jacobian matrix $\mathbf{J} = \begin{bmatrix} \frac{\partial x}{\partial u} & \frac{\partial x}{\partial v} \\ \frac{\partial y}{\partial u} & \frac{\partial y}{\partial v} \end{bmatrix} = \begin{bmatrix} 2u & -2v \\ 2v & 2u \end{bmatrix}$. Therefore,

$dx\,dy = \left||\mathbf{J}|\right|\,du\,dv = 4(u^2 + v^2)\,du\,dv$.

(e) The Jacobian matrix $\mathbf{J} = \begin{bmatrix} \frac{\partial x}{\partial u} & \frac{\partial x}{\partial v} \\ \frac{\partial y}{\partial u} & \frac{\partial y}{\partial v} \end{bmatrix} =$

$$\begin{bmatrix} \frac{((u+1)^2+v^2)(2)-2u(2(u+1))}{((u+1)^2+v^2)^2} & \frac{-4uv}{((u+1)^2+v^2)^2} \\ \frac{((u+1)^2+v^2)(-2u)-(1-(u^2+v^2))(2(u+1))}{((u+1)^2+v^2)^2} & \frac{((u+1)^2+v^2)(-2v)-(1-(u^2+v^2))(2v)}{((u+1)^2+v^2)^2} \end{bmatrix}.$$ Simplifying yields

$$\mathbf{J} = \frac{2}{((u+1)^2+v^2)^2} \begin{bmatrix} ((u+1)^2+v^2) - 2u(u+1) & -2uv \\ -u\left((u+1)^2+v^2\right) - (u+1)\left(1-(u^2+v^2)\right) & -2uv-2v \end{bmatrix}$$

$$= \frac{2}{((u+1)^2+v^2)^2} \begin{bmatrix} -u^2+v^2+1 & -2uv \\ -u^2-2u-1+v^2 & -2uv-2v \end{bmatrix}.$$

Now, for a 2×2 matrix $\mathbf{A}$, $|k\mathbf{A}| = k^2|\mathbf{A}|$, and so,

$|\mathbf{J}| = \frac{4}{((u+1)^2+v^2)^4}\left(\left(-u^2+v^2+1\right)(-2uv-2v) - (-2uv)\left(-u^2-2u-1+v^2\right)\right)$

$= \frac{4}{((u+1)^2+v^2)^4}\left(2u^3v - 2uv^3 - 2uv + 2u^2v - 2v^3 - 2v - 2u^3v - 4u^2v - 2uv + 2uv^3\right)$

$= \frac{4}{((u+1)^2+v^2)^4}\left(-4uv - 2u^2v - 2v^3 - 2v\right) = \frac{-8}{((u+1)^2+v^2)^4}\left(u^2v + 2uv + v^3 + v\right)$

$= \frac{-8}{((u+1)^2+v^2)^4}\left(\left((u+1)^2+v^2\right)v\right) = \frac{-8v}{((u+1)^2+v^2)^3}$.

Hence, $dx\,dy = \left||\mathbf{J}|\right|\, du\,dv = \frac{8|v|}{((u+1)^2+v^2)^3} du\,dv$.

(2) (a) The Jacobian matrix $\mathbf{J} = \begin{bmatrix} \frac{\partial x}{\partial u} & \frac{\partial x}{\partial v} & \frac{\partial x}{\partial w} \\ \frac{\partial y}{\partial u} & \frac{\partial y}{\partial v} & \frac{\partial y}{\partial w} \\ \frac{\partial z}{\partial u} & \frac{\partial z}{\partial v} & \frac{\partial z}{\partial w} \end{bmatrix} = \begin{bmatrix} 1 & 1 & 0 \\ 0 & 1 & 1 \\ 1 & 0 & 1 \end{bmatrix}$. Thus, $|\mathbf{J}| = (1)(1)(1) + (1)(1)(1) + (0)(0)(0) - (0)(1)(1) - (1)(1)(0) - (1)(0)(1) = 2$. Therefore, $dx\,dy\,dz = \left||\mathbf{J}|\right|\, du\,dv\,dw = 2\, du\,dv\,dw$.

(c) The Jacobian matrix $\mathbf{J} = \begin{bmatrix} \frac{\partial x}{\partial u} & \frac{\partial x}{\partial v} & \frac{\partial x}{\partial w} \\ \frac{\partial y}{\partial u} & \frac{\partial y}{\partial v} & \frac{\partial y}{\partial w} \\ \frac{\partial z}{\partial u} & \frac{\partial z}{\partial v} & \frac{\partial z}{\partial w} \end{bmatrix} =$

$$\frac{1}{(u^2+v^2+w^2)^2} \begin{bmatrix} (u^2+v^2+w^2) - 2u^2 & -2uv & -2uw \\ -2uv & (u^2+v^2+w^2) - 2v^2 & -2vw \\ -2uw & -2vw & (u^2+v^2+w^2) - 2w^2 \end{bmatrix}.$$

To make this expression simpler, let $A = (u^2+v^2+w^2)$.

Now, for a 3×3 matrix $\mathbf{M}$, $|k\mathbf{M}| = k^3|\mathbf{M}|$, and so by basketweaving,

$|\mathbf{J}| = \frac{1}{A^6}\left(\left(A-2u^2\right)\left(A-2v^2\right)\left(A-2w^2\right) + (-2uv)(-2vw)(-2uw) + (-2uw)(-2uv)(-2vw)\right.$
$\left. - (-2uw)(A-2v^2)(-2uw) - (A-2u^2)(-2vw)(-2vw) - (-2uv)(-2uv)(A-2w^2)\right)$

$= \frac{1}{A^6}\left(A^3 - 2u^2A^2 - 2v^2A^2 - 2w^2A^2 + 4u^2v^2A + 4u^2w^2A + 4v^2w^2A - 8u^2v^2w^2\right.$
$\left. -8u^2v^2w^2 - 8u^2v^2w^2 - 4u^2w^2A + 8u^2v^2w^2 - 4v^2w^2A + 8u^2v^2w^2 - 4u^2v^2A + 8u^2v^2w^2\right)$

$= \frac{1}{A^6}\left(A^3 - 2(u^2+v^2+w^2)A^2\right) = \frac{1}{A^6}\left(-A^3\right) = -\frac{1}{A^3}$.

Therefore, $dx\,dy\,dz = \left||\mathbf{J}|\right|\, du\,dv\,dw = \frac{1}{(u^2+v^2+w^2)^3}\, du\,dv\,dw$.

(4) (a) For the given region R, $r^2 = x^2 + y^2$ ranges from 1 to 9, and so r ranges from 1 to 3. Since R is restricted to the first quadrant, θ ranges from 0 to $\frac{\pi}{2}$. Therefore, using $x = r\cos\theta$, $y = r\sin\theta$, and $dx\,dy = r\,dr\,d\theta$ yields $\iint_R (x+y)\,dx\,dy = \iint_R (r\cos\theta + r\sin\theta)\,r\,dr\,d\theta =$

$$\int_0^{\frac{\pi}{2}}\int_1^3 (r\cos\theta + r\sin\theta)\,r\,dr\,d\theta = \int_0^{\frac{\pi}{2}} \frac{r^3}{3}(\cos\theta+\sin\theta)\Big|_1^3\,d\theta = \int_0^{\frac{\pi}{2}} \frac{26}{3}(\cos\theta+\sin\theta)\,d\theta$$

$$= \frac{26}{3}\left(\sin\theta - \cos\theta\right)\Big|_0^{\frac{\pi}{2}} = \frac{52}{3}.$$

(c) For the given sphere R, ρ ranges from 0 to 1. Since R is restricted to the upper half of the xy-plane, ϕ ranges from 0 to $\frac{\pi}{2}$. However, θ ranges from 0 to 2π, since R is the entire upper half of the sphere; that is, all the way around the z-axis. Therefore, using $z = \rho\cos\phi$ and $dx\,dy\,dz = (\rho^2\sin\phi)\,d\rho\,d\phi\,d\theta$, we get

$$\iiint_R z\;dx\,dy\,dz = \iiint_R \rho\cos\phi(\rho^2\sin\phi)\,d\rho\,d\phi\,d\theta = \int_0^{2\pi}\int_0^{\frac{\pi}{2}}\int_0^1 \rho^3\cos\phi\sin\phi\;d\rho\,d\phi\,d\theta$$

$$= \int_0^{2\pi}\int_0^{\frac{\pi}{2}}\left(\frac{\rho^4}{4}\cos\phi\sin\phi\right)\Big|_0^1 d\phi\,d\theta = \frac{1}{4}\int_0^{2\pi}\int_0^{\frac{\pi}{2}}\cos\phi\sin\phi\;d\phi\,d\theta = \frac{1}{4}\int_0^{2\pi}\frac{\sin^2\phi}{2}\Big|_0^{\frac{\pi}{2}} d\theta = \frac{1}{4}\int_0^{2\pi}\frac{1}{2}\,d\theta =$$

$$\frac{1}{8}\theta\Big|_0^{2\pi} = \frac{\pi}{4}.$$

(e) Since $r^2 = x^2 + y^2$, the condition $x^2 + y^2 \le 4$ on the region R means that r ranges from 0 to 2 and θ ranges from 0 to 2π. Therefore, using $x^2 + y^2 = r^2$ and $dx\,dy\,dz = r\,dr\,d\theta\,dz$ produces

$$\iiint_R \left(x^2+y^2+z^2\right)\,dx\,dy\,dz = \int_{-3}^5\int_0^{2\pi}\int_0^2\left(r^2+z^2\right)r\;dr\,d\theta\,dz = \int_{-3}^5\int_0^{2\pi}\left(\frac{r^4}{4}+\frac{r^2z^2}{2}\right)\Big|_0^2 d\theta\,dz =$$

$$\int_{-3}^5\int_0^{2\pi}\left(4+2z^2\right)\,d\theta\,dz = \int_{-3}^5\left(4+2z^2\right)\theta\Big|_0^{2\pi} dz = \int_{-3}^5\left(8\pi+4\pi z^2\right)\,dz =$$

$$\left(8\pi z + \frac{4}{3}\pi z^3\right)\Big|_{-3}^5 = \left(40\pi + \frac{500}{3}\pi\right) - \left(-24\pi - 36\pi\right) = \frac{800}{3}\pi.$$

(5) (a) True. This is because all of the partial derivatives in the Jacobian matrix will be constants, since they are derivatives of linear functions. Hence, the determinant of the Jacobian will be a constant.

(b) True. The determinant of the Jacobian matrix $\mathbf{J} = \begin{bmatrix} \frac{\partial x}{\partial u} & \frac{\partial x}{\partial v} \\ \frac{\partial y}{\partial u} & \frac{\partial y}{\partial v} \end{bmatrix}$ is (-1); however,

$dx\,dy = \Big||\mathbf{J}|\Big|\;du\,dv = |-1|\;du\,dv = du\,dv.$

(c) True. This is explained and illustrated in Section 8.7 of the textbook just after Example 3, in Figure 8.18, and again in Example 4.

(d) False. The scaling factor is the absolute value of the determinant of the Jacobian matrix, which will not equal the determinant of the Jacobian matrix when that determinant is negative. For a counterexample, see the solution to part (b) of this Exercise, above.

Section 8.7

(1) (c) For this parabola, the coefficients a, b, and c are $a = 3$, $b = 1$, and $c = -2\sqrt{3}$. Since, $a \ne b$,

$\theta = \frac{1}{2}\arctan\left(\frac{c}{a-b}\right) = \frac{1}{2}\arctan\left(-\sqrt{3}\right) = -\frac{\pi}{6}$. Hence, $\mathbf{P} = \begin{bmatrix} \cos\theta & -\sin\theta \\ \sin\theta & \cos\theta \end{bmatrix} = \begin{bmatrix} \frac{\sqrt{3}}{2} & \frac{1}{2} \\ -\frac{1}{2} & \frac{\sqrt{3}}{2} \end{bmatrix}$.

Therefore, $x = \frac{\sqrt{3}}{2}u + \frac{1}{2}v$ and $y = -\frac{1}{2}u + \frac{\sqrt{3}}{2}v$. Next, we substitute these expressions for x and y

into the original equation for the conic section, producing

$3\left(\frac{\sqrt{3}}{2}u+\frac{1}{2}v\right)^2+\left(-\frac{1}{2}u+\frac{\sqrt{3}}{2}v\right)^2-2\sqrt{3}\left(\frac{\sqrt{3}}{2}u+\frac{1}{2}v\right)\left(-\frac{1}{2}u+\frac{\sqrt{3}}{2}v\right)-(1+12\sqrt{3})\left(\frac{\sqrt{3}}{2}u+\frac{1}{2}v\right)+$

$(12-\sqrt{3})\left(-\frac{1}{2}u+\frac{\sqrt{3}}{2}v\right)+26=0$. Expanding the terms on the left side of the equation yields

$\frac{9}{4}u^2+\frac{3\sqrt{3}}{2}uv+\frac{3}{4}v^2+\frac{1}{4}u^2-\frac{\sqrt{3}}{2}uv+\frac{3}{4}v^2+\frac{3}{2}u^2-\sqrt{3}uv-\frac{3}{2}v^2$

$-\left(\frac{\sqrt{3}}{2}+18\right)u-\left(\frac{1}{2}+6\sqrt{3}\right)v+\left(-6+\frac{\sqrt{3}}{2}\right)u+\left(6\sqrt{3}-\frac{3}{2}\right)v+26=0.$

Combining like terms, rearranging terms, and dividing by 2 yields the following equation in uv-coordinates: $v=2u^2-12u+13$, which can be expressed, after completing the square on the right side, as $(v+5)=2(u-3)^2$. From this form, we see that the conic section is a parabola and that its vertex in uv-coordinates is $(3,-5)$. Its vertex in xy-coordinates is $\mathbf{P}\begin{bmatrix}3\\-5\end{bmatrix}\approx\begin{bmatrix}0.0981\\-5.830\end{bmatrix}$; that is, the point $(0.0981,-5.830)$. Figures 12 and 13 on the next page show the graph of the parabola in each of the two coordinate systems.

(d) For this ellipse, the coefficients a, b, and c are $a=29$, $b=36$, and $c=-24$. Since, $a\neq b$, $\theta=\frac{1}{2}\arctan\left(\frac{c}{a-b}\right)=\frac{1}{2}\arctan\left(\frac{24}{7}\right)\approx 0.6435$ radians (about $36°52'$). Let us compute exact values for $\sin\theta$ and $\cos\theta$. Now $2\theta=\arctan(\frac{24}{7})$. So, using this right triangle,

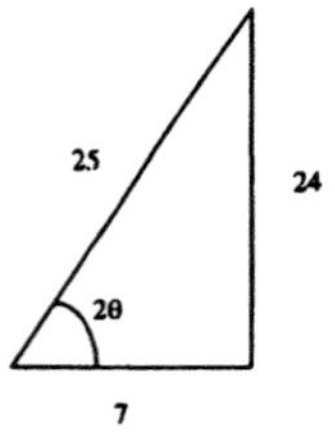

we see that $\cos 2\theta=\frac{7}{25}$, and so $\cos\theta=\sqrt{\frac{1+\cos 2\theta}{2}}=\sqrt{\frac{1+\frac{7}{25}}{2}}=\frac{4}{5}$ and $\sin\theta=\sqrt{1-\cos^2\theta}=\sqrt{1-\frac{16}{25}}=\frac{3}{5}$. Hence, $\mathbf{P}=\begin{bmatrix}\cos\theta & -\sin\theta\\ \sin\theta & \cos\theta\end{bmatrix}=\begin{bmatrix}\frac{4}{5} & -\frac{3}{5}\\ \frac{3}{5} & \frac{4}{5}\end{bmatrix}$. Therefore, $x=\frac{4}{5}u-\frac{3}{5}v$ and $y=\frac{3}{5}u+\frac{4}{5}v$. Next, we substitute these expressions for x and y into the original equation for the conic section, producing

$29\left(\frac{4}{5}u-\frac{3}{5}v\right)^2+36\left(\frac{3}{5}u+\frac{4}{5}v\right)^2-24\left(\frac{4}{5}u-\frac{3}{5}v\right)\left(\frac{3}{5}u+\frac{4}{5}v\right)-118\left(\frac{4}{5}u-\frac{3}{5}v\right)+24\left(\frac{3}{5}u+\frac{4}{5}v\right)-55=0.$

Expanding the terms on the left side of the equation yields

$\frac{464}{25}u^2-\frac{696}{25}uv+\frac{261}{25}v^2+\frac{324}{25}u^2+\frac{864}{25}uv+\frac{576}{25}v^2-\frac{288}{25}u^2-\frac{168}{25}uv+\frac{288}{25}v^2$

$-\frac{472}{5}u+\frac{354}{5}v+\frac{72}{5}u+\frac{96}{5}v-55=0.$

Combining like terms, rearranging terms, and dividing by 5 yields the following equation in uv-coordinates: $4u^2-16u+9v^2+18v=11$, which can be expressed, after completing the squares in both u and v, as $\frac{(u-2)^2}{9}+\frac{(v+1)^2}{4}=1$. From this form, we can see that the conic section is an ellipse and that its center in uv-coordinates is $(2,-1)$. Its center in xy-coordinates is

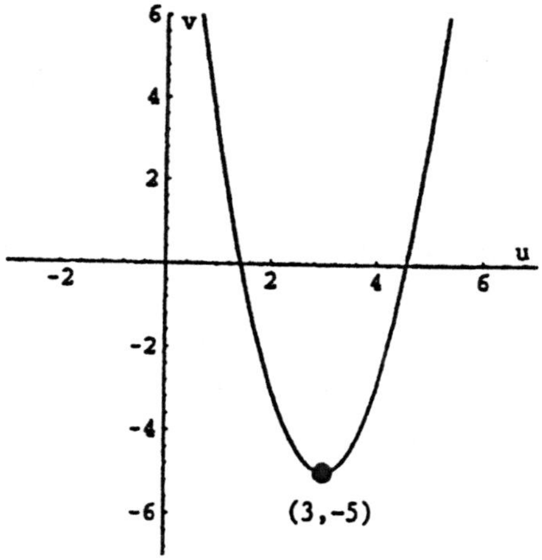

Figure 12: Rotated Parabola

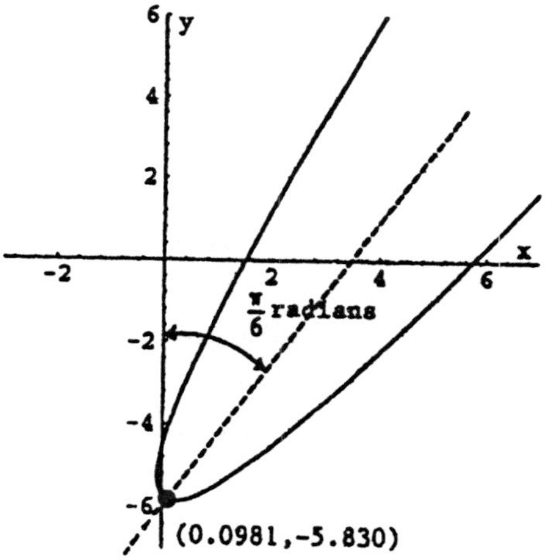

Figure 13: Original Parabola

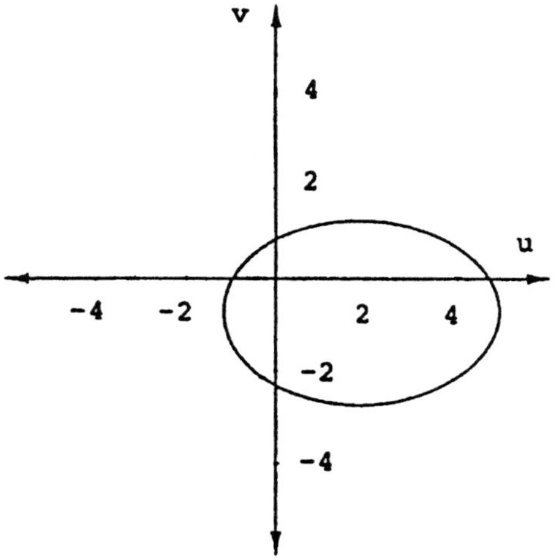

Figure 14: Rotated Ellipse

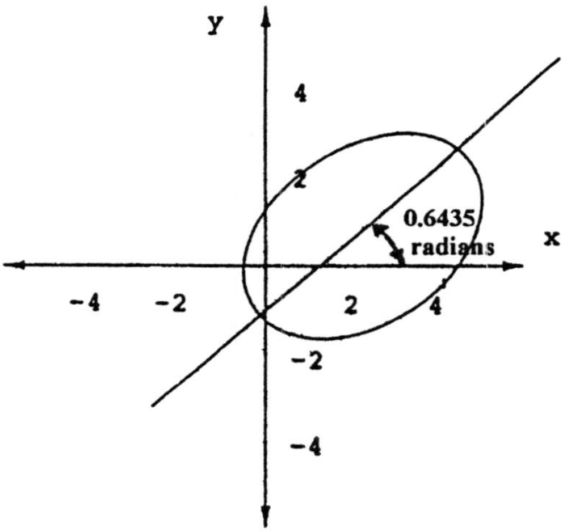

Figure 15: Original Ellipse

$\mathbf{P}\begin{bmatrix} 2 \\ -1 \end{bmatrix} = \begin{bmatrix} \frac{11}{5} \\ \frac{2}{5} \end{bmatrix}$; that is, the point $\left(\frac{11}{5}, \frac{2}{5}\right)$. Figures 14 and 15 show the graph of the ellipse in each of the two coordinate systems.

(f) (All answers are rounded to four significant digits.) For this hyperbola, the coefficients a, b, and c are $a = 8$, $b = -5$, and $c = 16$. Since, $a \neq b$, $\theta = \frac{1}{2}\arctan\left(\frac{c}{a-b}\right) = \frac{1}{2}\arctan\left(\frac{16}{13}\right) \approx 0.4442$

radians (about 25°27′). Hence, $\mathbf{P} = \begin{bmatrix} \cos\theta & -\sin\theta \\ \sin\theta & \cos\theta \end{bmatrix} = \begin{bmatrix} 0.9029 & -0.4298 \\ 0.4298 & 0.9029 \end{bmatrix}$. Therefore, $x = 0.9029u - 0.4298v$ and $y = 0.4298u + 0.9029v$. Next, we substitute these expressions for x and y into the original equation for the conic section, producing

$$8\,(0.9029u - 0.4298v)^2 - 5\,(0.4298u + 0.9029v)^2$$
$$+16\,(0.9029u - 0.4298v)\,(0.4298u + 0.9029v) - 37 = 0.$$

Expanding the terms on the left side of this equation yields

$$6.522u^2 - 6.209uv + 1.478v^2 - 0.9236u^2 - 3.881uv - 4.076v^2 + 6.209u^2 + 10.09uv - 6.209v^2 - 37 = 0.$$

Combining and rearranging terms and dividing by 37 yields the following equation in uv-coordinates: $\frac{u^2}{(1.770)^2} - \frac{v^2}{(2.050)^2} = 1$. From this form, we can see that the conic section is a hyperbola and that its center in uv-coordinates is $(0,0)$. Its center in xy-coordinates is $\mathbf{P}\begin{bmatrix} 0 \\ 0 \end{bmatrix} = \begin{bmatrix} 0 \\ 0 \end{bmatrix}$; that is, the origin. Figures 16 and 17 show the graph of the hyperbola in each of the two coordinate systems.

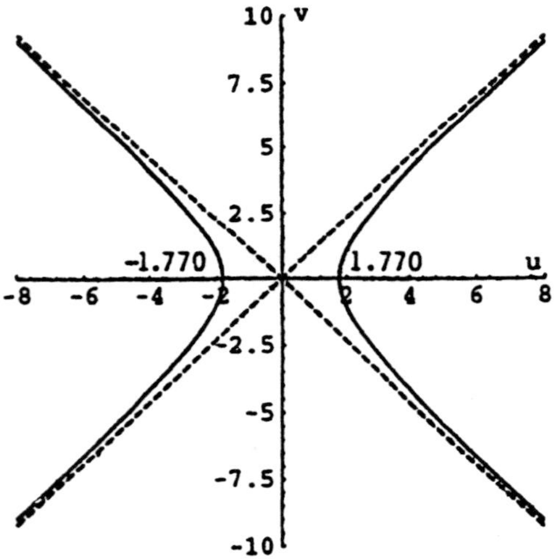

Figure 16: Rotated Hyperbola

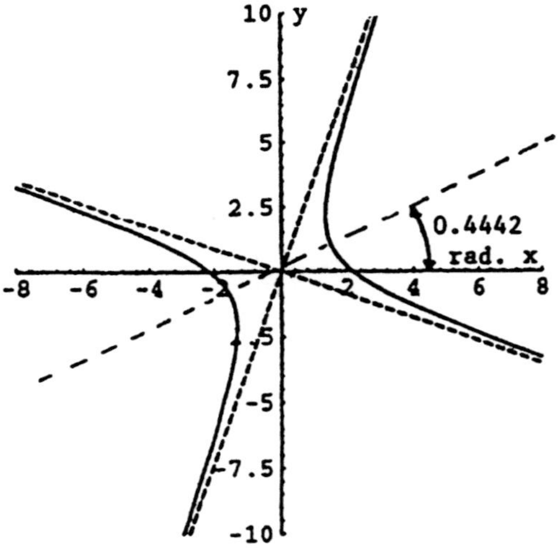

Figure 17: Original Hyperbola

(2) (a) True. This is because the coefficients of x^2 and y^2 are equal.

(b) False. Since the entire plane is rotated about the origin by the change in coordinates, the center of a hyperbola is rotated to a new position, unless the center is at the origin, in which case the center stays fixed. For example. the hyperbola $xy - x - y = 0$, which changes to $\frac{(u-\sqrt{2})^2}{2} - \frac{v^2}{2} = 1$ after a 45° counterclockwise rotation, has its center move from $(1,1)$ in xy-coordinates to $(\sqrt{2}, 0)$ in uv-coordinates.

(c) True. This is explained in Section 8.7 of the textbook, just after Example 1.

(d) False. For example, the conic section $x^2 - y = 0$ is a parabola with vertex at the origin whose graph is symmetric with respect to the y-axis, but not with respect to the x-axis. In general, if there is no xy term, any axis of symmetry must be parallel to either the x- or y-axis, but it does not have to *be* one of these axes. For example, the ellipse $(x-1)^2 + (y-1)^2 = 1$ (which, when expanded, has no xy term) has the lines $x = 1$ and $y = 1$ as horizontal and vertical axes of symmetry.

Section 8.8

(1) In each part we use the following list of vertices for the graphic in Figure 8.34(a) in Section 8.8 of the textbook: $(5,3)$, $(5,7)$, $(8,7)$, $(10,5)$ and $(8,3)$.

(a) A translation in the direction of the vector $[4,-2]$ is computed by merely adding this vector to each of the vertices in the graphic. This produces the new vertices $(9,1)$, $(9,5)$, $(12,5)$, $(14,3)$ and $(12,1)$.

(c) A reflection through a line of slope m through the origin is computed by multiplying the vector corresponding to each vertex by the matrix $\frac{1}{1+m^2}\begin{bmatrix} 1-m^2 & 2m \\ 2m & m^2-1 \end{bmatrix}$. In this problem, $m = 3$, and so the matrix used is $\mathbf{A} = \begin{bmatrix} -0.8 & 0.6 \\ 0.6 & 0.8 \end{bmatrix}$. Now, $\mathbf{A}\begin{bmatrix} 5 \\ 3 \end{bmatrix} = \begin{bmatrix} -2.2 \\ 5.4 \end{bmatrix}$, $\mathbf{A}\begin{bmatrix} 5 \\ 7 \end{bmatrix} = \begin{bmatrix} 0.2 \\ 8.6 \end{bmatrix}$, $\mathbf{A}\begin{bmatrix} 8 \\ 7 \end{bmatrix} = \begin{bmatrix} -2.2 \\ 10.4 \end{bmatrix}$, $\mathbf{A}\begin{bmatrix} 10 \\ 5 \end{bmatrix} = \begin{bmatrix} -5 \\ 10 \end{bmatrix}$ and $\mathbf{A}\begin{bmatrix} 8 \\ 3 \end{bmatrix} = \begin{bmatrix} -4.6 \\ 7.2 \end{bmatrix}$. Therefore, rounding each of these results to the nearest whole number, the vertices for the reflected graphic are $(-2,5)$, $(0,9)$, $(-2,10)$, $(-5,10)$ and $(-5,7)$.

(2) In each part we use the following list of vertices for the graphic in Figure 8.34(b) in Section 8.8 of the textbook: $(6,6)$, $(8,4)$, $(10,6)$, $(14,2)$ and $(18,6)$.

(b) A rotation about the origin through an angle θ is computed by multiplying the vector corresponding to each vertex by $\begin{bmatrix} \cos\theta & -\sin\theta \\ \sin\theta & \cos\theta \end{bmatrix}$. In this problem, $\theta = 120°$, and so the matrix is

$$\mathbf{A} = \begin{bmatrix} -\frac{1}{2} & -\frac{\sqrt{3}}{2} \\ \frac{\sqrt{3}}{2} & -\frac{1}{2} \end{bmatrix} \approx \begin{bmatrix} -0.5 & -0.866 \\ 0.866 & -0.5 \end{bmatrix}.$$ Now, $\mathbf{A}\begin{bmatrix} 6 \\ 6 \end{bmatrix} \approx \begin{bmatrix} -8.20 \\ 2.20 \end{bmatrix}$, $\mathbf{A}\begin{bmatrix} 8 \\ 4 \end{bmatrix} \approx \begin{bmatrix} -7.46 \\ 4.93 \end{bmatrix}$, $\mathbf{A}\begin{bmatrix} 10 \\ 6 \end{bmatrix} \approx \begin{bmatrix} -10.20 \\ 5.66 \end{bmatrix}$, $\mathbf{A}\begin{bmatrix} 14 \\ 2 \end{bmatrix} \approx \begin{bmatrix} -8.73 \\ 11.12 \end{bmatrix}$ and $\mathbf{A}\begin{bmatrix} 18 \\ 6 \end{bmatrix} \approx \begin{bmatrix} -14.20 \\ 12.59 \end{bmatrix}$.

Therefore, rounding each of these results to the nearest whole number, the vertices for the rotated graphic are $(-8,2)$, $(-7,5)$, $(-10,6)$, $(-9,11)$ and $(-14,13)$. The new graphic is illustrated in Figure 18 on the next page.

(d) A scaling about the origin with scale factors c in the x-direction and d in the y-direction is computed by multiplying the vector corresponding to each vertex by the matrix $\begin{bmatrix} c & 0 \\ 0 & d \end{bmatrix}$. In this problem, $c = \frac{1}{2}$ and $d = 3$, and so the matrix is $\mathbf{A} = \begin{bmatrix} \frac{1}{2} & 0 \\ 0 & 3 \end{bmatrix}$. Now, $\mathbf{A}\begin{bmatrix} 6 \\ 6 \end{bmatrix} = \begin{bmatrix} 3 \\ 18 \end{bmatrix}$,

$\mathbf{A}\begin{bmatrix} 8 \\ 4 \end{bmatrix} = \begin{bmatrix} 4 \\ 12 \end{bmatrix}$, $\mathbf{A}\begin{bmatrix} 10 \\ 6 \end{bmatrix} = \begin{bmatrix} 5 \\ 18 \end{bmatrix}$, $\mathbf{A}\begin{bmatrix} 14 \\ 2 \end{bmatrix} = \begin{bmatrix} 7 \\ 6 \end{bmatrix}$ and $\mathbf{A}\begin{bmatrix} 18 \\ 6 \end{bmatrix} = \begin{bmatrix} 9 \\ 18 \end{bmatrix}$. Therefore, the vertices for the scaled graphic are $(3,18)$, $(4,12)$, $(5,18)$, $(7,6)$ and $(9,18)$. The new graphic is illustrated in Figure 19.

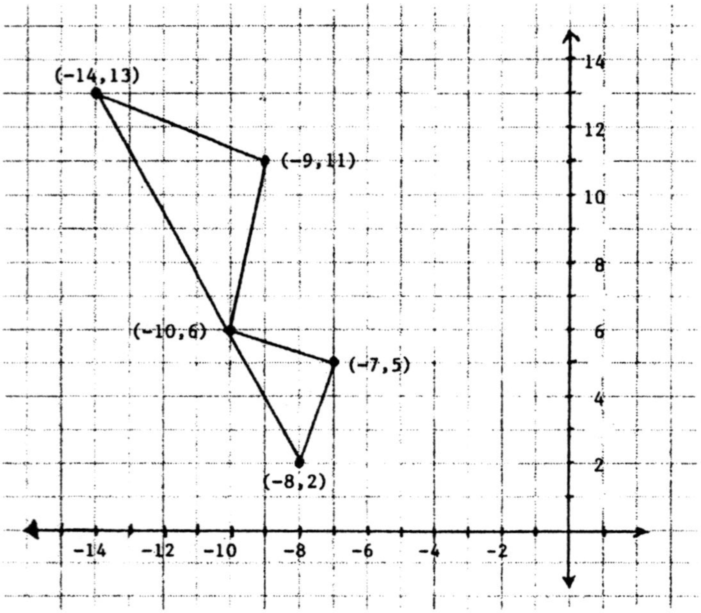

Figure 18

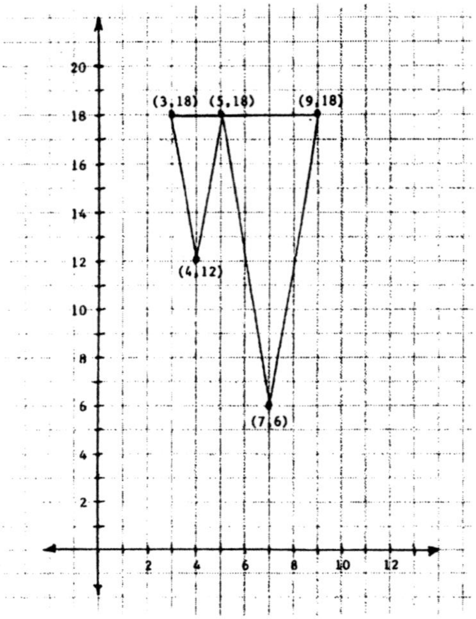

Figure 19

(3) In each part we use the following list of vertices for the graphic in Figure 8.34(c) in Section 8.8 of the textbook: $(2,4)$, $(2,10)$, (6.7), $(8,10)$ and $(10,4)$. In homogeneous coordinates, we express these points as the vectors $[2,4,1]$, $[2,10,1]$, $[6,7,1]$, $[8,10,1]$ and $[10,4,1]$.

(a) When using homogeneous coordinates, a rotation about the origin through an angle θ is computed by multiplying the vector corresponding to each vertex by $\mathbf{A} = \begin{bmatrix} \cos\theta & -\sin\theta & 0 \\ \sin\theta & \cos\theta & 0 \\ 0 & 0 & 1 \end{bmatrix}$. In this problem, $\theta = 45^\circ$, and so the matrix is $\mathbf{A} = \begin{bmatrix} \frac{\sqrt{2}}{2} & -\frac{\sqrt{2}}{2} & 0 \\ \frac{\sqrt{2}}{2} & \frac{\sqrt{2}}{2} & 0 \\ 0 & 0 & 1 \end{bmatrix} \approx \begin{bmatrix} 0.707 & -0.707 & 0 \\ 0.707 & 0.707 & 0 \\ 0 & 0 & 1 \end{bmatrix}$. Also, a reflection through a line of slope m through the origin is computed by multiplying the vector corresponding to each vertex by the matrix $\mathbf{B} = \frac{1}{1+m^2}\begin{bmatrix} 1-m^2 & 2m & 0 \\ 2m & m^2-1 & 0 \\ 0 & 0 & 1+m^2 \end{bmatrix}$. In this problem, $m = \frac{1}{2}$, and so the matrix used is $\mathbf{B} = \begin{bmatrix} 0.6 & 0.8 & 0 \\ 0.8 & -0.6 & 0 \\ 0 & 0 & 1 \end{bmatrix}$. The composition of the rotation followed by the reflection is therefore represented by the matrix

$\mathbf{BA} \approx \begin{bmatrix} 0.990 & 0.141 & 0 \\ 0.141 & -0.990 & 0 \\ 0 & 0 & 1 \end{bmatrix}$. Now $\mathbf{BA}\begin{bmatrix} 2 \\ 4 \\ 1 \end{bmatrix} \approx \begin{bmatrix} 2.545 \\ -3.676 \\ 1 \end{bmatrix}$, $\mathbf{BA}\begin{bmatrix} 2 \\ 10 \\ 1 \end{bmatrix} \approx \begin{bmatrix} 3.394 \\ -9.615 \\ 1 \end{bmatrix}$,

$\mathbf{BA}\begin{bmatrix} 6 \\ 7 \\ 1 \end{bmatrix} \approx \begin{bmatrix} 6.930 \\ -6.080 \\ 1 \end{bmatrix}$, $\mathbf{BA}\begin{bmatrix} 8 \\ 10 \\ 1 \end{bmatrix} \approx \begin{bmatrix} 9.332 \\ -8.767 \\ 1 \end{bmatrix}$ and $\mathbf{BA}\begin{bmatrix} 10 \\ 4 \\ 1 \end{bmatrix} \approx \begin{bmatrix} 10.464 \\ -2.545 \\ 1 \end{bmatrix}$.

Therefore, since each of the product vectors has a 1 in the 3rd coordinate, each can be converted from homogeneous coordinates back to regular coordinates by merely dropping that 1. Doing this and rounding each of these results to the nearest whole number we get the following vertices for the moved graphic: $(3,-4)$, $(3,-10)$, $(7,-6)$, $(9,-9)$ and $(10,-3)$.

(c) When using homogeneous coordinates, a scaling about the origin with scale factors c in the x-direction and d in the y-direction is computed by multiplying the vector corresponding to each vertex by $\mathbf{A} = \begin{bmatrix} c & 0 & 0 \\ 0 & d & 0 \\ 0 & 0 & 1 \end{bmatrix}$. In this problem, $c = 3$ and $d = \frac{1}{2}$, and so $\mathbf{A} = \begin{bmatrix} 3 & 0 & 0 \\ 0 & \frac{1}{2} & 0 \\ 0 & 0 & 1 \end{bmatrix}$. Also, a reflection through a line of slope m through the origin is computed by multiplying the vector corresponding to each vertex by the matrix $\mathbf{B} = \frac{1}{1+m^2}\begin{bmatrix} 1-m^2 & 2m & 0 \\ 2m & m^2-1 & 0 \\ 0 & 0 & 1+m^2 \end{bmatrix}$. In this problem, $m = 2$, and so $\mathbf{B} = \begin{bmatrix} -0.6 & 0.8 & 0 \\ 0.8 & 0.6 & 0 \\ 0 & 0 & 1 \end{bmatrix}$. The composition of the scaling followed by the reflection is therefore represented by the matrix $\mathbf{BA} = \begin{bmatrix} -1.8 & 0.4 & 0 \\ 2.4 & 0.3 & 0 \\ 0 & 0 & 1 \end{bmatrix}$. Now,

$\mathbf{BA}\begin{bmatrix} 2 \\ 4 \\ 1 \end{bmatrix} = \begin{bmatrix} -2 \\ 6 \\ 1 \end{bmatrix}$, $\mathbf{BA}\begin{bmatrix} 2 \\ 10 \\ 1 \end{bmatrix} = \begin{bmatrix} 0.4 \\ 7.8 \\ 1 \end{bmatrix}$, $\mathbf{BA}\begin{bmatrix} 6 \\ 7 \\ 1 \end{bmatrix} = \begin{bmatrix} -8 \\ 16.5 \\ 1 \end{bmatrix}$,

$\mathbf{BA}\begin{bmatrix} 8 \\ 10 \\ 1 \end{bmatrix} = \begin{bmatrix} -10.4 \\ 22.2 \\ 1 \end{bmatrix}$ and $\mathbf{BA}\begin{bmatrix} 10 \\ 4 \\ 1 \end{bmatrix} = \begin{bmatrix} -16.4 \\ 25.2 \\ 1 \end{bmatrix}$. Therefore, since each of the product vectors has a 1 in the 3rd coordinate, each can be converted from homogeneous coordinates back to regular coordinates by merely dropping that 1. Doing this and rounding each of these results to the nearest whole number we get the following vertices for the moved graphic: $(-2,6)$, $(0,8)$, $(-8,17)$, $(-10,22)$ and $(-16,25)$.

(4) In each part we use the following list of vertices for the graphic in Figure 8.35(a) in Section 8.8 of the textbook: $(5,4)$, $(5,9)$, $(8,6)$, $(8,9)$, $(8,11)$ and $(11,4)$. In homogeneous coordinates, we express these points as the vectors $[5,4,1]$, $[5,9,1]$, $[8,6,1]$, $[8,9,1]$, $[8,11,1]$ and $[11,4,1]$.

(a) Since the given rotation is not about the origin, we use the Similarity Method. First, we must consider the translation along the vector $[-8,-9]$ that moves the center of the rotation, $(8,9)$, to the origin. The matrix for this translation in homogeneous coordinates is $\mathbf{T}_1 = \begin{bmatrix} 1 & 0 & -8 \\ 0 & 1 & -9 \\ 0 & 0 & 1 \end{bmatrix}$. Next, we need to compute the matrix for a rotation of 120° about the origin. When using homogeneous coordinates, a rotation about the origin through an angle θ is computed by multiplying the vector corresponding to each vertex by the matrix $\mathbf{A} = \begin{bmatrix} \cos\theta & -\sin\theta & 0 \\ \sin\theta & \cos\theta & 0 \\ 0 & 0 & 1 \end{bmatrix}$. In this problem, $\theta = 120°$, and so the matrix used is $\mathbf{A} = \begin{bmatrix} -\frac{1}{2} & -\frac{\sqrt{3}}{2} & 0 \\ \frac{\sqrt{3}}{2} & -\frac{1}{2} & 0 \\ 0 & 0 & 1 \end{bmatrix} \approx \begin{bmatrix} -0.5 & -0.866 & 0 \\ 0.866 & -0.5 & 0 \\ 0 & 0 & 1 \end{bmatrix}$. Finally, we need to translate along the vector [8, 9] to bring the origin back to the point (8, 9). The matrix for this translation is $\mathbf{T}_2 = \begin{bmatrix} 1 & 0 & 8 \\ 0 & 1 & 9 \\ 0 & 0 & 1 \end{bmatrix}$. Hence, the rotation of 120° about (8, 9) is represented by the matrix $\mathbf{B} = \mathbf{T}_2\mathbf{A}\mathbf{T}_1 \approx \begin{bmatrix} -0.500 & -0.866 & 19.794 \\ 0.866 & -0.500 & 6.572 \\ 0 & 0 & 1 \end{bmatrix}$. Now $\mathbf{B}\begin{bmatrix} 5 \\ 4 \\ 1 \end{bmatrix} \approx \begin{bmatrix} 13.830 \\ 8.902 \\ 1 \end{bmatrix}$,

$\mathbf{B}\begin{bmatrix} 5 \\ 9 \\ 1 \end{bmatrix} \approx \begin{bmatrix} 9.500 \\ 6.402 \\ 1 \end{bmatrix}$, $\mathbf{B}\begin{bmatrix} 8 \\ 6 \\ 1 \end{bmatrix} \approx \begin{bmatrix} 10.598 \\ 10.500 \\ 1 \end{bmatrix}$, $\mathbf{B}\begin{bmatrix} 8 \\ 9 \\ 1 \end{bmatrix} = \begin{bmatrix} 8 \\ 9 \\ 1 \end{bmatrix}$,

$\mathbf{B}\begin{bmatrix} 8 \\ 11 \\ 1 \end{bmatrix} \approx \begin{bmatrix} 6.268 \\ 8.000 \\ 1 \end{bmatrix}$ and $\mathbf{B}\begin{bmatrix} 11 \\ 4 \\ 1 \end{bmatrix} \approx \begin{bmatrix} 10.830 \\ 14.098 \\ 1 \end{bmatrix}$. Therefore, since each of the product vectors has a 1 in the 3rd coordinate, each can be converted from homogeneous coordinates back to regular coordinates by merely dropping that 1. Doing this and rounding these results to the nearest whole number we get the following vertices for the moved graphic: (14, 9), (10, 6), (11, 11), (8, 9), (6, 8) and (11, 14).

(c) Since the given scaling is not about the origin, we use the Similarity Method. First, we must consider the translation along the vector [−8, −4] that moves the center of the scaling, (8, 4), to the origin. The matrix for this translation in homogeneous coordinates is $\mathbf{T}_1 = \begin{bmatrix} 1 & 0 & -8 \\ 0 & 1 & -4 \\ 0 & 0 & 1 \end{bmatrix}$.

Next we need to find the matrix for a scaling about the origin with factors 2 in the x-direction and $\frac{1}{3}$ in the y-direction. When using homogeneous coordinates, a scaling about the origin with scale factors c in the x-direction and d in the y-direction is computed by multiplying the vector corresponding to each vertex by the matrix $\mathbf{A} = \begin{bmatrix} c & 0 & 0 \\ 0 & d & 0 \\ 0 & 0 & 1 \end{bmatrix}$. In this problem, $c = 2$ and $d = \frac{1}{3}$, and so the matrix is $\mathbf{A} = \begin{bmatrix} 2 & 0 & 0 \\ 0 & \frac{1}{3} & 0 \\ 0 & 0 & 1 \end{bmatrix}$. Finally, we need to translate along the vector [8, 4] to

bring the origin back to the point (8, 4). The matrix for this translation is $\mathbf{T}_2 = \begin{bmatrix} 1 & 0 & 8 \\ 0 & 1 & 4 \\ 0 & 0 & 1 \end{bmatrix}$.
Hence, the scaling with factor 2 in the x-direction and $\frac{1}{3}$ in the y-direction about the point (8, 4) is represented by the matrix $\mathbf{B} = \mathbf{T}_2\mathbf{A}\mathbf{T}_1 = \begin{bmatrix} 2 & 0 & -8 \\ 0 & \frac{1}{3} & \frac{8}{3} \\ 0 & 0 & 1 \end{bmatrix}$. Now, $\mathbf{B}\begin{bmatrix} 5 \\ 4 \\ 1 \end{bmatrix} = \begin{bmatrix} 2 \\ 4 \\ 1 \end{bmatrix}$,

$\mathbf{B}\begin{bmatrix} 5 \\ 9 \\ 1 \end{bmatrix} = \begin{bmatrix} 2 \\ \frac{17}{3} \\ 1 \end{bmatrix}$, $\mathbf{B}\begin{bmatrix} 8 \\ 6 \\ 1 \end{bmatrix} = \begin{bmatrix} 8 \\ \frac{14}{3} \\ 1 \end{bmatrix}$, $\mathbf{B}\begin{bmatrix} 8 \\ 9 \\ 1 \end{bmatrix} = \begin{bmatrix} 8 \\ \frac{17}{3} \\ 1 \end{bmatrix}$,

$\mathbf{B}\begin{bmatrix} 8 \\ 11 \\ 1 \end{bmatrix} = \begin{bmatrix} 8 \\ \frac{19}{3} \\ 1 \end{bmatrix}$ and $\mathbf{B}\begin{bmatrix} 11 \\ 4 \\ 1 \end{bmatrix} = \begin{bmatrix} 14 \\ 4 \\ 1 \end{bmatrix}$. Therefore, since each of the product vectors has a 1 in the 3rd coordinate, each can be converted from homogeneous coordinates back to regular coordinates by merely dropping that 1. Doing this and rounding each of these results to the nearest whole number we get the following vertices for the moved graphic: (2, 4), (2, 6), (8, 5), (8, 6), (8, 6) and (14, 4).

(5) (b) We use the following list of vertices for the graphic in Figure 8.35(b) in Section 8.8 of the textbook: (7, 3), (7, 6), (10, 8), (13, 6) and (13, 3). In homogeneous coordinates, we express these points as the vectors [7, 3, 1], [7, 6, 1], [10, 8, 1], [13, 6, 1] and [13, 3, 1].
Since the given reflection is not about a line through the origin, we use the Similarity Method. First, we consider the translation along the vector [0, 10] that moves the y-intercept of the line of reflection, (0, −10), to the origin. The matrix for this translation in homogeneous coordinates is $\mathbf{T}_1 = \begin{bmatrix} 1 & 0 & 0 \\ 0 & 1 & 10 \\ 0 & 0 & 1 \end{bmatrix}$. Next, we find the matrix for the reflection about the line $y = 4x$ (since we have shifted the y-intercept to the origin). A reflection through a line of slope m through the origin is computed by multiplying the vector corresponding to each vertex by $\mathbf{A} = \frac{1}{1+m^2}\begin{bmatrix} 1-m^2 & 2m & 0 \\ 2m & m^2-1 & 0 \\ 0 & 0 & 1+m^2 \end{bmatrix}$. In this problem, $m = 4$, and so $\mathbf{A} = \frac{1}{17}\begin{bmatrix} -15 & 8 & 0 \\ 8 & 15 & 0 \\ 0 & 0 & 17 \end{bmatrix}$.
Finally, we need to translate along the vector [0, −10] to bring the origin back to the point (0, −10). The matrix for this translation is $\mathbf{T}_2 = \begin{bmatrix} 1 & 0 & 0 \\ 0 & 1 & -10 \\ 0 & 0 & 1 \end{bmatrix}$. Hence, the reflection about the line $y = 4x - 10$ is represented by the matrix $\mathbf{B} = \mathbf{T}_2\mathbf{A}\mathbf{T}_1 = \frac{1}{17}\begin{bmatrix} -15 & 8 & 80 \\ 8 & 15 & -20 \\ 0 & 0 & 17 \end{bmatrix}$. Now

$\mathbf{B}\begin{bmatrix} 7 \\ 3 \\ 1 \end{bmatrix} \approx \begin{bmatrix} -0.059 \\ 4.765 \\ 1 \end{bmatrix}$, $\mathbf{B}\begin{bmatrix} 7 \\ 6 \\ 1 \end{bmatrix} \approx \begin{bmatrix} 1.353 \\ 7.412 \\ 1 \end{bmatrix}$, $\mathbf{B}\begin{bmatrix} 10 \\ 8 \\ 1 \end{bmatrix} \approx \begin{bmatrix} -0.353 \\ 10.588 \\ 1 \end{bmatrix}$,

$$\mathbf{B}\begin{bmatrix} 13 \\ 6 \\ 1 \end{bmatrix} \approx \begin{bmatrix} -3.941 \\ 10.235 \\ 1 \end{bmatrix} \text{ and } \mathbf{B}\begin{bmatrix} 13 \\ 3 \\ 1 \end{bmatrix} \approx \begin{bmatrix} -5.353 \\ 7.588 \\ 1 \end{bmatrix}.$$

Therefore, since each of the product vectors has a 1 in the 3rd coordinate, each can be converted from homogeneous coordinates back to regular coordinates by merely dropping that 1. Doing this and rounding each of these results to the nearest whole number we get the following vertices for the moved graphic: $(0,5)$, $(1,7)$, $(0,11)$, $(-4,10)$ and $(-5,8)$.

(6) In each part we use the following list of vertices for the graphic in Figure 8.35(c) in Section 8.8 of the textbook: $(5,3)$, $(6,6)$, $(8,9)$, $(9,4)$, $(9,7)$ and $(12,8)$. In homogeneous coordinates, we express these points as the vectors $[5,3,1]$, $[6,6,1]$, $[8,9,1]$, $[9,4,1]$, $[9,7,1]$ and $[12,8,1]$.

(a) We begin with the rotation. Since the given rotation is not about the origin, we use the Similarity Method. First, we consider the translation along the vector $[-12,-8]$ that moves the center of the rotation, $(12,8)$, to the origin. The matrix for this translation in homogeneous coordinates is $\mathbf{T}_1 = \begin{bmatrix} 1 & 0 & -12 \\ 0 & 1 & -8 \\ 0 & 0 & 1 \end{bmatrix}$. Next, we need to compute the matrix for a rotation of 60° about the origin.

When using homogeneous coordinates, a rotation about the origin through an angle θ is computed by multiplying the vector corresponding to each vertex by the matrix $\mathbf{A} = \begin{bmatrix} \cos\theta & -\sin\theta & 0 \\ \sin\theta & \cos\theta & 0 \\ 0 & 0 & 1 \end{bmatrix}$.

In this problem, $\theta = 60°$, and so $\mathbf{A} = \begin{bmatrix} \frac{1}{2} & -\frac{\sqrt{3}}{2} & 0 \\ \frac{\sqrt{3}}{2} & \frac{1}{2} & 0 \\ 0 & 0 & 1 \end{bmatrix} \approx \begin{bmatrix} 0.5 & -0.866 & 0 \\ 0.866 & 0.5 & 0 \\ 0 & 0 & 1 \end{bmatrix}$. To finish the rotation, we need to translate along the vector $[12,8]$ to bring the origin back to the point $(12,8)$. The matrix for this is $\mathbf{T}_2 = \begin{bmatrix} 1 & 0 & 12 \\ 0 & 1 & 8 \\ 0 & 0 & 1 \end{bmatrix}$. Hence, the rotation of 60° about $(12,8)$ is represented by $\mathbf{B} = \mathbf{T}_2\mathbf{A}\mathbf{T}_1$.

Next, we compute the matrix for the reflection about $y = \frac{1}{2}x + 6$. Since the given reflection is not about a line through the origin, we use the Similarity Method. First, we must consider the translation along the vector $[0,-6]$ that moves the y-intercept of the line of reflection, $(0,6)$, to the origin. The matrix for this translation in homogeneous coordinates is $\mathbf{T}_3 = \begin{bmatrix} 1 & 0 & 0 \\ 0 & 1 & -6 \\ 0 & 0 & 1 \end{bmatrix}$.

Next, we need to find the matrix for the reflection about the line $y = \frac{1}{2}x$ (since we have shifted the y-intercept to the origin). A reflection through a line of slope m through the origin is computed by multiplying the vector corresponding to each vertex by $\mathbf{C} = \frac{1}{1+m^2}\begin{bmatrix} 1-m^2 & 2m & 0 \\ 2m & m^2-1 & 0 \\ 0 & 0 & 1+m^2 \end{bmatrix}$.

In this problem, $m = \frac{1}{2}$, and so $\mathbf{C} = \begin{bmatrix} 0.6 & 0.8 & 0 \\ 0.8 & -0.6 & 0 \\ 0 & 0 & 1 \end{bmatrix}$. Finally, we need to translate along the vector [0, 6] to bring the origin back to the point (0, 6). The matrix for this translation is $\mathbf{T}_4 = \begin{bmatrix} 1 & 0 & 0 \\ 0 & 1 & 6 \\ 0 & 0 & 1 \end{bmatrix}$. Hence, the reflection about the line $y = \frac{1}{2}x + 6$ is represented by the matrix $\mathbf{D} = \mathbf{T}_4\mathbf{C}\mathbf{T}_3$.

Finally, the composition of the rotation and the reflection is represented by the matrix

$\mathbf{DB} = \mathbf{T}_4\mathbf{C}\mathbf{T}_3\mathbf{T}_2\mathbf{A}\mathbf{T}_1 \approx \begin{bmatrix} 0.993 & -0.120 & -2.157 \\ -0.120 & -0.993 & 23.778 \\ 0 & 0 & 1 \end{bmatrix}$. Now $\mathbf{DB}\begin{bmatrix} 5 \\ 3 \\ 1 \end{bmatrix} \approx \begin{bmatrix} 2.448 \\ 20.201 \\ 1 \end{bmatrix}$,

$\mathbf{DB}\begin{bmatrix} 6 \\ 6 \\ 1 \end{bmatrix} \approx \begin{bmatrix} 3.082 \\ 17.103 \\ 1 \end{bmatrix}$, $\mathbf{DB}\begin{bmatrix} 8 \\ 9 \\ 1 \end{bmatrix} \approx \begin{bmatrix} 4.709 \\ 13.886 \\ 1 \end{bmatrix}$, $\mathbf{DB}\begin{bmatrix} 9 \\ 4 \\ 1 \end{bmatrix} \approx \begin{bmatrix} 6.300 \\ 18.730 \\ 1 \end{bmatrix}$,

$\mathbf{DB}\begin{bmatrix} 9 \\ 7 \\ 1 \end{bmatrix} \approx \begin{bmatrix} 5.941 \\ 15.752 \\ 1 \end{bmatrix}$ and $\mathbf{DB}\begin{bmatrix} 12 \\ 8 \\ 1 \end{bmatrix} = \begin{bmatrix} 8.8 \\ 14.4 \\ 1 \end{bmatrix}$. Therefore, since each of the product vectors has a 1 in the 3rd coordinate, each can be converted from homogeneous coordinates back to regular coordinates by merely dropping that 1. Doing this and rounding each of these results to the nearest whole number we get the following vertices for the moved graphic: (2, 20), (3, 17), (5, 14), (6, 19), (6, 16) and (9, 14). The new graphic is illustrated in Figure 20.

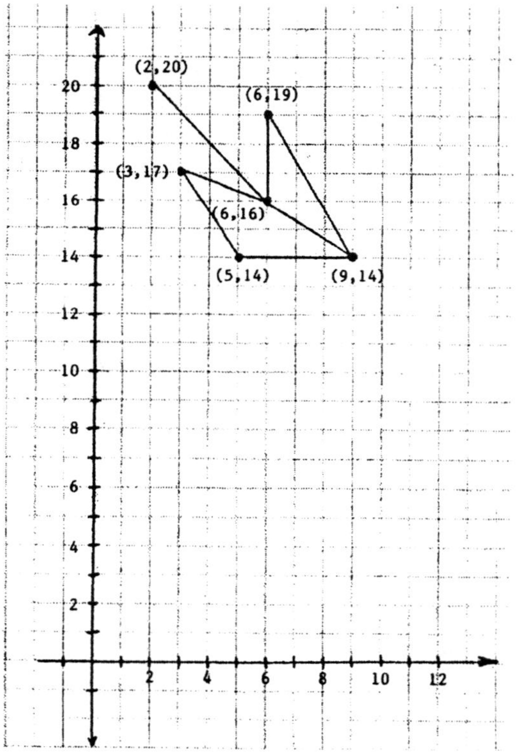

Figure 20

(c) We begin with the scaling. Since the given scaling is not about the origin, we use the Similarity Method. First, we must consider the translation along the vector $[-9,-4]$ that moves the center of the scaling, $(9,4)$, to the origin. The matrix for this translation in homogeneous coordinates is $\mathbf{T}_1 = \begin{bmatrix} 1 & 0 & -9 \\ 0 & 1 & -4 \\ 0 & 0 & 1 \end{bmatrix}$. Next we need to find the matrix for a scaling about the origin with factors $\frac{1}{3}$ in the x-direction and 2 in the y-direction. When using homogeneous coordinates, a scaling about the origin with scale factors c in the x-direction and d in the y-direction is computed by multiplying the vector corresponding to each vertex by $\mathbf{A} = \begin{bmatrix} c & 0 & 0 \\ 0 & d & 0 \\ 0 & 0 & 1 \end{bmatrix}$. In this problem, $c = \frac{1}{3}$ and $d = 2$, and so $\mathbf{A} = \begin{bmatrix} \frac{1}{3} & 0 & 0 \\ 0 & 2 & 0 \\ 0 & 0 & 1 \end{bmatrix}$. Finally, we need to translate along the vector $[9,4]$ to bring the origin back to the point $(9,4)$. The matrix for this is $\mathbf{T}_2 = \begin{bmatrix} 1 & 0 & 9 \\ 0 & 1 & 4 \\ 0 & 0 & 1 \end{bmatrix}$. Hence, the scaling with factor $\frac{1}{3}$ in the x-direction and 2 in the y-direction about the point $(9,4)$ is represented by the matrix $\mathbf{B} = \mathbf{T}_2\mathbf{A}\mathbf{T}_1$.

Next, we compute the matrix for the rotation. To do this, we first must consider the translation along the vector $[-2,-9]$ that moves the center of the rotation, $(2,9)$, to the origin. The matrix for this translation in homogeneous coordinates is $\mathbf{T}_3 = \begin{bmatrix} 1 & 0 & -2 \\ 0 & 1 & -9 \\ 0 & 0 & 1 \end{bmatrix}$. Next, we need to find the matrix for a rotation of 150° about the origin. When using homogeneous coordinates, a rotation about the origin through an angle θ is computed by multiplying the vector corresponding to each vertex by the matrix $\mathbf{C} = \begin{bmatrix} \cos\theta & -\sin\theta & 0 \\ \sin\theta & \cos\theta & 0 \\ 0 & 0 & 1 \end{bmatrix}$. In this problem, $\theta = 150°$, and so the matrix is $\mathbf{C} = \begin{bmatrix} -\frac{\sqrt{3}}{2} & -\frac{1}{2} & 0 \\ \frac{1}{2} & -\frac{\sqrt{3}}{2} & 0 \\ 0 & 0 & 1 \end{bmatrix} \approx \begin{bmatrix} -0.866 & -0.5 & 0 \\ 0.5 & -0.866 & 0 \\ 0 & 0 & 1 \end{bmatrix}$. To finish the rotation, we need to translate along the vector $[2,9]$ to bring the origin back to the point $(2,9)$. The matrix for this translation is $\mathbf{T}_4 = \begin{bmatrix} 1 & 0 & 2 \\ 0 & 1 & 9 \\ 0 & 0 & 1 \end{bmatrix}$. Hence, the rotation of 150° about $(2,9)$ is represented by the matrix $\mathbf{D} = \mathbf{T}_4\mathbf{C}\mathbf{T}_3$.

Finally, the composition of the rotation and the reflection is represented by the matrix $\mathbf{DB} = \mathbf{T}_4\mathbf{C}\mathbf{T}_3\mathbf{T}_2\mathbf{A}\mathbf{T}_1 \approx \begin{bmatrix} -0.289 & -1 & 5.036 \\ 0.167 & -1.732 & 22.258 \\ 0 & 0 & 1 \end{bmatrix}$. Now $\mathbf{DB}\begin{bmatrix} 5 \\ 3 \\ 1 \end{bmatrix} \approx \begin{bmatrix} 0.593 \\ 17.896 \\ 1 \end{bmatrix}$,

$$\mathbf{DB}\begin{bmatrix} 6 \\ 6 \\ 1 \end{bmatrix} \approx \begin{bmatrix} -2.696 \\ 12.866 \\ 1 \end{bmatrix}, \mathbf{DB}\begin{bmatrix} 8 \\ 9 \\ 1 \end{bmatrix} \approx \begin{bmatrix} -6.274 \\ 8.003 \\ 1 \end{bmatrix}, \mathbf{DB}\begin{bmatrix} 9 \\ 4 \\ 1 \end{bmatrix} \approx \begin{bmatrix} -1.562 \\ 16.830 \\ 1 \end{bmatrix},$$

$\mathbf{DB}\begin{bmatrix} 9 \\ 7 \\ 1 \end{bmatrix} \approx \begin{bmatrix} -4.562 \\ 11.634 \\ 1 \end{bmatrix}$ and $\mathbf{DB}\begin{bmatrix} 12 \\ 8 \\ 1 \end{bmatrix} \approx \begin{bmatrix} -6.428 \\ 10.402 \\ 1 \end{bmatrix}$. Therefore, since each of the product vectors obtained has a 1 in the 3rd coordinate, each can be converted from homogeneous coordinates back to regular coordinates by dropping that 1. Doing this and rounding each of these results to the nearest whole number yields the following vertices for the moved graphic: $(1, 18)$, $(-3, 13)$, $(-6, 8)$, $(-2, 17)$, $(-5, 12)$ and $(-6, 10)$. The new graphic is illustrated in Figure 21.

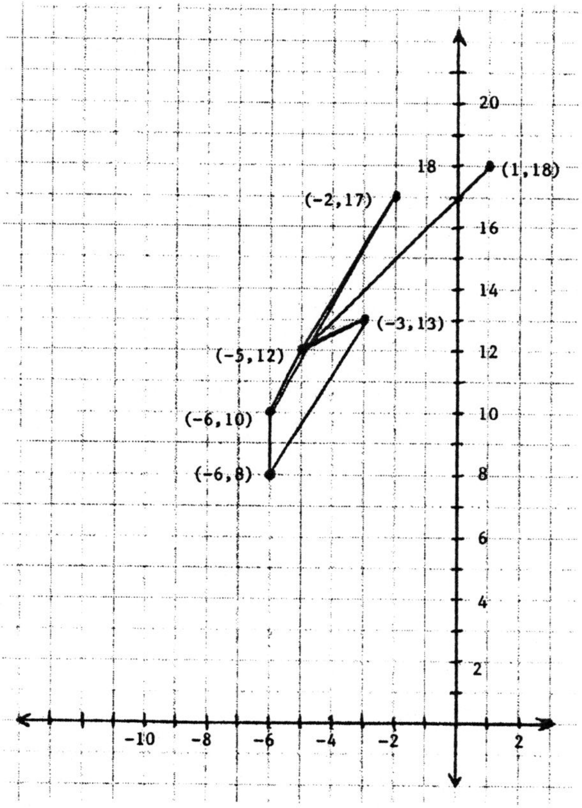

Figure 21

(9) (b) We use the following list of vertices for the graphic in Figure 8.35(b) in Section 8.8 of the textbook: $(7, 3)$, $(7, 6)$, $(10, 8)$, $(13, 6)$ and $(13, 3)$. In homogeneous coordinates, we express these points as the vectors $[7, 3, 1]$, $[7, 6, 1]$, $[10, 8, 1]$, $[13, 6, 1]$ and $[13, 3, 1]$.
Now according to part (b) of Exercise 7 in Section 8.8, a reflection about the line $y = mx + b$ is represented by the matrix $\mathbf{B} = \frac{1}{1+m^2}\begin{bmatrix} 1-m^2 & 2m & -2mb \\ 2m & m^2-1 & 2b \\ 0 & 0 & 1+m^2 \end{bmatrix}$. In this problem, the line is

$y = 4x - 10$, and so $m = 4$ and $b = -10$. Hence, $\mathbf{B} = \frac{1}{17}\begin{bmatrix} -15 & 8 & 80 \\ 8 & 15 & -20 \\ 0 & 0 & 17 \end{bmatrix}$. Now,

$\mathbf{B}\begin{bmatrix} 7 \\ 3 \\ 1 \end{bmatrix} \approx \begin{bmatrix} -0.059 \\ 4.765 \\ 1 \end{bmatrix}$, $\mathbf{B}\begin{bmatrix} 7 \\ 6 \\ 1 \end{bmatrix} \approx \begin{bmatrix} 1.353 \\ 7.412 \\ 1 \end{bmatrix}$, $\mathbf{B}\begin{bmatrix} 10 \\ 8 \\ 1 \end{bmatrix} \approx \begin{bmatrix} -0.353 \\ 10.588 \\ 1 \end{bmatrix}$,

$\mathbf{B}\begin{bmatrix} 13 \\ 6 \\ 1 \end{bmatrix} \approx \begin{bmatrix} -3.941 \\ 10.235 \\ 1 \end{bmatrix}$ and $\mathbf{B}\begin{bmatrix} 13 \\ 3 \\ 1 \end{bmatrix} \approx \begin{bmatrix} -5.353 \\ 7.588 \\ 1 \end{bmatrix}$. Therefore, since each of the product vectors has a 1 in the 3rd coordinate, each can be converted from homogeneous coordinates back to regular coordinates by merely dropping that 1. Doing this and rounding each of these results to the nearest whole number we get the following vertices for the moved graphic: $(0,5)$, $(1,7)$, $(0,11)$, $(-4,10)$ and $(-5,8)$.

(11) (b) Consider the reflection about the y-axis, whose matrix is given in Section 8.8 of the textbook as $\mathbf{A} = \begin{bmatrix} -1 & 0 & 0 \\ 0 & 1 & 0 \\ 0 & 0 & 1 \end{bmatrix}$, and a counterclockwise rotation of 90° about the origin, whose matrix is $\mathbf{B} = \begin{bmatrix} 0 & -1 & 0 \\ 1 & 0 & 0 \\ 0 & 0 & 1 \end{bmatrix}$. Then $\mathbf{AB} = \begin{bmatrix} 0 & 1 & 0 \\ 1 & 0 & 0 \\ 0 & 0 & 1 \end{bmatrix}$, but $\mathbf{BA} = \begin{bmatrix} 0 & -1 & 0 \\ -1 & 0 & 0 \\ 0 & 0 & 1 \end{bmatrix}$, and so $\mathbf{A}$ and $\mathbf{B}$ do not commute. In particular, starting from the point $(1,0)$, performing the rotation and then the reflection yields $(0,1)$. However, performing the reflection followed by the rotation produces $(0,-1)$.

(13) (a) False. As explained in Section 8.8 of the textbook, the 3rd coordinate of a vector in homogeneous coordinates must be nonzero.

(b) False. Vectors in homogeneous coordinates that are nonzero scalar multiples of each other represent the same pixel. For example, $[1,1,1]$ and $[2,2,2]$, as homogeneous coordinates, both represent the pixel at the point $(1,1)$.

(c) False. If $\mathbf{A}$ is a matrix for a rotation in homogeneous coordinates, then any nonzero scalar multiple of $\mathbf{A}$ also represents this same rotation.

(d) True. As explained in Section 8.8 of the textbook, *isometries* are the composition of translations, rotations and reflections only. However, any *similarity* can be expressed as a composition of translations, rotations, reflections, and scalings. But, scalings that are not the identity or a reflection are not isometries.

(e) True. This is because they do not send the origin to itself, violating part (1) of Theorem 5.1.

(f) False. Rotations about the origin and reflections about lines through the origin are linear transformations. However, no other rotations or reflections are linear transformations since they do not send the origin to itself, and therefore violate part (1) of Theorem 5.1.

Section 8.9

(1) (a) Now $\mathbf{A} = \begin{bmatrix} 13 & -28 \\ 6 & -13 \end{bmatrix}$, and so $p_{\mathbf{A}}(x) = \begin{vmatrix} x-13 & 28 \\ -6 & x+13 \end{vmatrix} = x^2 - 1 = (x-1)(x+1)$. Hence, **A** has 2 eigenvalues, $\lambda_1 = 1$ and $\lambda_2 = -1$. Next we solve for the corresponding eigenvectors.

For $\lambda_1 = 1$, we row reduce $[(1\mathbf{I}_2 - \mathbf{A})\,|\mathbf{0}] =$

$\left[\begin{array}{cc|c} -12 & 28 & 0 \\ -6 & 14 & 0 \end{array}\right]$ to obtain $\left[\begin{array}{cc|c} 1 & -\frac{7}{3} & 0 \\ 0 & 0 & 0 \end{array}\right]$, producing the eigenvector $[7,3]$.

For $\lambda_2 = -1$, we row reduce $[(-1\mathbf{I}_2 - \mathbf{A})\,|\mathbf{0}] =$

$\left[\begin{array}{cc|c} -14 & 28 & 0 \\ -6 & 12 & 0 \end{array}\right]$ to obtain $\left[\begin{array}{cc|c} 1 & -2 & 0 \\ 0 & 0 & 0 \end{array}\right]$, producing the eigenvector $[2,1]$.

Therefore, by Theorem 8.7, the general form for the solution to $\mathbf{F}'(t) = \mathbf{AF}(t)$ is

$\mathbf{F}(t) = b_1 e^t \begin{bmatrix} 7 \\ 3 \end{bmatrix} + b_2 e^{-t} \begin{bmatrix} 2 \\ 1 \end{bmatrix}$.

(c) Now $\mathbf{A} = \begin{bmatrix} 1 & 4 & 4 \\ -1 & 2 & 2 \\ 1 & 1 & 1 \end{bmatrix}$, and so $p_{\mathbf{A}}(x) = \begin{vmatrix} x-1 & -4 & -4 \\ 1 & x-2 & -2 \\ -1 & -1 & x-1 \end{vmatrix} = x^3 - 4x^2 + 3x$

$= x(x-1)(x-3)$. Hence, **A** has 3 eigenvalues, $\lambda_1 = 0$, $\lambda_2 = 1$ and $\lambda_3 = 3$. Next we solve for the corresponding eigenvectors.

For $\lambda_1 = 0$, we row reduce $[(0\mathbf{I}_3 - \mathbf{A})\,|\mathbf{0}] = [-\mathbf{A}\,|\mathbf{0}]$

$\left[\begin{array}{ccc|c} -1 & -4 & -4 & 0 \\ 1 & -2 & -2 & 0 \\ -1 & -1 & -1 & 0 \end{array}\right]$ to obtain $\left[\begin{array}{ccc|c} 1 & 0 & 0 & 0 \\ 0 & 1 & 1 & 0 \\ 0 & 0 & 0 & 0 \end{array}\right]$, producing the eigenvector $[0,-1,1]$.

For $\lambda_2 = 1$, we row reduce $[(1\mathbf{I}_3 - \mathbf{A})\,|\mathbf{0}] =$

$\left[\begin{array}{ccc|c} 0 & -4 & -4 & 0 \\ 1 & -1 & -2 & 0 \\ -1 & -1 & 0 & 0 \end{array}\right]$ to obtain $\left[\begin{array}{ccc|c} 1 & 0 & -1 & 0 \\ 0 & 1 & 1 & 0 \\ 0 & 0 & 0 & 0 \end{array}\right]$, producing the eigenvector $[1,-1,1]$.

For $\lambda_3 = 3$, we row reduce $[(3\mathbf{I}_3 - \mathbf{A})\,|\mathbf{0}] =$

$\left[\begin{array}{ccc|c} 2 & -4 & -4 & 0 \\ 1 & 1 & -2 & 0 \\ -1 & -1 & 2 & 0 \end{array}\right]$ to obtain $\left[\begin{array}{ccc|c} 1 & 0 & -2 & 0 \\ 0 & 1 & 0 & 0 \\ 0 & 0 & 0 & 0 \end{array}\right]$, producing the eigenvector $[2,0,1]$.

Therefore, by Theorem 8.7, the general form for the solution to $\mathbf{F}'(t) = \mathbf{AF}(t)$ is

$\mathbf{F}(t) = b_1 \begin{bmatrix} 0 \\ -1 \\ 1 \end{bmatrix} + b_2 e^t \begin{bmatrix} 1 \\ -1 \\ 1 \end{bmatrix} + b_3 e^{3t} \begin{bmatrix} 2 \\ 0 \\ 1 \end{bmatrix}$.

(d) Now $\mathbf{A} = \begin{bmatrix} -5 & -6 & 15 \\ -6 & -5 & 15 \\ -6 & -6 & 16 \end{bmatrix}$, and so $p_{\mathbf{A}}(x) = \begin{vmatrix} x+5 & 6 & -15 \\ 6 & x+5 & -15 \\ 6 & 6 & x-16 \end{vmatrix} = x^3 - 6x^2 + 9x - 4 =$

$(x-1)^2(x-4)$. Hence, **A** has 2 eigenvalues, $\lambda_1 = 1$ and $\lambda_2 = 4$. Next we solve for the corresponding eigenvectors.

For $\lambda_1 = 1$, we row reduce $[(1\mathbf{I}_3 - \mathbf{A})\,|\mathbf{0}] = \left[\begin{array}{ccc|c} 6 & 6 & -15 & 0 \\ 6 & 6 & -15 & 0 \\ 6 & 6 & -15 & 0 \end{array}\right]$ to obtain $\left[\begin{array}{ccc|c} 1 & 1 & -\frac{5}{2} & 0 \\ 0 & 0 & 0 & 0 \\ 0 & 0 & 0 & 0 \end{array}\right]$,

producing the eigenvectors $[-1,1,0]$ and $[5,0,2]$.

For $\lambda_2 = 4$, we row reduce $[(4\mathbf{I}_3 - \mathbf{A})\,|\mathbf{0}] =$

$\left[\begin{array}{ccc|c} 9 & 6 & -15 & 0 \\ 6 & 9 & -15 & 0 \\ 6 & 6 & -12 & 0 \end{array}\right]$ to obtain $\left[\begin{array}{ccc|c} 1 & 0 & -1 & 0 \\ 0 & 1 & -1 & 0 \\ 0 & 0 & 0 & 0 \end{array}\right]$, producing the eigenvector $[1,1,1]$.

Therefore, by Theorem 8.7, the general form for the solution to $\mathbf{F}'(t) = \mathbf{AF}(t)$ is

$\mathbf{F}(t) = b_1e^t \begin{bmatrix} -1 \\ 1 \\ 0 \end{bmatrix} + b_2e^t \begin{bmatrix} 5 \\ 0 \\ 2 \end{bmatrix} + b_3e^{4t} \begin{bmatrix} 1 \\ 1 \\ 1 \end{bmatrix}$. (There are other possible answers. For example, the first two vectors in the sum could be any basis for the two-dimensional eigenspace corresponding to the eigenvalue 1.)

(2) In each part, we use the technique described in Section 8.9 of the textbook between Examples 4 and 5.

(a) The characteristic equation is $x^2 + x - 6 = 0$, found by using the coefficients of y'', y', and y, in the given differential equation. Since $x^2 + x - 6 = (x-2)(x+3)$, the characteristic values (the roots of the characteristic equation) are $\lambda_1 = 2$ and $\lambda_2 = -3$. Hence, the general solution for the differential equation is $y = b_1e^{2t} + b_2e^{-3t}$.

(c) The characteristic equation is $x^4 - 6x^2 + 8 = 0$, found by using the coefficients of y'''', y''' y'', y' and y, in the given differential equation. Since $x^4 - 6x^2 + 8 = (x^2-4)(x^2-2) = (x-2)(x+2)\left(x-\sqrt{2}\right)\left(x+\sqrt{2}\right)$, the characteristic values (the roots of the characteristic equation) are $\lambda_1 = 2$, $\lambda_2 = -2$, $\lambda_3 = \sqrt{2}$ and $\lambda_4 = -\sqrt{2}$. Hence, the general solution for the differential equation is $y = b_1e^{2t} + b_2e^{-2t} + b_3e^{\sqrt{2}t} + b_4e^{-\sqrt{2}t}$.

(4) (b) First we find the general solution to $\mathbf{F}'(t) = \mathbf{AF}(t)$. Now $p_{\mathbf{A}}(x) = \begin{vmatrix} x+11 & 6 & -16 \\ 4 & x+1 & -4 \\ 12 & 6 & x-17 \end{vmatrix} =$

$x^3 - 5x^2 - x + 5 = (x-5)(x^2-1) = (x-5)(x-1)(x+1)$. Hence, $\mathbf{A}$ has 3 eigenvalues, $\lambda_1 = 5$, $\lambda_2 = 1$ and $\lambda_3 = -1$. Next we solve for the corresponding eigenvectors.

For $\lambda_1 = 5$, we row reduce $[(5\mathbf{I}_3 - \mathbf{A})\,|\mathbf{0}] =$

$\left[\begin{array}{ccc|c} 16 & 6 & -16 & 0 \\ 4 & 6 & -4 & 0 \\ 12 & 6 & -12 & 0 \end{array}\right]$ to obtain $\left[\begin{array}{ccc|c} 1 & 0 & -1 & 0 \\ 0 & 1 & 0 & 0 \\ 0 & 0 & 0 & 0 \end{array}\right]$, producing the eigenvector $[1,0,1]$.

For $\lambda_2 = 1$, we row reduce $[(1\mathbf{I}_3 - \mathbf{A})\,|\mathbf{0}] =$

$\left[\begin{array}{ccc|c} 12 & 6 & -16 & 0 \\ 4 & 2 & -4 & 0 \\ 12 & 6 & -16 & 0 \end{array}\right]$ to obtain $\left[\begin{array}{ccc|c} 1 & \frac{1}{2} & 0 & 0 \\ 0 & 0 & 1 & 0 \\ 0 & 0 & 0 & 0 \end{array}\right]$, producing the eigenvector $[-1,2,0]$.

For $\lambda_3 = -1$, we row reduce $[((-1)\mathbf{I}_3 - \mathbf{A})\,|\mathbf{0}] =$

$\left[\begin{array}{ccc|c} 10 & 6 & -16 & 0 \\ 4 & 0 & -4 & 0 \\ 12 & 6 & -18 & 0 \end{array}\right]$ to obtain $\left[\begin{array}{ccc|c} 1 & 0 & -1 & 0 \\ 0 & 1 & -1 & 0 \\ 0 & 0 & 0 & 0 \end{array}\right]$, producing the eigenvector $[1,1,1]$.

Therefore, by Theorem 8.7, the general form for the solution to $\mathbf{F}'(t) = \mathbf{AF}(t)$ is

$$\mathbf{F}(t) = b_1e^{5t}\begin{bmatrix}1\\0\\1\end{bmatrix} + b_2e^{t}\begin{bmatrix}-1\\2\\0\end{bmatrix} + b_3e^{-t}\begin{bmatrix}1\\1\\1\end{bmatrix}.$$

Now we want to find the particular solution such that $\mathbf{F}(0) = [1,-4,0]$. Plugging 0 into the general solution yields $\mathbf{F}(0) = [b_1, 0, b_1] + [-b_2, 2b_2, 0] + [b_3, b_3, b_3] = [b_1 - b_2 + b_3, 2b_2 + b_3, b_1 + b_3]$. Setting this equal to $[1,-4,0]$ produces the system

$$\begin{cases} b_1 - b_2 + b_3 = 1 \\ \quad\;\; 2b_2 + b_3 = -4 \\ b_1 \qquad\;\; + b_3 = 0 \end{cases}.$$

Row reducing $\left[\begin{array}{ccc|c} 1 & -1 & 1 & 1 \\ 0 & 2 & 1 & -4 \\ 1 & 0 & 1 & 0 \end{array}\right]$ to obtain $\left[\begin{array}{ccc|c} 1 & 0 & 0 & 2 \\ 0 & 1 & 0 & -1 \\ 0 & 0 & 1 & -2 \end{array}\right]$

shows that the unique solution to the differential equation with the given initial condition is

$$\mathbf{F}(t) = 2e^{5t}\begin{bmatrix}1\\0\\1\end{bmatrix} - e^{t}\begin{bmatrix}-1\\2\\0\end{bmatrix} - 2e^{-t}\begin{bmatrix}1\\1\\1\end{bmatrix}.$$

(5) (b) Base Step ($n = 1$): $\mathbf{A} = [-a_0]$, $x\mathbf{I}_1 - \mathbf{A} = [x + a_0]$, and so $p_{\mathbf{A}}(x) = x + a_0$.
Inductive Step: Assume true for k. Prove true for $k+1$. For the case $k+1$,

$$\mathbf{A} = \begin{bmatrix} 0 & 1 & 0 & 0 & \cdots & 0 \\ 0 & 0 & 1 & 0 & \cdots & 0 \\ \vdots & \vdots & \vdots & \vdots & \ddots & \vdots \\ 0 & 0 & 0 & 0 & \cdots & 1 \\ -a_0 & -a_1 & -a_2 & -a_3 & \cdots & -a_k \end{bmatrix}.$$

Then

$$(x\mathbf{I}_{k+1} - \mathbf{A}) = \begin{bmatrix} x & -1 & 0 & 0 & \cdots & 0 & 0 \\ 0 & x & -1 & 0 & \cdots & 0 & 0 \\ \vdots & \vdots & \vdots & \vdots & \ddots & \vdots & \vdots \\ 0 & 0 & 0 & 0 & \cdots & x & -1 \\ a_0 & a_1 & a_2 & a_3 & \cdots & a_{k-1} & x + a_k \end{bmatrix}.$$

Using a cofactor expansion on the first column yields

$$|x\mathbf{I}_{k+1} - \mathbf{A}| = x\begin{vmatrix} x & -1 & 0 & \cdots & 0 & 0 \\ \vdots & \vdots & \vdots & \ddots & \vdots & \vdots \\ 0 & 0 & 0 & \cdots & x & -1 \\ a_1 & a_2 & a_3 & \cdots & a_{k-1} & x + a_k \end{vmatrix}$$

$$+ (-1)^{(k+1)+1}a_0\begin{vmatrix} -1 & 0 & \cdots & 0 & 0 \\ x & -1 & \cdots & 0 & 0 \\ \vdots & \vdots & \ddots & \vdots & \vdots \\ 0 & 0 & \cdots & x & -1 \end{vmatrix}.$$

The first determinant equals $|x\mathbf{I}_k - \mathbf{B}|$, where

$$\mathbf{B} = \begin{bmatrix} 0 & 1 & 0 & \cdots & 0 & 0 \\ 0 & 0 & 1 & \cdots & 0 & 0 \\ \vdots & \vdots & \vdots & \ddots & \vdots & \\ 0 & 0 & 0 & \cdots & 0 & 1 \\ -a_1 & -a_2 & -a_3 & \cdots & -a_{k-1} & -a_k \end{bmatrix}.$$

Now, using the inductive hypothesis on the first determinant, and Theorems 3.2 and 3.9 on the second determinant (since its corresponding matrix is lower triangular) yields
$p_{\mathbf{A}}(x) = x(x^k + a_k x^{k-1} + \cdots + a_2 x + a_1) + (-1)^{k+2} a_0 (-1)^k = x^{k+1} + a_k x^k + \cdots + a_1 x + a_0$.

(7) (a) True. If $\mathbf{F}(t) = \mathbf{0}$, then $\mathbf{F}'(t)$ also equals $\mathbf{0}$. The equation $\mathbf{0} = \mathbf{A}\mathbf{0}$ is clearly satisfied for every matrix $\mathbf{A}$.

(b) True. Example 6 in Section 4.1 of the textbook shows that the set of all real-valued functions on $\mathbb{R}$ forms a vector space. An analogous proof shows that the set $\mathcal{V}$ of all $\mathbb{R}^n$-valued functions on $\mathbb{R}$ is also a vector space. It is easy to show that the subset of continuously differentiable functions that are solutions to $\mathbf{F}'(t) = \mathbf{A}\mathbf{F}(t)$ is closed under addition and scalar multiplication, and so Theorem 4.2 shows that this set is a subspace of $\mathcal{V}$, and is thus a vector space.

(c) True. Since $\mathbf{A} = \begin{bmatrix} 1 & 2 \\ 0 & 3 \end{bmatrix}$ is upper triangular, it is easy to see that the eigenvalues for $\mathbf{A}$ are $\lambda_1 = 1$ and $\lambda_2 = 3$. It is also easy to verify (by direct multiplication) that $[1,0]$ is an eigenvector for λ_1 and that $[1,1]$ is an eigenvector for λ_2. Hence, by Theorem 8.7, the solution set is $\left\{ b_1 e^t \begin{bmatrix} 1 \\ 0 \end{bmatrix} + b_2 e^{3t} \begin{bmatrix} 1 \\ 1 \end{bmatrix} \middle| \; b_1, b_2 \in \mathbb{R} \right\}$, which agrees with the given solution.

(d) False. The fact that the given matrix is not diagonalizable only means that the methods of Section 8.9 can not be used to solve the problem. For this particular matrix, complex eigenvalues need to be used, and will produce the general solution $\mathbf{F}(t) = [a \sin t + b \cos t, \; -b \sin t + a \cos t]$, which you can easily verify satisfies $\mathbf{F}'(t) = \mathbf{A}\mathbf{F}(t)$. (Note for those who have some experience with complex functions: using the method of Section 8.9 and complex eigenvalues to solve this problem will produce the solution $\mathbf{F}(t) = b_1 e^{it} \begin{bmatrix} -i \\ 1 \end{bmatrix} + b_2 e^{-it} \begin{bmatrix} i \\ 1 \end{bmatrix}$, which equals $\begin{bmatrix} a \sin t + b \cos t \\ -b \sin t + a \cos t \end{bmatrix}$, where $a = b_1 + b_2$ and $b = i(-b_1 + b_2)$.)

Section 8.10

(1) (a) Using the given matrix $\mathbf{A}$, $\mathbf{A}^T\mathbf{A} = \begin{bmatrix} 21 & 9 \\ 9 & 11 \end{bmatrix}$ and $\mathbf{A}^T\mathbf{b} = \begin{bmatrix} 26 \\ 19 \end{bmatrix}$. Thus, to solve $\mathbf{A}^T\mathbf{A}\mathbf{v} = \mathbf{A}^T\mathbf{b}$, we row reduce $\left[\begin{array}{cc|c} 21 & 9 & 26 \\ 9 & 11 & 19 \end{array}\right]$ obtaining $\left[\begin{array}{cc|c} 1 & 0 & \frac{23}{30} \\ 0 & 1 & \frac{11}{10} \end{array}\right]$.

Therefore, the unique least-squares solution is $\mathbf{v} = [\frac{23}{30}, \frac{11}{10}]$. Now, $||\mathbf{A}\mathbf{v} - \mathbf{b}|| =$

$$\left\| \begin{bmatrix} 2 & 3 \\ 1 & -1 \\ 4 & 1 \end{bmatrix} \begin{bmatrix} \frac{23}{30} \\ \frac{11}{10} \end{bmatrix} - \begin{bmatrix} 5 \\ 0 \\ 4 \end{bmatrix} \right\| = \left\| \begin{bmatrix} \frac{29}{6} \\ -\frac{1}{3} \\ \frac{25}{6} \end{bmatrix} - \begin{bmatrix} 5 \\ 0 \\ 4 \end{bmatrix} \right\| = ||[-\tfrac{1}{6}, -\tfrac{1}{3}, \tfrac{1}{6}]|| = \tfrac{\sqrt{6}}{6} \approx 0.408.$$ Note

that $||\mathbf{Az}-\mathbf{b}|| = \left\|\begin{bmatrix} 2 & 3 \\ 1 & -1 \\ 4 & 1 \end{bmatrix}\begin{bmatrix} 1 \\ 1 \end{bmatrix} - \begin{bmatrix} 5 \\ 0 \\ 4 \end{bmatrix}\right\| = \left\|\begin{bmatrix} 5 \\ 0 \\ 5 \end{bmatrix} - \begin{bmatrix} 5 \\ 0 \\ 4 \end{bmatrix}\right\| = ||[0,0,1]|| = 1$, which is larger than $||\mathbf{Av}-\mathbf{b}||$.

(c) Using the given matrix $\mathbf{A}$, $\mathbf{A}^T\mathbf{A} = \begin{bmatrix} 14 & 8 & 6 \\ 8 & 5 & 9 \\ 6 & 9 & 75 \end{bmatrix}$ and $\mathbf{A}^T\mathbf{b} = \begin{bmatrix} 18 \\ 7 \\ -35 \end{bmatrix}$. Thus, to solve $\mathbf{A}^T\mathbf{A}\mathbf{v} = \mathbf{A}^T\mathbf{b}$, we row reduce

$$\left[\begin{array}{ccc|c} 14 & 8 & 6 & 18 \\ 8 & 5 & 9 & 7 \\ 6 & 9 & 75 & -35 \end{array}\right] \text{ obtaining } \left[\begin{array}{ccc|c} 1 & 0 & -7 & \frac{17}{3} \\ 0 & 1 & 13 & -\frac{23}{3} \\ 0 & 0 & 0 & 0 \end{array}\right].$$

Therefore, there is an infinite number of least-squares solutions, all of the form $\mathbf{v} = \left[7c+\frac{17}{3}, -13c-\frac{23}{3}, c\right]$. Using $c=0$ and $c=\frac{1}{3}$ produces the two particular least-squares solutions $\left[\frac{17}{3}, -\frac{23}{3}, 0\right]$ and $\left[8, -12, \frac{1}{3}\right]$. With $\mathbf{v}$ as the first of these vectors, $||\mathbf{Av}-\mathbf{b}|| =$

$$\left\|\begin{bmatrix} 2 & 1 & -1 \\ 3 & 2 & 5 \\ 1 & 0 & -7 \end{bmatrix}\begin{bmatrix} \frac{17}{3} \\ -\frac{23}{3} \\ 0 \end{bmatrix} - \begin{bmatrix} 3 \\ 2 \\ 6 \end{bmatrix}\right\| = \left\|\begin{bmatrix} \frac{11}{3} \\ \frac{5}{3} \\ \frac{17}{3} \end{bmatrix} - \begin{bmatrix} 3 \\ 2 \\ 6 \end{bmatrix}\right\| = \left\|\left[\tfrac{2}{3}, -\tfrac{1}{3}, -\tfrac{1}{3}\right]\right\| = \frac{\sqrt{6}}{3} \approx 0.816.$$

Using the other particular solution we computed for $\mathbf{v}$, or any of the infinite number of least squares solutions instead, produces the same value of $\frac{\sqrt{6}}{3}$ for $||\mathbf{Av}-\mathbf{b}||$. Note that $||\mathbf{Az}-\mathbf{b}|| =$

$$\left\|\begin{bmatrix} 2 & 1 & -1 \\ 3 & 2 & 5 \\ 1 & 0 & -7 \end{bmatrix}\begin{bmatrix} 1 \\ 1 \\ -1 \end{bmatrix} - \begin{bmatrix} 3 \\ 2 \\ 6 \end{bmatrix}\right\| = \left\|\begin{bmatrix} 4 \\ 0 \\ 8 \end{bmatrix} - \begin{bmatrix} 3 \\ 2 \\ 6 \end{bmatrix}\right\| = ||[1,-2,2]|| = 3,$$ which is larger than $||\mathbf{Av}-\mathbf{b}||$.

(2) (a) Using the given matrix $\mathbf{A}$, $\mathbf{A}^T\mathbf{A} = \begin{bmatrix} 62 & -4 & 40 \\ -4 & 5 & -8 \\ 40 & -8 & 32 \end{bmatrix}$ and $\mathbf{A}^T\mathbf{b} = \begin{bmatrix} 29 \\ -3 \\ 20 \end{bmatrix}$. Thus, to solve

$\mathbf{A}^T\mathbf{A}\mathbf{v} = \mathbf{A}^T\mathbf{b}$, we row reduce $\left[\begin{array}{ccc|c} 62 & -4 & 40 & 29 \\ -4 & 5 & -8 & -3 \\ 40 & -8 & 32 & 20 \end{array}\right]$ obtaining $\left[\begin{array}{ccc|c} 1 & 0 & \frac{4}{7} & \frac{19}{42} \\ 0 & 1 & -\frac{8}{7} & -\frac{5}{21} \\ 0 & 0 & 0 & 0 \end{array}\right]$.

Therefore, there is an infinite number of least-squares solutions, all of the form $\mathbf{v} = \left[-\frac{4}{7}c+\frac{19}{42}, \frac{8}{7}c-\frac{5}{21}, c\right]$. Now, setting $-\frac{4}{7}c+\frac{19}{42} \geq 0$ yields $c \leq \frac{19}{24}$. And, setting $\frac{8}{7}c-\frac{5}{21} \geq 0$ produces $\frac{5}{24} \leq c$, which also implies $c \geq 0$. Hence, the desired solution set is $\left\{\left[-\frac{4}{7}c+\frac{19}{42}, \frac{8}{7}c-\frac{5}{21}, c\right] \;\middle|\; \frac{5}{24} \leq c \leq \frac{19}{24}\right\}$.

(3) (b) First, $(\lambda'\mathbf{I}_3 - \mathbf{C}) = \begin{bmatrix} -\frac{3}{2} & 3 & 2 \\ 5 & -\frac{7}{2} & -4 \\ -11 & 12 & \frac{21}{2} \end{bmatrix}$. We start with the homogeneous system $(\lambda'\mathbf{I}_3 - \mathbf{C})\mathbf{x} = \mathbf{0}$ and add the equation $x_1+x_2+x_3 = 1$. This produces the system $\mathbf{Ax} = \mathbf{b}$, with

$\mathbf{A} = \begin{bmatrix} -\frac{3}{2} & 3 & 2 \\ 5 & -\frac{7}{2} & -4 \\ -11 & 12 & \frac{21}{2} \\ 1 & 1 & 1 \end{bmatrix}$ and $\mathbf{b} = \begin{bmatrix} 0 \\ 0 \\ 0 \\ 1 \end{bmatrix}$. We find a least-squares solution for this system.

Now $\mathbf{A}^T\mathbf{A} = \begin{bmatrix} \frac{597}{4} & -153 & -\frac{275}{2} \\ -153 & \frac{665}{4} & 147 \\ -\frac{275}{2} & 147 & \frac{525}{4} \end{bmatrix}$ and $\mathbf{A}^T\mathbf{b} = \begin{bmatrix} 1 \\ 1 \\ 1 \end{bmatrix}$. Thus, to solve $\mathbf{A}^T\mathbf{A}\mathbf{v} = \mathbf{A}^T\mathbf{b}$, we row reduce $\left[\begin{array}{ccc|c} \frac{597}{4} & -153 & -\frac{275}{2} & 1 \\ -153 & \frac{665}{4} & 147 & 1 \\ -\frac{275}{2} & 147 & \frac{525}{4} & 1 \end{array}\right]$ obtaining $\left[\begin{array}{ccc|c} 1 & 0 & 0 & 0.46 \\ 0 & 1 & 0 & -0.36 \\ 0 & 0 & 1 & 0.90 \end{array}\right]$, where we have rounded to two places after the decimal point. Hence, the approximate eigenvector is $\mathbf{v} \approx [0.46, -0.36, 0.90]$. Direct computation produces $(\lambda'\mathbf{I}_3 - \mathbf{C})\mathbf{v} \approx [0.03, -0.04, 0.07]$, which is close to the zero vector. (Note that $\|(\lambda'\mathbf{I}_3 - \mathbf{C})\mathbf{v}\| \approx 0.086$, which is of the same order of magnitude as the difference between $\lambda' = \frac{3}{2}$ and the actual eigenvalue for $\mathbf{C}$ nearest to λ', which is $\sqrt{2} \approx 1.414$.)

(6) Let $\mathbf{b} = [b_1, \ldots, b_n]$ and $\mathbf{A}$ be as in Theorem 8.2, and let $f(x) = d_0 + d_1x + \ldots + d_tx^t$, and $\mathbf{z} = [d_0, d_1, \ldots, d_t]$. A short computation shows that $||\mathbf{Az} - \mathbf{b}||^2 = S_f$, the sum of the squares of the vertical distances illustrated in Figure 8.9, just before Theorem 8.2 in Section 8.3 of the textbook. Hence, minimizing $||\mathbf{Az} - \mathbf{b}||$ over all possible $(t+1)$-vectors $\mathbf{z}$ gives the coefficients of a degree t least-squares polynomial for the given points $(a_1, b_1), \ldots, (a_n, b_n)$. However, Theorem 8.8 shows that such a minimal solution is found by solving $(\mathbf{A}^T\mathbf{A})\mathbf{v} = \mathbf{A}^T\mathbf{b}$, thus proving the first claim of Theorem 8.2. (Notice that such a solution always exists because parts (1) and (3) of Theorem 8.8 indicates that $(\mathbf{A}^T\mathbf{A})\mathbf{v} = \mathbf{A}^T\mathbf{b}$ and $\mathbf{A}\mathbf{v} = \mathbf{proj}_{\mathcal{W}}\mathbf{b}$ are equivalent systems, and the second system is clearly consistent since $\mathcal{W}$ is the subspace $\{\mathbf{Ax} \mid \mathbf{x} \in \mathbb{R}^t\}$.) Finally, when $\mathbf{A}^T\mathbf{A}$ is row equivalent to $\mathbf{I}_{t+1}$, the uniqueness condition holds by Theorems 2.14 and 2.15.

(7) (a) False. The least-squares solution of an inconsistent system $\mathbf{Ax} = \mathbf{b}$ is, instead, any vector $\mathbf{v}$ such that the distance between $\mathbf{Av}$ and $\mathbf{b}$ is minimized.

(b) True. If $\mathbf{A}$ is an $m \times n$ matrix, then $\mathbf{A}^T\mathbf{A}$ is $n \times n$, which is square. Also, $(\mathbf{A}^T\mathbf{A})^T = \mathbf{A}^T(\mathbf{A}^T)^T = \mathbf{A}^T\mathbf{A}$, and so $\mathbf{A}^T\mathbf{A}$ is symmetric.

(c) True. If $\mathcal{W}$ is the subspace $\{\mathbf{Ax} \mid \mathbf{x} \in \mathbb{R}^n\}$, then $\mathbf{proj}_{\mathcal{W}}\mathbf{b}$ is a well-defined vector in $\mathcal{W}$. Hence, by the definition of $\mathcal{W}$, there must exist a vector $\mathbf{v}$ in $\mathbb{R}^n$ with $\mathbf{Av} = \mathbf{proj}_{\mathcal{W}}\mathbf{b}$. Theorem 8.8 asserts that such a vector $\mathbf{v}$ is a least-squares solution to $\mathbf{Ax} = \mathbf{b}$.

(d) True. By part (1) of Theorem 8.8, both $\mathbf{Av}_1$ and $\mathbf{Av}_2$ must equal $\mathbf{proj}_{\mathcal{W}}\mathbf{b}$, and so they must equal each other.

(e) True. This is illustrated in Example 3 in Section 8.10 of the textbook.

Section 8.11

(1) (a) First, $\nabla f = [3x^2 + 2x + 2y - 3, 2x + 2y]$. We find critical points by setting $\nabla f = \mathbf{0}$. Setting the 2nd coordinate of ∇f equal to zero yields $y = -x$. Plugging this into $3x^2 + 2x + 2y - 3 = 0$ produces $3x^2 - 3 = 0$, which has solutions $x = 1$ and $x = -1$. This gives the two critical points $(1, -1)$ and $(-1, 1)$. Next, $\mathbf{H} = \begin{bmatrix} 6x + 2 & 2 \\ 2 & 2 \end{bmatrix}$. At the first critical point, $\mathbf{H}\Big|_{(1,-1)} = \begin{bmatrix} 8 & 2 \\ 2 & 2 \end{bmatrix}$, which is

positive definite because its determinant is 12, which is positive, and its $(1,1)$ entry is 8, which is also positive. (See the comment just before Example 3 in Section 8.11 for easily verified necessary and sufficient conditions for a 2×2 matrix to be either positive definite or negative definite.) So, f has a local minimum at $(1,-1)$. At the second critical point, $\mathbf{H}\Big|_{(-1,1)} = \begin{bmatrix} -4 & 2 \\ 2 & 2 \end{bmatrix}$. We refer to this matrix as $\mathbf{A}$. Now $\mathbf{A}$ is neither positive definite nor negative definite, since its determinant is negative. Also note that $p_{\mathbf{A}}(x) = x^2 + 2x - 12$, which, by the Quadratic Formula, has roots $-1 \pm \sqrt{13}$. Since one of these is positive and the other is negative, we see that $(-1,1)$ is not a local extreme value.

(c) First, $\nabla f = [4x+2y+2, 2x+2y-2]$. We find critical points by setting $\nabla f = \mathbf{0}$. This corresponds to a linear system with two equations and two variables which has the unique solution $(-2,3)$. Next, $\mathbf{H} = \begin{bmatrix} 4 & 2 \\ 2 & 2 \end{bmatrix}$. $\mathbf{H}$ is positive definite since its determinant is 4, which is positive, and its $(1,1)$ entry is 4, which is positive. Hence, the critical point $(-2,3)$ is a local minimum.

(e) First, $\nabla f = [4x + 2y + 2z,\ 2x + 4y^3 + 12y^2z + 12yz^2 - 2y + 4z^3 - 4z,$ $2x+4y^3+12y^2z+12yz^2-4y+4z^3-2z]$. We find critical points by setting $\nabla f = \mathbf{0}$. To make things easier, notice that at the critical point, $\frac{\partial f}{\partial y} = \frac{\partial f}{\partial z}$. This yields $y = z$. With this, and $\frac{\partial f}{\partial x} = 0$, we obtain $x = -y$. Substituting for y and z into $\frac{\partial f}{\partial y} = 0$ and solving, produces $8x - 32x^3 = 0$. Hence, $8x(1-4x^2) = 8x(1-2x)(1+2x) = 0$. This yields the three critical points $(0,0,0)$, $(-\frac{1}{2}, \frac{1}{2}, \frac{1}{2})$, and $(\frac{1}{2}, -\frac{1}{2}, -\frac{1}{2})$. Next, $\mathbf{H} = \begin{bmatrix} 4 & 2 & 2 \\ 2 & 12y^2+24yz+12z^2-2 & 12y^2+24yz+12z^2-4 \\ 2 & 12y^2+24yz+12z^2-4 & 12y^2+24yz+12z^2-2 \end{bmatrix}$.

At the first critical point, $\mathbf{H}\Big|_{(0,0,0)} = \begin{bmatrix} 4 & 2 & 2 \\ 2 & -2 & -4 \\ 2 & -4 & -2 \end{bmatrix}$. We refer to this matrix as $\mathbf{A}$. Then $p_{\mathbf{A}}(x) = x^3 - 36x + 64 = (x-2)(x^2+2x-32)$. Hence, the eigenvalues of $\mathbf{A}$ are 2, and $-1 \pm \sqrt{33}$. Since $\mathbf{A}$ has both positive and negative eigenvalues, $(0,0,0)$ is not an extreme point.

At the second critical point, $\mathbf{H}\Big|_{(-\frac{1}{2},\frac{1}{2},\frac{1}{2})} = \begin{bmatrix} 4 & 2 & 2 \\ 2 & 10 & 8 \\ 2 & 8 & 10 \end{bmatrix}$. We refer to this matrix as $\mathbf{B}$. Then $p_{\mathbf{B}}(x) = x^3 - 24x^2 + 108x - 128 = (x-2)(x^2 - 22x + 64)$. Hence, the eigenvalues of $\mathbf{B}$ are all positive (2 and $11 \pm \sqrt{57}$). So f has a local minimum at $(-\frac{1}{2}, \frac{1}{2}, \frac{1}{2})$ by Theorem 8.12.

At the third critical point, $\mathbf{H}\Big|_{(\frac{1}{2},-\frac{1}{2},-\frac{1}{2})} = \mathbf{H}\Big|_{(-\frac{1}{2},\frac{1}{2},\frac{1}{2})}$, and so f also has a local minimum at $(\frac{1}{2}, -\frac{1}{2}, -\frac{1}{2})$.

(4) (a) True. We know from Calculus that if a real-valued function f on $\mathbb{R}^n$ has continuous second partial derivatives then $\frac{\partial^2 f}{\partial x_i \partial x_j} = \frac{\partial^2 f}{\partial x_j \partial x_i}$, for all i, j.

(b) False. For example, the matrix $\begin{bmatrix} 1 & 0 \\ 0 & -1 \end{bmatrix}$ represents neither a positive definite nor negative definite quadratic form since it has both a positive and a negative eigenvalue.

(c) True. In part (a) we noted that the Hessian matrix is symmetric. Theorems 6.17 and 6.19 then establish that the matrix is orthogonally diagonalizable, hence diagonalizable.

(d) True. We established that a symmetric 2×2 matrix $\mathbf{A}$ with $|\mathbf{A}| > 0$ and $a_{11} > 0$ represents a positive definite quadratic form (see the comment just before Example 3 in Section 8.11 of the textbook). In this problem, $|\mathbf{A}| = 1 > 0$ and $a_{11} = 5 > 0$.

(e) False. The eigenvalues of the given matrix are clearly 3, -9, and 4. Hence, the given matrix has both a positive and a negative eigenvalue, and so can not represent a positive definite quadratic form.

Chapter 9

Section 9.1

(1) (a) For the first system, row reduce

$$\left[\begin{array}{cc|c} 5 & -2 & 10 \\ 5 & -1.995 & 17.5 \end{array}\right] \text{ to obtain } \left[\begin{array}{cc|c} 1 & 0 & 602 \\ 0 & 1 & 1500 \end{array}\right].$$

Hence, the solution to the first system is $(602, 1500)$. For the second system, row reduce

$$\left[\begin{array}{cc|c} 5 & -2 & 10 \\ 5 & -1.99 & 17.5 \end{array}\right] \text{ to obtain } \left[\begin{array}{cc|c} 1 & 0 & 302 \\ 0 & 1 & 750 \end{array}\right].$$

Hence, the solution to the second system is $(302, 750)$. These systems are ill-conditioned, because a very small change in the coefficient of y leads to a very large change in the solution.

(2) Answers to this problem may differ significantly from the following if you round at different places in the algorithm. We rounded the original numbers in the system, and directly after performing each row operation.

(a) Without partial pivoting, we perform the following sequence of row operations:

$\langle 1\rangle \leftarrow \frac{1}{0.00072}\langle 1\rangle$ (new $\langle 1\rangle$ is $[1, \ -5990, \ -1370]$)

$\langle 2\rangle \leftarrow -2.31\langle 1\rangle + \langle 2\rangle$ (new $\langle 2\rangle$ is $[0, \ 3960, \ 3030]$)

$\langle 2\rangle \leftarrow \frac{1}{3960}\langle 2\rangle$ (new $\langle 2\rangle$ is $[0, \ 1, \ 0.765]$)

$\langle 1\rangle \leftarrow 5990\langle 2\rangle + \langle 1\rangle$ (new $\langle 1\rangle$ is $[1, \ 0, \ 3210]$)

giving the solution $(3210, 0.765)$.

With partial pivoting, we perform the following sequence of row operations:

$\langle 1\rangle \leftrightarrow \langle 2\rangle$

$\langle 1\rangle \leftarrow \frac{1}{2.31}\langle 1\rangle$ (new $\langle 1\rangle$ is $[1, \ -4280, \ -56.7]$)

$\langle 2\rangle \leftarrow -0.00072\langle 1\rangle + \langle 2\rangle$ (new $\langle 2\rangle$ is $[0, \ -1.23, \ -0.944]$)

$\langle 2\rangle \leftarrow -\frac{1}{1.23}\langle 2\rangle$ (new $\langle 2\rangle$ is $[0, \ 1, \ 0.767]$)

$\langle 1\rangle \leftarrow 4280\langle 2\rangle + \langle 1\rangle$ (new $\langle 1\rangle$ is $[1, \ 0, \ 3230]$)

giving the solution $(3230, 0.767)$. The actual solution is $(3214, 0.765)$.

(c) Without partial pivoting, we perform the following sequence of row operations:

$\langle 1\rangle \leftarrow \frac{1}{0.00032}\langle 1\rangle$ (new $\langle 1\rangle$ is $[1, \ 722, \ 397, \ -108]$)

$\langle 2\rangle \leftarrow 241\langle 1\rangle + \langle 2\rangle$ (new $\langle 2\rangle$ is $[0, \ 174000, \ 95700, \ -26600]$)

$\langle 3\rangle \leftarrow -49\langle 1\rangle + \langle 3\rangle$ (new $\langle 3\rangle$ is $[0, \ -35300, \ -19500, \ 5580]$)

$\langle 2\rangle \leftarrow \frac{1}{174000}\langle 2\rangle$ (new $\langle 2\rangle$ is $[0, \ 1, \ 0.55, \ -0.153]$)

$\langle 1\rangle \leftarrow -722\langle 2\rangle + \langle 1\rangle$ (new $\langle 1\rangle$ is $[1, \ 0, \ -0.1, \ 2.47]$)

$\langle 3\rangle \leftarrow 35300\langle 2\rangle + \langle 3\rangle$ (new $\langle 3\rangle$ is $[0, \ 0, \ -85, \ 179]$)

$\langle 3\rangle \leftarrow -\frac{1}{85}\langle 3\rangle$ (new $\langle 3\rangle$ is $[0, \ 0, \ 1, \ -2.11]$)

$\langle 1\rangle \leftarrow 0.1\langle 3\rangle + \langle 1\rangle$ (new $\langle 1\rangle$ is $[1, \ 0, \ 0, \ 2.26]$)

$\langle 2\rangle \leftarrow -0.55\langle 3\rangle + \langle 2\rangle$ (new $\langle 2\rangle$ is $[0, \ 1, \ 0, \ 1.01]$)

giving the solution $(2.26, 1.01, -2.11)$.

With partial pivoting, we perform the following sequence of row operations:

$\langle 1\rangle \leftrightarrow \langle 2\rangle$
$\langle 1\rangle \leftarrow -\frac{1}{241}\langle 1\rangle$ (new $\langle 1\rangle$ is $[1,\ 0.9,\ 0.0332,\ 2.39]$)
$\langle 2\rangle \leftarrow -0.00032\langle 1\rangle + \langle 2\rangle$ (new $\langle 2\rangle$ is $[0,\ 0.231,\ 0.127,\ -0.0354]$)
$\langle 3\rangle \leftarrow -49\langle 1\rangle + \langle 3\rangle$ (new $\langle 3\rangle$ is $[0,\ 0.9,\ 0.773,\ 166]$)
$\langle 2\rangle \leftrightarrow \langle 3\rangle$
$\langle 2\rangle \leftarrow \frac{1}{0.9}\langle 2\rangle$ (new $\langle 2\rangle$ is $[0,\ 1,\ 0.859,\ 184]$)
$\langle 1\rangle \leftarrow -0.9\langle 2\rangle + \langle 1\rangle$ (new $\langle 1\rangle$ is $[1,\ 0,\ -0.740,\ -163]$)
$\langle 3\rangle \leftarrow -0.231\langle 2\rangle + \langle 3\rangle$ (new $\langle 3\rangle$ is $[0,\ 0,\ -0.0714,\ -42.5]$)
$\langle 3\rangle \leftarrow -\frac{1}{0.0714}\langle 3\rangle$ (new $\langle 3\rangle$ is $[0,\ 0,\ 1,\ 595]$)
$\langle 1\rangle \leftarrow 0.74\langle 3\rangle + \langle 1\rangle$ (new $\langle 1\rangle$ is $[1,\ 0,\ 0,\ 277]$)
$\langle 2\rangle \leftarrow -0.859\langle 3\rangle + \langle 2\rangle$ (new $\langle 2\rangle$ is $[0,\ 1,\ 0,\ -327]$)
giving the solution $(277, -327, 595)$. The actual solution is $(267, -315, 573)$.

(3) Answers to this problem may differ significantly from the following if you round at different places in the algorithm. We rounded directly after performing each row operation.

(a) Without partial pivoting, we perform the following sequence of row operations:
$\langle 1\rangle \leftarrow \frac{1}{0.00072}\langle 1\rangle$ (new $\langle 1\rangle$ is $[1,\ -5989,\ -1368]$)
$\langle 2\rangle \leftarrow -2.31\langle 1\rangle + \langle 2\rangle$ (new $\langle 2\rangle$ is $[0,\ 3959,\ 3029]$)
$\langle 2\rangle \leftarrow \frac{1}{3959}\langle 2\rangle$ (new $\langle 2\rangle$ is $[0,\ 1,\ 0.7651]$)
$\langle 1\rangle \leftarrow 5989\langle 2\rangle + \langle 1\rangle$ (new $\langle 1\rangle$ is $[1,\ 0,\ 3214]$)
giving the solution $(3214, 0.7651)$.
With partial pivoting, we perform the following sequence of row operations:
$\langle 1\rangle \leftrightarrow \langle 2\rangle$
$\langle 1\rangle \leftarrow \frac{1}{2.31}\langle 1\rangle$ (new $\langle 1\rangle$ is $[1,\ -4275,\ -56.62]$)
$\langle 2\rangle \leftarrow -0.00072\langle 1\rangle + \langle 2\rangle$ (new $\langle 2\rangle$ is $[0,\ -1.234,\ -0.9438]$)
$\langle 2\rangle \leftarrow -\frac{1}{1.234}\langle 2\rangle$ (new $\langle 2\rangle$ is $[0,\ 1,\ 0.7648]$)
$\langle 1\rangle \leftarrow 4275\langle 2\rangle + \langle 1\rangle$ (new $\langle 1\rangle$ is $[1,\ 0,\ 3213]$)
giving the solution $(3213, 0.7648)$. The actual solution is $(3214, 0.765)$.

(c) Without partial pivoting, we perform the following sequence of row operations:
$\langle 1\rangle \leftarrow \frac{1}{0.00032}\langle 1\rangle$ (new $\langle 1\rangle$ is $[1,\ 723.1,\ 396.9,\ -108]$)
$\langle 2\rangle \leftarrow 241\langle 1\rangle + \langle 2\rangle$ (new $\langle 2\rangle$ is $[0,\ 174100,\ 95640,\ -26600]$)
$\langle 3\rangle \leftarrow -49\langle 1\rangle + \langle 3\rangle$ (new $\langle 3\rangle$ is $[0,\ -35390,\ -19450,\ 5575]$)
$\langle 2\rangle \leftarrow \frac{1}{174100}\langle 2\rangle$ (new $\langle 2\rangle$ is $[0,\ 1,\ 0.5493,\ -0.1528]$)
$\langle 1\rangle \leftarrow -723.1\langle 2\rangle + \langle 1\rangle$ (new $\langle 1\rangle$ is $[1,\ 0,\ -0.2988,\ 2.490]$)
$\langle 3\rangle \leftarrow 35390\langle 2\rangle + \langle 3\rangle$ (new $\langle 3\rangle$ is $[0,\ 0,\ -10.27,\ 167.4]$)
$\langle 3\rangle \leftarrow -\frac{1}{10.27}\langle 3\rangle$ (new $\langle 3\rangle$ is $[0,\ 0,\ 1,\ -16.30]$)
$\langle 1\rangle \leftarrow 0.2988\langle 3\rangle + \langle 1\rangle$ (new $\langle 1\rangle$ is $[1,\ 0,\ 0,\ -2.380]$)
$\langle 2\rangle \leftarrow -0.5493\langle 3\rangle + \langle 2\rangle$ (new $\langle 2\rangle$ is $[0,\ 1,\ 0,\ 8.801]$)
giving the solution $(-2.38, 8.801, -16.30)$.

With partial pivoting, we perform the following sequence of row operations:

$\langle 1\rangle \leftrightarrow \langle 2\rangle$

$\langle 1\rangle \leftarrow -\frac{1}{241}\langle 1\rangle$ (new $\langle 1\rangle$ is [1, 0.9004, 0.0332, 2.390])

$\langle 2\rangle \leftarrow -0.00032\langle 1\rangle + \langle 2\rangle$ (new $\langle 2\rangle$ is [0, 0.2311, 0.1270, $-$0.03532])

$\langle 3\rangle \leftarrow -49\langle 1\rangle + \langle 3\rangle$ (new $\langle 3\rangle$ is [0, 0.8804, 0.7732, 166.1])

$\langle 2\rangle \leftrightarrow \langle 3\rangle$

$\langle 2\rangle \leftarrow \frac{1}{0.8804}\langle 2\rangle$ (new $\langle 2\rangle$ is [0, 1, 0.8782, 188.7])

$\langle 1\rangle \leftarrow -0.9004\langle 2\rangle + \langle 1\rangle$ (new $\langle 1\rangle$ is [1, 0, $-$0.7575, $-$167.5])

$\langle 3\rangle \leftarrow -0.2311\langle 2\rangle + \langle 3\rangle$ (new $\langle 3\rangle$ is [0, 0, $-$0.07595, $-$43.64])

$\langle 3\rangle \leftarrow -\frac{1}{0.07595}\langle 3\rangle$ (new $\langle 3\rangle$ is [0, 0, 1, 574.6])

$\langle 1\rangle \leftarrow 0.7575\langle 3\rangle + \langle 1\rangle$ (new $\langle 1\rangle$ is [1, 0, 0, 267.8])

$\langle 2\rangle \leftarrow -0.8782\langle 3\rangle + \langle 2\rangle$ (new $\langle 2\rangle$ is [0, 1, 0, $-$315.9])

giving the solution $(267.8, -315.9, 574.6)$. The actual solution is $(267, -315, 573)$.

(4) (a) First we express the given system as

$$\begin{cases} x_1 = -\frac{1}{5}x_2 + \frac{26}{5} \\ x_2 = -\frac{3}{7}x_1 - 6 \end{cases}.$$

Starting with the initial values $x_1 = 0$ and $x_2 = 0$, we plug these in the right side of these equations to obtain the new values $x_1 = \frac{26}{5}$ and $x_2 = -6$. These are the values listed in the table, below, in the row labeled "After 1 Step." These new values are again put in the right side to solve for x_1 and x_2. After rounding to 3 decimal places, this yields $x_1 = 6.400$ and $x_2 = -8.229$, the numbers in the row labeled "After 2 Steps." We continue in this manner, producing the results given in the following table:

	x_1	x_2
Initial Values	0.000	0.000
After 1 Step	5.200	−6.000
After 2 Steps	6.400	−8.229
After 3 Steps	6.846	−8.743
After 4 Steps	6.949	−8.934
After 5 Steps	6.987	−8.978
After 6 Steps	6.996	−8.994
After 7 Steps	6.999	−8.998
After 8 Steps	7.000	−9.000
After 9 Steps	7.000	−9.000

After 9 steps we get the same result as we had after 8 steps, and so we are done. The solution obtained for the system is $(7, -9)$.

(c) First we express the given system as

$$\begin{cases} x_1 = -\frac{1}{7}x_2 + \frac{2}{7}x_3 - \frac{62}{7} \\ x_2 = \frac{1}{6}x_1 - \frac{1}{6}x_3 + \frac{9}{2} \\ x_3 = \frac{1}{3}x_1 - \frac{1}{6}x_2 - \frac{13}{3} \end{cases}.$$

Starting with the initial values $x_1 = 0$, $x_2 = 0$ and $x_3 = 0$, we plug these in the right side of these equations to obtain the new values $x_1 = -\frac{62}{7}$, $x_2 = \frac{9}{2}$ and $x_3 = -\frac{13}{3}$. These values, rounded to 3 places after the decimal point, are listed in the table, below, in the row labeled "After 1 Step." These new values are again put in the right side to solve for x_1, x_2 and x_3. After rounding to 3 decimal places, this yields $x_1 = -10.738$, $x_2 = 3.746$ and $x_3 = -8.036$, the numbers in the row labeled "After 2 Steps." We continue in this manner, producing the results given in the following table:

	x_1	x_2	x_3
Initial Values	0.000	0.000	0.000
After 1 Step	−8.857	4.500	−4.333
After 2 Steps	−10.738	3.746	−8.036
After 3 Steps	−11.688	4.050	−8.537
After 4 Steps	−11.875	3.975	−8.904
After 5 Steps	−11.969	4.005	−8.954
After 6 Steps	−11.988	3.998	−8.991
After 7 Steps	−11.997	4.001	−8.996
After 8 Steps	−11.999	4.000	−8.999
After 9 Steps	−12.000	4.000	−9.000
After 10 Steps	−12.000	4.000	−9.000

After 10 steps we get the same result as we had after 9 steps, and so we are done. The solution obtained for the system is $(-12, 4, -9)$.

(5) (a) First we express the given system as

$$\begin{cases} x_1 = -\frac{1}{5}x_2 + \frac{26}{5} \\ x_2 = -\frac{3}{7}x_1 - 6 \end{cases}.$$

We start with the initial values $x_1 = 0$ and $x_2 = 0$. We plug $x_2 = 0$ into the right side of the first equation to obtain the new value $x_1 = \frac{26}{5}$. Then, we substitute $x_1 = \frac{26}{5}$ into the right side of the second equation yielding $x_2 = -\frac{288}{35}$. After rounding to 3 places after the decimal point, we listed these results in the table, below, in the row labeled "After 1 Step." Next, we plug $x_2 = -8.229$ into the right side of the first equation to solve for x_1, which, after rounding to 3 decimal places, yields $x_1 = 6.846$. Substituting $x_1 = 6.846$ into the right side of the second equation produces $x_2 = -8.934$. The two new results are the numbers in the row labeled "After 2 Steps." We continue in this manner, producing the results given in the following table:

	x_1	x_2
Initial Values	0.000	0.000
After 1 Step	5.200	−8.229
After 2 Steps	6.846	−8.934
After 3 Steps	6.987	−8.994
After 4 Steps	6.999	−9.000
After 5 Steps	7.000	−9.000
After 6 Steps	7.000	−9.000

After 6 steps we get the same result as we had after 5 steps, and so we are done. The solution obtained for the system is $(7, -9)$.

(c) First we express the given system as

$$\begin{cases} x_1 = -\frac{1}{7}x_2 + \frac{2}{7}x_3 - \frac{62}{7} \\ x_2 = \frac{1}{6}x_1 - \frac{1}{6}x_3 + \frac{9}{2} \\ x_3 = \frac{1}{3}x_1 - \frac{1}{6}x_2 - \frac{13}{3} \end{cases}.$$

We start with the initial values $x_1 = 0$, $x_2 = 0$ and $x_3 = 0$. We plug $x_2 = x_3 = 0$ into the right side of the first equation to obtain the new value $x_1 = -\frac{62}{7} \approx -8.857$. Then, we substitute $x_1 = -8.857$ and $x_3 = 0$ into the right side of the second equation, yielding $x_2 \approx 3.024$. Finally, we plug $x_1 = -8.857$ and $x_2 = 3.024$ into the right side of the third equation, yielding $x_3 \approx -7.790$. We have listed these results in the table, below, in the row labeled "After 1 Step." Next, we plug $x_2 = 3.024$ and $x_3 = -7.790$ into the right side of the first equation to solve for x_1, which yields $x_1 \approx -11.515$. Substituting $x_1 = 11.515$ and $x_3 = -7.790$ into the right side of the second equation produces $x_2 \approx 3.879$. Finally, we plug $x_1 = -11.515$ and $x_2 = 3.879$ into the right side of the third equation, yielding $x_3 \approx -8.818$. The three new results are the numbers in the row labeled "After 2 Steps." We continue in this manner, producing the results given in the following table:

	x_1	x_2	x_3
Initial Values	0.000	0.000	0.000
After 1 Step	−8.857	3.024	−7.790
After 2 Steps	−11.515	3.879	−8.818
After 3 Steps	−11.931	3.981	−8.974
After 4 Steps	−11.990	3.997	−8.996
After 5 Steps	−11.998	4.000	−8.999
After 6 Steps	−12.000	4.000	−9.000
After 7 Steps	−12.000	4.000	−9.000

After 7 steps we get the same result as we had after 6 steps, and so we are done. The solution obtained for the system is $(-12, 4, -9)$.

(6) (a) **A** satisfies all of the criteria, so it is strictly diagonally dominant. (In row 1, $|-3| > |1|$; that is, $|a_{11}| > |a_{12}|$. In row 2, $|4| > |-2|$; that is, $|a_{22}| > |a_{21}|$.)

(b) **A** is not strictly diagonally dominant because $|a_{22}| < |a_{21}|$. (Note that $|a_{11}| = |a_{12}|$, which also prevents **A** from being strictly diagonally dominant.)

(c) **A** satisfies all of the criteria, so it is strictly diagonally dominant. (In row 1, $|a_{11}| > |a_{12}| + |a_{13}|$; in row 2, $|a_{22}| > |a_{21}| + |a_{23}|$; in row 3, $|a_{33}| > |a_{31}| + |a_{32}|$.)

(d) **A** is not strictly diagonally dominant because $|a_{22}| < |a_{21}| + |a_{23}|$.

(e) **A** is not strictly diagonally dominant because $|a_{22}| = |a_{21}| + |a_{23}|$.

(7) (a) Put the third equation first, and move the other two down. Then express the equations as follows:

$$\begin{cases} x_1 = \frac{1}{8}x_2 - \frac{3}{8}x_3 + \frac{25}{8} \\ x_2 = -\frac{2}{13}x_1 - \frac{1}{13}x_3 \\ x_3 = -\frac{1}{15}x_1 + \frac{2}{15}x_2 + \frac{26}{15} \end{cases}.$$

We start with the initial values $x_1 = 0$, $x_2 = 0$ and $x_3 = 0$. We plug $x_2 = x_3 = 0$ into the right side of the first equation to obtain the new value $x_1 = \frac{25}{8} = 3.125$. Then, we substitute $x_1 = 3.125$ and $x_3 = 0$ into the right side of the second equation, yielding $x_2 \approx -0.481$. Finally, we plug $x_1 = 3.125$ and $x_2 = -0.481$ into the right side of the third equation, yielding $x_3 \approx 1.461$.

We have listed these results in the table, below, in the row labeled "After 1 Step." Next, we plug $x_2 = -0.481$ and $x_3 = 1.461$ into the right side of the first equation to solve for x_1, which yields $x_1 \approx 2.517$. Substituting $x_1 = 2.517$ and $x_3 = 1.461$ into the right side of the second equation produces $x_2 \approx -0.500$. Finally, we plug $x_1 = 2.517$ and $x_2 = -0.500$ into the right side of the third equation yielding $x_3 \approx 1.499$. The three new results are the numbers in the row labeled "After 2 Steps." We continue in this manner, producing the results given in the following table:

	x_1	x_2	x_3
Initial Values	0.000	0.000	0.000
After 1 Step	3.125	−0.481	1.461
After 2 Steps	2.517	−0.500	1.499
After 3 Steps	2.500	−0.500	1.500
After 4 Steps	2.500	−0.500	1.500

After 4 steps we get the same result as we had after 3 steps, and so we are done. The solution obtained for the system is $(2.5, -0.5, 1.5)$.

(c) Put the second equation first, the fourth equation second, the first equation third, and the third equation fourth. Then express the equations as follows:

$$\begin{cases} x_1 = -\frac{2}{9}x_2 + \frac{1}{9}x_3 - \frac{1}{9}x_4 + \frac{49}{9} \\ x_2 = -\frac{1}{17}x_1 - \frac{3}{17}x_3 + \frac{2}{17}x_4 - \frac{86}{17} \\ x_3 = -\frac{1}{13}x_1 - \frac{1}{13}x_2 - \frac{2}{13}x_4 + \frac{120}{13} \\ x_4 = -\frac{1}{7}x_1 + \frac{3}{14}x_2 - \frac{1}{14}x_3 - \frac{55}{7} \end{cases}.$$

We starting with the initial values $x_1 = 0$, $x_2 = 0$, $x_3 = 0$ and $x_4 = 0$. We plug $x_2 = x_3 = x_4 = 0$ into the right side of the first equation to obtain the new value $x_1 = \frac{49}{9} \approx 5.444$. Then, we substitute $x_1 = 5.444$, $x_3 = 0$, and $x_4 = 0$ into the right side of the second equation, yielding $x_2 \approx -5.379$. Next, we substitute $x_1 = 5.444$, $x_2 = -5.379$, and $x_4 = 0$ into the right side of the third equation, yielding $x_3 \approx 9.226$. Finally, we plug $x_1 = 5.444$, $x_2 = -5.379$, and $x_3 = 9.226$ into the right side of the fourth equation, yielding $x_4 \approx -10.447$. We have listed these results in the table, below, in the row labeled "After 1 Step." Next, we plug $\text{x}_2 = -5.379$, $x_3 = 9.226$, and $x_4 = -10.447$ into the right side of the first equation to solve for x_1, which yields $x_1 \approx 8.826$. Substituting $x_1 = 8.826$, $x_3 = 9.226$ and $x_4 = -10.447$ into the right side of the second equation produces $x_2 \approx -8.435$. Next, substituting $x_1 = 8.826$, $x_2 = -8.435$ and $x_4 = -10.447$ into the right side of the third equation produces $x_3 \approx 10.808$. Finally, we plug $x_1 = 8.826$, $x_2 = -8.435$, and $x_3 = 10.808$ into the right side of the fourth equation, yielding $x_4 \approx -11.698$. The four new results are the numbers in the row labeled "After 2 Steps." We continue in this manner, producing the results given in the following table:

	x_1	x_2	x_3	x_4
Initial Values	0.000	0.000	0.000	0.000
After 1 Step	5.444	−5.379	9.226	−10.447
After 2 Steps	8.826	−8.435	10.808	−11.698
After 3 Steps	9.820	−8.920	10.961	−11.954
After 4 Steps	9.973	−8.986	10.994	−11.993
After 5 Steps	9.995	−8.998	10.999	−11.999
After 6 Steps	9.999	−9.000	11.000	−12.000
After 7 Steps	10.000	−9.000	11.000	−12.000
After 8 Steps	10.000	−9.000	11.000	−12.000

After 8 steps we get the same result as we had after 7 steps, and so we are done. The solution obtained for the system is $(10, -9, 11, -12)$.

(8) First we express the given system as

$$\begin{cases} x_1 = 5x_2 + x_3 + 16 \\ x_2 = 6x_1 - 2x_3 - 13 \\ x_3 = -7x_1 - x_2 + 12 \end{cases}.$$

Using the Jacobi Method, we start with the initial values $x_1 = 0$, $x_2 = 0$ and $x_3 = 0$, and plug these in the right side of these equations to obtain the new values $x_1 = 16$, $x_2 = -13$ and $x_3 = 12$. These values are listed in the table, below, in the row labeled "After 1 Step." These new values are put in the right side to again solve for x_1, x_2 and x_3 yielding $x_1 = -37$, $x_2 = 59$ and $x_3 = -87$, the numbers in the row labeled "After 2 Steps." We continue in this manner for 6 steps, producing the results given in the following table:

	x_1	x_2	x_3
Initial Values	0.0	0.0	0.0
After 1 Step	16.0	−13.0	12.0
After 2 Steps	−37.0	59.0	−87.0
After 3 Steps	224.0	−61.0	212.0
After 4 Steps	−77.0	907.0	−1495.0
After 5 Steps	3056.0	2515.0	−356.0
After 6 Steps	12235.0	19035.0	−23895.0

Note that the numbers are diverging rather than converging to a single solution.

Next, we use the Gauss-Seidel Method. We start with the initial values $x_1 = 0$, $x_2 = 0$ and $x_3 = 0$. We plug $x_2 = x_3 = 0$ into the right side of the first equation to obtain the new value $x_1 = 16$. Then, we substitute $x_1 = 16$ and $x_3 = 0$ into the right side of the second equation, yielding $x_2 = 83$. Finally, we plug $x_1 = 16$ and $x_2 = 83$ into the right side of the third equation, yielding $x_3 = -183$. We have listed these results in the table, below, in the row labeled "After 1 Step." Next, we plug $x_2 = 83.0$ and $x_3 = -183.0$ into the right side of the first equation to solve for x_1, which yields $x_1 = 248$. Substituting $x_1 = 248$ and $x_3 = -183$ into the right side of the second equation produces $x_2 = 1841$. Finally, we plug $x_1 = 248$ and $x_2 = 1841$ into the right side of the third equation, yielding $x_3 = -3565$. The three new results are the numbers in the row labeled "After 2 Steps." We continue in this manner for four steps, producing the results given in the following table:

	x_1	x_2	x_3
Initial Values	0.0	0.0	0.0
After 1 Step	16.0	83.0	−183.0
After 2 Steps	248.0	1841.0	−3565.0
After 3 Steps	5656.0	41053.0	−80633.0
After 4 Steps	124648.0	909141.0	−1781665.0

Again, the values are diverging rather than converging.

To find the actual solution, we use Gaussian elimination. We reduce

$$\left[\begin{array}{ccc|c} 1 & -5 & -1 & 16 \\ 6 & -1 & -2 & 13 \\ 7 & 1 & 1 & 12 \end{array}\right] \text{ to } \left[\begin{array}{ccc|c} 1 & -5 & -1 & 16 \\ 0 & 1 & \frac{4}{29} & -\frac{83}{29} \\ 0 & 0 & 1 & 1 \end{array}\right].$$

The last row yields $x_3 = 1$. The second row gives $x_2 = -\frac{4}{29}(1) - \frac{83}{29} = -3$, and the first row produces

$x_1 = 5(-3) + 1(1) + 16 = 2$. Hence, the actual solution for the system is $(2, -3, 1)$.

(10) (a) True. This is the definition of roundoff error given in Section 9.1 of the textbook.

(b) False. An ill-conditioned system is a system for which a very small change in the coefficients in the system leads to a very large change in the solution set.

(c) False. The method of partial pivoting specifies that one should use row swaps to ensure that each pivot element is as *large* as possible in absolute value.

(d) True. Roundoff errors do not accumulate in iterative methods since each iteration is using the previous iteration as a new starting point. Thus, in some sense, each iteration starts the solution process over again.

(e) False. This is done in the Gauss-Seidel Method, not the Jacobi Method.

(f) False. The values obtained should be $x = 6$ and $y = 1$, because the new value $x = 6$ is used in the computation of the new value of y.

Section 9.2

(1) (a) First, use the row operations $\langle 1\rangle \leftarrow \frac{1}{2}\langle 1\rangle$, $\langle 2\rangle \leftarrow 6\langle 1\rangle + \langle 2\rangle$ and $\langle 2\rangle \leftarrow \frac{1}{5}\langle 2\rangle$ to obtain the upper triangular matrix $\mathbf{U} = \begin{bmatrix} 1 & -2 \\ 0 & 1 \end{bmatrix}$. Using the formulas for the variables k_{ij} given in the proof of Theorem 9.1 in Section 9.2 of the textbook, $k_{11} = 2$, $k_{21} = -6$, and $k_{22} = 5$. Then, using the formulas for $\mathbf{L}$ and $\mathbf{D}$ given in that same proof, $\mathbf{L} = \begin{bmatrix} 1 & 0 \\ \frac{-6}{2} & 1 \end{bmatrix} = \begin{bmatrix} 1 & 0 \\ -3 & 1 \end{bmatrix}$, and $\mathbf{D} = \begin{bmatrix} 2 & 0 \\ 0 & 5 \end{bmatrix}$.

(c) First, use the row operations $\langle 1\rangle \leftarrow -\langle 1\rangle$, $\langle 2\rangle \leftarrow -2\langle 1\rangle + \langle 2\rangle$, $\langle 3\rangle \leftarrow -2\langle 1\rangle + \langle 3\rangle$, $\langle 2\rangle \leftarrow \frac{1}{2}\langle 2\rangle$, $\langle 3\rangle \leftarrow -8\langle 2\rangle + \langle 3\rangle$ and $\langle 3\rangle \leftarrow \frac{1}{3}\langle 3\rangle$ to obtain the upper triangular matrix $\mathbf{U} = \begin{bmatrix} 1 & -4 & 2 \\ 0 & 1 & -4 \\ 0 & 0 & 1 \end{bmatrix}$.

Using the formulas for the variables k_{ij} given in the proof of Theorem 9.1 in Section 9.2 of the textbook, $k_{11} = -1$, $k_{21} = 2$, $k_{31} = 2$, $k_{22} = 2$, $k_{32} = 8$, and $k_{33} = 3$. Then, using the formulas for $\mathbf{L}$ and $\mathbf{D}$ given in that same proof, $\mathbf{L} = \begin{bmatrix} 1 & 0 & 0 \\ \frac{2}{-1} & 1 & 0 \\ \frac{2}{-1} & \frac{8}{2} & 1 \end{bmatrix} = \begin{bmatrix} 1 & 0 & 0 \\ -2 & 1 & 0 \\ -2 & 4 & 1 \end{bmatrix}$, and

$\mathbf{D} = \begin{bmatrix} -1 & 0 & 0 \\ 0 & 2 & 0 \\ 0 & 0 & 3 \end{bmatrix}$.

(e) First, use the row operations $\langle 1\rangle \leftarrow -\frac{1}{3}\langle 1\rangle$, $\langle 2\rangle \leftarrow -4\langle 1\rangle + \langle 2\rangle$, $\langle 3\rangle \leftarrow -6\langle 1\rangle + \langle 3\rangle$, $\langle 4\rangle \leftarrow 2\langle 1\rangle + \langle 4\rangle$, $\langle 2\rangle \leftarrow -\frac{3}{2}\langle 2\rangle$, $\langle 3\rangle \leftarrow -\langle 2\rangle + \langle 3\rangle$, $\langle 4\rangle \leftarrow -\frac{4}{3}\langle 2\rangle + \langle 4\rangle$, and $\langle 3\rangle \leftarrow 2\langle 3\rangle$ to obtain the upper triangular matrix $\mathbf{U} = \begin{bmatrix} 1 & -\frac{1}{3} & -\frac{1}{3} & \frac{1}{3} \\ 0 & 1 & \frac{5}{2} & -\frac{11}{2} \\ 0 & 0 & 1 & 3 \\ 0 & 0 & 0 & 1 \end{bmatrix}$. Using the formulas for the variables k_{ij} given in the proof of Theorem 9.1 in Section 9.2 of the textbook, $k_{11} = -3$, $k_{21} = 4$, $k_{31} = 6$, $k_{41} = -2$, $k_{22} = -\frac{2}{3}$,

$k_{32} = 1$, $k_{42} = \frac{4}{3}$, $k_{33} = \frac{1}{2}$, $k_{43} = 0$ and $k_{44} = 1$. Then, using the formulas for **L** and **D** given in that same proof, $\mathbf{L} = \begin{bmatrix} 1 & 0 & 0 & 0 \\ \frac{4}{-3} & 1 & 0 & 0 \\ \frac{6}{-3} & \frac{1}{(-2/3)} & 1 & 0 \\ \frac{-2}{-3} & \frac{(4/3)}{(-2/3)} & \frac{0}{(1/2)} & 1 \end{bmatrix} = \begin{bmatrix} 1 & 0 & 0 & 0 \\ -\frac{4}{3} & 1 & 0 & 0 \\ -2 & -\frac{3}{2} & 1 & 0 \\ \frac{2}{3} & -2 & 0 & 1 \end{bmatrix}$, and

$\mathbf{D} = \begin{bmatrix} -3 & 0 & 0 & 0 \\ 0 & -\frac{2}{3} & 0 & 0 \\ 0 & 0 & \frac{1}{2} & 0 \\ 0 & 0 & 0 & 1 \end{bmatrix}$.

(3) (a) The coefficient matrix for the system $\mathbf{AX} = \mathbf{B}$ is $\mathbf{A} = \begin{bmatrix} -1 & 5 \\ 2 & -13 \end{bmatrix}$, with $\mathbf{B} = \begin{bmatrix} -9 \\ 21 \end{bmatrix}$. We need to find the **KU** decomposition for **A**. To do this, we use the row operations $\langle 1 \rangle \leftarrow -\langle 1 \rangle$, $\langle 2 \rangle \leftarrow -2\langle 1 \rangle + \langle 2 \rangle$ and $\langle 2 \rangle \leftarrow -\frac{1}{3}\langle 2 \rangle$ to obtain the upper triangular matrix $\mathbf{U} = \begin{bmatrix} 1 & -5 \\ 0 & 1 \end{bmatrix}$. Using the formulas for the variables k_{ij} given in the proof of Theorem 9.1 in Section 9.2 of the textbook, $k_{11} = -1$, $k_{21} = 2$, and $k_{22} = -3$, yielding the lower triangular matrix $\mathbf{K} = \begin{bmatrix} -1 & 0 \\ 2 & -3 \end{bmatrix}$. Now, $\mathbf{AX} = \mathbf{B}$ is equivalent to $\mathbf{K}(\mathbf{UX}) = \mathbf{B}$. Letting $\mathbf{Y} = \mathbf{UX}$, we must first solve $\mathbf{KY} = \mathbf{B}$. This is equivalent to the system

$$\begin{cases} -y_1 & & = -9 \\ 2y_1 & - \ 3y_2 & = 21 \end{cases},$$

whose solution is easily seen to be $y_1 = 9$, $y_2 = -1$; that is, $\mathbf{Y} = [9, -1]$. Finally, we find **X** by solving the system $\mathbf{UX} = \mathbf{Y}$, which is

$$\begin{cases} x_1 & - \ 5x_2 & = 9 \\ & x_2 & = -1 \end{cases}.$$

Back substitution gives the solution $x_2 = -1$, $x_1 = 4$. Hence, the solution set for the original system is $\{(4, -1)\}$.

(c) The coefficient matrix for the system $\mathbf{AX} = \mathbf{B}$ is $\mathbf{A} = \begin{bmatrix} -1 & 3 & -2 \\ 4 & -9 & -7 \\ -2 & 11 & -31 \end{bmatrix}$, with $\mathbf{B} = \begin{bmatrix} -13 \\ 28 \\ -68 \end{bmatrix}$.

We need to find the **KU** decomposition for **A**. To do this, we use the row operations $\langle 1 \rangle \leftarrow -\langle 1 \rangle$, $\langle 2 \rangle \leftarrow -4\langle 1 \rangle + \langle 2 \rangle$, $\langle 3 \rangle \leftarrow 2\langle 1 \rangle + \langle 3 \rangle$, $\langle 2 \rangle \leftarrow \frac{1}{3}\langle 2 \rangle$, $\langle 3 \rangle \leftarrow -5\langle 2 \rangle + \langle 3 \rangle$ and $\langle 3 \rangle \leftarrow -\frac{1}{2}\langle 3 \rangle$ to obtain the upper triangular matrix $\mathbf{U} = \begin{bmatrix} 1 & -3 & 2 \\ 0 & 1 & -5 \\ 0 & 0 & 1 \end{bmatrix}$. Using the formulas for the variables k_{ij} given in the proof of Theorem 9.1 in Section 9.2 of the textbook, $k_{11} = -1$, $k_{21} = 4$, $k_{31} = -2$, $k_{22} = 3$, $k_{32} = 5$, and $k_{33} = -2$, yielding the lower triangular matrix $\mathbf{K} = \begin{bmatrix} -1 & 0 & 0 \\ 4 & 3 & 0 \\ -2 & 5 & -2 \end{bmatrix}$. Now,

AX = **B** is equivalent to **K**(**UX**) = **B**. Letting **Y** = **UX**, we must first solve **KY** = **B**. This is equivalent to the system

$$\begin{cases} -y_1 & & & = & -13 \\ 4y_1 & + & 3y_2 & & = & 28 \\ -2y_1 & + & 5y_2 & - & 2y_3 & = & -68 \end{cases},$$

whose solution is easily seen to be $y_1 = 13$, $y_2 = \frac{1}{3}(-4(13)+28) = -8$, $y_3 = -\frac{1}{2}(2(13)-5(-8)-68) = 1$; that is, $\mathbf{Y} = [13, -8, 1]$. Finally, we find **X** by solving the system **UX** = **Y**, which is

$$\begin{cases} x_1 & - & 3x_2 & + & 2x_3 & = & 13 \\ & & x_2 & - & 5x_3 & = & -8 \\ & & & & x_3 & = & 1 \end{cases}.$$

Back substitution gives the solution $x_3 = 1$, $x_2 = 5(1) - 8 = -3$, $x_1 = 3(-3) - 2(1) + 13 = 2$. Hence, the solution set for the original system is $\{(2, -3, 1)\}$.

(4) (a) False. Nonsingular matrices that require row swaps to reduce to $\mathbf{I}_n$ may not have **LDU** decompositions. For example, $\begin{bmatrix} 0 & 1 \\ 1 & 0 \end{bmatrix}$ has no **LDU** decomposition, as shown in Exercise 2 of Section 9.2.

(b) True. The method for doing this is given in the proof of Theorem 9.1.

(c) False. The given row operation adds a multiple of row 3 to row 2, which is above row 3. This is not allowed for lower type (II) operations.

(d) False. First solve for **Y** in **KY** = **B**, then solve for **X** in **UX** = **Y**.

Section 9.3

(1) In each part, we only print m digits after the decimal point for all computations, even though more digits are actually used and computed. We write all vectors as row vectors to conserve space, even when correct notation calls for a column vector. Also, let **A** represent the given matrix.

(a) We begin with $\mathbf{u}_0 = [1, 0]$. Then, $\mathbf{w}_1 = \mathbf{A}\mathbf{u}_0 = [2, 36]$, and so $\|\mathbf{w}_1\| = \sqrt{2^2 + 36^2} \approx 36.06$. Using this gives $\mathbf{u}_1 = \frac{\mathbf{w}_1}{\|\mathbf{w}_1\|} \approx \frac{1}{36.06}[2, 36] \approx [0.06, 1.00]$. These results are displayed in the first line of the table, below. Continuing in this manner for a total of 9 iterations produces the following:

k	$\mathbf{w}_k = \mathbf{A}\mathbf{u}_{k-1}$	$\|\mathbf{w}_k\|$	$\mathbf{u}_k = \frac{\mathbf{w}_k}{\|\mathbf{w}_k\|}$
1	[2, 36]	36.06	[0.06, 1.00]
2	[36.06, 24.96]	43.85	[0.82, 0.57]
3	[22.14, 42.69]	48.09	[0.46, 0.89]
4	[32.88, 36.99]	49.49	[0.66, 0.75]
5	[28.24, 41.11]	49.87	[0.57, 0.82]
6	[30.81, 39.34]	49.97	[0.62, 0.79]
7	[29.58, 40.30]	49.99	[0.59, 0.81]
8	[30.21, 39.84]	50.00	[0.60, 0.80]
9	[29.90, 40.08]	50.00	[0.60, 0.80]

Since $\mathbf{u}_8 = \mathbf{u}_9$, we stop here. Hence, the Power Method produces the eigenvector [0.60, 0.80]. Since the vectors $\mathbf{u}_8$ and $\mathbf{u}_9$ are equal (rather than being the negations of each other), the sign of the corresponding eigenvalue is positive, and so the eigenvalue is $\|\mathbf{w}_9\| = 50$.

(c) We begin with $\mathbf{u}_0 = [0,1,0]$. Then, $\mathbf{w}_1 = \mathbf{A}\mathbf{u}_0 = [3,0,1]$, and so $\|\mathbf{w}_1\| = \sqrt{3^2+0^2+1^2} \approx 3.16$. Using this gives $\mathbf{u}_1 = \frac{\mathbf{w}_1}{\|\mathbf{w}_1\|} \approx \frac{1}{3.16}[3,0,1] \approx [0.95, 0.00, 0.32]$. These results are displayed in the first line of the table, below. Continuing in this manner for a total of 7 iterations produces the following:

k	$\mathbf{w}_k = \mathbf{A}\mathbf{u}_{k-1}$	$\|\mathbf{w}_k\|$	$\mathbf{u}_k = \frac{\mathbf{w}_k}{\|\mathbf{w}_k\|}$
1	$[3,0,1]$	3.16	$[0.95, 0.00, 0.32]$
2	$[1.58, 1.26, 3.16]$	3.75	$[0.42, 0.34, 0.84]$
3	$[1.01, 1.26, 2.44]$	2.93	$[0.34, 0.43, 0.83]$
4	$[1.15, 1.18, 2.30]$	2.83	$[0.41, 0.42, 0.81]$
5	$[1.25, 1.22, 2.45]$	3.01	$[0.42, 0.41, 0.81]$
6	$[1.23, 1.23, 2.47]$	3.02	$[0.41, 0.41, 0.82]$
7	$[1.22, 1.23, 2.45]$	3.00	$[0.41, 0.41, 0.82]$

Since $\mathbf{u}_6 = \mathbf{u}_7$, we stop here. Hence, the Power Method produces the eigenvector $[0.41, 0.41, 0.82]$. Since the vectors $\mathbf{u}_6$ and $\mathbf{u}_7$ are equal (rather than being negations of each other), the sign of the corresponding eigenvalue is positive, and so the eigenvalue is $\|\mathbf{w}_7\| = 3.0$.

(e) We begin with $\mathbf{u}_0 = [3,8,2,3]$. Then, $\mathbf{w}_1 = \mathbf{A}\mathbf{u}_0 = [17,40,14,19]$, and so $\|\mathbf{w}_1\| = \sqrt{17^2+40^2+14^2+19^2} \approx 49.457$. Using this gives $\mathbf{u}_1 = \frac{\mathbf{w}_1}{\|\mathbf{w}_1\|} \approx \frac{1}{49.457}[17,40,14,19] \approx [0.344, 0.809, 0.283, 0.384]$. These results are displayed in the first line of the table, below. Continuing in this manner for a total of 15 iterations produces the following:

k	$\mathbf{w}_k = \mathbf{A}\mathbf{u}_{k-1}$	$\|\mathbf{w}_k\|$	$\mathbf{u}_k = \frac{\mathbf{w}_k}{\|\mathbf{w}_k\|}$
1	$[17,40,14,19]$	49.457	$[0.344, 0.809, 0.283, 0.384]$
2	$[2.123, 4.448, 2.143, 2.285]$	5.840	$[0.364, 0.762, 0.367, 0.391]$
3	$[1.825, 4.224, 1.392, 1.928]$	5.179	$[0.352, 0.816, 0.269, 0.372]$
4	$[1.935, 4.468, 1.440, 2.010]$	5.461	$[0.354, 0.818, 0.264, 0.368]$
5	$[1.878, 4.471, 1.232, 1.929]$	5.363	$[0.350, 0.834, 0.230, 0.360]$
6	$[1.893, 4.537, 1.190, 1.929]$	5.413	$[0.350, 0.838, 0.220, 0.356]$
7	$[1.878, 4.553, 1.116, 1.903]$	5.397	$[0.348, 0.844, 0.207, 0.353]$
8	$[1.879, 4.575, 1.084, 1.896]$	5.406	$[0.347, 0.846, 0.201, 0.351]$
9	$[1.874, 4.585, 1.053, 1.886]$	5.403	$[0.347, 0.848, 0.195, 0.349]$
10	$[1.873, 4.593, 1.036, 1.881]$	5.405	$[0.346, 0.850, 0.192, 0.348]$
11	$[1.871, 4.598, 1.022, 1.876]$	5.405	$[0.346, 0.851, 0.189, 0.347]$
12	$[1.870, 4.602, 1.013, 1.874]$	5.405	$[0.346, 0.851, 0.187, 0.347]$
13	$[1.869, 4.605, 1.007, 1.872]$	5.405	$[0.346, 0.852, 0.186, 0.346]$
14	$[1.869, 4.606, 1.003, 1.871]$	5.405	$[0.346, 0.852, 0.185, 0.346]$
15	$[1.868, 4.607, 1.000, 1.870]$	5.405	$[0.346, 0.852, 0.185, 0.346]$

Since $\mathbf{u}_{14} = \mathbf{u}_{15}$, we stop here. Hence, the Power Method produces the eigenvector $[0.346, 0.852, 0.185, 0.346]$. Since the vectors $\mathbf{u}_{14}$ and $\mathbf{u}_{15}$ are equal (rather than being negations of each other), the sign of the corresponding eigenvalue is positive, and so the eigenvalue is $\|\mathbf{w}_{15}\| = 5.405$.

(3) (b) Let $\lambda_1, \ldots \lambda_n$ be the eigenvalues of $\mathbf{A}$ with $|\lambda_1| > |\lambda_j|$, for $2 \le j \le n$. Let $\{\mathbf{v}_1, \ldots, \mathbf{v}_n\}$ be a basis of unit eigenvectors for $\mathbb{R}^n$ corresponding to $\lambda_1, \ldots, \lambda_n$, respectively. Suppose the initial vector in the Power Method is $\mathbf{u}_0 = a_{01}\mathbf{v}_1 + \cdots + a_{0n}\mathbf{v}_n$ and the ith iteration yields $\mathbf{u}_i = a_{i1}\mathbf{v}_1 + \cdots + a_{in}\mathbf{v}_n$.

Thus, for $2 \le j \le n$, $\lambda_j \ne 0$, and $a_{0j} \ne 0$, we have

$$\frac{|a_{i1}|}{|a_{ij}|} = \left|\frac{\lambda_1}{\lambda_j}\right|^i \frac{|a_{01}|}{|a_{0j}|}.$$

Proof: The Power Method states that each $\mathbf{u}_i$ is derived from $\mathbf{u}_{i-1}$ as $\mathbf{u}_i = c_i\mathbf{A}\mathbf{u}_{i-1}$, where c_i is a normalizing constant. A proof by induction shows that $\mathbf{u}_i = k_i\mathbf{A}^i\mathbf{u}_0$ for some constant k_i. (Base Step $i = 0$: $\mathbf{u}_0 = \mathbf{I}_n\mathbf{u}_0 = 1\mathbf{A}^0\mathbf{u}_0$, with $k_0 = 1$. Inductive Step: $\mathbf{u}_i = c_i\mathbf{A}\mathbf{u}_{i-1} = c_i\mathbf{A}(k_{i-1}\mathbf{A}^{i-1}\mathbf{u}_0) = c_ik_{i-1}\mathbf{A}^i\mathbf{u}_0$, with $k_i = c_ik_{i-1}$.) Now, if $\mathbf{u}_0 = a_{01}\mathbf{v}_1 + \cdots + a_{0n}\mathbf{v}_n$, then $\mathbf{u}_i = k_i\mathbf{A}^i\mathbf{u}_0 = a_{01}k_i\mathbf{A}^i\mathbf{v}_1 + \cdots + a_{0n}k_i\mathbf{A}^i\mathbf{v}_n$. But because $\mathbf{v}_j$ is an eigenvector for $\mathbf{A}$ corresponding to the eigenvalue λ_j, $\mathbf{A}^i\mathbf{v}_j = \lambda_j^i\mathbf{v}_j$. Therefore $\mathbf{u}_i = a_{01}k_i\lambda_1^i\mathbf{v}_1 + \cdots + a_{0n}k_i\lambda_n^i\mathbf{v}_n$. Since we also have $\mathbf{u}_i = a_{i1}\mathbf{v}_1 + \cdots + a_{in}\mathbf{v}_n$, the uniqueness of expression of a vector with respect to a basis shows that $a_{ij} = a_{0j}k_i\lambda_j^i$. Hence,

$$\frac{|a_{i1}|}{|a_{ij}|} = \frac{|a_{01}k_i\lambda_1^i|}{|a_{0j}k_i\lambda_j^i|} = \left|\frac{\lambda_1}{\lambda_j}\right|^i \frac{|a_{01}|}{|a_{0j}|}.$$

(4) (a) False. The eigenvalue could be $-\|\mathbf{A}\mathbf{u}_{k-1}\|$. This occurs when $\mathbf{u}_k = -\mathbf{u}_{k-1}$. For an example, consider applying the Power Method to the matrix $\begin{bmatrix} -2 & 1 \\ 0 & 1 \end{bmatrix}$, starting with $\mathbf{u}_0 = [1,0]$. Then $\mathbf{w}_1 = [-2,0]$, and $\|\mathbf{w}_1\| = 2$. Hence, $\mathbf{u}_1 = [-1,0] = -\mathbf{u}_0$, and the Power Method yields the eigenvector $[-1,0]$. The eigenvalue corresponding to this eigenvector is -2. However, $\|\mathbf{A}\mathbf{u}_0\| = \|\mathbf{w}_1\| = 2$.

(b) True. Each iteration of the Power Method will just repeatedly produce the initial vector.

(c) True. Starting with $\mathbf{u}_0 = [1,0,0,0]$, the Power Method will produce $\mathbf{u}_i = \frac{1}{2}[1,1,1,1]$ for all $i \ge 1$ and $\mathbf{w}_i = [2,2,2,2]$ for all $i \ge 2$. Note that $\|[2,2,2,2]\| = 4$.

(d) False. The dominant eigenvalue in this case is -3, *not* 2, because $|-3| > |2|$. Hence, the Power Method will produce an eigenvector corresponding to the eigenvalue -3. (Note: The Power Method converges here since $\mathbf{A}$ is diagonalizable and there is a dominant eigenvalue.)

Chapter 10

Section 10.1

(1) In each part, let $\mathbf{E}$ represent the given matrix.

(a) The operation (III): $\langle 2\rangle \longleftrightarrow \langle 3\rangle$ was performed on $\mathbf{I}_3$ to obtain $\mathbf{E}$. The inverse operation is (III): $\langle 2\rangle \longleftrightarrow \langle 3\rangle$. Since this row operation is its own inverse, the matrix $\mathbf{E}$ is its own inverse.

(b) The operation (I): $\langle 2\rangle \leftarrow -2\langle 2\rangle$ was performed on $\mathbf{I}_3$ to obtain $\mathbf{E}$. The inverse operation is (I): $\langle 2\rangle \longleftarrow -\frac{1}{2}\langle 2\rangle$. Hence, $\mathbf{E}^{-1} = \begin{bmatrix} 1 & 0 & 0 \\ 0 & -\frac{1}{2} & 0 \\ 0 & 0 & 1 \end{bmatrix}$, the matrix obtained by performing the inverse row operation on $\mathbf{I}_3$.

(e) The operation (II): $\langle 3\rangle \leftarrow -2\langle 4\rangle + \langle 3\rangle$ was performed on $\mathbf{I}_4$ to obtain $\mathbf{E}$. The inverse operation is (II): $\langle 3\rangle \longleftarrow 2\langle 4\rangle + \langle 3\rangle$. Hence, $\mathbf{E}^{-1} = \begin{bmatrix} 1 & 0 & 0 & 0 \\ 0 & 1 & 0 & 0 \\ 0 & 0 & 1 & 2 \\ 0 & 0 & 0 & 1 \end{bmatrix}$, the matrix obtained by performing the inverse row operation on $\mathbf{I}_4$.

(2) In each part, let $\mathbf{A}$ represent the given matrix.

(a) We list the row operations needed to convert $\mathbf{A}$ to reduced row echelon form, the resulting matrix after each step, the inverse of the row operation used, and the elementary matrix corresponding to each inverse operation:

Row Operation	Resultant Matrix	Inverse Row Operation	Elementary Matrix for Inverse Operation
$\langle 1\rangle \leftarrow \frac{1}{4}\langle 1\rangle$	$\begin{bmatrix} 1 & \frac{9}{4} \\ 3 & 7 \end{bmatrix}$	$\langle 1\rangle \leftarrow 4\langle 1\rangle$	$\mathbf{F}_4 = \begin{bmatrix} 4 & 0 \\ 0 & 1 \end{bmatrix}$
$\langle 2\rangle \leftarrow -3\langle 1\rangle + \langle 2\rangle$	$\begin{bmatrix} 1 & \frac{9}{4} \\ 0 & \frac{1}{4} \end{bmatrix}$	$\langle 2\rangle \leftarrow 3\langle 1\rangle + \langle 2\rangle$	$\mathbf{F}_3 = \begin{bmatrix} 1 & 0 \\ 3 & 1 \end{bmatrix}$
$\langle 2\rangle \leftarrow 4\langle 2\rangle$	$\begin{bmatrix} 1 & \frac{9}{4} \\ 0 & 1 \end{bmatrix}$	$\langle 2\rangle \leftarrow \frac{1}{4}\langle 2\rangle$	$\mathbf{F}_2 = \begin{bmatrix} 1 & 0 \\ 0 & \frac{1}{4} \end{bmatrix}$
$\langle 1\rangle \leftarrow -\frac{9}{4}\langle 2\rangle + \langle 1\rangle$	$\begin{bmatrix} 1 & 0 \\ 0 & 1 \end{bmatrix}$	$\langle 1\rangle \leftarrow \frac{9}{4}\langle 2\rangle + \langle 1\rangle$	$\mathbf{F}_1 = \begin{bmatrix} 1 & \frac{9}{4} \\ 0 & 1 \end{bmatrix}$

Then $\mathbf{A} = \mathbf{F}_4\mathbf{F}_3\mathbf{F}_2\mathbf{F}_1\mathbf{I}_2$.

(c) We list the row operations needed to convert $\mathbf{A}$ to reduced row echelon form, the resulting matrix after each step, the inverse of the row operation used, and the elementary matrix corresponding to each inverse operation:

Row Operation	Resultant Matrix	Inverse Row Operation	Elementary Matrix for Inverse Operation
$\langle 1\rangle \leftrightarrow \langle 2\rangle$	$\begin{bmatrix} -3 & 0 & 0 & -2 \\ 0 & 0 & 5 & 0 \\ 0 & 6 & -10 & -1 \\ 3 & 0 & 0 & 3 \end{bmatrix}$	$\langle 1\rangle \leftrightarrow \langle 2\rangle$	$\mathbf{F}_9 = \begin{bmatrix} 0 & 1 & 0 & 0 \\ 1 & 0 & 0 & 0 \\ 0 & 0 & 1 & 0 \\ 0 & 0 & 0 & 1 \end{bmatrix}$
$\langle 1\rangle \leftarrow -\frac{1}{3}\langle 1\rangle$	$\begin{bmatrix} 1 & 0 & 0 & \frac{2}{3} \\ 0 & 0 & 5 & 0 \\ 0 & 6 & -10 & -1 \\ 3 & 0 & 0 & 3 \end{bmatrix}$	$\langle 1\rangle \leftarrow -3\langle 1\rangle$	$\mathbf{F}_8 = \begin{bmatrix} -3 & 0 & 0 & 0 \\ 0 & 1 & 0 & 0 \\ 0 & 0 & 1 & 0 \\ 0 & 0 & 0 & 1 \end{bmatrix}$
$\langle 4\rangle \leftarrow -3\langle 1\rangle + \langle 4\rangle$	$\begin{bmatrix} 1 & 0 & 0 & \frac{2}{3} \\ 0 & 0 & 5 & 0 \\ 0 & 6 & -10 & -1 \\ 0 & 0 & 0 & 1 \end{bmatrix}$	$\langle 4\rangle \leftarrow 3\langle 1\rangle + \langle 4\rangle$	$\mathbf{F}_7 = \begin{bmatrix} 1 & 0 & 0 & 0 \\ 0 & 1 & 0 & 0 \\ 0 & 0 & 1 & 0 \\ 3 & 0 & 0 & 1 \end{bmatrix}$
$\langle 2\rangle \leftrightarrow \langle 3\rangle$	$\begin{bmatrix} 1 & 0 & 0 & \frac{2}{3} \\ 0 & 6 & -10 & -1 \\ 0 & 0 & 5 & 0 \\ 0 & 0 & 0 & 1 \end{bmatrix}$	$\langle 2\rangle \leftrightarrow \langle 3\rangle$	$\mathbf{F}_6 = \begin{bmatrix} 1 & 0 & 0 & 0 \\ 0 & 0 & 1 & 0 \\ 0 & 1 & 0 & 0 \\ 0 & 0 & 0 & 1 \end{bmatrix}$
$\langle 2\rangle \leftarrow \frac{1}{6}\langle 2\rangle$	$\begin{bmatrix} 1 & 0 & 0 & \frac{2}{3} \\ 0 & 1 & -\frac{5}{3} & -\frac{1}{6} \\ 0 & 0 & 5 & 0 \\ 0 & 0 & 0 & 1 \end{bmatrix}$	$\langle 2\rangle \leftarrow 6\langle 2\rangle$	$\mathbf{F}_5 = \begin{bmatrix} 1 & 0 & 0 & 0 \\ 0 & 6 & 0 & 0 \\ 0 & 0 & 1 & 0 \\ 0 & 0 & 0 & 1 \end{bmatrix}$
$\langle 3\rangle \leftarrow \frac{1}{5}\langle 3\rangle$	$\begin{bmatrix} 1 & 0 & 0 & \frac{2}{3} \\ 0 & 1 & -\frac{5}{3} & -\frac{1}{6} \\ 0 & 0 & 1 & 0 \\ 0 & 0 & 0 & 1 \end{bmatrix}$	$\langle 3\rangle \leftarrow 5\langle 3\rangle$	$\mathbf{F}_4 = \begin{bmatrix} 1 & 0 & 0 & 0 \\ 0 & 1 & 0 & 0 \\ 0 & 0 & 5 & 0 \\ 0 & 0 & 0 & 1 \end{bmatrix}$
$\langle 2\rangle \leftarrow \frac{5}{3}\langle 3\rangle + \langle 2\rangle$	$\begin{bmatrix} 1 & 0 & 0 & \frac{2}{3} \\ 0 & 1 & 0 & -\frac{1}{6} \\ 0 & 0 & 1 & 0 \\ 0 & 0 & 0 & 1 \end{bmatrix}$	$\langle 2\rangle \leftarrow -\frac{5}{3}\langle 3\rangle + \langle 2\rangle$	$\mathbf{F}_3 = \begin{bmatrix} 1 & 0 & 0 & 0 \\ 0 & 1 & -\frac{5}{3} & 0 \\ 0 & 0 & 1 & 0 \\ 0 & 0 & 0 & 1 \end{bmatrix}$
$\langle 1\rangle \leftarrow -\frac{2}{3}\langle 4\rangle + \langle 1\rangle$	$\begin{bmatrix} 1 & 0 & 0 & 0 \\ 0 & 1 & 0 & -\frac{1}{6} \\ 0 & 0 & 1 & 0 \\ 0 & 0 & 0 & 1 \end{bmatrix}$	$\langle 1\rangle \leftarrow \frac{2}{3}\langle 4\rangle + \langle 1\rangle$	$\mathbf{F}_2 = \begin{bmatrix} 1 & 0 & 0 & \frac{2}{3} \\ 0 & 1 & 0 & 0 \\ 0 & 0 & 1 & 0 \\ 0 & 0 & 0 & 1 \end{bmatrix}$
$\langle 2\rangle \leftarrow \frac{1}{6}\langle 4\rangle + \langle 2\rangle$	$\begin{bmatrix} 1 & 0 & 0 & 0 \\ 0 & 1 & 0 & 0 \\ 0 & 0 & 1 & 0 \\ 0 & 0 & 0 & 1 \end{bmatrix}$	$\langle 2\rangle \leftarrow -\frac{1}{6}\langle 4\rangle + \langle 2\rangle$	$\mathbf{F}_1 = \begin{bmatrix} 1 & 0 & 0 & 0 \\ 0 & 1 & 0 & -\frac{1}{6} \\ 0 & 0 & 1 & 0 \\ 0 & 0 & 0 & 1 \end{bmatrix}$

Then $\mathbf{A} = \mathbf{F}_9\mathbf{F}_8\mathbf{F}_7\mathbf{F}_6\mathbf{F}_5\mathbf{F}_4\mathbf{F}_3\mathbf{F}_2\mathbf{F}_1\mathbf{I}_4$.

(7) $\mathbf{A}$ is nonsingular if and only if $\text{rank}(\mathbf{A}) = n$ (by Theorem 2.14) if and only if $\mathbf{A}$ is row equivalent to

$\mathbf{I}_n$ (by the definition of rank) if and only if $\mathbf{A} = \mathbf{E}_k \cdots \mathbf{E}_1\mathbf{I}_n = \mathbf{E}_k \cdots \mathbf{E}_1$ for some elementary matrices $\mathbf{E}_1, \ldots, \mathbf{E}_k$ (by Theorem 10.3).

(10) (a) True. An elementary matrix is formed by performing a single row operation on $\mathbf{I}_n$. Since $\mathbf{I}_n$ is square, the result of any row operation performed on it will also be square.

(b) False. The equation $\mathbf{B} = \mathbf{EA}$ is true only if $\mathbf{B}$ can be obtained from $\mathbf{A}$ using a single row operation – namely the row operation corresponding to the elementary matrix $\mathbf{E}$. However, it may take several row operations to convert one matrix $\mathbf{A}$ to a row equivalent matrix $\mathbf{B}$. For example, $\mathbf{I}_2$ and $5\mathbf{I}_2$ are row equivalent, but it takes at least two row operations to convert $\mathbf{I}_2$ to $5\mathbf{I}_2$.

(c) False. The inverse of $\mathbf{E}_1\mathbf{E}_2\cdots\mathbf{E}_k$ is $\mathbf{E}_k^{-1}\cdots\mathbf{E}_2^{-1}\mathbf{E}_1^{-1}$, *not* $\mathbf{E}_k\cdots\mathbf{E}_2\mathbf{E}_1$. For example, if $\mathbf{E}_1 = \begin{bmatrix} 2 & 0 \\ 0 & 1 \end{bmatrix}$ and $\mathbf{E}_2 = \begin{bmatrix} 1 & 0 \\ 0 & 2 \end{bmatrix}$, then $\mathbf{E}_1\mathbf{E}_2 = 2\mathbf{I}_2$, and so $(\mathbf{E}_1\mathbf{E}_2)^{-1} = \frac{1}{2}\mathbf{I}_2$. However, $\mathbf{E}_2\mathbf{E}_1 = 2\mathbf{I}_2$.

(d) True. If $\mathbf{A}$ is nonsingular, then $\mathbf{A}^{-1}$ is also nonsingular. The given statement then follows from Corollary 10.4.

(e) True. The proof of Theorem 10.2 shows that the elementary matrix corresponding to the reverse of a row operation R is the inverse of the elementary matrix corresponding to R. The given statement follows from this fact and Theorem 10.1.

Section 10.2

(1) (a) To test for linear independence, we must determine whether $ae^x + be^{2x} + ce^{3x} = 0$ has any nontrivial solutions. Using this equation we substitute the following values for x:

$$\begin{cases} \text{Letting } x = 0 \implies & a + b + c = 0 \\ \text{Letting } x = 1 \implies & a(e) + b(e^2) + c(e^3) = 0 \\ \text{Letting } x = 2 \implies & a(e^2) + b(e^4) + c(e^6) = 0 \end{cases}$$

But $\left[\begin{array}{ccc|c} 1 & 1 & 1 & 0 \\ e & e^2 & e^3 & 0 \\ e^2 & e^4 & e^6 & 0 \end{array}\right]$ row reduces to $\left[\begin{array}{ccc|c} 1 & 0 & 0 & 0 \\ 0 & 1 & 0 & 0 \\ 0 & 0 & 1 & 0 \end{array}\right]$,

showing that the system has only the trivial solution $a = b = c = 0$. Hence, the given set is linearly independent.

(c) To test for linear independence, we must check for nontrivial solutions to

$a\left(\frac{5x-1}{1+x^2}\right) + b\left(\frac{3x+1}{2+x^2}\right) + c\left(\frac{7x^3-3x^2+17x-5}{x^4+3x^2+2}\right) = 0.$

Using this equation we substitute the following values for x:

$$\begin{cases} \text{Letting } x = 0 \implies & -a + \frac{1}{2}b - \frac{5}{2}c = 0 \\ \text{Letting } x = 1 \implies & 2a + \frac{4}{3}b + \frac{8}{3}c = 0 \\ \text{Letting } x = -1 \implies & -3a - \frac{2}{3}b - \frac{16}{3}c = 0 \end{cases}.$$

But $\left[\begin{array}{ccc|c} -1 & \frac{1}{2} & -\frac{5}{2} & 0 \\ 2 & \frac{4}{3} & \frac{8}{3} & 0 \\ -3 & -\frac{2}{3} & -\frac{16}{3} & 0 \end{array}\right]$ row reduces to $\left[\begin{array}{ccc|c} 1 & 0 & 2 & 0 \\ 0 & 1 & -1 & 0 \\ 0 & 0 & 0 & 0 \end{array}\right]$,

which has nontrivial solution $a = -2$, $b = 1$, $c = 1$. Algebraic simplification verifies that

$(-2)\left(\frac{5x-1}{1+x^2}\right)+(1)\left(\frac{3x+1}{2+x^2}\right)+(1)\left(\frac{7x^3-3x^2+17x-5}{x^4+3x^2+2}\right)=0$

is a functional identity. Hence, since a nontrivial linear combination of the elements of S produces $\mathbf{0}$, the set S is linearly dependent.

(4) (a) First, to simplify matters, we eliminate any functions that we can see by inspection are linear combinations of other functions in S. The familiar trigonometric identities $\sin^2 x+\cos^2 x=1$ and $\sin x\cos x=\frac{1}{2}\sin 2x$ show that the last two functions in S are linear combinations of earlier functions. Thus, the subset $S_1=\{\sin 2x,\cos 2x,\sin^2 x,\cos^2 x\}$ has the same span as S. However, the identity $\cos^2 x=\cos 2x+\sin^2 x$ (more commonly stated as $\cos 2x=\cos^2 x-\sin^2 x$) shows that $B=\{\sin 2x,\cos 2x,\sin^2 x\}$ also has the same span. We suspect that we have now eliminated all of the functions that we can, and that B is linearly independent, making it a basis for span(S). We verify the linear independence of B by considering the equation

$a\sin 2x+b\cos 2x+c\sin^2 x=0.$

Using this equation we substitute the following values for x:

$$\begin{cases}\text{Letting } x=0 \implies & & b & & = 0\\ \text{Letting } x=\frac{\pi}{6} \implies & \frac{\sqrt{3}}{2}a & +\ \frac{1}{2}b & +\ \frac{1}{4}c & = 0\\ \text{Letting } x=\frac{\pi}{4} \implies & a & & +\ \frac{1}{2}c & = 0\end{cases}.$$

Next, row reduce $\left[\begin{array}{ccc|c}0&1&0&0\\ \frac{\sqrt{3}}{2}&\frac{1}{2}&\frac{1}{4}&0\\ 1&0&\frac{1}{2}&0\end{array}\right]$ to $\left[\begin{array}{ccc|c}1&0&0&0\\ 0&1&0&0\\ 0&0&1&0\end{array}\right]$.

Since the system has only the trivial solution, B is linearly independent, and is thus the desired basis for span(S).

(c) First, to simplify matters, we eliminate any functions that we can see by inspection are linear combinations of other functions in S. We start with the familiar trigonometric identities $\sin(a+b)=\sin a\cos b+\sin b\cos a$ and $\cos(a+b)=\cos a\cos b-\sin a\sin b$. Hence,

$\sin(x+1)=(\sin x)(\cos 1)+(\sin 1)(\cos x)$,
$\cos(x+1)=(\cos x)(\cos 1)-(\sin x)(\sin 1)$,
$\sin(x+2)=(\sin x)(\cos 2)+(\sin 2)(\cos x)$ and
$\cos(x+2)=(\cos x)(\cos 2)-(\sin x)(\sin 2)$.

Therefore, each of the elements of S is a linear combination of $\sin x$ and $\cos x$. That is, $\text{span}(S)\subseteq\text{span}(\{\sin x,\cos x\})$. Thus, $\dim(\text{span}(S))\leq 2$. So, if we can find two linearly independent vectors in S, they form a basis for span(S). We claim that the subset $B=\{\sin(x+1),\cos(x+1)\}$ of S is linearly independent, and hence is a basis contained in S for which we are looking.

To show that B is linearly independent, we plug two values for x into $a\sin(x+1)+b\cos(x+1)=0$:

$$\begin{cases}\text{Letting } x=0 \implies & (\sin 1)a & +\ (\cos 1)b & = 0\\ \text{Letting } x=-1 \implies & & b & = 0\end{cases}.$$

Using back substitution and the fact that $\sin 1\neq 0$ shows that $a=b=0$, proving that B is linearly independent.

(5) (a) Since $\mathbf{v}=5e^x+0e^{2x}+(-7)e^{3x}$, $[\mathbf{v}]_B=[5,0,-7]$ by the definition of $[\mathbf{v}]_B$.

(c) We start with the familiar trigonometric identity $\sin(a+b)=\sin a\cos b+\sin b\cos a$, which yields $\sin(x+1)=(\sin x)(\cos 1)+(\sin 1)(\cos x)$ and

$\sin(x+2) = (\sin x)(\cos 2) + (\sin 2)(\cos x)$.

Dividing the first equation by $\cos 1$ and the second by $\cos 2$, and then subtracting and rearranging terms yields $\frac{1}{\cos 1}\sin(x+1) - \frac{1}{\cos 2}\sin(x+2) = \left(\frac{\sin 1}{\cos 1} - \frac{\sin 2}{\cos 2}\right)\cos x$. Now $\frac{\sin 1}{\cos 1} - \frac{\sin 2}{\cos 2} = \frac{\sin 1 \cos 2 - \sin 2 \cos 1}{\cos 1 \cos 2} = \frac{\sin(-1)}{\cos 1 \cos 2} = -\frac{\sin 1}{\cos 1 \cos 2}$. Hence, $\cos x = -\frac{\cos 2}{\sin 1}\sin(x+1) + \frac{\cos 1}{\sin 1}\sin(x+2)$. Therefore, the definition of $[\mathbf{v}]_B$ yields $[\mathbf{v}]_B = [-\frac{\cos 2}{\sin 1}, \frac{\cos 1}{\sin 1}]$, or approximately $[0.4945, 0.6421]$.

An alternate approach is to solve for a and b in the equation $\cos x = a\sin(x+1) + b\sin(x+2)$. Plug in the following values for x:

$$\begin{cases} \text{Letting } x=-1 \implies & \cos 1 = & & (\sin 1)b \\ \text{Letting } x=-2 \implies & \cos 2 = & -(\sin 1)a & \end{cases},$$

producing the same results for a and b as above.

(6) (a) True. In *any* vector space, the definition of linear dependence asserts that $S = \{\mathbf{v}_1, \mathbf{v}_2\}$ is linearly dependent if and only if either of the vectors in S can be expressed as a linear combination of the other. Since there are only two vectors in the set, neither of which is zero, this happens precisely when $\mathbf{v}_1$ is a nonzero constant multiple of $\mathbf{v}_2$.

(b) True. Since the polynomials all have different degrees, none of them is a linear combination of the others.

(c) False. By Theorem 4.7, the set of vectors would be linearly independent, *not* linearly dependent.

(d) False. It is possible that the three values chosen for x do not produce equations proving linear independence, but choosing a different set of three values might. This principle is illustrated directly after Example 1 in Section 10.2 of the textbook. We are only assured of linear dependence when we have found values of a, b, and c, not all zero, such that $a\mathrm{f}_1(x) + b\mathrm{f}_2(x) + c\mathrm{f}_3(x) = 0$ is a functional identity.

Section 10.3

(1) In each part, we use the fact that the entries for the upper triangular matrix $\mathbf{C}$ are given, for $i \le j$, by $c_{ij} =$ (coefficient of $x_i x_j$) in the formula for Q. Then, the entries of the symmetric matrix $\mathbf{A}$ are given, for $i < j$, by $a_{ij} = a_{ji} = \frac{1}{2}c_{ij}$, and, on the diagonal, $a_{ii} = c_{ii}$.

(a) Using the formulas given above produces $\mathbf{C} = \begin{bmatrix} 8 & 12 \\ 0 & -9 \end{bmatrix}$ and $\mathbf{A} = \begin{bmatrix} 8 & 6 \\ 6 & -9 \end{bmatrix}$.

(c) Using the formulas given above yields $\mathbf{C} = \begin{bmatrix} 5 & 4 & -3 \\ 0 & -2 & 5 \\ 0 & 0 & 0 \end{bmatrix}$ and $\mathbf{A} = \begin{bmatrix} 5 & 2 & -\frac{3}{2} \\ 2 & -2 & \frac{5}{2} \\ -\frac{3}{2} & \frac{5}{2} & 0 \end{bmatrix}$.

(2) (a) The entries for the symmetric matrix $\mathbf{A}$ representing Q are given, for $i \ne j$, by $a_{ij} = \frac{1}{2}$(coefficient of $x_i x_j$) in the expression for Q, and, on the diagonal, $a_{ii} =$ (the coefficient of x_i^2). Hence, $\mathbf{A} = \begin{bmatrix} 43 & -24 \\ -24 & 57 \end{bmatrix}$. We need to orthogonally diagonalize $\mathbf{A}$.

Now $p_{\mathbf{A}}(x) = (x-43)(x-57) - 24^2 = x^2 - 100x + 1875 = (x-75)(x-25)$. Thus, $\mathbf{A}$ has eigenvalues $\lambda_1 = 75$ and $\lambda_2 = 25$. Using row reduction to find particular solutions to the systems $[75\mathbf{I}_2 - \mathbf{A}|\mathbf{0}]$ and $[25\mathbf{I}_2 - \mathbf{A}|\mathbf{0}]$ as in the Diagonalization Method of Section 3.4 produces the basis $\{[-3, 4]\}$ for E_{75} and $\{[4, 3]\}$ for E_{25}. Since E_{75} and E_{25} are 1-dimensional, the given bases are already orthogonal. To create orthonormal bases, we need to normalize the vectors, yielding $\{\frac{1}{5}[-3,4]\}$ for E_{75} and $\{\frac{1}{5}[4,3]\}$ for E_{25}. Thus, an orthonormal basis B for $\mathbb{R}^2$ is the set $B = (\frac{1}{5}[-3,4], \frac{1}{5}[4,3])$. Now $\mathbf{P}$ is the matrix whose columns are the vectors in B, and $\mathbf{D}$ is the diagonal matrix whose main diagonal entries are the eigenvalues 75 and 25. Hence, $\mathbf{P} = \frac{1}{5}\begin{bmatrix} -3 & 4 \\ 4 & 3 \end{bmatrix}$ and $\mathbf{D} = \begin{bmatrix} 75 & 0 \\ 0 & 25 \end{bmatrix}$.

Next, direct computation using the given formula for Q yields $Q([1,-8]) = 4075$. However, we can also calculate $Q([1,-8])$ as $[1,-8]_B^T\mathbf{D}[1,-8]_B$, where

$$[1,-8]_B = \mathbf{P}^{-1}\begin{bmatrix} 1 \\ -8 \end{bmatrix} = \mathbf{P}^T\begin{bmatrix} 1 \\ -8 \end{bmatrix} = \frac{1}{5}\begin{bmatrix} -3 & 4 \\ 4 & 3 \end{bmatrix}\begin{bmatrix} 1 \\ -8 \end{bmatrix} = \frac{1}{5}\begin{bmatrix} -35 \\ -20 \end{bmatrix} = \begin{bmatrix} -7 \\ -4 \end{bmatrix}.$$

Then $Q([1,-8]) = \begin{bmatrix} -7 & -4 \end{bmatrix}\begin{bmatrix} 75 & 0 \\ 0 & 25 \end{bmatrix}\begin{bmatrix} -7 \\ -4 \end{bmatrix} = \begin{bmatrix} -7 & -4 \end{bmatrix}\begin{bmatrix} -525 \\ -100 \end{bmatrix} = 4075.$

(c) The entries for the symmetric matrix $\mathbf{A}$ representing Q are given, for $i \neq j$, by $a_{ij} = \frac{1}{2}$(coefficient of x_ix_j) in the expression for Q, and, on the diagonal, a_{ii} = (the coefficient of x_i^2). Hence,

$\mathbf{A} = \begin{bmatrix} 18 & 48 & -30 \\ 48 & -68 & 18 \\ -30 & 18 & 1 \end{bmatrix}$. We need to orthogonally diagonalize $\mathbf{A}$.

Now $p_{\mathbf{A}}(x) = x^3 + 49x^2 - 4802x = x(x^2 + 49x - 4802)$. Thus, $\lambda_1 = 0$ is an eigenvalue for $\mathbf{A}$. Using the quadratic formula on the second factor of $p_{\mathbf{A}}(x)$ produces the eigenvalues $\lambda_2 = 49$ and $\lambda_3 = -98$. Then, using row reduction to find particular solutions to the systems $[0\mathbf{I}_3 - \mathbf{A}|\mathbf{0}]$, $[49\mathbf{I}_3 - \mathbf{A}|\mathbf{0}]$ and $[-98\mathbf{I}_3 - \mathbf{A}|\mathbf{0}]$ as in the Diagonalization Method of Section 3.4 produces the basis $\{[2,3,6]\}$ for E_0, $\{[-6,-2,3]\}$ for E_{49} and $\{[3,-6,2]\}$ for E_{-98}. Since E_0, E_{49} and E_{-98} are 1-dimensional, the given bases are already orthogonal. To create orthonormal bases, we need to normalize the vectors, yielding $\{\frac{1}{7}[2,3,6]\}$ for E_0, $\{\frac{1}{7}[-6,-2,3]\}$ for E_{49} and $\{\frac{1}{7}[3,-6,2]\}$ for E_{-98}. Thus, the orthonormal basis B for $\mathbb{R}^3$ is the set $B = (\frac{1}{7}[2,3,6], \frac{1}{7}[-6,-2,3], \frac{1}{7}[3,-6,2])$. Now $\mathbf{P}$ is the matrix whose columns are the vectors in B, and $\mathbf{D}$ is the diagonal matrix whose main diagonal entries are the eigenvalues 0, 49 and -98. Hence $\mathbf{P} = \frac{1}{7}\begin{bmatrix} 2 & -6 & 3 \\ 3 & -2 & -6 \\ 6 & 3 & 2 \end{bmatrix}$, and

$$\mathbf{D} = \begin{bmatrix} 0 & 0 & 0 \\ 0 & 49 & 0 \\ 0 & 0 & -98 \end{bmatrix}.$$

Next, direct computation using the given formula for Q yields $Q([4,-3,6]) = -3528$. However, we can also calculate $Q([4,-3,6])$ as $[4,-3,6]_B^T\mathbf{D}[4,-3,6]_B$, where

$$[4,-3,6]_B = \mathbf{P}^{-1}\begin{bmatrix} 4 \\ -3 \\ 6 \end{bmatrix} = \mathbf{P}^T\begin{bmatrix} 4 \\ -3 \\ 6 \end{bmatrix} = \frac{1}{7}\begin{bmatrix} 2 & 3 & 6 \\ -6 & -2 & 3 \\ 3 & -6 & 2 \end{bmatrix}\begin{bmatrix} 4 \\ -3 \\ 6 \end{bmatrix} = \frac{1}{7}\begin{bmatrix} 35 \\ 0 \\ 42 \end{bmatrix} = \begin{bmatrix} 5 \\ 0 \\ 6 \end{bmatrix}.$$

Then $Q([4,-3,6]) = \begin{bmatrix} 5 & 0 & 6 \end{bmatrix}\begin{bmatrix} 0 & 0 & 0 \\ 0 & 49 & 0 \\ 0 & 0 & -98 \end{bmatrix}\begin{bmatrix} 5 \\ 0 \\ 6 \end{bmatrix} = \begin{bmatrix} 5 & 0 & 6 \end{bmatrix}\begin{bmatrix} 0 \\ 0 \\ -588 \end{bmatrix} = -3528.$

(4) Yes. If $Q(\mathbf{x}) = \Sigma a_{ij}x_ix_j$, $1 \leq i \leq j \leq n$, then $\mathbf{x}^T\mathbf{C}_1\mathbf{x}$ and $\mathbf{C}_1$ upper triangular imply that the (i,j) entry for $\mathbf{C}_1$ is zero if $i > j$ and a_{ij} if $i \leq j$. A similar argument describes $\mathbf{C}_2$. Thus, $\mathbf{C}_1 = \mathbf{C}_2$.

(6) (a) True. This is discussed in the proof of Theorem 10.5.

(b) False. For this quadratic form, the coefficients of x^2 and y^2 are zero, which is not disallowed in the definition of a quadratic form.

(c) False. A quadratic form can be represented by many matrices. In particular, every quadratic form is represented by both an upper triangular matrix and a symmetric matrix (which are usually distinct). For example, the quadratic form $Q([x, y]) = 2xy$ can be expressed both as $\begin{bmatrix} x & y \end{bmatrix} \begin{bmatrix} 0 & 2 \\ 0 & 0 \end{bmatrix} \begin{bmatrix} x \\ y \end{bmatrix}$ and as $\begin{bmatrix} x & y \end{bmatrix} \begin{bmatrix} 0 & 1 \\ 1 & 0 \end{bmatrix} \begin{bmatrix} x \\ y \end{bmatrix}$.

(d) True. This is the conclusion of the Principal Axes Theorem (Theorem 10.7).

(e) True. This can be inferred from the proof of the Principal Axes Theorem (Theorem 10.7) and the Orthogonal Diagonalization Method from Section 6.3 of the textbook. This fact is utilized in the Quadratic Form Method in Section 10.3 of the textbook.

Appendices

Appendix B

(1) (a) f is not a function because it is not defined for every element of its stated domain. In particular, f is not defined for $x < 1$.

(c) h is not a function because two values are assigned to each $x \neq 1$.

(e) k is not a function because it is not defined for every element of its stated domain. In particular, k is not defined at $\theta = \frac{\pi}{2}n$, where n is an odd integer.

(f) l is a function. Its range is the set of all prime numbers. The range is contained in the set of prime numbers because the definition of l defines an image under l to be a prime number. The range contains *every* prime number, because, if p is prime, $l(p) = p$, and so p is in the range. The image of $2 = l(2) = 2$. The pre-image of $2 = \{0, 1, 2\}$, the set of all natural numbers less than or equal to 2.

(2) (a) The pre-image of $\{10, 20, 30\}$ is the set of all integers whose absolute value is 5, 10, or 15, since 2 times the absolute value must be 10, 20, or 30. This is the set $\{-15, -10, -5, 5, 10, 15\}$.

(c) The pre-image of the set of multiples of 4 is the set of all integers which, when doubled, have absolute values that are multiples of 4. This is the set of integers whose absolute value is a multiple of 2; that is, the set $\{\ldots, -8, -6, -4, -2, 0, 2, 4, 6, 8, \ldots\}$.

(3) $(g \circ f)(x) = g(f(x)) = g\left(\frac{5x-1}{4}\right) = \sqrt{3\left(\frac{5x-1}{4}\right)^2 + 2} = \sqrt{\frac{3}{16}\left(25x^2 - 10x + 1\right) + 2} = \frac{1}{4}\sqrt{75x^2 - 30x + 35}$.
$(f \circ g)(x) = f(g(x)) = f\left(\sqrt{3x^2+2}\right) = \frac{1}{4}(5\sqrt{3x^2+2} - 1)$.

(4) $(g \circ f)\left(\begin{bmatrix} x \\ y \end{bmatrix}\right) = g\left(f\left(\begin{bmatrix} x \\ y \end{bmatrix}\right)\right) = g\left(\begin{bmatrix} 3 & -2 \\ 1 & 4 \end{bmatrix}\begin{bmatrix} x \\ y \end{bmatrix}\right) = \begin{bmatrix} -4 & 4 \\ 0 & 2 \end{bmatrix}\left(\begin{bmatrix} 3 & -2 \\ 1 & 4 \end{bmatrix}\begin{bmatrix} x \\ y \end{bmatrix}\right) =$
$\left(\begin{bmatrix} -4 & 4 \\ 0 & 2 \end{bmatrix}\begin{bmatrix} 3 & -2 \\ 1 & 4 \end{bmatrix}\right)\begin{bmatrix} x \\ y \end{bmatrix} = \begin{bmatrix} -8 & 24 \\ 2 & 8 \end{bmatrix}\begin{bmatrix} x \\ y \end{bmatrix}$.
$(f \circ g)\left(\begin{bmatrix} x \\ y \end{bmatrix}\right) = f\left(g\left(\begin{bmatrix} x \\ y \end{bmatrix}\right)\right) = f\left(\begin{bmatrix} -4 & 4 \\ 0 & 2 \end{bmatrix}\begin{bmatrix} x \\ y \end{bmatrix}\right) = \begin{bmatrix} 3 & -2 \\ 1 & 4 \end{bmatrix}\left(\begin{bmatrix} -4 & 4 \\ 0 & 2 \end{bmatrix}\begin{bmatrix} x \\ y \end{bmatrix}\right) =$
$\left(\begin{bmatrix} 3 & -2 \\ 1 & 4 \end{bmatrix}\begin{bmatrix} -4 & 4 \\ 0 & 2 \end{bmatrix}\right)\begin{bmatrix} x \\ y \end{bmatrix} = \begin{bmatrix} -12 & 8 \\ -4 & 12 \end{bmatrix}\begin{bmatrix} x \\ y \end{bmatrix}$.

(8) The function f is not one-to-one because $f(x+1) = f(x+3) = 1$, which shows that two different domain elements have the same image. The function f is not onto because there is no pre-image for x^n. For $n \geq 3$, the pre-image of $\mathcal{P}_2$ is $\mathcal{P}_3$.

(10) The function f is one-to-one, because if $f(\mathbf{A}_1) = f(\mathbf{A}_2)$, then
$\mathbf{B}^{-1}\mathbf{A}_1\mathbf{B} = \mathbf{B}^{-1}\mathbf{A}_2\mathbf{B} \Rightarrow \mathbf{B}(\mathbf{B}^{-1}\mathbf{A}_1\mathbf{B})\mathbf{B}^{-1} = \mathbf{B}(\mathbf{B}^{-1}\mathbf{A}_2\mathbf{B})\mathbf{B}^{-1}$
$\Rightarrow (\mathbf{B}\mathbf{B}^{-1})\mathbf{A}_1(\mathbf{B}\mathbf{B}^{-1}) = (\mathbf{B}\mathbf{B}^{-1})\mathbf{A}_2(\mathbf{B}\mathbf{B}^{-1})$
$\Rightarrow \mathbf{I}_n\mathbf{A}_1\mathbf{I}_n = \mathbf{I}_n\mathbf{A}_2\mathbf{I}_n \Rightarrow \mathbf{A}_1 = \mathbf{A}_2$.
The function f is onto, because for any $\mathbf{C} \in \mathcal{M}_{nn}$, $f(\mathbf{B}\mathbf{C}\mathbf{B}^{-1}) = \mathbf{B}^{-1}(\mathbf{B}\mathbf{C}\mathbf{B}^{-1})\mathbf{B} = \mathbf{C}$.
Also, $f^{-1}(\mathbf{A}) = \mathbf{B}\mathbf{A}\mathbf{B}^{-1}$.

(12) (a) False. Functions may have two different elements of the domain mapped to the same element of the codomain. For example, $f : \mathbb{R} \to \mathbb{R}$ given by $f(x) = x^2$ is such a function, since f maps both 2 and -2 to 4.

(b) True. Since each element of X is assigned to exactly one element of Y, f is a function by definition. However, a function f having this property is not onto, since not every element of Y has a corresponding element of X.

(c) False. Since $f(5) = f(6)$, f is not one-to-one. Hence, f^{-1} does not exist. For example, the constant function $f : \mathbb{R} \to \mathbb{R}$ given by $f(x) = 7$ satisfies $f(5) = f(6)$, but has no inverse, and so $f^{-1}(5)$ is not defined. However, for any function f, $f^{-1}(\{5\})$ *is* defined and represents the set of all pre-images of 5. If f is the constant function $f : \mathbb{R} \to \mathbb{R}$ given by $f(x) = 7$, then $f^{-1}(\{5\})$ equals the empty set, while if $f : \mathbb{R} \to \mathbb{R}$ is given by $f(x) = 5$, then $f^{-1}(\{5\}) = \mathbb{R}$.

(d) False. For f to be onto, the codomain of f must equal the range of f, *not* the domain of f. For a specific counterexample, consider $f : \mathbb{R} \to \mathbb{R}$ given by $f(x) = 0$. Then the domain and codomain of f are both $\mathbb{R}$; however, f is not onto because no nonzero element of the codomain $\mathbb{R}$ has a pre-image.

(e) False. Instead, f is one-to-one if $f(x_1) = f(x_2)$ implies $x_1 = x_2$. Notice, by the way, that every function, by definition, has the property that $x_1 = x_2$ implies $f(x_1) = f(x_2)$. Hence, any function that is not one-to-one (such as $f : \mathbb{R} \to \mathbb{R}$ given by $f(x) = 0$) provides a specific counterexample to the given statement.

(f) False. Consider $f : \{0,1\} \to \{0,1\}$ given by $f(x) = x$, and $g : \{0,1,2\} \to \{0,1\}$ given by $g(0) = 0$, and $g(1) = g(2) = 1$. Then simple computation shows that $(g \circ f)(x)$ actually equals $f(x)$. Now, both f and $g \circ f$ are easily seen to be one-to-one, but g is not one-to-one because $g(1) = g(2)$.

(g) False. By Theorem B.2, in order for a function f to have an inverse, if must be both one-to-one *and* onto. Hence, any function that is one-to-one but not onto (for example, $f : \mathbb{R} \to \mathbb{R}$ given by $f(x) = \arctan(x)$) or that is onto but not one-to-one (for example, $g : \mathbb{R} \to \mathbb{R}$ given by $g(x) = x^3 - x$) provides a counterexample.

(h) False. Instead, by Theorem B.4, $(g \circ f)^{-1} = f^{-1} \circ g^{-1}$. For a specific counterexample to the statement in the problem, consider

$f : \{0,1,2\} \to \{0,1,2\}$ given by $f(0) = 1$, $f(1) = 2$ and $f(2) = 0$, and

$g : \{0,1,2\} \to \{0,1,2\}$ given by $g(0) = 1$, $g(1) = 0$ and $g(2) = 2$. Then, it is easily seen that

$f^{-1} : \{0,1,2\} \to \{0,1,2\}$ is given by $f^{-1}(0) = 2$, $f^{-1}(1) = 0$ and $f^{-1}(2) = 1$, and

$g^{-1} : \{0,1,2\} \to \{0,1,2\}$ is given by $g^{-1}(0) = 1$, $g^{-1}(1) = 0$ and $g^{-1}(2) = 2$. Also,

$g \circ f : \{0,1,2\} \to \{0,1,2\}$ is given by $(g \circ f)(0) = 0$, $(g \circ f)(1) = 2$ and $(g \circ f)(2) = 1$. Hence,

$(g \circ f)^{-1} : \{0,1,2\} \to \{0,1,2\}$ is given by $(g \circ f)^{-1}(0) = 0$, $(g \circ f)^{-1}(1) = 2$ and $(g \circ f)^{-1}(2) = 1$.

However, $g^{-1} \circ f^{-1} : \{0,1,2\} \to \{0,1,2\}$ is given by $(g^{-1} \circ f^{-1})(0) = 2$, $(g^{-1} \circ f^{-1})(1) = 1$ and $(g^{-1} \circ f^{-1})(2) = 0$. Therefore, $(g \circ f)^{-1} \neq g^{-1} \circ f^{-1}$.

Appendix C

(1) (a) $(6 - 3i) + (5 + 2i) = (6 + 5) + (-3 + 2)i = 11 - i$

(c) $4\Big((8 - 2i) - (3 + i)\Big) = 4\Big((8 - 3) + (-2 - 1)i\Big) = 4(5 - 3i) = 20 - 12i$

(e) $(5 + 3i)(3 + 2i) = \Big((5)(3) - (3)(2)\Big) + \Big((5)(2) + (3)(3)\Big)i = 9 + 19i$

(g) $(7-i)(-2-3i) = \Big((7)(-2)-(-1)(-3)\Big) + \Big((7)(-3)+(-1)(-2)\Big)i = -17-19i$

(i) $\overline{9-2i} = 9+2i$

(k) $\overline{(6+i)(2-4i)} = \overline{\Big((6)(2)-(1)(-4)\Big) + \Big((6)(-4)+(1)(2)\Big)i} = \overline{16-22i} = 16+22i$

(m) $|-2+7i| = \sqrt{(-2)^2+(7)^2} = \sqrt{53}$

(2) In each part we use the formula $\frac{1}{z} = \frac{\overline{z}}{|z|^2}$ and the fact that, for $z = a+bi$, $|z|^2 = \left(\sqrt{a^2+b^2}\right)^2 = a^2+b^2$.

(a) If $z = 6-2i$, then $\overline{z} = 6+2i$. Also, $|z|^2 = 6^2+(-2)^2 = 40$. Hence, $\frac{1}{z} = \frac{6+2i}{40} = \frac{3}{20}+\frac{1}{20}i$.

(c) If $z = -4+i$, then $\overline{z} = -4-i$. Also, $|z|^2 = (-4)^2+(-1)^2 = 17$. Hence, $\frac{1}{z} = \frac{-4-i}{17} = -\frac{4}{17}-\frac{1}{17}i$.

(3) In all parts, let $z_1 = a_1+b_1i$ and $z_2 = a_2+b_2i$.

(a) Part (1): $\overline{z_1+z_2} = \overline{(a_1+b_1i)+(a_2+b_2i)} = \overline{(a_1+a_2)+(b_1+b_2)i} = (a_1+a_2)-(b_1+b_2)i = (a_1-b_1i)+(a_2-b_2i) = \overline{z_1}+\overline{z_2}$.
Part (2): $\overline{(z_1z_2)} = \overline{(a_1+b_1i)(a_2+b_2i)}$
$= \overline{(a_1a_2-b_1b_2)+(a_1b_2+a_2b_1)i} = (a_1a_2-b_1b_2)-(a_1b_2+a_2b_1)i = (a_1a_2-(-b_1)(-b_2)) + (a_1(-b_2)+a_2(-b_1))i = (a_1-b_1i)(a_2-b_2i) = \overline{z_1}\,\overline{z_2}$.

(b) If $z_1 \neq 0$, then $\frac{1}{z_1}$ exists. Hence, $\frac{1}{z_1}(z_1z_2) = \frac{1}{z_1}(0)$, implying $z_2 = 0$.

(c) Part (4): $z_1 = \overline{z_1} \Leftrightarrow a_1+b_1i = a_1-b_1i \Leftrightarrow b_1i = -b_1i \Leftrightarrow 2b_1i = 0 \Leftrightarrow b_1 = 0 \Leftrightarrow z_1$ is real.
Part (5): $z_1 = -\overline{z_1} \Leftrightarrow a_1+b_1i = -(a_1-b_1i) \Leftrightarrow a_1+b_1i = -a_1+b_1i \Leftrightarrow a_1 = -a_1 \Leftrightarrow 2a_1 = 0 \Leftrightarrow a_1 = 0 \Leftrightarrow z_1$ is pure imaginary.

(5) (a) False. Instead, $z\overline{z} = |z|^2$. For example, if $z = 1+i$, then $\overline{z} = 1-i$ and $z\overline{z} = (1+i)(1-i) = 2$, but $|z| = \sqrt{1^2+1^2} = \sqrt{2}$.

(b) False. Every real number z (such as $z = 3$) has the property that $z = \overline{z}$.

(c) True. A pure imaginary number is of the form $z = 0+bi = bi$, whose conjugate is $\overline{z} = 0-bi = -bi$.

(d) True. The additive inverse of $z = a+bi$ is $-z = -a-bi$.

(e) False. The complex number 0 has no multiplicative inverse. However, every *nonzero* complex number z has the multiplicative inverse $\frac{1}{z} = \left(\frac{1}{|z|^2}\right)\overline{z}$.

Printed in the United States
116488LV00003B/35-48/A

9 780120 586226